LTE for 4G Mobile Broadband

Do you need to get up-to-speed quickly on Long-Term Evolution (LTE)?

Understand the new technologies of the LTE standard and how they contribute to improvements in system performance with this practical and valuable guide, written by an expert on LTE who was intimately involved in the drafting of the standard. In addition to a strong grounding in the technical details, you'll also get fascinating insights into why particular technologies were chosen in the development process.

Core topics covered include:

- Network architecture and protocols;
- OFDMA downlink access;
- Low-PAPR SC-FDMA uplink access;
- Transmit diversity and MIMO spatial multiplexing;
- Channel structure and bandwidths;
- Cell search, reference signals and random access;
- Turbo coding with contention-free interleaver;
- Scheduling, link adaptation, hybrid ARQ and power control;
- Uplink and downlink physical control signaling;
- Inter-cell interference mitigation techniques;
- Single-frequency network (SFN) broadcast;
- MIMO spatial channel model;
- Evaluation methodology and system performance.

With extensive references, a useful discussion of technologies that were not included in the standard, and end-of-chapter summaries that draw out and emphasize all the key points, this book is an essential resource for practitioners in the mobile cellular communications industry and for graduate students studying advanced wireless communications.

Farooq Khan is Technology Director at the Samsung Telecom R&D Center, Dallas, Texas, where he manages the design, performance evaluation, and standardization of next-generation wireless communications systems. Previously, he was a Member of Technical Staff at Bell Laboratories, where he conducted research on the evolution of cdma2000 and UMTS systems towards high-speed packet access (HSPA). He also worked at Ericsson Research in Sweden, contributing to the design and performance evaluation of EDGE and WCDMA technologies. He has authored more than 30 research papers and holds over 50 US patents, all in the area of wireless communications.

LTE for 4G Mobile Broadband

Air Interface Technologies and Performance

FAROOQ KHAN

Telecom R&D Center
Samsung Telecommunications, America

CAMBRIDGE
UNIVERSITY PRESS

CAMBRIDGE UNIVERSITY PRESS
Cambridge, New York, Melbourne, Madrid, Cape Town, Singapore, São Paulo,
Delhi, Dubai, Tokyo

Cambridge University Press
The Edinburgh Building, Cambridge CB2 8RU, UK

Published in the United States of America by Cambridge University Press, New York

www.cambridge.org
Information on this title: www.cambridge.org/9780521882217

First published 2009
Reprinted with corrections 2010

Printed in the United Kingdom at the University Press, Cambridge

A catalog record for this publication is available from the British Library

ISBN 978-0-521-88221-7 hardback

To my wonderful wife, Mobeena; our three precious children, Nemul, Haris and Alisha; my father and to the memory of my late mother.

Contents

Preface

The Global system for mobile communications (GSM) is the dominant wireless cellular standard with over 3.5 billion subscribers worldwide covering more than 85% of the global mobile market. Furthermore, the number of worldwide subscribers using high-speed packet access (HSPA) networks topped 70 million in 2008. HSPA is a 3G evolution of GSM supporting high-speed data transmissions using WCDMA technology. Global uptake of HSPA technology among consumers and businesses is accelerating, indicating continued traffic growth for high-speed mobile networks worldwide. In order to meet the continued traffic growth demands, an extensive effort has been underway in the 3G Partnership Project (3GPP) to develop a new standard for the evolution of GSM/HSPA technology towards a packet-optimized system referred to as Long-Term Evolution (LTE).

The goal of the LTE standard is to create specifications for a new radio-access technology geared to higher data rates, low latency and greater spectral efficiency. The spectral efficiency target for the LTE system is three to four times higher than the current HSPA system. These aggressive spectral efficiency targets require pushing the technology envelope by employing advanced air-interface techniques such as low-PAPR orthogonal uplink multiple access based on SC-FDMA (single-carrier frequency division multiple access) MIMO multiple-input multiple-output multi-antenna technologies, inter-cell interference mitigation techniques, low-latency channel structure and single-frequency network (SFN) broadcast. The researchers and engineers working on the standard come up with new innovative technology proposals and ideas for system performance improvement. Due to the highly aggressive standard development schedule, these researchers and engineers are generally unable to publish their proposals in conferences or journals, etc. In the standards development phase, the proposals go through extensive scrutiny with multiple sources evaluating and simulating the proposed technologies from system performance improvement and implementation complexity perspectives. Therefore, only the highest-quality proposals and ideas finally make into the standard.

The book provides detailed coverage of the air-interface technologies and protocols that withstood the scrutiny of the highly sophisticated technology evaluation process typically used in the 3GPP physical layer working group. We describe why certain technology choices were made in the standard development process and how each of the technology components selected contributes to the overall system performance improvement. As such, the book serves as a valuable reference for system designers and researchers not directly involved in the standard development phase.

I am indebted to many colleagues at Samsung, in particular to Zhouyue (Jerry) Pi, Jianzhong (Charlie) Zhang, Jiann-An Tsai, Juho Lee, Jin-Kyu Han and Joonyoung Cho. These colleagues and other valued friends, too numerous to be mentioned, have deeply influenced my understanding of wireless communications and standards. Without the unprecedented support of Phil Meyler, Sarah Matthews, Dawn Preston and their colleagues at Cambridge University Press, this monograph would never have reached the readers. Finally my sincere gratitude goes to the numerous researchers and engineers who contributed to the development of the LTE standard in 3GPP, without whom this book would not have materialized.

1 Introduction

The cellular wireless communications industry witnessed tremendous growth in the past decade with over four billion wireless subscribers worldwide. The first generation (1G) analog cellular systems supported voice communication with limited roaming. The second generation (2G) digital systems promised higher capacity and better voice quality than did their analog counterparts. Moreover, roaming became more prevalent thanks to fewer standards and common spectrum allocations across countries particularly in Europe. The two widely deployed second-generation (2G) cellular systems are GSM (global system for mobile communications) and CDMA (code division multiple access). As for the 1G analog systems, 2G systems were primarily designed to support voice communication. In later releases of these standards, capabilities were introduced to support data transmission. However, the data rates were generally lower than that supported by dial-up connections. The ITU-R initiative on IMT-2000 (international mobile telecommunications 2000) paved the way for evolution to 3G. A set of requirements such as a peak data rate of 2 Mb/s and support for vehicular mobility were published under IMT-2000 initiative. Both the GSM and CDMA camps formed their own separate 3G partnership projects (3GPP and 3GPP2, respectively) to develop IMT-2000 compliant standards based on the CDMA technology. The 3G standard in 3GPP is referred to as wideband CDMA (WCDMA) because it uses a larger 5 MHz bandwidth relative to 1.25 MHz bandwidth used in 3GPP2's cdma2000 system. The 3GPP2 also developed a 5 MHz version supporting three 1.25 MHz subcarriers referred to as cdma2000-3x. In order to differentiate from the 5 MHz cdma2000-3x standard, the 1.25 MHz system is referred to as cdma2000-1x or simply 3G-1x.

The first release of the 3G standards did not fulfill its promise of high-speed data transmissions as the data rates supported in practice were much lower than that claimed in the standards. A serious effort was then made to enhance the 3G systems for efficient data support. The 3GPP2 first introduced the HRPD (high rate packet data) [1] system that used various advanced techniques optimized for data traffic such as channel sensitive scheduling, fast link adaptation and hybrid ARQ, etc. The HRPD system required a separate 1.25 MHz carrier and supported no voice service. This was the reason that HRPD was initially referred to as cdma2000-1xEVDO (evolution data only) system. The 3GPP followed a similar path and introduced HSPA (high speed packet access) [2] enhancement to the WCDMA system. The HSPA standard reused many of the same data-optimized techniques as the HRPD system. A difference relative to HRPD, however, is that both voice and data can be carried on the same 5 MHz carrier in HSPA. The voice and data traffic are code multiplexed in the downlink. In parallel to HRPD, 3GPP2 also developed a joint voice data standard that was referred to as cdma2000-1xEVDV (evolution data voice) [3]. Like HSPA, the cdma2000-1xEVDV system supported both voice and data on the same carrier but it was never commercialized. In the later release of HRPD, VoIP (Voice over Internet Protocol) capabilities were introduced to provide both voice and data service on the same carrier. The two 3G standards namely HSPA and HRPD were finally able to fulfill the 3G promise and have been widely deployed in major cellular markets to provide wireless data access.

1.1   Beyond 3G systems

While HSPA and HRPD systems were being developed and deployed, IEEE 802 LMSC (LAN/MAN Standard Committee) introduced the IEEE 802.16e standard [4] for mobile broadband wireless access. This standard was introduced as an enhancement to an earlier IEEE 802.16 standard for fixed broadband wireless access. The 802.16e standard employed a different access technology named OFDMA (orthogonal frequency division multiple access) and claimed better data rates and spectral efficiency than that provided by HSPA and HRPD. Although the IEEE 802.16 family of standards is officially called WirelessMAN in IEEE, it has been dubbed WiMAX (worldwide interoperability for microwave access) by an industry group named the WiMAX Forum. The mission of the WiMAX Forum is to promote and certify the compatibility and interoperability of broadband wireless access products. The WiMAX system supporting mobility as in IEEE 802.16e standard is referred to as Mobile WiMAX. In addition to the radio technology advantage, Mobile WiMAX also employed a simpler network architecture based on IP protocols.

The introduction of Mobile WiMAX led both 3GPP and 3GPP2 to develop their own version of beyond 3G systems based on the OFDMA technology and network architecture similar to that in Mobile WiMAX. The beyond 3G system in 3GPP is called evolved universal terrestrial radio access (evolved UTRA) [5] and is also widely referred to as LTE (Long-Term Evolution) while 3GPP2's version is called UMB (ultra mobile broadband) [6] as depicted in Figure 1.1. It should be noted that all three beyond 3G systems namely Mobile WiMAX,

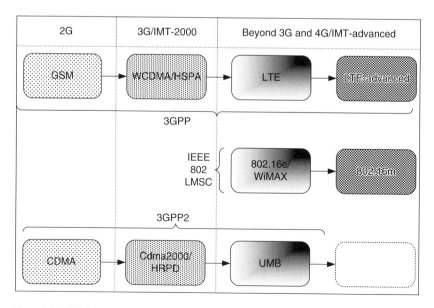

Figure 1.1. Cellular systems evolution.

Table 1.1. LTE system attributes.

Bandwidth		1.25–20 MHz
Duplexing		FDD, TDD, half-duplex FDD
Mobility		350 km/h
Multiple access	Downlink	OFDMA
	Uplink	SC-FDMA
MIMO	Downlink	$2 \times 2, 4 \times 2, 4 \times 4$
	Uplink	$1 \times 2, 1 \times 4$
Peak data rate in 20 MHz	Downlink	173 and 326 Mb/s for 2×2 and 4×4 MIMO, respectively
	Uplink	86 Mb/s with 1×2 antenna configuration
Modulation		QPSK, 16-QAM and 64-QAM
Channel coding		Turbo code
Other techniques		Channel sensitive scheduling, link adaptation, power control, ICIC and hybrid ARQ

LTE and UMB meet IMT-2000 requirements and hence they are also part of IMT-2000 family of standards.

1.2   Long-Term Evolution (LTE)

The goal of LTE is to provide a high-data-rate, low-latency and packet-optimized radio-access technology supporting flexible bandwidth deployments [7]. In parallel, new network architecture is designed with the goal to support packet-switched traffic with seamless mobility, quality of service and minimal latency [8].

The air-interface related attributes of the LTE system are summarized in Table 1.1. The system supports flexible bandwidths thanks to OFDMA and SC-FDMA access schemes. In addition to FDD (frequency division duplexing) and TDD (time division duplexing), half-duplex FDD is allowed to support low cost UEs. Unlike FDD, in half-duplex FDD operation a UE is not required to transmit and receive at the same time. This avoids the need for a costly duplexer in the UE. The system is primarily optimized for low speeds up to 15 km/h. However, the system specifications allow mobility support in excess of 350 km/h with some performance degradation. The uplink access is based on single carrier frequency division multiple access (SC-FDMA) that promises increased uplink coverage due to low peak-to-average power ratio (PAPR) relative to OFDMA.

The system supports downlink peak data rates of 326 Mb/s with 4×4 MIMO (multiple input multiple output) within 20 MHz bandwidth. Since uplink MIMO is not employed in the first release of the LTE standard, the uplink peak data rates are limited to 86 Mb/s within 20 MHz bandwidth. In addition to peak data rate improvements, the LTE system provides two to four times higher cell spectral efficiency relative to the Release 6 HSPA system. Similar improvements are observed in cell-edge throughput while maintaining same-site locations as deployed for HSPA. In terms of latency, the LTE radio-interface and network provides capabilities for less than 10 ms latency for the transmission of a packet from the network to the UE.

1.3   Evolution to 4G

The radio-interface attributes for Mobile WiMAX and UMB are very similar to those of LTE given in Table 1.1. All three systems support flexible bandwidths, FDD/TDD duplexing, OFDMA in the downlink and MIMO schemes. There are a few differences such as uplink in LTE is based on SC-FDMA compared to OFDMA in Mobile WiMAX and UMB. The performance of the three systems is therefore expected to be similar with small differences.

Similar to the IMT-2000 initiative, ITU-R Working Party 5D has stated requirements for IMT-advanced systems. Among others, these requirements include average downlink data rates of 100 Mbit/s in the wide area network, and up to 1 Gbit/s for local access or low-mobility scenarios. Also, at the World Radiocommunication Conference 2007 (WRC-2007), a maximum of a 428 MHz new spectrum is identified for IMT systems that also include a 136 MHz spectrum allocated on a global basis.

Both 3GPP and IEEE 802 LMSC are actively developing their own standards for submission to IMT-advanced. The goal for both LTE-advanced [9] and IEEE 802.16 m [10] standards is to further enhance system spectral efficiency and data rates while supporting backward compatibility with their respective earlier releases. As part of the LTE-advanced and IEEE 802.16 standards developments, several enhancements including support for a larger than 20 MHz bandwidth and higher-order MIMO are being discussed to meet the IMT-advanced requirements.

References

[1] 3GPP2 TSG C.S0024-0 v2.0, cdma2000 High Rate Packet Data Air Interface Specification.
[2] 3GPP TSG RAN TR 25.848 v4.0.0, Physical Layer Aspects of UTRA High Speed Downlink Packet Access.
[3] 3GPP2 TSG C.S0002-C v1.0, Physical Layer Standard for cdma2000 Spread Spectrum Systems, Release C.
[4] IEEE Std 802.16e-2005, Air Interface for Fixed and Mobile Broadband Wireless Access Systems.
[5] 3GPP TSG RAN TR 25.912 v7.2.0, Feasibility Study for Evolved Universal Terrestrial Radio Access (UTRA) and Universal Terrestrial Radio Access Network (UTRAN).
[6] 3GPP2 TSG C.S0084-001-0 v2.0, Physical Layer for Ultra Mobile Broadband (UMB) Air Interface Specification.
[7] 3GPP TSG RAN TR 25.913 v7.3.0, Requirements for Evolved Universal Terrestrial Radio Access (UTRA) and Universal Terrestrial Radio Access Network (UTRAN).
[8] 3GPP TSG RAN TR 23.882 v1.15.1, 3GPP System Architecture Evolution: Report on Technical Options and Conclusions.
[9] 3GPP TSG RAN TR 36.913 v8.0.0, Requirements for Further Advancements for E-UTRA (LTE-Advanced).
[10] IEEE 802.16m-07/002r4, TGm System Requirements Document (SRD).

2 Network architecture and protocols

The LTE network architecture is designed with the goal of supporting packet-switched traffic with seamless mobility, quality of service (QoS) and minimal latency. A packet-switched approach allows for the supporting of all services including voice through packet connections. The result in a highly simplified flatter architecture with only two types of node namely evolved Node-B (eNB) and mobility management entity/gateway (MME/GW). This is in contrast to many more network nodes in the current hierarchical network architecture of the 3G system. One major change is that the radio network controller (RNC) is eliminated from the data path and its functions are now incorporated in eNB. Some of the benefits of a single node in the access network are reduced latency and the distribution of the RNC processing load into multiple eNBs. The elimination of the RNC in the access network was possible partly because the LTE system does not support macro-diversity or soft-handoff.

In this chapter, we discuss network architecture designs for both unicast and broadcast traffic, QoS architecture and mobility management in the access network. We also briefly discuss layer 2 structure and different logical, transport and physical channels along with their mapping.

2.1   Network architecture

All the network interfaces are based on IP protocols. The eNBs are interconnected by means of an X2 interface and to the MME/GW entity by means of an S1 interface as shown in Figure 2.1. The S1 interface supports a many-to-many relationship between MME/GW and eNBs [1].

The functional split between eNB and MME/GW is shown in Figure 2.2. Two logical gateway entities namely the serving gateway (S-GW) and the packet data network gateway (P-GW) are defined. The S-GW acts as a local mobility anchor forwarding and receiving packets to and from the eNB serving the UE. The P-GW interfaces with external packet data networks (PDNs) such as the Internet and the IMS. The P-GW also performs several IP functions such as address allocation, policy enforcement, packet filtering and routing.

The MME is a signaling only entity and hence user IP packets do not go through MME. An advantage of a separate network entity for signaling is that the network capacity for signaling and traffic can grow independently. The main functions of MME are idle-mode UE reachability including the control and execution of paging retransmission, tracking area list management, roaming, authentication, authorization, P-GW/S-GW selection, bearer management including dedicated bearer establishment, security negotiations and NAS signaling, etc.

Evolved Node-B implements Node-B functions as well as protocols traditionally implemented in RNC. The main functions of eNB are header compression, ciphering and reliable delivery of packets. On the control side, eNB incorporates functions such as admission

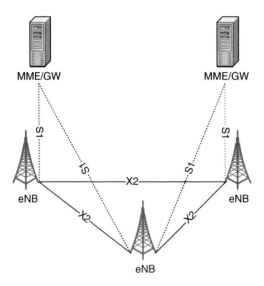

Figure 2.1. Network architecture.

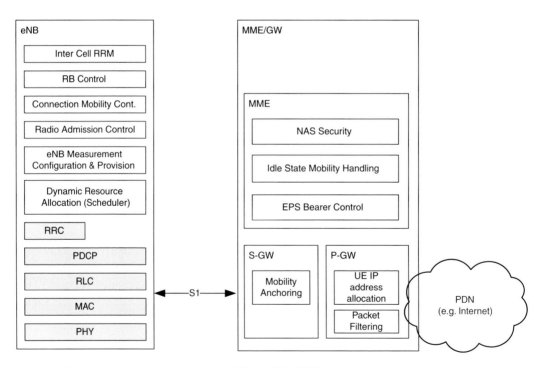

Figure 2.2. Functional split between eNB and MME/GW.

control and radio resource management. Some of the benefits of a single node in the access network are reduced latency and the distribution of RNC processing load into multiple eNBs.

The user plane protocol stack is given in Figure 2.3. We note that packet data convergence protocol (PDCP) and radio link control (RLC) layers traditionally terminated in RNC on

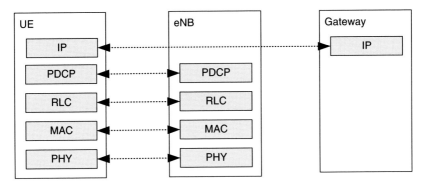

Figure 2.3. User plane protocol.

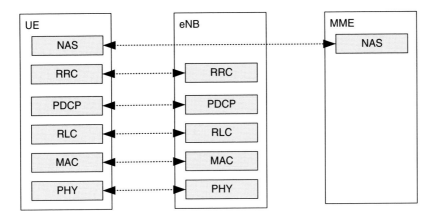

Figure 2.4. Control plane protocol stack.

the network side are now terminated in eNB. The functions performed by these layers are described in Section 2.2.

Figure 2.4 shows the control plane protocol stack. We note that RRC functionality traditionally implemented in RNC is now incorporated into eNB. The RLC and MAC layers perform the same functions as they do for the user plane. The functions performed by the RRC include system information broadcast, paging, radio bearer control, RRC connection management, mobility functions and UE measurement reporting and control. The non-access stratum (NAS) protocol terminated in the MME on the network side and at the UE on the terminal side performs functions such as EPS (evolved packet system) bearer management, authentication and security control, etc.

The S1 and X2 interface protocol stacks are shown in Figures 2.5 and 2.6 respectively. We note that similar protocols are used on these two interfaces. The S1 user plane interface (S1-U) is defined between the eNB and the S-GW. The S1-U interface uses GTP-U (GPRS tunneling protocol – user data tunneling) [2] on UDP/IP transport and provides non-guaranteed delivery of user plane PDUs between the eNB and the S-GW. The GTP-U is a relatively simple IP based tunneling protocol that permits many tunnels between each set of end points. The S1 control plane interface (S1-MME) is defined as being between the eNB and the MME. Similar to the user plane, the transport network layer is built on IP transport and for the reliable

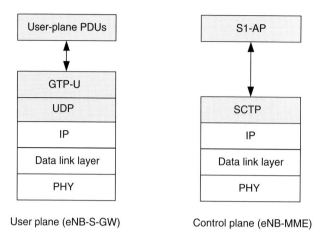

Figure 2.5. S1 interface user and control planes.

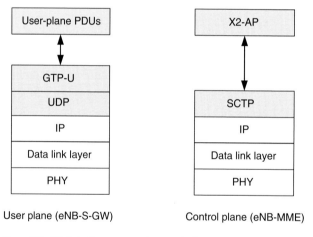

Figure 2.6. X2 interface user and control planes.

transport of signaling messages SCTP (stream control transmission protocol) is used on top of IP. The SCTP protocol operates analogously to TCP ensuring reliable, in-sequence transport of messages with congestion control [3]. The application layer signaling protocols are referred to as S1 application protocol (S1-AP) and X2 application protocol (X2-AP) for S1 and X2 interface control planes respectively.

2.2   QoS and bearer service architecture

Applications such as VoIP, web browsing, video telephony and video streaming have special QoS needs. Therefore, an important feature of any all-packet network is the provision of a QoS mechanism to enable differentiation of packet flows based on QoS requirements. In EPS, QoS flows called EPS bearers are established between the UE and the P-GW as shown in Figure 2.7. A radio bearer transports the packets of an EPS bearer between a UE and an eNB. Each IP flow (e.g. VoIP) is associated with a different EPS bearer and the network can

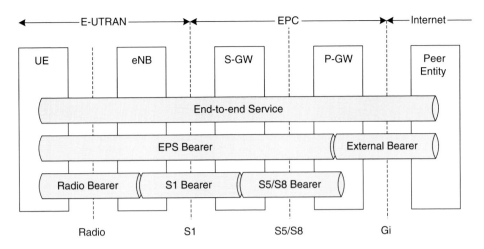

Figure 2.7. EPS bearer service architecture.

prioritize traffic accordingly. When receiving an IP packet from the Internet, P-GW performs packet classification based on certain predefined parameters and sends it an appropriate EPS bearer. Based on the EPS bearer, eNB maps packets to the appropriate radio QoS bearer. There is one-to-one mapping between an EPS bearer and a radio bearer.

2.3   Layer 2 structure

The layer 2 of LTE consists of three sublayers namely medium access control, radio link control (RLC) and packet data convergence protocol (PDCP). The service access point (SAP) between the physical (PHY) layer and the MAC sublayer provide the transport channels while the SAP between the MAC and RLC sublayers provide the logical channels. The MAC sublayer performs multiplexing of logical channels on to the transport channels.

The downlink and uplink layer 2 structures are given in Figures 2.8 and 2.9 respectively. The difference between downlink and uplink structures is that in the downlink, the MAC sublayer also handles the priority among UEs in addition to priority handling among the logical channels of a single UE. The other functions performed by the MAC sublayers in both downlink and uplink include mapping between the logical and the transport channels, multiplexing of RLC packet data units (PDU), padding, transport format selection and hybrid ARQ (HARQ).

The main services and functions of the RLC sublayers include segmentation, ARQ in-sequence delivery and duplicate detection, etc. The in-sequence delivery of upper layer PDUs is not guaranteed at handover. The reliability of RLC can be configured to either acknowledge mode (AM) or un-acknowledge mode (UM) transfers. The UM mode can be used for radio bearers that can tolerate some loss. In AM mode, ARQ functionality of RLC retransmits transport blocks that fail recovery by HARQ. The recovery at HARQ may fail due to hybrid ARQ NACK to ACK error or because the maximum number of retransmission attempts is reached. In this case, the relevant transmitting ARQ entities are notified and potential retransmissions and re-segmentation can be initiated.

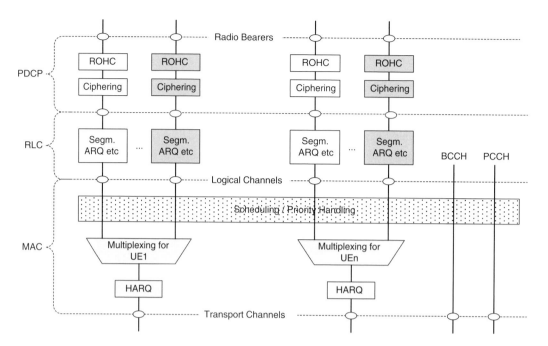

Figure 2.8. Downlink layer 2 structure.

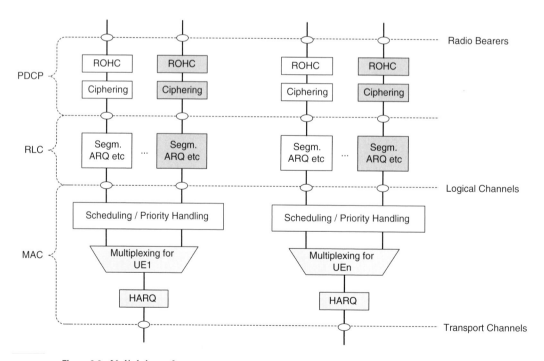

Figure 2.9. Uplink layer 2 structure.

The PDCP layer performs functions such as header compression and decompression, ciphering and in-sequence delivery and duplicate detection at handover for RLC AM, etc. The header compression and decompression is performed using the robust header compression (ROHC) protocol [4].

2.3.1 Downlink logical, transport and physical channels

The relationship between downlink logical, transport and physical channels is shown in Figure 2.10. A logical channel is defined by the type of information it carriers. The logical channels are further divided into control channels and traffic channels. The control channels carry control-plane information, while traffic channels carry user-plane information.

In the downlink, five control channels and two traffic channels are defined. The downlink control channel used for paging information transfer is referred to as the paging control channel (PCCH). This channel is used when the network has no knowledge about the location cell of the UE. The channel that carries system control information is referred to as the broadcast control channel (BCCH). Two channels namely the common control channel (CCCH) and the dedicated control channel (DCCH) can carry information between the network and the UE. The CCCH is used for UEs that have no RRC connection while DCCH is used for UEs that have an RRC connection. The control channel used for the transmission of MBMS control information is referred to as the multicast control channel (MCCH). The MCCH is used by only those UEs receiving MBMS.

The two traffic channels in the downlink are the dedicated traffic channel (DTCH) and the multicast traffic channel (MTCH). A DTCH is a point-to-point channel dedicated to a single UE for the transmission of user information. An MTCH is a point-to-multipoint channel used for the transmission of user traffic to UEs receiving MBMS.

The paging control channel is mapped to a transport channel referred to as paging channel (PCH). The PCH supports discontinuous reception (DRX) to enable UE power saving. A DRX cycle is indicated to the UE by the network. The BCCH is mapped to either a transport

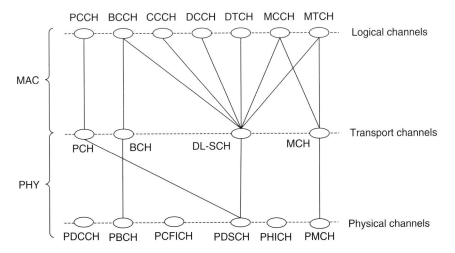

Figure 2.10. Downlink logical, transport and physical channels mapping.

channel referred to as a broadcast channel (BCH) or to the downlink shared channel (DL-SCH). The BCH is characterized by a fixed pre-defined format as this is the first channel UE receives after acquiring synchronization to the cell. The MCCH and MTCH are either mapped to a transport channel called a multicast channel (MCH) or to the downlink shared channel (DL-SCH). The MCH supports MBSFN combining of MBMS transmission from multiple cells. The other logical channels mapped to DL-SCH include CCCH, DCCH and DTCH. The DL-SCH is characterized by support for adaptive modulation/coding, HARQ, power control, semi-static/dynamic resource allocation, DRX, MBMS transmission and multi-antenna technologies. All the four-downlink transport channels have the requirement to be broadcast in the entire coverage area of a cell.

The BCH is mapped to a physical channel referred to as physical broadcast channel (PBCH), which is transmitted over four subframes with 40 ms timing interval. The 40 ms timing is detected blindly without requiring any explicit signaling. Also, each subframe transmission of BCH is self-decodable and UEs with good channel conditions may not need to wait for reception of all the four subframes for PBCH decoding. The PCH and DL-SCH are mapped to a physical channel referred to as physical downlink shared channel (PDSCH). The multicast channel (MCH) is mapped to physical multicast channel (PMCH), which is the multi-cell MBSFN transmission channel.

The three stand-alone physical control channels are the physical control format indicator channel (PCFICH), the physical downlink control channel (PDCCH) and the physical hybrid ARQ indicator channel (PHICH). The PCFICH is transmitted every subframe and carries information on the number of OFDM symbols used for PDCCH. The PDCCH is used to inform the UEs about the resource allocation of PCH and DL-SCH as well as modulation, coding and hybrid ARQ information related to DL-SCH. A maximum of three or four OFDM symbols can be used for PDCCH. With dynamic indication of number of OFDM symbols used for PDCCH via PCFICH, the unused OFDM symbols among the three or four PDCCH OFDM symbols can be used for data transmission. The PHICH is used to carry hybrid ARQ ACK/NACK for uplink transmissions.

2.3.2   Uplink logical, transport and physical channels

The relationship between uplink logical, transport and physical channels is shown in Figure 2.11. In the uplink two control channels and a single traffic channel is defined. As for the downlink, common control channel (CCCH) and dedicated control channel (DCCH) are used to carry information between the network and the UE. The CCCH is used for UEs having no RRC connection while DCCH is used for UEs having an RRC connection. Similar to downlink, dedicated traffic channel (DTCH) is a point-to-point channel dedicated to a single UE for transmission of user information. All the three uplink logical channels are mapped to a transport channel named uplink shared channel (UL-SCH). The UL-SCH supports adaptive modulation/coding, HARQ, power control and semi-static/dynamic resource allocation.

Another transport channel defined for the uplink is referred to as the random access channel (RACH), which can be used for transmission of limited control information from a UE with possibility of collisions with transmissions from other UEs. The RACH is mapped to physical random access channel (PRACH), which carries the random access preamble.

The UL-SCH transport channel is mapped to physical uplink shared channel (PUSCH). A stand-alone uplink physical channel referred to as physical uplink control channel (PUCCH) is used to carry downlink channel quality indication (CQI) reports, scheduling request (SR) and hybrid ARQ ACK/NACK for downlink transmissions.

2.4   Protocol states and states transitions

In the LTE system, two radio resource control (RRC) states namely RRC IDLE and RRC CONNECTED states are defined as depicted in Figure 2.12. A UE moves from RRC IDLE state to RRC CONNECTED state when an RRC connection is successfully established. A UE can move back from RRC CONNECTED to RRC IDLE state by releasing the RRC connection. In the RRC IDLE state, UE can receive broadcast/multicast data, monitors a paging channel to detect incoming calls, performs neighbor cell measurements and cell selection/reselection and acquires system information. Furthermore, in the RRC IDLE state, a UE specific DRX (discontinuous reception) cycle may be configured by upper layers to enable UE power savings. Also, mobility is controlled by the UE in the RRC IDLE state.

In the RRC CONNECTED state, the transfer of unicast data to/from UE, and the transfer of broadcast/multicast data to UE can take place. At lower layers, the UE may be configured with a UE specific DRX/DTX (discontinuous transmission). Furthermore, UE monitors control channels associated with the shared data channel to determine if data is scheduled for it, provides channel quality feedback information, performs neighbor cell measurements and measurement reporting and acquires system information. Unlike the RRC IDLE state, the mobility is controlled by the network in this state.

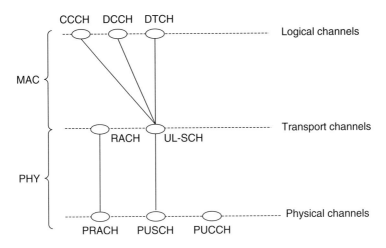

Figure 2.11. Uplink logical, transport and physical channels mapping.

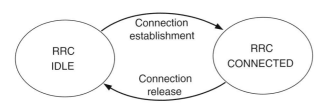

Figure 2.12. UE states and state transitions.

2.5   Seamless mobility support

An important feature of a mobile wireless system such as LTE is support for seamless mobility across eNBs and across MME/GWs. Fast and seamless handovers (HO) are particularly important for delay-sensitive services such as VoIP. The handovers occur more frequently across eNBs than across core networks because the area covered by MME/GW serving a large number of eNBs is generally much larger than the area covered by a single eNB. The signaling on X2 interface between eNBs is used for handover preparation. The S-GW acts as anchor for inter-eNB handovers.

In the LTE system, the network relies on the UE to detect the neighboring cells for handovers and therefore no neighbor cell information is signaled from the network. For the search and measurement of inter-frequency neighboring cells, only the carrier frequencies need to be indicated. An example of active handover in an RRC CONNECTED state is shown in Figure 2.13 where a UE moves from the coverage area of the source eNB (eNB1) to the coverage area of the target eNB (eNB2). The handovers in the RRC CONNECTED state are network controlled and assisted by the UE. The UE sends a radio measurement report to the source eNB1 indicating that the signal quality on eNB2 is better than the signal quality on eNB1. As preparation for handover, the source eNB1 sends the coupling information and the UE context to the target eNB2 (HO request) [6] on the X2 interface. The target eNB2 may perform admission control dependent on the received EPS bearer QoS information. The target eNB configures the required resources according to the received EPS bearer QoS information and reserves a C-RNTI (cell radio network temporary identifier) and optionally a RACH

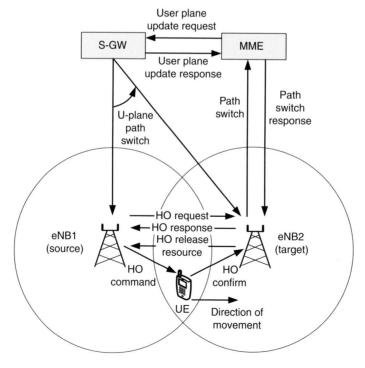

Figure 2.13. Active handovers.

preamble. The C-RNTI provides a unique UE identification at the cell level identifying the RRC connection. When eNB2 signals to eNB1 that it is ready to perform the handover via HO response message, eNB1 commands the UE (HO command) to change the radio bearer to eNB2. The UE receives the HO command with the necessary parameters (i.e. new C-RNTI, optionally dedicated RACH preamble, possible expiry time of the dedicated RACH preamble, etc.) and is commanded by the source eNB to perform the HO. The UE does not need to delay the handover execution for delivering the HARQ/ARQ responses to source eNB.

After receiving the HO command, the UE performs synchronization to the target eNB and accesses the target cell via the random access channel (RACH) following a contention-free procedure if a dedicated RACH preamble was allocated in the HO command or following a contention-based procedure if no dedicated preamble was allocated. The network responds with uplink resource allocation and timing advance to be applied by the UE. When the UE has successfully accessed the target cell, the UE sends the HO confirm message (C-RNTI) along with an uplink buffer status report indicating that the handover procedure is completed for the UE. After receiving the HO confirm message, the target eNB sends a path switch message to the MME to inform that the UE has changed cell. The MME sends a user plane update message to the S-GW. The S-GW switches the downlink data path to the target eNB and sends one or more "end marker" packets on the old path to the source eNB and then releases any user-plane/TNL resources towards the source eNB. Then S-GW sends a user plane update response message to the MME. Then the MME confirms the path switch message from the target eNB with the path switch response message. After the path switch response message is received from the MME, the target eNB informs success of HO to the source eNB by sending release resource message to the source eNB and triggers the release of resources. On receiving the release resource message, the source eNB can release radio and C-plane related resources associated with the UE context.

During handover preparation U-plane tunnels can be established between the source eNB and the target eNB. There is one tunnel established for uplink data forwarding and another one for downlink data forwarding for each EPS bearer for which data forwarding is applied. During handover execution, user data can be forwarded from the source eNB to the target eNB. Forwarding of downlink user data from the source to the target eNB should take place in order as long as packets are received at the source eNB or the source eNB buffer is exhausted.

For mobility management in the RRC IDLE state, concept of tracking area (TA) is introduced. A tracking area generally covers multiple eNBs as depicted in Figure 2.14. The tracking area identity (TAI) information indicating which TA an eNB belongs to is broadcast as part of system information. A UE can detect change of tracking area when it receives a different TAI than in its current cell. The UE updates the MME with its new TA information as it moves across TAs. When P-GW receives data for a UE, it buffers the packets and queries the MME for the UE's location. Then the MME will page the UE in its most current TA. A UE can be registered in multiple TAs simultaneously. This enables power saving at the UE under conditions of high mobility because it does not need to constantly update its location with the MME. This feature also minimizes load on TA boundaries.

2.6 Multicast broadcast system architecture

In the LTE system, the MBMS either use a single-cell transmission or a multi-cell transmission. In single-cell transmission, MBMS is transmitted only in the coverage of a specific cell and therefore combining MBMS transmission from multiple cells is not supported. The single-cell

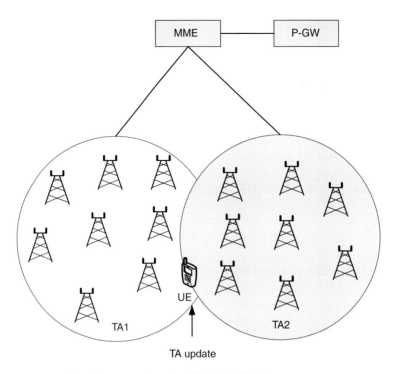

Figure 2.14. Tracking area update for UE in RRC IDLE state.

MBMS transmission is performed on DL-SCH and hence uses the same network architecture as the unicast traffic. The MTCH and MCCH are mapped on DL-SCH for point-to-multipoint transmission and scheduling is done by the eNB. The UEs can be allocated dedicated uplink feedback channels identical to those used in unicast transmission, which enables HARQ ACK/NACK and CQI feedback. The HARQ retransmissions are made using a group (service specific) RNTI (radio network temporary identifier) in a time frame that is co-ordinated with the original MTCH transmission. All UEs receiving MBMS are able to receive the retransmissions and combine with the original transmissions at the HARQ level. The UEs that are allocated a dedicated uplink feedback channel are in RRC CONNECTED state. In order to avoid unnecessary MBMS transmission on MTCH in a cell where there is no MBMS user, network can detect presence of users interested in the MBMS service by polling or through UE service request.

The multi-cell transmission for the evolved multimedia broadcast multicast service (eMBMS) is realized by transmitting identical waveform at the same time from multiple cells. In this case, MTCH and MCCH are mapped on to MCH for point-to-multipoint transmission. This multi-cell transmission mode is referred to as multicast broadcast single frequency network (MBSFN) as described in detail in Chapter 17. An MBSFN transmission from multiple cells within an MBSFN area is seen as a single transmission by the UE. An MBSFN area comprises a group of cells within an MBSFN synchronization area of a network that are co-ordinated to achieve MBSFN transmission. An MBSFN synchronization area is defined as an area of the network in which all eNBs can be synchronized and perform MBSFN transmission. An MBMS service area may consist of multiple MBSFN areas. A cell within an MBSFN synchronization area may form part of multiple SFN areas each characterized by

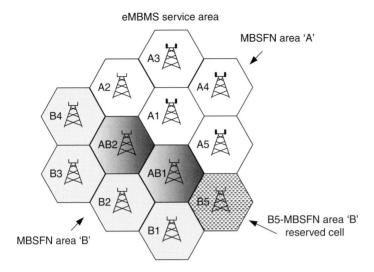

Figure 2.15. The eMBMS service area and MBSFN areas.

different content and set of participating cells. An example of MBMS service area consisting of two MBSFN areas, area A and area B, is depicted in Figure 2.15. The MBSFNA area consists of cells A1–A5, cell AB1 and AB2. The MBSFN area consists of cells B1–B5, cell AB1 and AB2. The cells AB1 and AB2 are part of both MBSFN area A and area B. The cell B5 is part of area B but does not contribute to MBSFN transmission. Such a cell is referred to as MBSFN area reserved cell. The MBSFN area reserved cell may be allowed to transmit for other services on the resources allocated for the MBSFN but at a restricted power. The MBSFN synchronization area, the MBSFN area and reserved cells can be semi-statically configured by O&M.

The MBMS architecture for multi-cell transmission is depicted in Figure 2.16. The multi-cell multicast coordination entity (MCE) is a logical entity, which means it can also be part of another network element such as eNB. The MCE performs functions such as the allocation of the radio resources used by all eNBs in the MBSFN area as well as determining the radio configuration including the modulation and coding scheme. The MBMS GW is also a logical entity whose main function is sending/broadcasting MBMS packets with the SYNC protocol to each eNB transmitting the service. The MBMS GW hosts the PDCP layer of the user plane and uses IP multicast for forwarding MBMS user data to eNBs.

The eNBs are connected to eMBMS GW via a pure user plane interface M1. As M1 is a pure user plane interface, no control plane application part is defined for this interface. Two control plane interfaces M2 and M3 are defined. The application part on M2 interface conveys radio configuration data for the multi-cell transmission mode eNBs. The application part on M3 interface between MBMS GW and MCE performs MBMS session control signaling on EPS bearer level that includes procedures such as session start and stop.

An important requirement for multi-cell MBMS service transmission is MBMS content synchronization to enable MBSFN operation. The eMBMS user plane architecture for content synchronization is depicted in Figure 2.17. A SYNC protocol layer is defined on the transport network layer (TNL) to support the content synchronization mechanism. The SYNC protocol carries additional information that enables eNBs to identify the timing for

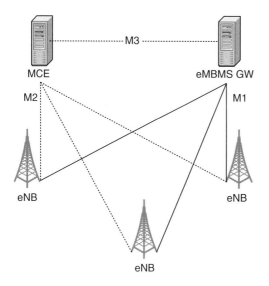

Figure 2.16. eMBMS logical architecture.

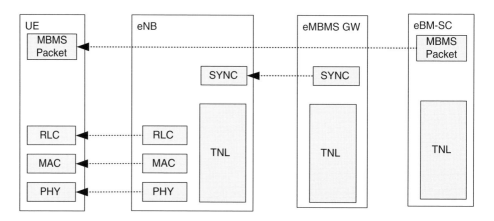

Figure 2.17. The eMBMS user plane architecture for content synchronization.

radio frame transmission as well as detect packet loss. The eNBs participating in multi-cell MBMS transmission are required to comply with content synchronization mechanism. An eNB transmitting only in single-cell service is not required to comply with the stringent timing requirements indicated by SYNC protocol. In case PDCP is used for header compression, it is located in eMBMS GW.

The UEs receiving MTCH transmissions and taking part in at least one MBMS feedback scheme need to be in an RRC CONNECTED state. On the other hand, UEs receiving MTCH transmissions without taking part in an MBMS feedback mechanism can be in either an RRC IDLE or an RRC CONNECTED state. For receiving single-cell transmission of MTCH, a UE may need to be in RRC CONNECTED state. The signaling by which a UE is triggered to move to RRC CONNECTED state solely for single-cell reception purposes is carried on MCCH.

2.7   Summary

The LTE system is based on highly simplified network architecture with only two types of nodes namely eNode-B and MME/GW. Fundamentally, it is a flattened architecture that enables simplified network design while still supporting seamless mobility and advanced QoS mechanisms. This is a major change relative to traditional wireless networks with many more network nodes using hierarchical network architecture. The simplification of network was partly possible because LTE system does not support macro-diversity or soft-handoff and hence does not require a RNC in the access network for macro-diversity combining. Many of the other RNC functions are incorporated into the eNB. The QoS logical connections are provided between the UE and the gateway enabling differentiation of IP flows and meeting the requirements for low-latency applications.

A separate architecture optimized for multi-cell multicast and broadcast is provided, which consists of two logical nodes namely the multicast co-ordination entity (MCE) and the MBMS gateway. The MCE allocates radio resources as well as determines the radio configuration to be used by all eNBs in the MBSFN area. The MBMS gateway broadcasts MBMS packets with the SYNC protocol to each eNB transmitting the service. The MBMS gateway uses IP multicast for forwarding MBMS user data to eNBs.

The layer 2 and radio resource control protocols are designed to enable reliable delivery of data, ciphering, header compression and UE power savings.

References

[1] 3GPP TS 36.300 V8.4.0, Evolved Universal Terrestrial Radio Access Network (E-UTRA): Overall Description.
[2] 3GPP TS 29.060 V8.3.0, GPRS Tunneling Protocol (GTP) Across the Gn and Gp Interface.
[3] IETF RFC 4960, Stream Control Transmission Protocol.
[4] IETF RFC 3095, RObust Header Compression (ROHC): Framework and Four Profiles: RTP, UDP, ESP, and uncompressed.
[5] 3GPP TS 36.331 V8.1.0, Radio Resource Control (RRC) Protocol Specification.
[6] 3GPP TR 23.882 V1.15.1, 3GPP System Architecture Evolution (SAE): Report on Technical Options and Conclusions.

3 Downlink access

The current 3G systems use a wideband code division multiple access (WCDMA) scheme within a 5 MHz bandwidth in both the downlink and the uplink. In WCDMA, multiple users potentially using different orthogonal Walsh codes [1] are multiplexed on to the same carrier. In a WCDMA downlink (Node-B to UE link), the transmissions on different Walsh codes are orthogonal when they are received at the UE. This is due to the fact that the signal is transmitted from a fixed location (base station) on the downlink and all the Walsh codes are received synchronized. Therefore, in the absence of multi-paths, transmissions on different codes do not interfere with each other. However, in the presence of multi-path propagation, which is typical in cellular environments, the Walsh codes are no longer orthogonal and interfere with each other resulting in inter-user and/or inter-symbol interference (ISI). The multi-path interference can possibly be eliminated by using an advanced receiver such as linear minimum mean square error (LMMSE) receiver. However, this comes at the expense of significant increase in receiver complexity.

The multi-path interference problem of WCDMA escalates for larger bandwidths such as 10 and 20 MHz required by LTE for support of higher data rates. This is because chip rate increases for larger bandwidths and hence more multi-paths can be resolved due to shorter chip times. Note that LMMSE receiver complexity increases further for larger bandwidths due to increase of multi-path intensity. Another possibility is to employ multiple 5 MHz WCDMA carriers to support 10 and 20 MHz bandwidths. However, transmitting and receiving multiple carriers add to the Node-B and UE complexity. Another concern against employing WCDMA for LTE was lack of flexible bandwidth support as bandwidths supported can only be multiples of 5 MHz and also bandwidths smaller than 5 MHz cannot be supported.

Taking into account the LTE requirements and scalability and complexity issues associated with WCDMA, it was deemed necessary to employ a new access scheme in the LTE downlink.

3.1 OFDM

Orthogonal frequency division multiplexing (OFDM) approach was first proposed more than four decades ago by R. W. Chang [3]. The scheme was soon analyzed by Saltzberg in [4]. The basic principle of OFDM is to divide the available spectrum into narrow-band parallel channels referred to as subcarriers and transmit information on these parallel channels at a reduced signaling rate. The goal is to let each channel experience almost flat-fading simplifying the channel equalization process. The name OFDM comes from the fact that the frequency responses of the subchannels are overlapping and orthogonal. An example of five OFDM subchannels or subcarriers at frequencies f_1, f_2, f_3, f_4 and

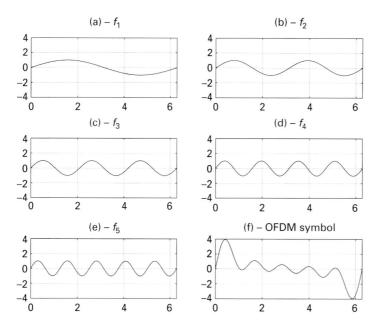

Figure 3.1. An illustration of subcarriers and OFDM symbol.

f_5 is shown in Figure 3.1. The subchannel frequency $f_k = k\Delta f$, where Δf is the subcarrier spacing. Each subcarrier is modulated by a data symbol and an OFDM symbol is formed by simply adding the modulated subcarrier signals. The modulation symbol in Figure 3.1 is obtained by assuming that all subcarriers are modulated by data symbols 1's. An interesting observation to make is that the OFDM symbol signal has much larger signal amplitude variations than the individual subcarriers. This characteristic of OFDM signal leads to larger signal peakiness as discussed in more detail in Chapter 5.

The orthogonality of OFDM subcarriers can be lost when the signal passes through a time-dispersive radio channel due to inter-OFDM symbol interference. However, a cyclic extension of the OFDM signal can be performed [6] to avoid this interference. In cyclic prefix extension, the last part of the OFDM signal is added as cyclic prefix (CP) in the beginning of the OFDM signal as shown in Figure 3.2. The cyclic prefix length is generally chosen to accommodate the maximum delay spread of the wireless channel. The addition of the cyclic prefix makes the transmitted OFDM signal periodic and helps in avoiding inter-OFDM symbol and inter-subcarrier interference as explained later.

The baseband signal within an OFDM symbol can be written as:

$$s(t) = \sum_{k=0}^{(N-1)} X(k) \times e^{j2\pi k \Delta f t}, \tag{3.1}$$

where N represents the number of subcarriers, $X(k)$ complex modulation symbol transmitted on the kth subcarrier $e^{j2\pi k \Delta f t}$ and Δf subcarrier spacing as shown in Figure 3.3.

The OFDM receiver model is given in Figure 3.4. At the receiver, the estimate of the complex modulation symbol $X(m)$ is obtained by multiplying the received signal with $e^{-j2\pi m \Delta f t}$ and integrating over an OFDM symbol duration as below:

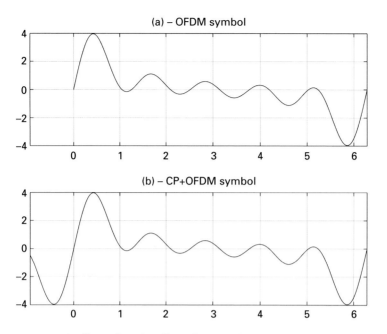

Figure 3.2. An illustration of cyclic prefix extension of OFDM symbol.

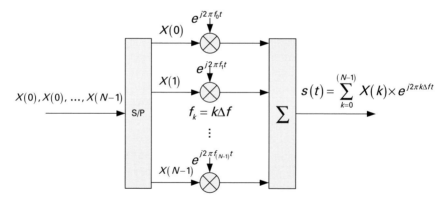

Figure 3.3. OFDM continuous-time transmitter model using banks of subcarrier oscillators.

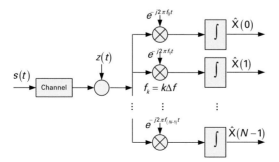

Figure 3.4. OFDM continuous-time receiver model using banks of subcarrier oscillators.

$$\hat{X}(m) = \frac{1}{T_s} \int_0^{T_s} [s(t) + z(t)] e^{-j2\pi m \Delta f t} \, dt$$

$$= \frac{1}{T_s} \int_0^{T_s} s(t) e^{-j2\pi m \Delta f t} \, dt + \frac{1}{T_s} \int_0^{T_s} z(t)^{-j2\pi m \Delta f t} \, dt. \tag{3.2}$$

We assume perfect time and frequency synchronization and ignore the effect of wireless channel. Under these assumptions, the only source of signal degradation is AWGN component $z(t)$. By letting

$$\tilde{z} = \frac{1}{T_s} \int_0^{T_s} z(t)^{-j2\pi m \Delta f t} \, dt, \tag{3.3}$$

we can rewrite (3.2) as:

$$\hat{X}(m) = \frac{1}{T_s} \sum_{k=0}^{(N-1)} X(k) \int_0^{T_s} e^{-j2\pi k \Delta f t} e^{-j2\pi m \Delta f t} \, dt + \tilde{z}$$

$$= \frac{1}{T_s} \sum_{k=0}^{(N-1)} X(k) \int_0^{T_s} e^{-j2\pi \Delta f (k-m) t} \, dt + \tilde{z} \tag{3.4}$$

$$= X(m) + \tilde{z}.$$

We note that under the assumptions of perfect time and frequency synchronization and also that the wireless channel does not cause any time or frequency dispersion, the transmitted data message is perfectly recovered with the only source of degradation being the noise. This is guaranteed by the mutual orthogonality of OFDM subcarriers over the OFDM symbol duration T_s as below:

$$\frac{1}{T_s} \int_0^{T_s} e^{-j2\pi \Delta f (k-m) t} \, dt = \begin{cases} 1 & k = m \\ 0 & k \neq m. \end{cases} \tag{3.5}$$

We can represent the OFDM baseband transmit signal in the following form:

$$s(t) = \sum_{k=0}^{(N-1)} \sum_{n=-\infty}^{\infty} X_n(k) \, e^{j2\pi k \Delta f t} \cdot \text{rect}(t - nT_s), \tag{3.6}$$

where n is the OFDM symbol number and the rectangular filter is defined as:

$$\text{rect}(t) = \begin{cases} \frac{1}{\sqrt{T_s}} & 0 \leq t < T_0 \\ 0 & \text{elsewhere.} \end{cases} \tag{3.7}$$

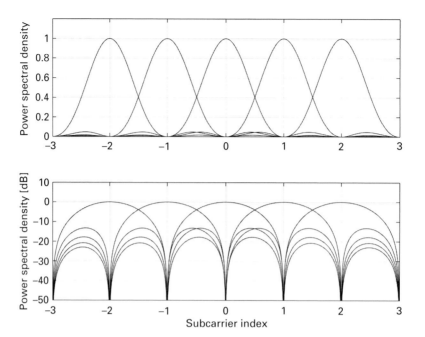

Figure 3.5. OFDM power spectral density.

The power spectral density for the kth subcarrier is given as:

$$\chi_k\left(f\right) = \left[\frac{\sin\left(\pi\left(\frac{f}{\Delta f} - k\right)\right)}{\pi\left(\frac{f}{\Delta f} - k\right)}\right]^2. \tag{3.8}$$

The application of rectangular pulse in OFDM results in a sinc-square shape power spectral density as shown in Figure 3.5. This allows minimal subcarrier separation with overlapping spectra where a signal peak for a given subcarrier corresponds to spectrum nulls for the remaining subcarriers.

Until now we discussed OFDM transmitter and receiver implementation using banks of subcarrier oscillators. A more efficient processing for OFDM using discrete Fourier transform (DFT) and hence eliminating the need for banks of subcarrier oscillators was presented by Weinstein and Ebert [5].

By assuming N times sampling of the OFDM symbol at time instants of $t = \frac{m}{N}T_s$, we can rewrite (3.1) as below:

$$s\left(\frac{m}{N}T_s\right) = \sum_{k=0}^{(N-1)} X\left(k\right) \times e^{j2\pi k\frac{m}{N}} \quad m = 0, 1, \ldots, (N-1). \tag{3.9}$$

In (3.9), we used $\Delta f = 1/T_s$. We can represent $s\left(\frac{m}{N}T_s\right)$ as $s\left(m\right)$ as it depends upon m and then (3.9) can be written as:

$$s\left(m\right) = N \cdot \text{IDFT}\left\{X\left(k\right)\right\} \quad k, m = 0, 1, \ldots, (N-1), \tag{3.10}$$

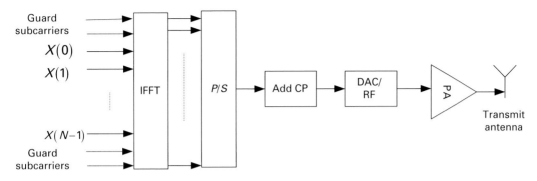

Figure 3.6. A digital implementation of baseband OFDM transmitter.

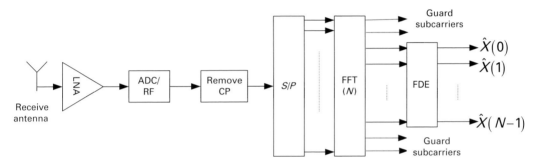

Figure 3.7. A digital implementation of baseband OFDM receiver.

where IDFT is the inverse discrete Fourier transform operator. The number of OFDM subcarriers in an OFDM system is generally selected as power of 2, which allows using more efficient FFT (Fast Fourier Transform) and IFFT (Inverse FFT) algorithms. The complex modulations symbols $X(k) \, k = 0, 1, \ldots, (N - 1)$ are mapped to the input of IFFT. No information is transmitted on the guard subcarriers. A cyclic prefix is added after IFFT operation and the resulting sequence is up-converted to RF, amplified and transmitted as shown in Figure 3.6. In the receiver side, the received signal is filtered, amplified and down-converted from RF as shown in Figure 3.7. The cyclic prefix samples are discarded and an FFT operation is performed on the received samples sequence. A frequency-domain equalization (FDE) operation is performed using channel estimates obtained from received pilots or reference signals and the estimates of the transmitted complex modulation symbols are obtained.

Let $X(k) \in C^N$ be a vector of complex modulation symbols transmitted on N subcarriers. The time-domain samples at the output of the IFFT are then given as:

$$s(n) = W^H X(k) \quad k, n = 0, 1, \ldots, (N - 1), \tag{3.11}$$

where W is a $N \times N$ Fast Fourier Transform (FFT) matrix with entries given by:

$$[W]_{k,n} = e^{-j2\pi kn/N} \quad k, n = 0, 1, \ldots, (N - 1) \tag{3.12}$$

$$
W = \begin{bmatrix}
1 & 1 & 1 & \cdots & 1 \\
1 & e^{-j2\pi/N} & e^{-j4\pi/N} & \cdots & e^{-j2\pi(N-1)/N} \\
\vdots & \vdots & \vdots & \vdots & \vdots \\
1 & e^{-j2\pi(N-2)/N} & e^{-j4\pi(N-2)/N} & \cdots & e^{-j2\pi(N-1)(N-2)/N} \\
1 & e^{-j2\pi(N-1)/N} & e^{-j4\pi(N-1)/N} & \cdots & e^{-j2\pi(N-1)(N-1)/N}
\end{bmatrix}. \tag{3.13}
$$

Since DFT matrix W is a unitary matrix $\left(WW^H = W^H W = I_N\right)$, its inverse W^{-1} referred to as inverse FFT (IFFT) matrix can be obtained by simply taking conjugate transpose or Hermitian transpose of W as below:

$$
W^{-1} = \left[W^H\right]_{k,n} = e^{j2\pi kn/N} \quad k,n = 0,1,\ldots,(N-1). \tag{3.14}
$$

In order to avoid interference between OFDM symbols, a length G cyclic prefix is inserted in front of the time-domain samples and the resulting time-domain signal sequence s' is given as:

$$
s' = C_T \times s \tag{3.15}
$$

where C_T is a $(N+G) \times N$ matrix that represents the cyclic prefix addition operation at the transmitter and is defined as:

$$
C_T \equiv \begin{bmatrix} 0_{G\times(N-G)} & I_G \\ & I_N \end{bmatrix}. \tag{3.16}
$$

The time-domain samples sequence after cyclic prefix addition is then given as:

$$
\begin{aligned}
s'(n) &= \begin{bmatrix} 0_{G\times(N-G)} & I_G \\ & I_N \end{bmatrix} \begin{bmatrix} s(0) \\ s(1) \\ \vdots \\ s(N-1) \end{bmatrix} \\
&= \begin{bmatrix} s(N-G) & \cdots & s(N-1) & s(0) & \cdots & s(N-1) \end{bmatrix}^T.
\end{aligned} \tag{3.17}
$$

This signal is transmitted and goes through a linear channel before arriving at the receiver. The multi-path propagation channel can be modeled as a finite impulse response (FIR) filter of order L with tap coefficients $[h_0, h_1, \ldots, h_L]^T$. The time-domain received signal is then given as:

$$
y(n) = \sum_{i=0}^{L} h_i s'(n-i) + z(n) \quad n = 0,1,\ldots,(N-1) \tag{3.18}
$$

where $z(n) \sim N\left(0,\sigma^2\right)$ is additive white Gaussian noise (AWGN). The convolution operation in Equation (3.18) can be constructed as a matrix multiplication with matrix $\tilde{H} \in C^{(N+L)\times(N+L)}$ given below:

$$\tilde{H} = \begin{bmatrix} h_0 & 0 & 0 & 0 & 0 & 0 & 0 \\ h_1 & h_0 & 0 & \vdots & 0 & 0 & 0 \\ \vdots & h_1 & \ddots & 0 & \vdots & 0 & 0 \\ h_L & \vdots & \ddots & h_0 & 0 & \vdots & 0 \\ 0 & h_L & \vdots & h_1 & h_0 & 0 & \vdots \\ \vdots & \vdots & \ddots & \vdots & \ddots & \ddots & 0 \\ 0 & 0 & \cdots & h_L & \cdots & h_1 & h_0 \end{bmatrix}. \tag{3.19}$$

We note that the above time-domain channel matrix is in familiar lower triangular Toeplitz form. A Toeplitz matrix allows constructing a convolution operation as a matrix multiplication. A cyclic prefix length of at least L samples is necessary for avoiding inter OFDM-symbol interference in a channel with time dispersion of L samples. Let us assume that the length of the channel impulse response and the length of the cyclic prefix are equal $G = L$. This is a reasonable assumption as the cyclic prefix length is generally selected to cover the maximum time dispersion in a channel.

The received time-domain signal can then be written as:

$$y = \tilde{H}s' + z = \tilde{H}C_T W^H X + z. \tag{3.20}$$

The first operation performed on the received signal is the removal of the cyclic prefix that can be expressed as multiplication of the received signal with a $N \times (N + G)$ matrix C_R given below

$$C_R = [0_{N \times G} \; I_N]. \tag{3.21}$$

The received signal is then written as:

$$y' = C_R[\tilde{H}s' + z(n)] = C_R \tilde{H} C_T W^H X + z'(n). \tag{3.22}$$

Let us have a closer look at the matrix $C_R \tilde{H} C_T$:

$$C_R \tilde{H} C_T = \begin{bmatrix} 0_{N \times G} & I_N \end{bmatrix} \times \begin{bmatrix} h_0 & 0 & 0 & 0 & 0 & 0 & 0 \\ h_1 & h_0 & 0 & \vdots & 0 & 0 & 0 \\ \vdots & h_1 & \ddots & 0 & \vdots & 0 & 0 \\ h_{(G-1)} & \vdots & \ddots & h_0 & 0 & \vdots & 0 \\ 0 & h_{(G-1)} & \vdots & h_1 & h_0 & 0 & \vdots \\ \vdots & \vdots & \ddots & \vdots & \ddots & \ddots & 0 \\ 0 & 0 & \cdots & h_{(G-1)} & \cdots & h_1 & h_0 \end{bmatrix}$$

$$\times \begin{bmatrix} 0_{G \times (N-G)} & I_G \\ & I_N \end{bmatrix}$$

$$
= \begin{bmatrix}
h_0 & 0 & 0 & 0 & h_{(G-1)} & \cdots & h_1 \\
h_1 & h_0 & 0 & \vdots & 0 & 0 & h_2 \\
\vdots & h_1 & \ddots & 0 & \vdots & 0 & 0 \\
h_{(G-1)} & \vdots & \ddots & h_0 & 0 & \vdots & h_{(G-1)} \\
0 & h_{(G-1)} & \vdots & h_1 & h_0 & 0 & \vdots \\
\vdots & \vdots & \ddots & \vdots & \ddots & \ddots & 0 \\
0 & 0 & \cdots & h_{(G-1)} & h_{(G-2)} & \cdots & h_0
\end{bmatrix}. \tag{3.23}
$$

We note that the use of a cyclic prefix in OFDM changes the Toeplitz-like channel matrix into a circulant matrix. Equivalently, the use of a cyclic prefix transforms the linear convolution in the channel to a circular convolution. The frequency-domain received signal is obtained by performing FFT operation on the received time-domain signal as below:

$$
Y = W C_R \left[\tilde{H} s' + z(n) \right] = W \overbrace{C_R \tilde{H} C_T}^{\text{circulant}} \underbrace{W^H}_{\text{diagonal}} X + z'', \tag{3.24}
$$

where $z'' = W C_R z$. We know that a circulant matrix can be diagonalized by the DFT matrix. The equivalent diagonal channel matrix $H = W C_R \tilde{H} C_T W^H$ that includes the affects of cyclic prefix and FFT can be written as:

$$
H = W C_R \tilde{H} C_T W^H = \begin{bmatrix}
H(0) & 0 & \cdots & 0 & 0 \\
0 & H(1) & 0 & \cdots & 0 \\
\vdots & \ddots & \ddots & \ddots & \vdots \\
0 & 0 & \cdots & H(N-2) & 0 \\
0 & 0 & \cdots & 0 & H(N-1)
\end{bmatrix}. \tag{3.25}
$$

With the above definition of equivalent channel matrix H, the received frequency-domain signal is simply given as:

$$
Y = HX + z''. \tag{3.26}
$$

We note that in an OFDM system, the modulation symbols can be transmitted in an ISI-free fashion in a multi-path propagation channel. This is achieved with a low complexity 1-tap equalizer per subcarrier thanks to diagonal structure of the equivalent channel matrix H.

3.2   Downlink capacity comparison

A key difference between OFDM and WCDMA with Rake receiver is that the latter suffers from ISI (Inter Symbol Interference) in multi-path dispersive fading channels. In the absence of multi-paths, which is the case for a single-path flat-fading channel, the performance of OFDM and WCDMA is similar. The performance difference between OFDM and WCDMA depends upon the extent of multi-path interference. In general, the use of larger bandwidths in multi-path dispersive channels leads to more multi-path interference and therefore favoring OFDM over WCDMA. In this section, we compare the performance of OFDM and WCDMA for the case where the WCDMA performance is limited by the multi-path interference. This case is representative of larger bandwidths and highly dispersive channels where OFDM is expected to provide the greatest advantage over WCDMA.

3.2.1 Wideband CDMA capacity

In a WCDMA system using a Rake receiver, the signal-to-interference-plus-noise ratio (SINR) for the signal received at nth multi-path component ρ_n can be expressed as:

$$\rho_n = \frac{P_n}{\left(fP + \sum_{i=0,i\neq n}^{N-1} P_i + N_0 \right)}, \tag{3.27}$$

where P_n is the received power on the nth multi-path component from the cell of interest. Also, f represents the ratio between other-cell and own-cell signal. For simplicity of analysis, let us assume that equal power of P/N is received on each of the N multi-path components where P is the total received power on all the multi-path components. With this assumption, Equation (3.75) can be simplified as:

$$\rho_n = \frac{P/N}{\left(fP + (N-1)\dfrac{P}{N} + N_0 \right)}. \tag{3.28}$$

In deriving (3.27) and (3.28), we assumed that a single-user is scheduled at a time in a TDM (time-division multiplexed) fashion using all the cell resources. Therefore, there is no need to account for the interference among users in the same cell. Further assuming a maximum-ratio-combining (MRC) for signals received at different Rake fingers (multi-paths), the average received signal-to-noise ratio can be expressed as the sum:

$$\rho_{\text{WCDMA}} = \sum_{n=0}^{(N-1)} \rho_n = \sum_{n=0}^{(N-1)} \left(\frac{P/N}{\left(fP+(N-1)\dfrac{P}{N}+N_0 \right)} \right) \tag{3.29}$$

$$= \frac{P}{fP+\left(1-\dfrac{1}{N}\right)P+N_0} = \frac{\rho}{f\rho+\left(1-\dfrac{1}{N}\right)\rho+1}$$

where $\rho = P/N_0$ is the SINR when all the power is received on a single-path and there is no interference from the other cells. This should actually be referred to as SNR rather than SINR as the interference is zero. For a single-path frequency-flat fading channel ($N = 1$), the SINR in a WCDMA system becomes:

$$\rho_{\text{WCDMA}}^1 = \frac{P}{f \cdot P + N_0}. \tag{3.30}$$

For a very large number of multi-paths, that is $N \gg 1$

$$\rho_{\text{WCDMA}} = \frac{\rho}{\rho \cdot (f+1) + 1}. \tag{3.31}$$

It can be noted that the maximum achievable SINR in a WCDMA system is limited to zero dB when the WCDMA signal is received with a large number of multi-path components ($N \gg 1$). The capacity limit for a WCDMA system can then be written as:

$$C_{\text{WCDMA}} = \log_2(1 + \rho_{\text{WCDMA}}) \quad \text{[b/s/Hz]}. \tag{3.32}$$

With maximum SINR limited to zero dB, the peak data rate achievable with WCDMA will be limited to 1 b/s/Hz when the signal is received with an arbitrarily large number of multi-path components.

3.2.2   OFDMA capacity

In OFDM, there is no multi-path interference due to use of a cyclic prefix and 1-tap equalization of OFDM subcarriers. Therefore, the sources of SINR degradation in an OFDMA system are the other-cell interference and the background noise. The SINR in an OFDM system is then approximated as:

$$\rho_{OFDM} = \frac{P}{f \cdot P + N_0}. \tag{3.33}$$

By comparing the above equation with Equation (3.30), we note that SINRs in an OFDM system and a WCDMA system are the same for a single-path frequency-flat fading channel. The capacity limit of an OFDM system is given as:

$$C_{\text{OFDM}} = \log_2 \left(1 + \frac{P}{fP + N_0} \right) = \log_2 \left(1 + \frac{\rho}{\rho \cdot f + 1} \right) \quad \text{[b/s/Hz]}. \tag{3.34}$$

We also need to take into account the cyclic prefix overhead for the OFDMA case. Therefore, the capacity of an OFDM system is scaled-down to account for CP overhead as below:

$$C_{\text{OFDMA}} = \left(1 - \frac{\Delta}{T_s} \right) \cdot \log_2 \left(1 + \frac{\rho}{\rho \cdot f + 1} \right) \quad \text{[b/s/Hz]}, \tag{3.35}$$

where T_s is the OFDM symbol duration and Δ is the cyclic prefix duration.

It can be noted that the SINR in an OFDM system degrades with increasing f. In general, f is larger for cell-edge users experiencing high interference from neighboring cells and lower for cell-center users receiving little interference from the neighboring cells. Therefore, users closer to the cell with low f are expected to benefit more from OFDMA than users at the cell edge. The performance of cell edge users is generally dominated by interference from a neighboring cell rather than the multi-path interference. Therefore, OFDMA is expected to provide relatively smaller gains for the cell-edge users.

3.2.3 Capacity comparison results

Figures 3.8 and 3.9 provide the performance comparison of WCDMA and OFDMA for the case of $\rho = 10$ and 0 dB respectively. Note that ρ is defined as the SINR when all the power is received on a single-path and there is no interference from the other cells. In these results, we assumed 10% CP overhead for the OFDMA case. The performance of WCDMA is provided for three different cases where the number of equal-power multi-path components is 2, 4 and very large ($N \gg 1$). Note that the WCDMA performance is similar to that of OFDM for the case of single-path ($N = 1$) channel. It can be noted that OFDMA provides greater capacity advantage for the cases when the other-cell interference f is relatively small. This is usually the case for the users closer to the base station. For these good users, the performance is dominated by the multi-path interference for WCDMA and, therefore, eliminating multi-path interference by employing OFDMA improves capacity significantly. However, for the weak users at the cell edge with relatively larger f, the performance is dominated by interference from a neighboring cell rather than the own-cell multi-path interference. Therefore, OFDMA provides little advantage for the weak users in the cell. It can also be observed that the performance difference between OFDM and WCDMA is small for smaller ρ as shown in Figure 3.9. In general, multi-path interference has greater impact on users with good channel conditions compared to the users with relatively weaker channel conditions. The potential performance gains of OFDMA over WCDMA are also summarized in Table 3.1.

3.3 Effect of frequency selectivity on OFDM performance

In case of OFDMA, an interesting aspect is the frequency selective fading introduced across the OFDMA signal spectrum when signal is received over multiple paths. Different subcarriers

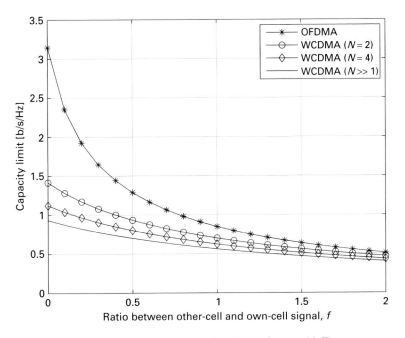

Figure 3.8. Capacity limits for OFDMA and WCDMA for $\rho = 10$ dB.

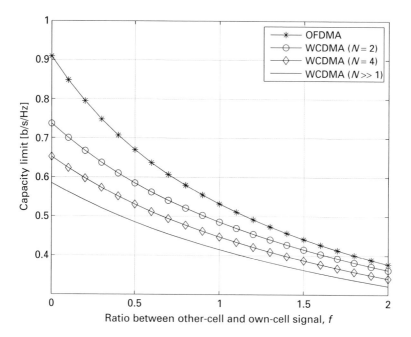

Figure 3.9. Capacity limits for OFDMA and WCDMA for $\rho = 0$ dB.

Table 3.1. Capacity gains of OFDMA over WCDMA.

		OFDMA gains over WCDMA		
	f	$(N = 2)$	$(N = 4)$	$(N \gg 1)$
$\rho = 10$ dB	0.0	122%	180%	237%
	1.0	21%	36%	51%
	2.0	9%	18%	27%
$\rho = 0$ dB	0.0	23%	40%	55%
	1.0	10%	19%	28%
	2.0	4%	11%	17%

then experience different fading due to frequency selective effect. In this section, we evaluate the effect of frequency selective fading on OFDM capacity performance. Let N_{sc} be the total number of subcarriers over which the transmission is performed. We also assume that all subcarriers are allocated equal power with $P_{\text{sc}} = P/N_{\text{sc}}$ where P is the total power across the whole bandwidth and P_{sc} is the power per subcarrier.

The capacity limit of an OFDM system in a frequency-selective channel can then be written as:

$$C_{\text{OFDM}} = \left(1 - \frac{\Delta}{T_s}\right) \cdot \frac{1}{N_{\text{sc}}} \sum_{i=1}^{N_{\text{sc}}} \log_2 \left(1 + \frac{|H_c(i)|^2 P_{\text{sc}}}{f \times |H_{\text{int}}(i)|^2 \times P_{\text{sc}} + N_0}\right) \text{ [b/s/Hz]}, \quad (3.36)$$

where $|H_c(i)|^2$ and $|H_{\text{int}}(i)|^2$ represent channel gains from the cell of interest and the interferer respectively. Note that we assumed that the channel transfer function H_c is perfectly known at the receiver. By defining average SNR per subcarrier as $\rho_{sc} = P_{sc}/N_0$, Equation (3.36) can be simplified as below:

$$C_{\text{OFDM}} = \left(1 - \frac{\Delta}{T_s}\right) \cdot \frac{1}{N_{sc}} \sum_{i=1}^{N_{sc}} \log_2\left(1 + \frac{|H_c(i)|^2 \rho_{sc}}{f \times |H_{\text{int}}(i)|^2 \times \rho_{sc} + 1}\right) \text{ [b/s/Hz]}. \quad (3.37)$$

Let us ignore the other-cell interference for a moment, that is ($f = 0$), and rewrite the above equation as:

$$C_{\text{OFDM}} = \left(1 - \frac{\Delta}{T_s}\right) \cdot E\left[\log_2\left(1 + |H_c(i)|^2 \rho_{sc}\right)\right] \text{ [b/s/Hz]}. \quad (3.38)$$

At low SNR $|H_c(i)|^2 \rho_{sc} \ll 1$ and the above expression can be approximated as:

$$C_{\text{OFDM}} \approx \left(1 - \frac{\Delta}{T_s}\right) \cdot E\left[|H_c(i)|^2 \rho_{sc} \log_2 e\right] \text{ [b/s/Hz]}. \quad (3.39)$$

Here we used the approximation $\log_2(1 + x) \approx x \log_2 e$ for x small.
As $E\left[|H_c(i)|^2\right] = 1$, we can rewrite (3.87) as:

$$C_{\text{OFDM}} \approx \left(1 - \frac{\Delta}{T_s}\right) \cdot \rho_{sc} \log_2 e \text{ [b/s/Hz]}. \quad (3.40)$$

We note that capacity at low SNR is a linear function of SNR. This is in contrast to high SNR where capacity is a concave function of SNR.

At very high SNR, $|H_c(i)|^2 \rho_{sc} \gg 1$ and Equation (3.86) can be simplified as:

$$C_{\text{OFDM}} \approx \left(1 - \frac{\Delta}{T_s}\right) \cdot E\left[\log_2\left(|H_c(i)|^2 \rho_{sc}\right)\right] \text{ [b/s/Hz]} \quad (3.41)$$

$$C_{\text{OFDM}} \approx \left(1 - \frac{\Delta}{T_s}\right) \cdot \left(\log_2(\rho_{sc}) + E\left[\log_2\left(|H_c(i)|^2\right)\right]\right) \text{ [b/s/Hz]}, \quad (3.42)$$

where $E\left[\log_2\left(|H_c(i)|^2\right)\right]$ represents penalty due to frequency selectivity at high SNR. For a Rayleigh fading channel,

$$E\left[\log_2\left(|H_c(i)|^2\right)\right] = -0.83 \text{ [b/s/Hz]}. \quad (3.43)$$

This means an SNR penalty of 2.5 dB $\left(10 \times \log_{10}\left[2^{0.83}\right]\right)$ relative to the case of a flat-fading (AWGN) channel. Note that this SNR penalty is at very high SNR where $|H_c(i)|^2 \rho_{sc} \gg 1$. Let us define effective SNR as below:

$$SNR_{\text{eff}} = 2^{C_{\text{OFDM}}} - 1, \quad (3.44)$$

where C is the Ergodic capacity defined as

$$C_{\text{OFDM}} = E\left[\log_2\left(1 + |H_c(i)|^2 \rho_{sc}\right)\right] \text{ [b/s/Hz]}. \quad (3.45)$$

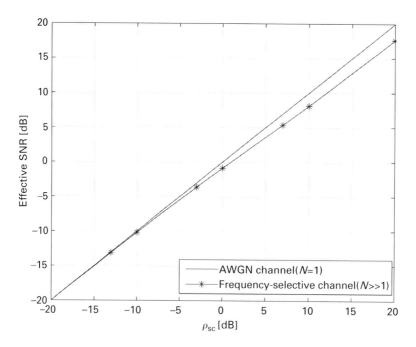

Figure 3.10. Effective SNR for a freqeuncy-flat and Rayleigh fading i.i.d channel.

Note that Equation (3.45) is the same as Equation (3.39) with the OFDM cyclic prefix overhead dropped for simplicity. We note that in a flat-fading channel, $SNR_{eff} = \rho_{sc}$ because $|H_c(i)|^2 = 1$ for all subcarriers. In a frequency selective channel $SNR_{eff} \leq \rho_{sc}$ and the difference between SNR_{eff} and ρ_{sc} depends upon the channel selectivity and the ρ_{sc} itself. We quantified this difference as 2.5 dB in a Rayleigh fading channel at very high ρ_{sc}. We plot SNR_{eff} versus ρ_{sc} for a flat-fading channel and an $i.i.d.$ Rayleigh fading channel in Figure 3.10. In the $i.i.d.$ Rayleigh fading channel, the fading is independent from one subcarrier to the other subcarrier and is representative of a case with a very large number of multi-paths. We note that SNR penalty due to frequency-selectivity is smaller at lower SNR and approaches to 2.5 dB at very large SNR. At very low SNR, there is no difference between SNR_{eff} versus ρ_{sc} as predicted by Equation (3.40). We observe an effective SNR penalty of 0.89, 1.88 and 2.36 dB, for $\rho_{sc} = 0, 10, 20$ dB respectively.

Let us turn our attention to the capacity in (3.36) and refer to it as capacity with frequency selective interference (FSI), C_{OFDM}^{FSI}:

$$C_{OFDM}^{FSI} = \left(1 - \frac{\Delta}{T_s}\right) \cdot \frac{1}{N_{sc}} \sum_{i=1}^{N_{sc}} \log_2\left(1 + \frac{|H_c(i)|^2 \rho_{sc}}{f \times |H_{int}(i)|^2 \times \rho_{sc} + 1}\right) [\text{b/s/Hz}]. \qquad (3.46)$$

By frequency selective interference we mean that the interference channel is frequency selective with the varying channel gains $H_{int}(i)$. In a frequency-flat channel, $|H_{int}(i)|^2 = 1$ and (3.46) can be simplified as below:

$$C_{OFDM}^{NFSI} = \left(1 - \frac{\Delta}{T_s}\right) \cdot \frac{1}{N_{sc}} \sum_{i=1}^{N_{sc}} \log_2\left(1 + \frac{|H_c(i)|^2 \rho_{sc}}{f \times \rho_{sc} + 1}\right) [\text{b/s/Hz}]. \qquad (3.47)$$

Furthermore, in a frequency-flat (AWGN) channel, $|H_c(i)|^2 = 1$ and (3.47) can be written as:

$$C_{\text{OFDM}}^{\text{AWGN}} = \left(1 - \frac{\Delta}{T_s}\right) \log_2 \left(1 + \frac{1}{f + \frac{1}{\rho_{\text{sc}}}}\right) \text{[b/s/Hz]}. \qquad (3.48)$$

We note that the capacity relationship in (3.48) is the same as in (3.35) that we used for OFDM capacity comparison against WCDMA capacity. The effective SINR for the three cases in (3.46)–(3.48) are obtained using the relationship in (3.44). The effective SINR plots are shown in Figures 3.11 and 3.12 for the case of $(f = 0.1)$ and $(f = 1)$ respectively. We assume independent Rayleigh fading from subcarrier to subcarrier. Also, in the case of a frequency selective interferer, the interference is assumed to be independent Rayleigh fading from subcarrier to subcarrier and also independent of the signal fading. We note that a frequency selective (FS) interference helps to improve OFDM performance relative to the case of a non-frequency selective (NFS) interference. Even for relatively small interference $(f = 0.1)$, the performance with a frequency selective (FS) interference approaches the AWGN performance at very high SNR, ρ_{sc}. For relatively larger interference assuming interference power equal to the signal power $(f = 1.0)$, the performance with a frequency selective (FS) interference is better than the AWGN channel performance particularly at high ρ_{sc}. Therefore, we can conclude that a frequency-selective interference can actually result in OFDM performance improvement relative to a flat-fading interferer. Also, the assumption we made for OFDM capacity when comparing it against WCDMA capacity is pessimistic particularly for the higher SNR case and therefore actual gains of OFDM over WCDMA can even be larger than predicted in Section 3.2.

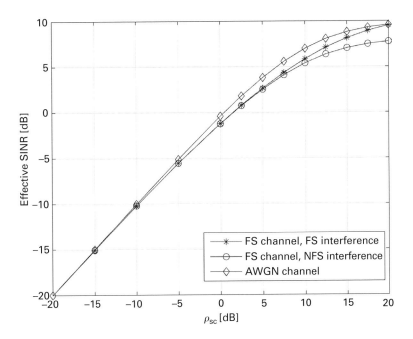

Figure 3.11. Effective SNR with and without a frequency selective interference $(f = 0.1)$.

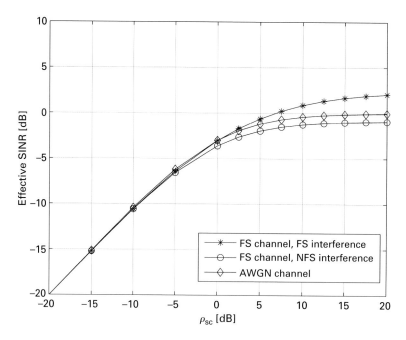

Figure 3.12. Effective SNR with and without a frequency selective interference ($f = 1.0$).

3.4   Single-carrier with FDE

An alternative approach following the same spirit as OFDM is to define a modulation format in time domain instead of in frequency domain and apply the frequency domain equalization (FDE) [6] technique similar to single-tap OFDM subcarrier equalization. Such an approach where modulation and demodulation is performed in the time domain is referred to as single-carrier FDE [7, 8] as shown in Figure 3.13. We know that in case of OFDM, an IDFT operation is performed at the transmitter while a DFT operation is performed at the receiver. In SC-FDE, however, both the IDFT and DFT operations are performed at the receiver. We can see this as the IDFT operation of OFDM transmitter is moved to the SC-FDE receiver. The frequency domain equalization is performed on the received symbols in the frequency-domain after the DFT operation. Note that in both OFDM and SC-FDE, equalization is always performed in the frequency domain. The difference is that modulation and demodulation in OFDM is performed in the frequency domain, while in SC-FDE these operations are performed in the time domain.

3.4.1   SNR analysis for OFDM

In case of OFDM, an IDFT operation is performed on the complex-valued modulation symbols $X = [X(0), \ldots, X(N-1)]^T$ transmitted in the frequency domain. Before equalization, the received frequency domain symbols $Y = [Y(0), \ldots, Y(N-1)]^T$ are given as:

$$Y = HX + Z, \tag{3.49}$$

where $Z = [Z(0), \ldots, Z(N-1)]^T$ are frequency-domain AWGN samples. Also the frequency-domain channel matrix H is a diagonal matrix given as:

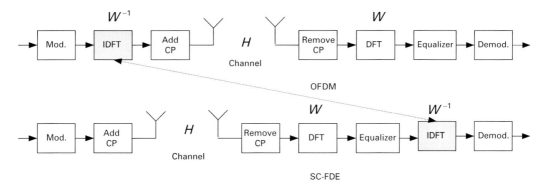

Figure 3.13. Block diagram of OFDM and single-carrier FDE.

$$H = \begin{bmatrix} H(0) & 0 & \cdots & 0 & 0 \\ 0 & H(1) & 0 & \cdots & 0 \\ \vdots & \ddots & \ddots & \ddots & \vdots \\ 0 & 0 & \cdots & H(N-2) & 0 \\ 0 & 0 & \cdots & 0 & H(N-1) \end{bmatrix}. \tag{3.50}$$

We rewrite Equation (3.49) in its expanded form as below:

$$\begin{bmatrix} y(0) \\ y(1) \\ \vdots \\ y(N-2) \\ y(N-1) \end{bmatrix} = \begin{bmatrix} H(0) & 0 & \cdots & 0 & 0 \\ 0 & H(1) & 0 & \cdots & 0 \\ \vdots & \ddots & \ddots & \ddots & \vdots \\ 0 & 0 & \cdots & H(N-2) & 0 \\ 0 & 0 & \cdots & 0 & H(N-1) \end{bmatrix}$$
$$\times \begin{bmatrix} X(0) \\ X(1) \\ \vdots \\ X(N-2) \\ X(N-1) \end{bmatrix} + \begin{bmatrix} Z(0) \\ Z(1) \\ \vdots \\ Z(N-2) \\ Z(N-1) \end{bmatrix}. \tag{3.51}$$

Assuming an MRC receiver, the channel equalization matrix is simply conjugate transpose H^H of the channel matrix H. The estimates of received modulation symbols in the frequency domain are:

$$\hat{X} = H^H (HX + Z) = H^H H X + Z', \tag{3.52}$$

where $Z' = H^H Z$ and the matrix $H^H H$ is given by:

$$H^H H = \begin{bmatrix} |H(0)|^2 & 0 & \cdots & 0 & 0 \\ 0 & |H(1)|^2 & 0 & \cdots & 0 \\ \vdots & \ddots & \ddots & \ddots & \vdots \\ 0 & 0 & \cdots & |H(N-2)|^2 & 0 \\ 0 & 0 & \cdots & 0 & |H(N-1)|^2 \end{bmatrix}. \tag{3.53}$$

We note that all the non-diagonal elements of the matrix $H^H H$ are zero, which means that the channel matrix H is always orthogonal. This is the case even when the channel gains $H(0), H(1), \ldots, H(N-1)$ are different as is the case in a multi-path frequency-selective channel. The received modulation symbols estimates are then given as:

$$
\begin{bmatrix} \hat{X}(0) \\ \hat{X}(1) \\ \vdots \\ \hat{X}(N-2) \\ \hat{X}(N-1) \end{bmatrix} = \begin{bmatrix} |H(0)|^2 & 0 & \cdots & 0 & 0 \\ 0 & |H(1)|^2 & 0 & \cdots & 0 \\ \vdots & & \ddots & \ddots & \vdots \\ 0 & 0 & \cdots & |H(N-2)|^2 & 0 \\ 0 & 0 & \cdots & 0 & |H(N-1)|^2 \end{bmatrix}
$$

$$
\times \begin{bmatrix} X(0) \\ X(1) \\ \vdots \\ X(N-2) \\ X(N-1) \end{bmatrix} + \begin{bmatrix} Z'(0) \\ Z'(1) \\ \vdots \\ Z'(N-2) \\ Z'(N-1) \end{bmatrix}
$$

$$
\begin{bmatrix} \hat{X}(0) \\ \hat{X}(1) \\ \vdots \\ \hat{X}(N-2) \\ \hat{X}(N-1) \end{bmatrix} = \begin{bmatrix} |H(0)|^2 X(0) \\ |H(1)|^2 X(1) \\ \vdots \\ |H(N-2)|^2 X(N-2) \\ |H(N-1)|^2 X(N-1) \end{bmatrix} + \begin{bmatrix} Z'(0) \\ Z'(1) \\ \vdots \\ Z'(N-2) \\ Z'(N-1) \end{bmatrix} \tag{3.54}
$$

where $Z'(k) = H^*(k)Z(k)$, $k = 0, 1, \ldots, (N-1)$.

The modulation symbol estimate on the kth subcarrier from Equation (3.54) is:

$$
\hat{X}(k) = \underbrace{|H(k)|^2 X(k)}_{\text{signal}} + \underbrace{Z'(k)}_{\text{noise}}. \tag{3.55}
$$

We note that there is no inter-subcarrier or inter-modulation-symbol symbol interference in OFDM. The SNR on the kth subcarrier for an OFDM system using MRC receiver is then simply:

$$
\gamma_{\text{OFDM-MRC}}(k) = \frac{P|H(k)|^2}{\sigma^2} = \gamma |H(k)|^2, \tag{3.56}
$$

where $\gamma = P/\sigma^2$ is the ratio of signal power to the AWGN noise power. We note that an OFDM system does not provide any frequency diversity within a modulation symbol as each modulation symbol experiences a single channel gain. In order to exploit the frequency diversity via channel coding, data symbols need to be coded and interleaved over the transmitted subcarriers.

In case of an MMSE receiver, the equalization weight on the kth subcarrier is given as:

$$g(k) = \frac{H(k)^*}{|H(k)|^2 + \sigma^2}. \tag{3.57}$$

The estimates of received modulation symbols in the frequency-domain are:

$$\hat{X} = G(HX + Z) = GHX + Z', \tag{3.58}$$

where $Z' = GZ$ and the matrix GH is given by:

$$GH = \begin{bmatrix} \frac{|H(0)|^2}{|H(0)|^2 + \sigma^2} & 0 & \cdots & 0 \\ 0 & \frac{|H(1)|^2}{|H(1)|^2 + \sigma^2} & \cdots & 0 \\ \vdots & \vdots & \cdots & \vdots \\ 0 & 0 & \cdots & \frac{|H(N-1)|^2}{|H(N-1)|^2 + \sigma^2} \end{bmatrix}. \tag{3.59}$$

The modulation symbol estimate on the kth subcarrier is:

$$\hat{X}(k) = \underbrace{\frac{|H(k)|^2}{|H(k)|^2 + \sigma^2} X(k)}_{\text{signal}} + \underbrace{\frac{H(k)^*}{|H(k)|^2 + \sigma^2} Z(k)}_{\text{noise}} \tag{3.60}$$

The SNR on the kth subcarrier for an OFDM system using an MMSE receiver is then simply:

$$\gamma_{\text{OFDM-MMSE}}(k) = \frac{P \left(\frac{|H(k)|^2}{|H(k)|^2 + \sigma^2} \right)^2}{\sigma^2 \left(\frac{|H(k)|}{|H(k)|^2 + \sigma^2} \right)^2} = \frac{P |H(k)|^2}{\sigma^2} = \gamma |H(k)|^2. \tag{3.61}$$

We note that an MMSE receiver provides the same performance as an MRC receiver in case of OFDM. This is because in OFDM the transmission on each subcarrier is orthogonal and there is no inter-carrier interference.

In case of a zero-forcing receiver, the equalization weight on the kth subcarrier is given as:

$$g(k) = \frac{1}{H(k)} = \frac{H(k)^*}{|H(k)|^2}. \tag{3.62}$$

The modulation symbol estimate on the kth subcarrier is:

$$\hat{X}(k) = \underbrace{X(k)}_{\text{signal}} + \underbrace{\frac{H(k)^*}{|H(k)|^2}Z(k)}_{\text{noise}}.$$

(3.63)

The SNR on the kth subcarrier for an OFDM system using a ZF receiver is then simply:

$$\gamma_{\text{OFDM-ZF}}(k) = \frac{P}{\sigma^2\left(\dfrac{|H(k)|}{|H(k)|^2}\right)^2} = \gamma\,|H(k)|^2.$$

(3.64)

We note that a ZF receiver provides the same performance as an MRC or MMSE receiver. Therefore, we can conclude that for 1-tap equalization in OFDM, any of the receivers among MRC, MMSE or ZF can be used for achieving the same performance.

$$\gamma_{\text{OFDM}}(k) = \gamma_{\text{OFDM-MRC}}(k) = \gamma_{\text{OFDM-MMSE}}(k) = \gamma_{\text{OFDM-ZF}}(k).$$

(3.65)

It should be noted, however, that when inter-cell interference or MIMO is considered an MMSE receiver could lead to superior performance due to its interference suppression capability.

3.4.2   SNR analysis for SC-FDE

In case of SC-FDE, the modulation symbols $x = [x(0), \ldots, x(N_{\text{FFT}} - 1)]^T$ are transmitted in the time domain. The frequency-domain symbols $X = [X(0), \ldots, X(N_{\text{FFT}} - 1)]^T$ are then given as:

$$X(k) = \sum_{n=0}^{N-1} x(n) \times e^{-j2\pi \frac{n}{N}k} \quad k = 0, 1, \ldots, (N-1).$$

(3.66)

The modulation symbol estimates in the time-domain $x = [x(0), \ldots, x(N_{\text{FFT}} - 1)]^T$ assuming an MRC receiver can be written as:

$$\hat{x} = W^H H^H H W x + W^H H^H Z = (HW)^H HWx + (HW)^H Z.$$

(3.67)

Let us have a closer look at the equivalent channel matrix (HW):

$$HW = \begin{bmatrix} H(0) & H(0) & \cdots & H(0) \\ H(1) & H(1)e^{-j2\pi/N} & \cdots & H(1)e^{-j2\pi(N-1)/N} \\ \vdots & \vdots & \vdots & \vdots \\ H(N-1) & H(N-1)e^{-j2\pi(N-1)/N} & \cdots & H(N-1)e^{-j2\pi(N-1)(N-1)/N} \end{bmatrix}.$$

(3.68)

The time-domain noise $z' = (HW)^H Z$ term can be expanded as below:

$$(HW)^H Z = \frac{1}{N} \begin{bmatrix} H(0)^* & H(1)^* & \dots & H(N-1)^* \\ H(0)^* & H(1)^* e^{j2\pi/N} & \dots & H(N-1)^* e^{j2\pi(N-1)/N} \\ \vdots & \vdots & \vdots & \vdots \\ H(0)^* & H(1)^* e^{j2\pi(N-1)/N} & \dots & H(N-1)^* e^{j2\pi(N-1)(N-1)/N} \end{bmatrix}$$

$$\times \begin{bmatrix} Z(0) \\ Z(1) \\ \vdots \\ Z(N-1) \end{bmatrix}$$

$$\begin{bmatrix} z'(0) \\ z'(1) \\ \vdots \\ z'(N-1) \end{bmatrix} = \frac{1}{N} \begin{bmatrix} H(0)^* Z(0) + H(1)^* Z(1) \\ + \dots + H(N-1)^* Z(N-1) \\ H(0)^* Z(0) + H(1)^* Z(1) e^{j2\pi/N} \\ + \dots + H(N-1)^* e^{j2\pi(N-1)/N} Z(N-1) \\ \vdots \\ H(0)^* Z(0) + H(1)^* Z(1) e^{j2\pi(N-1)/N} \\ + \dots + H(N-1)^* Z(N-1) e^{j2\pi(N-1)(N-1)/N} \end{bmatrix}. \quad (3.69)$$

From Equation (3.69), the noise $z'(n)$ experienced by the time-domain symbol $x(n)$ is given as:

$$z'(n) = \frac{1}{N} \sum_{k=0}^{N-1} H^*(k) Z(k) e^{j2\pi kn/N} \quad n = 0, 1, \dots, (N-1). \quad (3.70)$$

We note that with an MRC receiver, there is no enhancement to the noise experienced by the time-domain symbols.

Let us have a close look at the equivalent operation $(HW)^H HW$ on the time-domain data symbols $x = [x(0), \ldots, x(N_{\text{FFT}} - 1)]^T$:

$$(HW)^H HW = \frac{1}{N} \begin{bmatrix} \sum_{k=0}^{(N-1)} |H(k)|^2 & \sum_{k=0}^{(N-1)} |H(k)|^2 e^{-j2\pi k/N} & \cdots & \sum_{k=0}^{(N-1)} |H(k)|^2 e^{j2\pi k(1-N)/N} \\ \sum_{k=0}^{(N-1)} |H(k)|^2 e^{j2\pi k/N} & \sum_{k=0}^{(N-1)} |H(k)|^2 & \cdots & \sum_{k=0}^{(N-1)} |H(k)|^2 e^{j2\pi k(2-N)/N} \\ \vdots & \vdots & \vdots & \vdots \\ \sum_{k=0}^{(N-1)} |H(k)|^2 e^{j2\pi k(1-N)/N} & \sum_{k=0}^{(N-1)} |H(k)|^2 e^{j2\pi k(2-N)/N} & \cdots & \sum_{k=0}^{(N-1)} |H(k)|^2 \end{bmatrix}.$$

$$(3.71)$$

We note that this operation is not orthogonal which would result in inter-modulation-symbol interference. The estimates $\hat{x} = [\hat{x}(0), \ldots, \hat{x}(N_{\text{FFT}} - 1)]^T$ of time-domain data symbols are given as:

$$\begin{bmatrix} \hat{x}(0) \\ \hat{x}(1) \\ \vdots \\ \hat{x}(N-1) \end{bmatrix} = \frac{1}{N} \begin{bmatrix} x(0) \sum_{k=0}^{(N-1)} |H(k)|^2 + x(1) \sum_{k=0}^{(N-1)} |H(k)|^2 e^{-j2\pi k/N} \\ + \ldots + x(N-1) \sum_{k=0}^{(N-1)} |H(k)|^2 e^{j2\pi k(1-N)/N} \\ x(1) \sum_{k=0}^{(N-1)} |H(k)|^2 + x(0) \sum_{k=0}^{(N-1)} |H(k)|^2 e^{j2\pi k/N} \\ + \ldots + x(N-1) \sum_{k=0}^{(N-1)} |H(k)|^2 e^{j2\pi k(2-N)/N} \\ \vdots \\ x(N-1) \sum_{k=0}^{(N-1)} |H(k)|^2 + x(0) \sum_{k=0}^{(N-1)} |H(k)|^2 e^{-j2\pi k(1-N)/N} \\ + x(1) \sum_{k=0}^{(N-1)} |H(k)|^2 e^{-j2\pi k(2-N)/N} + \ldots \end{bmatrix}$$

$$+ \begin{bmatrix} z'(0) \\ z'(1) \\ \vdots \\ z'(N-1) \end{bmatrix}.$$

$$(3.72)$$

We note that each modulation symbol experiences frequency diversity (in a frequency-selective channel) due to an effective channel gain of $\sum_{k=0}^{(N-1)} |H(k)|^2$. On the other hand, we note the presence of inter-modulation-symbol interference terms in each time-domain symbol estimate. These interfering terms disappear when all the frequency-domain channel gains are the same $\left(|H(0)|^2 = |H(1)|^2 = \cdots = |H(N-1)|^2\right)$ as is the case in a frequency-flat fading channel. This can be observed from the fact that under equal frequency-domain channel gains assumption, the matrix HW becomes orthogonal:

$$(HW)^H \, HW = \begin{bmatrix} |H(0)|^2 & 0 & \cdots & 0 \\ 0 & |H(0)|^2 & \cdots & 0 \\ \vdots & \vdots & \vdots & \vdots \\ 0 & 0 & \cdots & |H(0)|^2 \end{bmatrix}. \tag{3.73}$$

This shows that SC-FDE produces no noise enhancement or inter-modulation-symbol interference and its performance is equivalent to OFDM performance in a frequency flat-fading channel.

We rewrite time-domain symbol estimate $\hat{x}(n)$ from Equation (3.72) as:

$$\hat{x}(n) = \underbrace{x(n)\frac{1}{N}\sum_{k=0}^{(N-1)}|H(k)|^2}_{\text{signal}} + \underbrace{\frac{1}{N}\sum_{m=0,m\neq n}^{N-1}\sum_{k=0}^{N-1}|H(k)|^2 \, x(m) \, e^{j2\pi k(n-m)/N}}_{\text{interference}} + \underbrace{z'(n)}_{\text{noise}}$$

$$\tag{3.74}$$

The noise term $z'(n)$ is given in Equation (3.70). Let us now focus on a linear MMSE receiver. With an MMSE equalizer, the equalization weight on subcarrier k is given as:

$$g(k) = \frac{H(k)^*}{|H(k)|^2 + \sigma^2}. \tag{3.75}$$

The time-domain symbol estimate $\hat{x}(n)$ is then given as:

$$\hat{x}(n) = x(n)\frac{1}{N}\sum_{k=0}^{(N-1)}\frac{|H(k)|^2}{|H(k)|^2 + \gamma^{-1}}$$

$$+ \frac{1}{N}\sum_{m=0,m\neq n}^{N-1}\sum_{k=0}^{N-1}\frac{|H(k)|^2}{|H(k)|^2 + \gamma^{-1}}x(m)\,e^{j2\pi k(n-m)/N} + z'(n), \tag{3.76}$$

where $\gamma = P/\sigma^2$ is the ratio of signal power to the AWGN noise power. The noise $z'(n)$ experienced by the time-domain symbol $x(n)$ is given as:

$$z'(n) = \frac{1}{N}\sum_{k=0}^{N-1}\frac{|H(k)|^2}{|H(k)|^2 + \gamma^{-1}}Z(k)e^{j2\pi kn/N} \quad n = 0, 1, \ldots, (N-1). \tag{3.77}$$

The SNR of estimated symbol $\hat{x}(n)$ assuming an MMSE equalizer is given as:

$$
\gamma_{\text{SC-FDE-MMSE}} = \frac{\left(\dfrac{1}{N}\sum\limits_{k=0}^{(N-1)}\dfrac{|H(k)|^2}{|H(k)|^2+\gamma^{-1}}\right)^2}{\dfrac{1}{N}\sum\limits_{m=0,m\neq n}^{N-1}\left|\sum\limits_{k=0}^{N-1}\dfrac{|H(k)|^2}{|H(k)|^2+\gamma^{-1}}e^{j2\pi k(n-m)/N}\right|^2+\gamma^{-1}\dfrac{1}{N}\sum\limits_{k=0}^{(N-1)}\dfrac{|H(k)|^2}{|H(k)|^2+\gamma^{-1}}}
$$

$$
= \frac{\dfrac{1}{N^2}\left(\sum\limits_{k=0}^{(N-1)}\dfrac{|H(k)|^2}{|H(k)|^2+\gamma^{-1}}\right)^2}{\dfrac{1}{N}\sum\limits_{m=0}^{N-1}\sum\limits_{k=0}^{N-1}\left|\dfrac{|H(k)|^2}{|H(k)|^2+\gamma^{-1}}e^{j2\pi k(n-m)/N}\right|^2-\dfrac{1}{N^2}\left(\sum\limits_{k=0}^{(N-1)}\dfrac{|H(k)|^2}{|H(k)|^2+\gamma^{-1}}\right)^2+\gamma^{-1}\dfrac{1}{N}\sum\limits_{k=0}^{(N-1)}\dfrac{|H(k)|^2}{|H(k)|^2+\gamma^{-1}}}
$$

$$
= \frac{\dfrac{1}{N^2}\left(\sum\limits_{k=0}^{(N-1)}\dfrac{|H(k)|^2}{|H(k)|^2+\gamma^{-1}}\right)^2}{\dfrac{1}{N}\sum\limits_{k=0}^{(N-1)}\left(\dfrac{|H(k)|^2}{|H(k)|^2+\gamma^{-1}}\right)^2-\dfrac{1}{N^2}\left(\sum\limits_{k=0}^{(N-1)}\dfrac{|H(k)|^2}{|H(k)|^2+\gamma^{-1}}\right)^2+\gamma^{-1}\dfrac{1}{N}\sum\limits_{k=0}^{(N-1)}\dfrac{|H(k)|^2}{|H(k)|^2+\gamma^{-1}}}
$$

$$
= \frac{\left(\sum\limits_{k=0}^{(N-1)}\dfrac{|H(k)|^2}{|H(k)|^2+\gamma^{-1}}\right)^2}{N\sum\limits_{k=0}^{(N-1)}\dfrac{|H(k)|^2\left(|H(k)|^2+\gamma^{-1}\right)}{\left(|H(k)|^2+\gamma^{-1}\right)^2}-\left(\sum\limits_{k=0}^{(N-1)}\dfrac{|H(k)|^2}{|H(k)|^2+\gamma^{-1}}\right)^2} = \frac{\beta^2}{N\beta-\beta^2} = \frac{1}{1-\beta/N}-1 \tag{3.78}
$$

where β is defined as:

$$
\beta = \sum_{k=0}^{(N-1)}\frac{\gamma\,|H(k)|^2}{\gamma\,|H(k)|^2+1}. \tag{3.79}
$$

By substituting β back in Equation (3.78), $\gamma_{\text{SC-FDE-MMSE}}$ is given as:

$$
\gamma_{\text{SC-FDE-MMSE}} = \frac{1}{1-\dfrac{1}{N}\sum\limits_{k=0}^{(N-1)}\dfrac{\gamma\,|H(k)|^2}{\gamma\,|H(k)|^2+1}}-1 = \frac{1}{\dfrac{1}{N}\sum\limits_{k=0}^{(N-1)}\dfrac{1}{\gamma\,|H(k)|^2+1}}-1. \tag{3.80}
$$

We note that $\gamma\,|H(k)|^2 = \gamma_{\text{OFDM}}(k)$ represents SNR on the kth subcarrier in an OFDM system, and therefore SNR in an OFDM and SC-FDE system with MMSE receiver are related by:

$$
\gamma_{\text{SC-FDE-MMSE}} = \frac{1}{\dfrac{1}{N}\sum\limits_{k=0}^{(N-1)}\dfrac{1}{\gamma_{\text{OFDM}}(k)+1}}-1. \tag{3.81}
$$

When all the frequency-domain channel gains are equal $\left(|H(0)|^2 = |H(1)|^2 = \cdots = |H(N-1)|^2\right)$:

$$
\gamma_{\text{SC-FDE-MMSE}} = \frac{1}{\dfrac{1}{N}\sum\limits_{k=0}^{(N-1)}\dfrac{1}{\gamma_{\text{OFDM}}(k)+1}}-1 = \gamma\,|H(0)|^2 = \gamma_{\text{OFDM}}. \tag{3.82}
$$

We noted that SC-FDE provides frequency diversity but results in SNR degradation due to inter-modulation-symbol interference relative to OFDM in a frequency-selective channel. Can we somehow eliminate inter-modulation symbol interference in SC-FDE? One possibility is to employ a ZF-receiver. With a ZF-receiver, the modulation symbol estimates in the time-domain $\hat{x} = \left[\hat{x}(0), \ldots, \hat{x}(N_{FFT} - 1)\right]^T$ can be written as:

$$\hat{x} = W^H H^{-1} H W x + W^H H^{-1} Z = x + W^H H^{-1} Z. \tag{3.83}$$

Let us have a closer look at the noise-scaling matrix $W^H H^{-1}$:

$$W^H H^{-1} = \frac{1}{N} \begin{bmatrix} 1 & 1 & \cdots & 1 \\ 1 & e^{j2\pi/N} & \cdots & e^{j2\pi(N-1)/N} \\ \vdots & \vdots & \cdots & \vdots \\ 1 & e^{j2\pi(N-1)/N} & \cdots & e^{j2\pi(N-1)(N-1)/N} \end{bmatrix}$$

$$\times \begin{bmatrix} \dfrac{H(0)^*}{|H(0)|^2} & 0 & \cdots & 0 \\ 0 & \dfrac{H(1)^*}{|H(1)|^2} & \cdots & 0 \\ \vdots & \vdots & \cdots & \vdots \\ 0 & 0 & \cdots & \dfrac{H(N-1)^*}{|H(N-1)|^2} \end{bmatrix}$$

$$W^H H^{-1} = \frac{1}{N} \begin{bmatrix} \dfrac{H(0)^*}{|H(0)|^2} & \dfrac{H(1)^*}{|H(1)|^2} & \cdots & \dfrac{H(N-1)^*}{|H(N-1)|^2} \\ \dfrac{H(0)^*}{|H(0)|^2} & \dfrac{H(1)^*}{|H(1)|^2}e^{j2\pi/N} & \cdots & \dfrac{H(N-1)^*}{|H(N-1)|^2}e^{j2\pi(N-1)/N} \\ \vdots & \vdots & \cdots & \vdots \\ \dfrac{H(0)^*}{|H(0)|^2} & \dfrac{H(1)^*}{|H(1)|^2}e^{j2\pi(N-1)/N} & \cdots & \dfrac{H(N-1)^*}{|H(N-1)|^2}e^{j2\pi(N-1)(N-1)/N} \end{bmatrix}.$$

$$\tag{3.84}$$

The time-domain noise samples $W^H H^{-1} Z$ are given as:

$$
\begin{bmatrix} z'(0) \\ z'(1) \\ \vdots \\ z'(N-1) \end{bmatrix} = \frac{1}{N} \begin{bmatrix} \dfrac{H(0)^*}{|H(0)|^2}Z(0) + \dfrac{H(1)^*}{|H(1)|^2}Z(1) + \dfrac{H(N-1)^*}{|H(N-1)|^2}Z(N-1) \\[2mm] \dfrac{H(0)^*}{|H(0)|^2}Z(0) + \dfrac{H(1)^*}{|H(1)|^2}e^{j2\pi/N}Z(1) \\[1mm] + \dfrac{H(N-1)^*}{|H(N-1)|^2}e^{j2\pi(N-1)/N}Z(N-1) \\[2mm] \vdots \\[2mm] \dfrac{H(0)^*}{|H(0)|^2}Z(0) + \dfrac{H(1)^*}{|H(1)|^2}e^{j2\pi(N-1)/N}Z(1) \\[1mm] + \dfrac{H(N-1)^*}{|H(N-1)|^2}e^{j2\pi(N-1)(N-1)/N}Z(N-1) \end{bmatrix}.
$$

$$(3.85)$$

Let us now write Equation (3.83) as below:

$$
\begin{bmatrix} \hat{x}(0) \\ \hat{x}(1) \\ \vdots \\ \hat{x}(N-1) \end{bmatrix} = \begin{bmatrix} x(0) \\ x(1) \\ \vdots \\ x(N-1) \end{bmatrix} + \begin{bmatrix} z'(0) \\ z'(1) \\ \vdots \\ z'(N-1) \end{bmatrix}.
\tag{3.86}
$$

We note that there is no inter-subcarrier or inter-modulation symbol interference in an SC-FDE system using a zero-forcing receiver. From Equation (3.85), the noise $z'(n)$ experienced by the time-domain symbol $x(n)$ is given as:

$$
z'(n) = \frac{1}{N} \sum_{k=0}^{N-1} \frac{H^*(k)}{|H(k)|^2} Z(k) e^{j2\pi kn/N} \quad n = 0, 1, \ldots, (N-1).
\tag{3.87}
$$

We note that a zero-forcing receiver can result in noise enhancement when some frequency-domain channel gains are highly attenuated which is generally the case in a frequency-selective channel. The noise enhancement negatively influences the link performance of an SC-FDE system and therefore the ZF receiver is generally not used in SC-FDE systems. When all the frequency-domain channel gains are equal $\left(|H(0)|^2 = |H(1)|^2 = \cdots = |H(N-1)|^2\right)$ as is the case in a frequency-flat fading channel, the noise terms in Equation (3.87) are scaled equally and there is no noise enhancement. Therefore, in a frequency-flat fading channel the performance of OFDM and SC-FDE is the same for the three receivers types considered, that is MRC, MMSE and ZF receiver.

The time-domain symbol estimate $\hat{x}(n)$ using a ZF-receiver is then given as:

$$\hat{x}(n) = \underbrace{x(n)}_{\text{signal}} + \underbrace{\frac{1}{N}\sum_{k=0}^{N-1}\frac{H^*(k)}{|H(k)|^2}Z(k)e^{j2\pi kn/N}}_{\text{noise}}. \tag{3.88}$$

The SNR on the nth time-domain symbol for an SC-FDE system using a ZF receiver is then simply:

$$
\begin{aligned}
\gamma_{\text{SC-FDE-ZF}} &= \frac{P}{\sigma^2 \dfrac{1}{N}\displaystyle\sum_{k=0}^{N-1}\left|\dfrac{H^*(k)}{|H(k)|^2}Z(k)e^{j2\pi kn/N}\right|^2} \\[2em]
&= \frac{\gamma}{\dfrac{1}{N}\displaystyle\sum_{k=0}^{N-1}\dfrac{1}{|H(k)|^2}} = \frac{1}{\dfrac{1}{N}\displaystyle\sum_{k=0}^{N-1}\dfrac{1}{\gamma\,|H(k)|^2}} = \frac{1}{\dfrac{1}{N}\displaystyle\sum_{k=0}^{N-1}\dfrac{1}{\gamma_{\text{OFDM}}(k)}}.
\end{aligned}
\tag{3.89}
$$

When all the frequency-domain channel gains are equal, that is $\left(|H(0)|^2 = |H(1)|^2 = \ldots = |H(N-1)|^2\right)$:

$$\gamma_{\text{SC-FDE-ZF}} = \frac{\gamma}{\dfrac{1}{N}\displaystyle\sum_{k=0}^{N-1}\dfrac{1}{|H(0)|^2}} = \gamma\,|H(0)|^2 = \gamma_{\text{OFDM}}. \tag{3.90}$$

3.4.3 SNR analysis results

We noted that in a frequency-flat fading channel, OFDM and SC-FDE provides similar performance. In a frequency-selective fading channel, however, the performance of both schemes degrades. In case of OFDM, the performance degrades because different OFDM symbol experience different SNR in a frequency-selective fading channel. The expected capacity of an OFDM system is given as:

$$C_{\text{OFDM}} = \frac{1}{N}\sum_{k=0}^{N-1}\log_2\left(1 + \gamma_{\text{OFDM}}(k)\right). \tag{3.91}$$

Let us define effective SNR, $\text{SNR}_{\text{eff-OFDM}}$ as below:

$$\text{SNR}_{\text{eff-OFDM}} = 2^{C_{\text{OFDM}}} - 1. \tag{3.92}$$

The effective SNR, $\text{SNR}_{\text{eff-OFDM}}$ can be seen as the SNR on each subcarrier in an equivalent flat-fading channel. It can be easily recognized that $\text{SNR}_{\text{eff-OFDM}}$ is less than or equal to average SNR in an OFDM system.

$$\text{SNR}_{\text{eff-OFDM}} \leq \frac{1}{N}\sum_{k=0}^{N-1}\gamma_{\text{OFDM}}(k). \tag{3.93}$$

The equality sign holds for a frequency-flat fading channel. In Section 3.3, we noted that the difference between $SNR_{eff\text{-}OFDM}$ and average SNR can be up to 2.5 dB in a Rayleigh fading channel at very high SNRs.

We also noticed that in an SC-FDE system, the performance degrades in a frequency-selective channel due to either noise enhancement for a ZF-receiver or loss of orthogonality for an MMSE-receiver that results in inter-modulation symbol interference. There is no additional capacity loss in an SC-FDE system because all the modulation symbols experience the same SNR i.e. effective SNR and average SNR are the same.

We plot post-receiver SNR_{eff} for OFDM and SC-FDE as a function of $\rho = P/N_0$ in Figure 3.14 for the case of a very large number of subcarriers fading independently according to a Rayleigh fading process. This would model a highly frequency selective channel with a large delay spread. The reference curve labeled as AWGN represents SNR experienced by OFDM and SC-FDE in a frequency-flat channel. We note that at very low SNR, OFDM and SC-FDE with MMSE perform similarly and achieve AWGN channel performance. This is because in case of OFDM, the loss due to varying SNR on different subcarriers is negligible at low SNR as we discussed in Section 3.3. In SC-FDE with MMSE, the loss is small at low SNR because noise and not inter-modulation-symbol interference is a dominant source of degradation. In contrast, SC-FDE with ZF-receiver gives very poor performance due to noise enhancement. At high SNR, OFDM experiences up to a maximum of 2.5 dB loss due to varying SNR on different subcarriers. The loss due to inter-modulation-symbol interference for SC-FDE with MMSE receiver is even larger and always more than the OFDM loss.

Until now, we have considered the case of a single receiver antenna. With multiple receive antennas both OFDM and SC-FDE systems would experience spatial diversity reducing the symbol SNR variations in the frequency domain. Assuming MRC combining of the signals

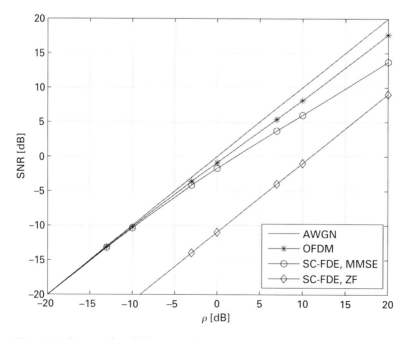

Figure 3.14. Post-receiver SNR comparison between OFDM and SC-FDE.

across receive antenna paths, the SNR for OFDM can be written as:

$$\gamma_{\text{OFDM}}(k) = \sum_{p=0}^{P} \gamma_{\text{OFDM}}(k,p),\tag{3.94}$$

where $(P-1)$ is the number of receiving antennas. Similarly, we can write SNR expressions for SC-FDE using MMSE and ZF-receivers as below:

$$\gamma_{\text{SC-FDE-MMSE}} = \cfrac{1}{\cfrac{1}{N}\displaystyle\sum_{k=0}^{(N-1)} \cfrac{1}{\displaystyle\sum_{p=0}^{P}\gamma_{\text{OFDM}}(k,p)+1}} - 1\tag{3.95}$$

$$\gamma_{\text{SC-FDE-ZF}} = \cfrac{1}{\cfrac{1}{N}\displaystyle\sum_{k=0}^{N-1} \cfrac{1}{\displaystyle\sum_{p=0}^{P}\gamma_{\text{OFDM}}(k,p)}}.\tag{3.96}$$

We again plot post-receiver SNR_{eff} for OFDM and SC-FDE for the case of a very large number of subcarriers fading independently for two and four receive antennas in Figures 3.15 and 3.16 respectively. As we expect, the performance gap closes between different schemes as the diversity order increases. This is because receive diversity reduces SNR variations in the frequency domain which leads to less loss due to varying symbol SNR in OFDM and also

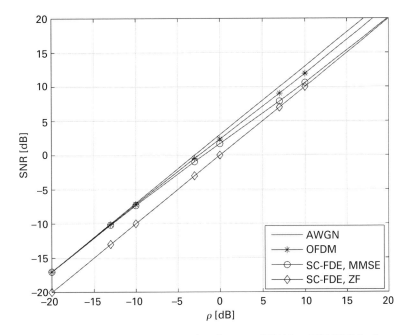

Figure 3.15. Post-receiver SNR comparison between OFDM and SC-FDE for 2-way receive diversity.

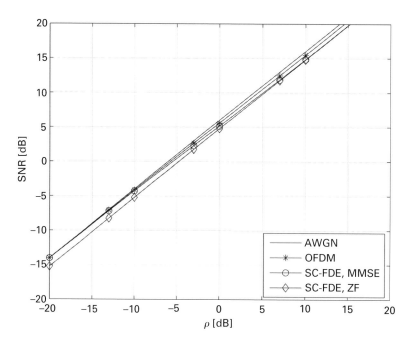

Figure 3.16. Post-receiver SNR comparison between OFDM and SC-FDE for 4-way receive diversity.

lesser loss due to noise enhancement or loss of orthogonality in SC-FDE. When the diversity order becomes very large, the performance approaches the AWGN channel and all the schemes converge providing the same performance. Even though we only considered receive diversity to show the effect of flattening frequency-domain SNR, similar behavior would be observed by using other schemes that provide diversity within a symbol such as the Alamouti transmit diversity scheme discussed in Chapter 6.

We looked at OFDM and SC-FDE schemes from a capacity perspective with the underlying assumption that a capacity achieving channel code is available. Most modern cellular systems including LTE employ very powerful codes based on the turbo coding principle. By using a strong channel code, OFDM can achieve frequency diversity via coding in a frequency-selective fading channel. In the presence of a weak code or in an uncoded system, the SC-FDE scheme can have a diversity advantage over OFDM. However, a weak code with a higher coding rate is generally used at a very high SNR only. This is because multiple modulation formats such as QPSK, 16-QAM and 64-QAM are generally available. By using adaptive modulation and coding (AMC), as the SNR increases, the system switches to a higher order modulation, while keeping a reasonable channel code rate. However, when the highest-level modulation format is reached there is no choice but to use higher coding rates to achieve higher data rates. We also noted that the SC-FDE loss relative to the OFDM increases with SNR. It is therefore not clear if the diversity gain of SC-FDE can compensate for its noise enhancement or orthogonality loss at high SNR.

Another aspect of SC-FDE to consider is complexity for MIMO processing. In case of a $P \times L$ MIMO system with P transmit and L receive antennas, the frequency-domain channel matrix H for OFDM is a block-diagonal matrix given as:

$$
H = \begin{bmatrix}
H_{L \times P}(0) & 0 & \cdots & 0 & 0 \\
0 & H_{L \times P}(1) & 0 & \cdots & 0 \\
\vdots & \ddots & \ddots & \ddots & \vdots \\
0 & 0 & \cdots & H_{L \times P}(N-2) & 0 \\
0 & 0 & \cdots & 0 & H_{L \times P}(N-1)
\end{bmatrix}
\tag{3.97}
$$

where the MIMO channel matrix $H_{L \times P}(k)$ for the kth subcarrier is given as:

$$
H_{L \times P}(k) = \begin{bmatrix}
H_{0,0}(k) & H_{0,1}(k) & \cdots & H_{0,(P-1)}(k) \\
H_{1,0}(k) & H_{1,1}(k) & \cdots & H_{1,(P-1)}(k) \\
\vdots & \vdots & \vdots & \vdots \\
H_{(L-1),0}(k) & H_{(L-1),1}(k) & \cdots & H_{(L-1),(P-1)}(k)
\end{bmatrix}
\tag{3.98}
$$

where $H_{l,p}(k)$ is the channel gain on the kth subcarrier from the pth transmit antenna to the lth receive antenna. In OFDM, the MIMO processing such as MMSE or MLD (Maximum Likelihood Detection) can be performed separately on each subcarrier, as there is no inter-carrier interference. The equivalent MIMO channel matrix HW for SC-FDE is given as:

$$
HW = \begin{bmatrix}
H_{L \times P}(0) & H_{L \times P}(0) & \cdots & H_{L \times P}(0) \\
H_{L \times P}(1) & H_{L \times P}(1)e^{-j2\pi/N} & \cdots & H_{L \times P}(1)e^{-j2\pi(N-1)/N} \\
\vdots & \vdots & \vdots & \vdots \\
H_{L \times P}(N-1) & H_{L \times P}(N-1)\,e^{-j2\pi(N-1)/N} & \cdots & H_{L \times P}(N-1)\,e^{-j2\pi(N-1)(N-1)/N}
\end{bmatrix} .
\tag{3.99}
$$

Since the equivalent MIMO channel matrix HW is not block-diagonal, we cannot perform MIMO processing on a per subcarrier basis leading to a more complex receiver.

3.5   Frequency diversity

In a frequency-selective channel, different modulations symbols transmitted using OFDM can experience different fading as shown in Figure 3.17. Since each modulation symbol is transmitted on a single subcarrier, diversity within a modulation symbol is not achieved. In contrast, in an SC-FDE system, a modulation symbol is transmitted over the whole bandwidth and therefore SC-FDE scheme captures diversity within a modulation symbol.

In the presence of inter-cell interference in an OFDM system, a subcarrier sees interference from a single subcarrier and hence a single modulation symbol transmitted in the interfering cell. If different transmission power levels are used on different subcarriers, interference diversity can be provided by transmitting a modulation symbol over multiple subcarriers. Several

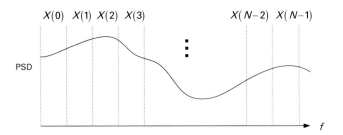

Figure 3.17. An illustration of frequency-selective fading in OFDM.

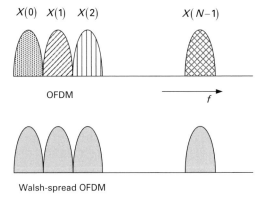

Figure 3.18. An illustration of Walsh-spread OFDM.

techniques that can provide frequency and interference-diversity in an OFDM system can be considered. We discuss below two schemes namely Walsh-spread OFDM also referred to as Multi-carrier CDMA (MC-CDMA) in the literature [11] and fast frequency hopping (FFH) OFDM [12,13] scheme. Another scheme referred to as Rotational OFDM is not discussed here and the interested reader is referred to reference [14] for details on Rotational OFDM and also to reference [15] for details on signal-space diversity concept.

3.5.1   Walsh-spread OFDM

The goal of Walsh-spread OFDM is to provide frequency as well as interference diversity within a modulation symbol. In OFDM, each subcarrier carries one modulation symbol as shown in Figure 3.18. In contrast, in Walsh-spread OFDM, using Walsh codes a modulation symbol is spread over all the subcarriers. Each subcarrier then contains some linear combination of the actual modulation symbols. We can expect to capture frequency diversity within each modulation symbol as each symbol is transmitted over all the subcarriers. Another expected advantage is that all modulation symbols transmitted in Walsh-spread OFDM would experience the same SNR. We know that in OFDM different modulation symbols experience different SNR in a frequency-selective channel.

Let $x = [x(0), \ldots, x(N-1)]^T$ be the N complex modulation symbols to be transmitted. Then the signal $[X(k), k = 0, 1, \ldots, N]$ transmitted on the kth subcarrier in Walsh-spread OFDM is given as:

$$
\begin{bmatrix} X(0) \\ X(1) \\ \vdots \\ X(N-2) \\ X(N-1) \end{bmatrix} = R_N \times \begin{bmatrix} x(0) \\ x(1) \\ \vdots \\ x(N-2) \\ x(N-1) \end{bmatrix}, \tag{3.100}
$$

where R_N is $N \times N$ Hadamard matrix given as:

$$
R_N = \begin{pmatrix} R_{N/2} & R_{N/2} \\ R_{N/2} & -R_{N/2} \end{pmatrix}. \tag{3.101}
$$

The first three Hadamard matrices R_1, R_2 and R_4 are given below:

$$
R_1 = [1] \quad R_2 = \begin{bmatrix} 1 & 1 \\ 1 & -1 \end{bmatrix}
$$

$$
R_4 = \begin{bmatrix} 1 & 1 & 1 & 1 \\ 1 & -1 & 1 & -1 \\ 1 & 1 & -1 & -1 \\ 1 & -1 & -1 & 1 \end{bmatrix}. \tag{3.102}
$$

We note that Hadamard matrices are symmetric square shaped matrices. Each column or row corresponds to a Walsh code of length N. Also, a Hadamard matrix is an orthogonal matrix, that is:

$$
H_N^H \times H_N = I_N. \tag{3.103}
$$

Each row is orthogonal to all the other rows and similarly each column is orthogonal to all the other columns.

The transmission and reception chain for Walsh-spread OFDM is given in Figure 3.19. The spreading is done before the IFDT operation at the transmitter. On the receiver side, the spreading takes place after the frequency-domain equalization (FDE).

The received signal vector can be written as:

$$
Y = HRx + Z. \tag{3.104}
$$

The modulation symbol estimates $x = [x(0), \ldots, x(N-1)]^T$ assuming an MRC receiver can be written as:

$$
\hat{x} = R^H H^H Y = R^H H^H HRx + R^H H^H Z = (HR)^H HRx + Z', \tag{3.105}
$$

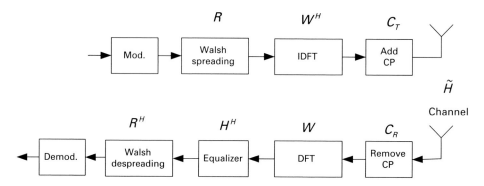

Figure 3.19. Walsh-spread OFDM transmitter and receiver.

where $H = WC_R\tilde{H}C_T W^H$ is a frequency-domain diagonal channel matrix as noted in (3.24).

As an example, let us consider transmission over two subcarriers with channel gains $H(1)$ and $H(2)$. The composite matrix HR is then given as:

$$HR = \begin{bmatrix} H(1) & 0 \\ 0 & H(2) \end{bmatrix} \frac{1}{\sqrt{2}} \begin{bmatrix} 1 & 1 \\ 1 & -1 \end{bmatrix} = \frac{1}{\sqrt{2}} \begin{bmatrix} H(1) & H(1) \\ H(2) & -H(2) \end{bmatrix}. \tag{3.106}$$

Let us look at the matrix $(HR)^H HR$ operating on the transmitted modulation symbols:

$$(HR)^H HR = \frac{1}{2} \begin{bmatrix} H^*(1) & H^*(2) \\ H^*(1) & -H^*(2) \end{bmatrix} \times \begin{bmatrix} H(1) & H(1) \\ H(2) & -H(2) \end{bmatrix}$$

$$= \begin{bmatrix} H^2(1) + H^2(2) & H^2(1) - H^2(2) \\ H^2(1) - H^2(2) & H^2(1) + H^2(2) \end{bmatrix}. \tag{3.107}$$

We note that this operation is only orthogonal when $H(1) = H(2)$. In more general terms, the operation is orthogonal when the channel gains on all the subcarriers used for modulation symbols transmissions are the same. This would be the case for a flat-fading channel.

The received modulation symbol estimates are:

$$\begin{bmatrix} \hat{x}(1) \\ \hat{x}(2) \end{bmatrix} = \frac{1}{2} \begin{bmatrix} H^2(1) + H^2(2) & H^2(1) - H^2(2) \\ H^2(1) - H^2(2) & H^2(1) + H^2(2) \end{bmatrix} \begin{bmatrix} x(1) \\ x(2) \end{bmatrix} + \begin{bmatrix} Z'(1) \\ Z'(2) \end{bmatrix}$$

$$\begin{bmatrix} \hat{x}(1) \\ \hat{x}(2) \end{bmatrix} = \frac{1}{2} \begin{bmatrix} [H^2(1) + H^2(2)]x(1) + [H^2(1) - H^2(2)]x(2) \\ [H^2(1) - H^2(2)]x(1) + [H^2(1) + H^2(2)]x(2) \end{bmatrix} + \begin{bmatrix} Z'(1) \\ Z'(2) \end{bmatrix}. \tag{3.108}$$

Let us look at the first modulation symbol estimate:

$$\hat{x}(1) = \underbrace{[H^2(1) + H^2(2)]x(1)}_{\text{signal}} + \underbrace{[H^2(1) - H^2(2)]x(2)}_{\text{interference}} + \underbrace{Z'(1)}_{\text{noise}}. \tag{3.109}$$

We note that the received modulation symbol experiences diversity when $H(1)$ and $H(2)$ are different. However, there is also an interference term introduced due to spreading of modulation symbols. This interference term disappears when the channel gains on the two subcarriers are the same. This would also mean that there is no diversity when the channel gains are the same. Therefore, on one hand Walsh-spread OFDM promises frequency diversity but on the other hand inter-modulation symbol interference is introduced when the channel can provide diversity. This situation is similar to SC-FDE scheme and similar tradeoffs apply for Walsh-spread OFDM.

3.5.2 Fast frequency-hopping OFDM

The fast frequency-hopping (FFH) scheme is also designed to provide frequency diversity as well as interference diversity within modulation symbols. This is achieved by hopping the subcarrier frequencies at the time-domain samples level within an OFDM symbol as shown in Figure 3.20 for the case of four subcarriers OFDM system. In conventional OFDM scheme, the subcarrier used for transmission of a modulation symbol is fixed for the duration of the OFDM symbol. Accordingly, the kth modulation symbol $X(k)$ is transmitted on the kth subcarrier during the four time-domain samples. By assuming a simple form of a cyclic fast frequency-hopping pattern, the kth modulation symbol $X(k)$ is transmitted on the subcarrier number mod $[k + n, (N - 1)]$ in the nth time-domain sample. When the number of modulation symbols to be transmitted is equal to the number of subcarriers, using FFH-OFDM each modulation symbol is transmitted over all the subcarriers therefore capturing frequency diversity within a modulation symbol.

In regular OFDM, the time-domain samples at the output of the IFFT are given as:

$$s(n) = W^H X(k) \quad n, k = 0, 1, \ldots, (N - 1), \tag{3.110}$$

where W^H is an $N \times N$ inverse fast fourier transform (IFFT) matrix with entries given by:

$$\left[W^H \right]_{n,k} = e^{j2\pi nk/N} \quad n, k = 0, 1, \ldots, (N - 1). \tag{3.111}$$

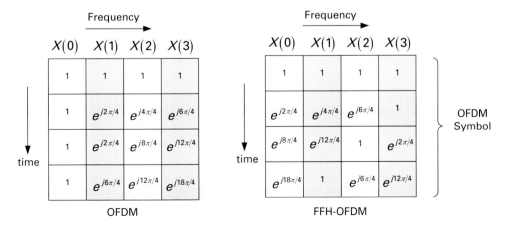

Figure 3.20. An illustration of FFH-OFDM.

In FFH-OFDM, a shuffled version of the IFFT matrix $\left[W_F^H\right]_{n,k}$ is used instead:

$$\left[W_F^H\right]_{n,k} = e^{j2\pi n[\Phi]_{n,k}/N} \qquad n,k = 0,1,\ldots(N-1). \tag{3.112}$$

The use of matrix W_F^H allows spreading modulation symbols over all the subcarriers. The matrix $[\Phi]_{n,k}$ forms the frequency-hopping pattern. In conventional OFDM, the matrix $[\Phi]_{n,k}$ takes the form:

$$[\Phi]_{n,k} = \begin{bmatrix} 0 & 1 & \cdots & (N-1) \\ 0 & 1 & \cdots & (N-1) \\ & & \vdots & \\ 0 & 1 & \cdots & (N-1) \end{bmatrix}. \tag{3.113}$$

Which means that the subcarrier frequencies at a certain time sample n are arranged in ascending order starting from 0 to $(N-1)$. We note that in this case $\left[W_F^H\right]_{n,k} = \left[W^H\right]_{n,k}$ as below.

$$W_F^H = W^H = \begin{bmatrix} 1 & 1 & 1 & \cdots & 1 \\ 1 & e^{j2\pi/N} & e^{j4\pi/N} & \cdots & e^{j2\pi(N-1)/N} \\ \vdots & \vdots & \vdots & \vdots & \vdots \\ 1 & e^{j2\pi(N-2)/N} & e^{j4\pi(N-2)/N} & \cdots & e^{j2\pi(N-1)(N-2)/N} \\ 1 & e^{j2\pi(N-1)/N} & e^{j4\pi(N-1)/N} & \cdots & e^{j2\pi(N-1)(N-1)/N} \end{bmatrix}. \tag{3.114}$$

In the case of simple cyclic shift hopping, the matrix $[\Phi]_{n,k}$ is given as:

$$[\Phi]_{n,k} = \mathrm{mod}\left[(f_n + k),(N-1)\right]$$

$$= \begin{bmatrix} 0 & 1 & \cdots & (N-1) \\ 1 & 2 & \cdots & 0 \\ \vdots & \vdots & \vdots & \vdots \\ (N-1) & 0 & \cdots & (N-2) \end{bmatrix}. \tag{3.115}$$

This means that the frequency of subcarrier with $k = 0$ is set to $f_n = 0,1,2,\ldots,(N-1)$ and the remaining frequencies arranged in ascending order with the modulo operation to enable cyclic hopping. With the definition of $[\Phi]_{n,k}$ in (3.115), the matrix W_F^H becomes:

$$
W_F^H = \begin{bmatrix}
1 & 1 & 1 & \cdots & 1 \\
e^{j2\pi/N} & e^{j4\pi/N} & e^{j6\pi/N} & \cdots & 1 \\
\vdots & \vdots & \vdots & \vdots & \vdots \\
e^{j2\pi(N-2)(N-2)/N} & e^{j2\pi(N-2)(N-1)/N} & 1 & \cdots & e^{j2\pi(N-2)(N-3)/N} \\
e^{j2\pi(N-1)(N-1)/N} & 1 & e^{j2\pi(N-1)/N} & \cdots & e^{j2\pi(N-1)(N-2)/N}
\end{bmatrix}.
$$

$$\text{(3.116)}$$

The time-domain samples in FFH-OFDM are then given as:

$$
s(n) = W_F^H X(k)
$$

$$
\begin{bmatrix}
s(0) \\
s(1) \\
\vdots \\
s(N-1)
\end{bmatrix}
= W_F^H \times
\begin{bmatrix}
X(0) \\
X(1) \\
\vdots \\
X(N-1)
\end{bmatrix}
$$

$$
\begin{bmatrix}
1 & 1 & 1 & \cdots & 1 \\
e^{j2\pi/N} & e^{j4\pi/N} & e^{j6\pi/N} & \cdots & 1 \\
\vdots & \vdots & \vdots & \vdots & \vdots \\
e^{j2\pi(N-2)(N-2)/N} & e^{j2\pi(N-2)(N-1)/N} & 1 & \cdots & e^{j2\pi(N-2)(N-3)/N} \\
e^{j2\pi(N-1)(N-1)/N} & 1 & e^{j2\pi(N-1)/N} & \cdots & e^{j2\pi(N-1)(N-2)/N}
\end{bmatrix}
$$

$$
\times
\begin{bmatrix}
X(0) \\
X(1) \\
\vdots \\
X(N-1)
\end{bmatrix}.
$$

$$\text{(3.117)}$$

From the frequency-shifting property of Fourier transform, we know that shifting in the frequency domain is equivalent to phase shift in the time domain as below:

$$
F[g(t)e^{j2\pi f_0 t}] = \int_{-\infty}^{\infty} g(t)e^{j2\pi f_0 t} e^{-j2\pi ft}\, dt
$$

$$
= \int_{-\infty}^{\infty} g(t)\, e^{-j2\pi(f-f_0)t}\, dt
$$

$$\text{(3.118)}$$

$$
= G(f - f_0).
$$

Therefore, cyclic hopping FFH-OFDM can be implemented by multiplying the time-domain samples at the output of IFFT in conventional OFDM with a phase shift value. The time-domain samples in FFH-OFDM are then given as:

$$s = \Delta W^H X, \tag{3.119}$$

where Δ is a diagonal matrix given as:

$$\Delta = \begin{bmatrix} 1 & 0 & \cdots & 0 \\ 0 & e^{j2\pi/N} & \cdots & 0 \\ \vdots & \vdots & \ddots & \vdots \\ 0 & 0 & \cdots & e^{j2\pi(N-1)/N} \end{bmatrix}. \tag{3.120}$$

The matrix diagonal matrix Δ relates the IFFT matrix W^H and the shuffled cyclic shifted IFFT matrix W_F^H as below:

$$W_F^H = \Delta W^H. \tag{3.121}$$

The FFH-OFDM scheme can also be seen as precoding of the modulation symbols with a unitary precoder matrix U before the IFFT operation as shown in Figure 3.21. The time-domain samples at the output of the IFFT are then given as:

$$s = W^H U X. \tag{3.122}$$

For the FFH scheme based on cyclic shift hopping, U is a circulant matrix given as:

$$U = W \Delta W^H = W W_F^H, \tag{3.123}$$

where Δ is a diagonal matrix from (3.120). The matrix U has the following structure:

$$U = \begin{bmatrix} X & 0 & X & \cdots & X \\ 0 & X & 0 & \cdots & 0 \\ \vdots & \ddots & \ddots & \ddots & \vdots \\ X & 0 & X & \cdots & X \\ 0 & X & 0 & \cdots & 0 \end{bmatrix}, \tag{3.124}$$

where X represents non-zero elements. Assuming precoding with matrix U, the time-domain samples at the output of the IFFT can be written as:

$$s = W^H U X = W^H W \Delta W^H X. \tag{3.125}$$

The received signal is then written as:

$$\begin{aligned} Y &= W C_R \tilde{H} C_T W^H U X + Z \\ &= H U X + Z, \end{aligned} \tag{3.126}$$

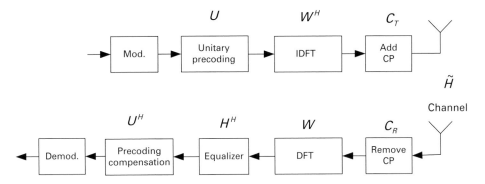

Figure 3.21. FFH-OFDM transmitter and receiver.

where $H = WC_R\tilde{H}C_T W^H$ is a frequency-domain diagonal channel matrix as noted in (3.24).

The modulation symbol estimates $\hat{X} = \left[\hat{X}(0), \ldots, \hat{X}(N-1)\right]^T$ assuming an MRC receiver can be written as:

$$\hat{X} = U^H H^H [Y]$$
$$= U^H H^H HUX + U^H H^H Z$$
$$= (HU)^H HUX + Z'. \tag{3.127}$$

Let us call the matrix HU as the equivalent channel matrix H_{eq}, then the modulation symbols estimates are given as:

$$\hat{X} = \left(H_{\text{eq}}\right)^H H_{\text{eq}}X + Z'. \tag{3.128}$$

As an example, let us consider transmission of four modulations symbols over four subcarriers and look at the shuffled IFFT matrix W_F^H:

$$W_F^H = \Delta W^H = \frac{1}{\sqrt{4}} \begin{bmatrix} 1 & 0 & 0 & 0 \\ 0 & j & 0 & 0 \\ 0 & 0 & 1 & 0 \\ 0 & 0 & 0 & j \end{bmatrix} \times \begin{bmatrix} 1 & 1 & 1 & 1 \\ 1 & j & -1 & -j \\ 1 & -1 & 1 & -1 \\ 1 & -j & -1 & j \end{bmatrix}$$

$$= \frac{1}{\sqrt{4}} \begin{bmatrix} 1 & 1 & 1 & 1 \\ j & -1 & -j & 1 \\ 1 & -1 & 1 & -1 \\ j & 1 & -j & -1 \end{bmatrix}. \tag{3.129}$$

We know that multiplying a matrix $[A]_{n,k}$ from the left with a diagonal matrix $[B]_{n,k} = b_n$ amounts to multiplying the nth row of $[A]_{n,k}$ with b_n.

We now construct the unitary matrix U as below:

$$U = WW_F^H = W \Delta W^H = \frac{1}{\sqrt{4}} \begin{bmatrix} 1 & 1 & 1 & 1 \\ 1 & -j & -1 & j \\ 1 & -1 & 1 & -1 \\ 1 & j & -1 & -j \end{bmatrix}$$

$$\times \frac{1}{\sqrt{4}} \begin{bmatrix} 1 & 1 & 1 & 1 \\ j & -1 & -j & 1 \\ 1 & -1 & 1 & -1 \\ j & 1 & -j & -1 \end{bmatrix} = \frac{1}{4} \begin{bmatrix} 1+j & 0 & 1-j & 0 \\ 0 & 1+j & 0 & 1-j \\ 1-j & 0 & 1+j & 0 \\ 0 & 1-j & 0 & 1+j \end{bmatrix}. \quad (3.130)$$

We note that the unitary precoding matrix U is orthogonal:

$$U^H U = I_N. \quad (3.131)$$

Let us now look at the equivalent channel matrix H_{eq}:

$$H_{\text{eq}} = HU = \frac{1}{\sqrt{4}} \begin{bmatrix} H(0) & 0 & 0 & 0 \\ 0 & H(1) & 0 & 0 \\ 0 & 0 & H(2) & 0 \\ 0 & 0 & 0 & H(3) \end{bmatrix} \times \begin{bmatrix} 1+j & 0 & 1-j & 0 \\ 0 & 1+j & 0 & 1-j \\ 1-j & 0 & 1+j & 0 \\ 0 & 1-j & 0 & 1+j \end{bmatrix}$$

$$= \begin{bmatrix} H(0)(1+j) & 0 & H(0)(1-j) & 0 \\ 0 & H(1)(1+j) & 0 & H(1)(1-j) \\ H(2)(1-j) & 0 & H(2)(1+j) & 0 \\ 0 & H(3)(1-j) & 0 & H(3)(1+j) \end{bmatrix}. \quad (3.132)$$

Since H is a diagonal matrix, the equivalent channel matrix $H_{\text{eq}} = HU$ has the same form as the unitary matrix U in (3.124).

Let us now have a close look at the equivalent channel matrix H_{eq}.

$$H_{eq}^H H_{eq} = (HU)^H HU = \begin{bmatrix} H(0)^*(1-j) & 0 & H(2)^*(1+j) & 0 \\ 0 & H(1)^*(1-j) & 0 & H(3)^*(1+j) \\ H(0)^*(1+j) & 0 & H(2)^*(1-j) & 0 \\ 0 & H(1)^*(1+j) & 0 & H(3)^*(1-j) \end{bmatrix}$$

$$\times \begin{bmatrix} H(0)(1+j) & 0 & H(0)(1-j) & 0 \\ 0 & H(1)(1+j) & 0 & H(1)(1-j) \\ H(2)(1-j) & 0 & H(2)(1+j) & 0 \\ 0 & H(3)(1-j) & 0 & H(3)(1+j) \end{bmatrix}$$

$$= \begin{bmatrix} |H(0)|^2 + |H(2)|^2 & 0 & -j\left[|H(0)|^2 - |H(2)|^2\right] & 0 \\ 0 & |H(1)|^2 + |H(3)|^2 & 0 & -j\left[|H(1)|^2 - |H(3)|^2\right] \\ j\left[|H(0)|^2 - |H(2)|^2\right] & 0 & |H(0)|^2 + |H(2)|^2 & 0 \\ 0 & j\left[|H(1)|^2 - |H(3)|^2\right] & 0 & |H(1)|^2 + |H(3)|^2 \end{bmatrix}.$$

$$(3.133)$$

We note that the equivalent channel matrix $H_{eq} = HU$ is only orthogonal when all the channel gains are equal, that is $H(k) = H(0), k = 0, 1, 2, 3$, which is the case when the channel is frequency-flat. We note that even when the channel gains are different on different subcarriers, odd columns are orthogonal to even columns and vice versa. The modulation symbols estimates are now written as:

$$\begin{bmatrix} \hat{X}(0) \\ \hat{X}(1) \\ \vdots \\ \hat{X}(N-1) \end{bmatrix} = (HU)^H HU \times \begin{bmatrix} X(0) \\ X(1) \\ \vdots \\ X(N-1) \end{bmatrix} + \begin{bmatrix} Z(0) \\ Z(1) \\ \vdots \\ Z(N-1) \end{bmatrix}. \qquad (3.134)$$

Let us look at the first modulation symbol estimate:

$$\hat{X}(1) = \underbrace{\left[|H(0)|^2 + |H(2)|^2\right] X(0)}_{\text{signal}} \underbrace{-j\left[|H(0)|^2 - |H(2)|^2\right] X(2)}_{\text{interference}} + \underbrace{Z'(0)}_{\text{noise}}. \qquad (3.135)$$

We note that the received modulation symbol experiences diversity when $H(0)$ and $H(2)$ are different. However, there is also an interference term introduced due to the fact that the equivalent channel matrix $H_{eq} = HU$ is not orthogonal. This interference term disappears when the channel gains are the same. This would also mean that there is no diversity when the channel gains are the same. Similar to Walsh-spread OFDM, FFH-OFDM promises frequency diversity but at the same time introduces inter-modulation symbol interference. We noted in (3.133) that even-numbered columns are orthogonal to odd-numbered columns while odd-numbered columns are orthogonal to other even-numbered columns. This would mean that even-numbered modulations symbols see interference from other even-numbered symbols and odd-numbered modulations symbols see interference from other odd-numbered symbols. This is in contrast to SC-FDE where each modulations symbol sees interference from all the other symbols when the channel gains on each subcarrier are different. At the same time, however, we note that the diversity order provided by FFH-OFDM is half the diversity order provided by SC-FDE.

3.6   OFDM/OQAM with pulse-shaping

The impulse response of a multi-path radio channel has a duration of T_m, which corresponds to the largest path delay. Also due to mobility, the channel introduces a frequency spread B_d referred to as the Doppler spread. Therefore, a signal transmitted over a mobile wireless channel experiences dispersion both in time and in frequency. The equivalent of time dispersion in frequency domain is coherence bandwidth B_c and an equivalent to frequency spread in time domain is coherence time T_c. The coherence bandwidth B_c represents the frequency interval over which the channel transfer function can be assumed constant. Similarly, coherence time refers to the time interval over which transfer function of the channel does not vary significantly.

Let us denote the time and frequency intervals occupied by the transmission signal as T_s and Δf respectively. We set the following conditions for the transmission signal time and frequency dispersions ΔT and ΔW respectively [16]:

$$B_d \ll \Delta f \Delta W \ll B_c$$
$$T_m \ll T_s \Delta T \ll T_c. \tag{3.136}$$

The bandwidth occupied by the transmission signal can be written as:

$$W = (N - 1) \, \Delta f + \Delta W. \tag{3.137}$$

We can also represent the efficiency of the modulation scheme as:

$$\eta = \frac{\beta N}{T_s W} = \frac{\beta N}{T_s \left[(N - 1) \, \Delta f + \Delta W \right]}$$

$$= \frac{\beta}{\dfrac{(N - 1)}{N} \Delta f T_s + \dfrac{\Delta W T_s}{N}} \quad \text{b/s/Hz}, \tag{3.138}$$

where β denotes the number of bits per symbol and N the number of channels each with frequency interval Δf. We note that the maximum efficiency can be achieved for $\Delta f \approx 1/T_s$ and $N \gg 1$.

We can represent the baseband transmit signal in the following generalized form:

$$s(t) = \sum_{k=0}^{(N-1)} \sum_{n=-\infty}^{\infty} X_n(k) \Psi_{k,n}(t), \tag{3.139}$$

where $X_n(k)$ represents the data symbol transmitted on the kth subcarrier in the nth time interval $[nT_s, (n + 1) \, T_s]$. Also, the basis functions $\Psi_{k,n}(t)$ are obtained by translation in time by nT_s and in frequency by $k \Delta f$ of a prototype function $g(t)$ as:

$$\Psi_{k,n}(t) = e^{j2\pi k \Delta f t} g \left(t - nT_s \right), \tag{3.140}$$

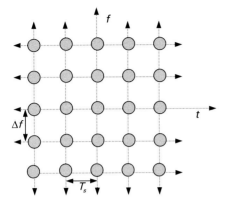

Figure 3.22. Two-dimensional time-frequency lattice.

where $g(t)$ is the transmitter pulse shaping filter. The family of basis functions $\Psi_{k,n}(t)$ forms a discrete two-dimensional time-frequency lattice as shown in Figure 3.22. If the original function $g(t)$ is centered on $(0, 0)$ in a time-frequency lattice, then the function $\Psi_{k,n}(t)$ will be centered on $(nT_s, k\Delta f)$. We can write the center of a normalized time function and its Fourier transform as below:

$$\hat{t} = \int_{-\infty}^{\infty} t \, |g(t)|^2 \, dt$$

$$\hat{f} = \int_{-\infty}^{\infty} f \, |G(f)|^2 \, df. \tag{3.141}$$

Since all basis functions $\Psi_{k,n}(t)$ are obtained from time-frequency translations of the same prototype function $g(t)$, they have the same time dispersion ΔT and frequency dispersion ΔW. These parameters for a function $g(t)$ centered on $(0, 0)$ in time-frequency lattice are given by:

$$(\Delta T)^2 = \int_{-\infty}^{\infty} t^2 \, |g(t)|^2 \, dt$$

$$(\Delta W)^2 = \int_{-\infty}^{\infty} f^2 \, |G(f)|^2 \, df. \tag{3.142}$$

Under the condition of orthogonality of the base functions, we have:

$$\int_{-\infty}^{\infty} \Psi_{k,n}(t)\Psi_{k,n}^*(t) \, dt = \delta_{k,k'} \, \delta_{n,n'}. \tag{3.143}$$

We can rewrite (3.143) as:

$$\int_{-\infty}^{\infty} \Psi_{0,0}(t)\Psi_{k,n}^*(t)\,dt = \delta_k\,\delta_n. \tag{3.144}$$

This is because time-frequency lattice translation is invariant relative to the signal $\Psi_{0,0}(t) = g(t)$.

The demodulation can be done by projecting the received signal $s(t)$ on the base functions as below:

$$
\begin{aligned}
\hat{X}_i(k) &= \int_{-\infty}^{\infty} s(t)\Psi_{k,i}^*(t)\,dt \\
&= \int_{-\infty}^{\infty} s(t)e^{-j2\pi k\Delta ft}g\,(t - iT_s)\,dt.
\end{aligned}
\tag{3.145}
$$

Therefore, we can consider modulation as pulse-shaping of the signal in the time-frequency and the demodulation as a "projection" of the received signal $s(t)$ on the discrete time-frequency lattice $(nT_s, k\Delta f)$.

We can find many orthogonal sets satisfying (3.144) but an important consideration for a mobile wireless channel is robustness against time and frequency dispersion. For this reason, the functions $\Psi_{k,n}(t)$ need to be well localized in time and frequency. We can characterize the time and frequency dispersion by the product $(\Delta T\Delta W)$. This product should be small to prevent the symbol energy smearing over to the neighboring symbols in a time-frequency dispersive mobile radio channel.

We know from Fourier analysis that the more concentrated a function $g(t)$ is in time, the more spread out its Fourier transform $g\,(f)$ is. When we squeeze a function in time, it spreads out in frequency and vice versa. Therefore, we cannot arbitrarily concentrate a function in both time and frequency. In fact, the dispersion product is lower bounded by the uncertainty principle [22]:

$$\Delta T\Delta W \geq \frac{1}{4\pi}. \tag{3.146}$$

In the conventional OFDM scheme, each subcarrier is modulated with a complex QAM data symbol as shown in Figure 3.3. Let us call the conventional OFDM scheme OFDM/QAM to differentiate it from the OFDM/OQAM scheme discussed here. In OFDM, we can represent the baseband transmision signal in the following form:

$$s(t) = \sum_{k=0}^{(N-1)}\sum_{n=-\infty}^{\infty} X_n(k)e^{j2\pi k\Delta ft} \cdot \mathrm{Rect}\,(t - nT_0), \tag{3.147}$$

where $T_0 = T_s + \Delta$ is the OFDM symbol duration including the cyclic prefix interval Δ and the rectangular filter is defined as:

$$\text{Rect}(t) = \begin{cases} \dfrac{1}{\sqrt{T_s}} & 0 \leq t < T_0 \\ 0 & \text{elsewhere.} \end{cases} \tag{3.148}$$

The efficiency of a conventional OFDM system using CP is given as:

$$\eta = \beta \left(1 - \frac{\Delta}{T_s}\right) \quad \text{b/s/Hz.} \tag{3.149}$$

The typical values for the cyclic prefix interval Δ are between $T_s/16$ and $T_s/4$, which represents a significant overhead in both bandwidth and power. We note that the cyclic prefix can be removed if the waveform (also called prototype function) modulating each sub carrier is very well localized in the time domain to limit the inter-symbol interference. Moreover, it can be chosen to be localized well in the frequency domain to limit the inter-carrier interference. Functions having these characteristics exist but the optimally localized ones only guarantee orthogonality on *real* values [17].

The OFDM/OQAM is another well-known parallel transmission scheme [18–21] where the QAM modulation on each subcarrier is replaced with Offset QAM (OQAM) modulation. In OQAM, a $T_s/2$ delay is introduced between in-phase and quadrature components of each QAM symbol as shown in Figure 3.23. Since the OFDM/OQAM introduces an offset between real and imaginary parts of the modulation symbols it allows using well-localized functions operating over *real* values. The orthogonality is then guaranteed over *real* values that lead to a "weak" orthogonality because the channel gain multiplying each subcarrier is a *complex* random variable.

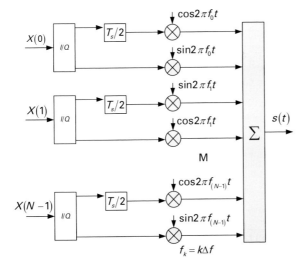

Figure 3.23. Time-continuous model for OFDM/OQAM.

The transmission signal for OFDM/OQAM can be written as:

$$s(t) = \sum_{k=0}^{(N-1)} \left[\sum_{n=-\infty}^{\infty} X_n^{\Re}(k) j^k e^{j2\pi k \Delta ft} g(t - nT_s) \right.$$
$$\left. + \sum_{n=-\infty}^{\infty} j X_n^I(k) j^k e^{j2\pi k \Delta ft} g\left(t - \frac{T_s}{2} - nT_s\right) \right], \quad (3.150)$$

where $X_n^{\Re}(k) = Re[X_n(k)]$ and $X_n^I(k) = Im[X_n(k)]$ denote the real and imaginary parts of the data modulation symbol $X_n(k)$.

The demodulation is performed by taking the real part of the projection of the received signal $s(t)$ on the base functions as below:

$$\hat{X}_n^{\Re}(k) = \int_{t=(n-1)T_s}^{nT_s} Re\left[s(t) j^{-k} e^{-j2\pi k \Delta ft}\right] g(t - nT_s) \, dt$$

$$\hat{X}_n^I(k) = \int_{t=(n-1)T_s}^{nT_s} Re\left[js(t) j^{-k} e^{-j2\pi k \Delta ft}\right] g\left(t - \frac{T_s}{2} - nT_s\right) dt. \quad (3.151)$$

We can rewrite (3.150) as:

$$s(t) = \sum_{k=0}^{(N-1)} \left[\sum_{n=-\infty}^{\infty} Z_n(k) j^{(k+n)} e^{j2\pi k \Delta ft} g\left(t - \frac{nT_s}{2}\right) \right], \quad (3.152)$$

where $Z_n(k) = Re[X_n(k)] \, or \, Im[X_n(k)]$.

The time-frequency lattice for the OFDM/OQAM scheme is shown in Figure 3.24. We note that in OFDM/OQAM one real symbol is sent every $T_s/2$ seconds. This means that one complex symbol is transmitted every T_s seconds providing the same spectral efficiency as an OFDM/QAM scheme.

The equality in (3.146) for optimal time-frequency localization is achieved for the Gaussian function, which means that the Gaussian function is maximally concentrated in time frequency. In addition, a Gaussian function equals its Fourier transform and hence strikes a balance between being concentrated and spread out. Let us consider the Gaussian function

$$\psi(x) = e^{-\pi x^2}. \quad (3.153)$$

Let us define the density of the discrete time-frequency lattice as $1/(\Delta fT_s)$. We know from (3.138) that maximum spectral efficiency is achieved for $(\Delta fT_s = 1)$. If we orthonormalize the Gaussian function in Equation (3.153) for a half-density lattice $\Delta f = T_s = \sqrt{2}$ and $1/(\Delta fT_s) = 1/2$, the resulting orthonormalized signal $\phi(x)$ is the isotropic orthogonal transform algorithm (IOTA) function. By isotropic we mean that the function has the same shape as its Fourier transform that would allow similar robustness against delay spread and

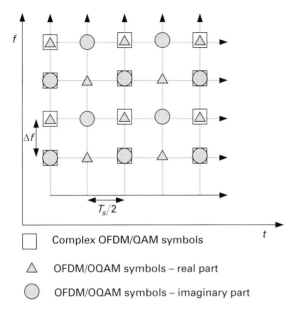

Figure 3.24. OFDM/OQAM time-frequency lattice.

Doppler spread. The IOTA orthonormalization operator is defined as [39]:

$$O\left[\psi\left(x\right)\right] = \frac{2^{1/4}\psi\left(x\right)}{\sqrt{\sum_{m}\left\|\psi\left(x - \frac{m}{\sqrt{2}}\right)\right\|^{2}}}.\tag{3.154}$$

The operator O in (3.154) orthogonalizes the function $O\left[\psi\left(x\right)\right]$ along the frequency axis. Similarly, we can have operator $W^{-1}OW$ that orthogonalizes the function along the time axis. The IOTA function can then be written as:

$$\phi\left(x\right) = W^{-1}OWO\left[\psi\left(x\right)\right]\tag{3.155}$$

where W is the Fourier transform operator.

3.7 Summary

The WCDMA scheme used in the current 3G systems suffers from multi-path interference limiting the achievable peak data rates and system capacity. The multi-path interference in WCDMA worsens for larger bandwidths as the chip rate increases. The multi-path interference problem can be partly resolved by employing the more complex LMMSE receiver. Another problem with WCDMA is its inflexibility in scaling for various system bandwidths required by LTE which include both larger and smaller than 5 MHz bandwidths used by WCDMA.

The OFDM approach can support bandwidths in multiples of subcarrier spacing thereby providing a high degree of flexibility in supporting scalable bandwidths. We also noted that the use of a cyclic prefix in OFDM allows using low complexity 1-tap equalizer per subcarrier thanks to diagonal structure of the equivalent channel matrix. In turn, the modulation symbols can be transmitted in an inter-symbol-interference-free fashion in a multi-path propagation channel.

The OFDM approach was selected as the multiple access scheme for the LTE downlink due to low complexity and the scalability advantages it offers over WCDMA. Other flavors of OFDM such as Walsh-spread OFDM (or MC-CDMA) and FFH-OFDM aim to provide frequency and interference diversity similar to the SC-FDE scheme. As the LTE system is optimized for low speeds, frequency-selective multi-user scheduling based on user channel quality information can be used for improved system performance and capacity. Also, frequency diversity can be exploited via channel coding as the data channel employs turbo coding with a base code rate of 1/3. We also observed that interference fluctuations in OFDM can actually lead to better performance relative to the case of interference diversity scheme that averages out the interference. It is for these reasons that schemes like Walsh-spread OFDM were not adopted for the data channel in the LTE system. The situation is different for some control channels that do not employ channel coding such as 1-bit hybrid ARQ ACK/NACK, which can benefit from frequency and interference diversity offered by Walsh-spread OFDM.

Another OFDM enhancement that we discussed is the scheme called OFDM/OQAM with pulse shaping. This scheme promises to eliminate the cyclic prefix overhead by using functions that are well localized in time and frequency. As the cyclic prefix overhead can be significant, substantial performance gains can be achieved by eliminating the cyclic prefix. The issues with the OFDM/OQAM pulse shaping scheme, however, are increased complexity and also challenges in channel estimation. Because of these concerns, the OFDM/OQAM pulse shaping scheme was not included in the LTE standard.

References

[1] Hanzo, L., Münster, M., Choi, B. J. and Keller, T., *OFDM and MC-CDMA for Broadband Multi-User Communications, WLANs and Broadcasting*, Chichester: Wiley, 2003.

[2] Bingham, J. A. C., "Multicarrier modulation for data transmission: an idea whose time has come," *IEEE Communications Magazine*, vol. 28, no. 5, pp. 5–14, May 1990.

[3] Chang, R.W., "Synthesis of band-limited orthogonal signals for multichannel data transmission," *Bell Systems Technical Journal*, vol. 45, pp. 1775–1796, Dec. 1966.

[4] Saltzberg, B., "Performance of an efficient parallel data transmission system," *IEEE Transactions on Communications*, vol. COM-15, no. 6, pp. 805–811, Dec. 1967.

[5] Weinstein, S. B. and Ebert, P. M., "Data transmission by frequency-division multiplexing using the discrete Fourier transform," *IEEE Transactions on communications*, vol. COM-19, no. 5, pp. 628–634, Oct. 1971.

[6] Peled, A. and Ruiz, A., "Frequency domain data transmission using reduced computational complexity algorithms," *IEEE International Conference on Acoustics, Speech, and Signal Processing ICASSP '80*, vol. 5, pp. 964–967, Apr. 1980.

[7] Walzman, T., "Automatic equalization and discrete Fourier transform techniques" (Ph.D. Thesis abstract), *IEEE Transactions on Information Theory*, vol. IT.18, pp. 455–455, May 1972.

[8] Sari, H., Karam, G. and Jeanclaude, I., "Transmission techniques for digital terrestrial TV broadcasting," *IEEE Communications Magazine*, vol. IT.33, pp. 100–109, Feb. 1995.

[9] Benvenuto, N. and Tomasin, S., "On the comparison between OFDM and single carrier modulation with a DFE using a frequency-domain feedforward filter," *IEEE Transactions on Communications*, vol. COM-50, pp. 947–955, June 2002.

[10] Shi, T., Zhou, S. and Yao, Y., "Capacity of single carrier systems with frequency-domain equalization," *Proceedings of the IEEE 6th Circuits and Systems Symposium on Emerging Technologies*, pp. 429–432, June 2004.

[11] Zervos, N. and Kalet, I., "Optimized decision feedback equalization versus optimized orthogonal frequency division multiplexing for high-speed data transmission over the local cable network," *IEEE International Conference on Communications, ICC 98*, pp. 1080–1085, June 1989.

[12] Scholand, T., Faber, T., Seebens, A., Lee, J., Cho, J., Cho, Y. and Lee, H.W., "A fast frequency hopping concept," *Electronics Letters*, vol. 41, no. 13, pp. 748–749, June 2005.

[13] Berens, F., Rüegg, A., Scholand, T., Hessamian-Alinejad, A. and Jung, P., "Fast frequency hopping diversity scheme for OFDM based UWB systems," *Electronics Letters*, vol. 43, no. 1, pp. 41–42, Jan. 2007.

[14] Miyazaki, N., Hatakawa, Y., Yamamoto, T., Ishikawa, H., Suzuki, T. and Takeuchi, K., "A study on rotational OFDM transmission with multi-dimensional demodulator and twin turbo decoder," *IEEE 64th Vehicular Technology Conference, 2006*, VTC-2006 Fall, pp. 1–5, Sept. 2006.

[15] Boutros, J. and Viterbo, E., "Signal space diversity: a power and bandwidth efficient diversity technique for the Rayleigh fading channel," *IEEE Transactions on Information Theory*, vol. IT-44, no. 4, pp. 1453–1467, July 1998.

[16] Haas, R. and Belfiore, J.-C., "Multiple carrier transmission with time-frequency well-localized impulses," *IEEE Second Symposium on Communications and Vehicular Technology in the Benelux*, pp. 187–193, Nov. 1994.

[17] Lacroix, D. and Javaudin, J.P., "A new channel estimation method for OFDM/OQAM," *Proceedings of 7th International OFDM Workshop*, Sep. 2002.

[18] Hirosaki, B., "An analysis of automatic equalizers for orthogonally multiplexed QAM systems," *IEEE Transactions on Communications*, vol. COM-28, no. 1, pp. 73–83, Jan. 1980.

[19] Lacroix, D., Goudard, N. and Alard, M., "OFDM with guard interval versus OFDM/offsetQAM for high data rate UMTS downlink transmission," *IEEE Vehicular Technology Conference 2001*, VTC 2001 Fall, vol. COM-s4, pp. 2682–2686, 2001.

[20] 3GPP TR 25.892 V2.0.0, Feasibility study for OFDM for UTRAN enhancement.

[21] Bolcskei, H., Duhamel, P. and Hleiss, R., "Design of pulse shaping OFDM/OQAM systems for high data-rate transmission over wireless channels," *IEEE International Conference on Communications, ICC '99*, pp. 559–564, 1999.

[22] Folland, G. and Sitaram, A., "The uncertainty principle: a mathematical survey", *Journal of Fourier Analysis and Applications*, pp. 207–238, 1997.

[23] Le Floch, B., Alard, M. and Berrou, C., "Coded orthogonal frequency division multiplex TV broadcasting," *Proceedings of the IEEE*, vol. 83, no. 6, pp. 982–996, June 1995.

4 Single-carrier FDMA

The design of an efficient multiple access and multiplexing scheme is more challenging on the uplink than on the downlink due to the many-to-one nature of the uplink transmissions. Another important requirement for uplink transmissions is low signal peakiness due to the limited transmission power at the user equipment (UE). The current 3G systems use the wideband code division multiple access (WCDMA) scheme both in the uplink and in the downlink. In a WCDMA downlink (Node-B to UE link) the transmissions on different Walsh codes are orthogonal when they are received at the UE. This is due to the fact that the signal is transmitted from a fixed location (base station) on the downlink and all the Walsh codes received are synchronized. Therefore, in the absence of multi-paths, transmissions on different codes do not interfere with each other. However, in the presence of multi-path propagation, which is typically the case in cellular environments, the Walsh codes are no longer orthogonal and interfere with each other resulting in inter-user and/or inter-symbol interference (ISI).

The problem is even more severe on the uplink because the received Walsh codes from multiple users are not orthogonal even in the absence of multi-paths. In the uplink (UE to Node-B link), the propagation times from UEs at different locations in the cell to the Node-B are different. The received codes are not synchronized when they arrive at the Node-B and therefore orthogonality cannot be guaranteed. In fact, in the current 3G systems based on WCDMA, the same Walsh code though scrambled with different PN (pseudo-noise) sequences is allocated to multiple users accessing the system making the uplink transmissions non-orthogonal to each other.

The simultaneous transmissions from multiple users interfering with each other contribute to the noise rise seen by each of the users. In general, the noise rise at the base station is kept below a certain threshold called the rise-over-thermal (RoT) threshold in order to guarantee desirable capacity and coverage. The RoT is defined as:

$$\text{RoT} = \frac{I_0 + N_0}{N_0}, \tag{4.1}$$

where I_0 is the received signal power density (including both the inter-cell and intra-cell power) and N_0 is the thermal noise (also referred to as the background noise) density. The RoT threshold limits the amount of power above the thermal noise at which mobiles can transmit. The RoT can be related to the uplink loading by the following expression:

$$\text{RoT} = \frac{I_0 + N_0}{N_0} = \frac{1}{1 - \eta_{\text{UL}}}, \tag{4.2}$$

where η_{UL} represents uplink loading in a fraction of the CDMA pole capacity [7]. In general uplink loading in a CDMA system is tightly controlled for system stability and coverage.

Assuming 70% loading, a noise rise of 5.2 dB is obtained. It can be noted that 5.2 dB is also the maximum SINR (or chip energy to noise-plus-interference spectral density, E_c/N_t) that can be observed if there is no inter-cell and intra-cell interference i.e. a single user transmitting in the system. However, in a realistic scenario, the maximum E_c/N_t seen by a user would be much smaller due to inter-cell, intra-cell and multi-path interference. Therefore, the noise rise constraint limits the maximum achievable E_c/N_t that in turn limits the maximum achievable data rate in a CDMA uplink.

The noise rise constraint is a result of the near-far problem in a non-orthogonal WCDMA uplink. However, if the uplink can somehow be made orthogonal (e.g. using synchronous WCDMA), the noise rise constraint can be relaxed. Synchronous WCDMA requires sub-chip level synchronization and thus puts very strict requirements on uplink timing control in a mobile environment. The subchip level synchronization required in synchronous CDMA makes it undesirable in a mobile wireless system. It is also possible to improve the performance of a WCDMA uplink by using a more complex successive interference cancellation (SIC) receiver.

In a WCDMA uplink the number of Walsh codes used for uplink transmission from a single user is limited to a few codes. This is to limit the uplink signal peakiness to improve the power amplifier efficiency. With a single code transmission, WCDMA provides low signal peakiness due to its single-carrier property.

The goal for the LTE uplink access scheme design is to provide low signal peakiness comparable to WCDMA signal peakiness while providing orthogonal access not requiring an SIC receiver. The two candidate technologies for orthogonal access are OFDMA (orthogonal frequency division multiple access) and SC-FDMA (single-carrier frequency division multiple access).

In this chapter, we first present details of the SC-FDMA scheme. We then discuss the capacity performance aspects of orthogonal and non-orthogonal uplink multiple access schemes. The signal peakiness comparison for various candidate schemes is performed in the next chapter. The capacity performance comparison and signal peakiness tradeoffs should help the reader understand why SC-FDMA was selected as the multiple access schemes for LTE uplink.

4.1   Single-carrier FDMA

Single-carrier FDMA scheme provides orthogonal access to multiple users simultaneously accessing the system [2]. Another attractive feature of SC-FDMA discussed in the next chapter is low signal peakiness due to the single carrier transmission property. In one flavor of SC-FDMA scheme referred to as IFDMA in [5], the user's data sequence is first repeated a predetermined number of times. Then the repeated data sequence is multiplied with a user-specific phase vector. Another way of looking at this approach is FFT precoding of the data sequence and then mapping of the FFT-precoded data sequence to uniformly spaced subcarriers at the input of IFFT. The uniform spacing is determined by the repetition factor Q. The multiplication of the repeated data sequence with a user-specific phase vector can be seen as frequency shift applied in order to map transmissions from multiple users on non-overlapping orthogonal subcarriers. It should be noted that each data modulation symbol is spread out on all the subcarriers used by the UE. This can provide a frequency-diversity benefit in a frequency-selective channel. However, there may be some impact on performance as well due to loss of orthogonality or noise enhancement when data subcarriers experience frequency selective fading. We will refer to the IFDMA scheme as distributed FDMA (DFDMA) from now on. The mapping of FFT-precoded data sequence to contiguous subcarriers results in a localized transmission in the frequency domain. Similar to distributed mapping or DFDMA, localized mapping also results in a low PAPR signal. The distributed and localized mapping of FFT pre-coded data sequence to OFDM subcarriers is sometimes collectively referred to as DFT-spread OFDM.

In case of distributed FDMA, the input samples to IFFT are given as [3]:

$$\tilde{X}_l = \begin{cases} X_{\frac{l}{Q}} & l = Q \cdot k, 0 \le k \le M - 1 \\ 0 & \text{otherwise.} \end{cases} \tag{4.3}$$

Let us define $n = M \times q + m$, where $0 \le q \le Q - 1$ and $0 \le m \le M - 1$. The time domain samples at the output of IDFT are then given as:

$$\tilde{x}_n = \frac{1}{N} \sum_{l=0}^{N-1} \tilde{X}_l e^{j2\pi \frac{n}{N} l}. \tag{4.4}$$

Using the relationship of Equation (4.3), the above equation can be simplified as:

$$\tilde{x}_n = \frac{1}{Q \cdot M} \sum_{k=0}^{M-1} X_k e^{j2\pi \frac{n}{M} k}. \tag{4.5}$$

Let us now introduce $n = M \times q + m$

$$\begin{aligned} \tilde{x}_n &= \frac{1}{Q \cdot M} \sum_{k=0}^{M-1} X_k e^{j2\pi \frac{M \times q + m}{M} k} \\ &= \frac{1}{Q} \left(\frac{1}{M} \sum_{k=0}^{M-1} X_k e^{j2\pi \frac{m}{M} k} \right) \\ &= \frac{1}{Q} x_m. \end{aligned} \tag{4.6}$$

It can be seen that the time-domain symbols at the output of size N IDFT are repetitions of time-domain symbols at the input of size M DFT [5].

In case of localized FDMA, the input samples to IFFT are given as:

$$\tilde{X}_l = \begin{cases} X_l & 0 \le l \le M - 1 \\ 0 & M \le l \le N - 1. \end{cases} \tag{4.7}$$

Let us define $n = Q \times m + q$, where $0 \le q \le Q - 1$ and $0 \le m \le M - 1$. The time-domain samples at the output of IDFT are then given as:

$$\tilde{x}_n = \frac{1}{N} \sum_{l=0}^{N-1} \tilde{X}_l e^{j2\pi \frac{n}{N} l}. \tag{4.8}$$

Using the relationship of Equation (4.7) and $n = Q \times m + q$, the above equation can be simplified as:

$$\tilde{x}_n = \frac{1}{Q \cdot M} \sum_{l=0}^{M-1} X_l e^{j2\pi \frac{Q \times m + q}{QM} k}.$$ (4.9)

Let us now introduce $q = 0$

$$\tilde{x}_n = \frac{1}{Q} \cdot \frac{1}{M} \sum_{l=0}^{M-1} X_l e^{j2\pi \frac{m}{M} k}$$

$$= \frac{1}{Q} x_m.$$ (4.10)

It can be seen that every Qth time-domain sample at the output of size N IDFT is the same as the time-domain sample at the input of size M DFT. For $q \neq 0$, the time-domain sample at the output of size N IDFT is the sum of time-domain samples at the input of size M DFT with different complex weighting. An example of DFDMA and LFDMA mapping for $M = 4, N = 8$ and $Q = \frac{N}{M} = 2$ is given in Figure 4.1.

The transmit and receive chains for LFDMA are given in Figures 4.2 and 4.3 respectively. All UEs use the same IDFT size of N. However, different UEs can use different DFT-precoding sizes. The size of the DFT precoder for a UE is proportional to the orthogonal subcarriers allocated to the UE for uplink transmission. Let M_i represent DFT-precoding size for the ith UE, then the following applies:

$$\sum_{i=1}^{K} M_i \leq (N - G),$$ (4.11)

where K represents the number of UEs transmitting simultaneously and G the number of guard subcarriers. In case of uplink multi-user MIMO (multiple input multiple output), different UEs can be allocated overlapping subcarriers and therefore the condition in Equation (4.11) does not apply.

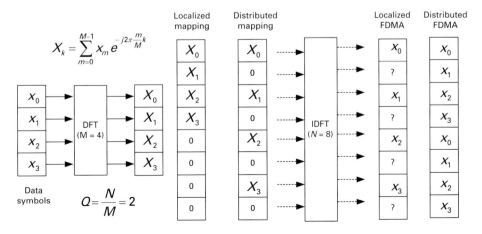

Figure 4.1. Subcarrier mapping for Localized and Distributed FDMA.

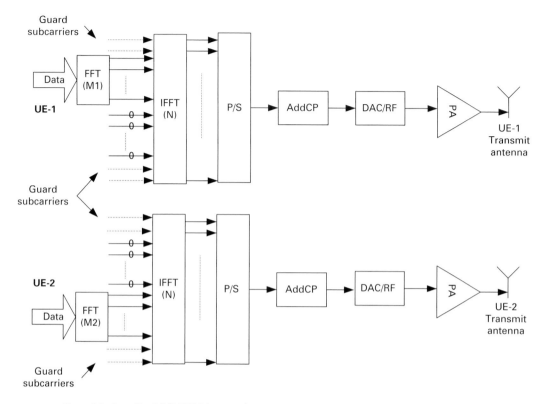

Figure 4.2. Localized SC-FDMA transmitter.

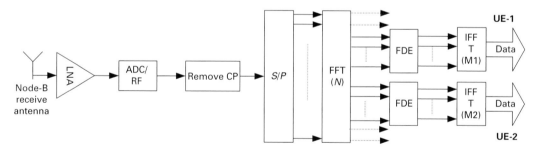

Figure 4.3. Localized SC-FDMA receiver.

A cyclic prefix is added after IDFT operation and the resulting sequence is up-converted to RF, amplified and transmitted. All UEs transmitting simultaneously with their data mapped to orthogonal subcarriers perform this operation independently. Each UE power amplifier then sees a single FFT-precoded transmission, which leads to a low signal peakiness single-carrier transmission. In Figure 4.2, for example, UE1 and UE2 use DFT sizes of M_1 and M_2 for transmission from their respective power amplifiers. If, for example, we map FFT-precoded data sequences of UE1 and UE2 to the same IDFT and transmit from a single power amplifier that will no longer be a low signal peakiness single-carrier transmission. This is one of the reasons why SC-FDMA is generally not considered for downlink transmissions. In the downlink, Node-B generally transmits simultaneous signals to multiple UEs on orthogonal subcarriers using a single common power amplifier.

In the receiver side, the received signal is filtered, amplified and down-converted from RF. The cyclic prefix samples are discarded and a size N DFT operation is performed on the received samples sequence. The symbols for each UE are separated by collecting data from the subcarriers allocated to a UE. A frequency-domain equalization (FDE) operation is performed using channel estimates obtained from pilots or reference signals received for each UE. An IDFT operation is then performed separately for each UE to recover the transmitted data sequence. It should be noted that data demodulation in SC-FDMA happens in the time domain after the IDFT operation.

4.2   Uplink capacity comparison

By using simple analytical models, we provide a channel capacity performance comparison of non-orthogonal uplink multiple access and orthogonal uplink multiple access schemes. The orthogonal schemes that we considered include OFDMA, SC-FDMA and TDMA while WCDMA is assumed as baseline for a non-orthogonal scheme. Before starting the discussion on capacity comparison, we present the power-limited capacity of a wireless link.

4.2.1   Power-limited capacity

The capacity of a wireless communication channel with constant transmission power can be expressed as:

$$C = W \cdot \log_2 \left(1 + \frac{\text{SNR}_{1.0\text{MHz}}}{W} \right) \text{ [Mb/s]}, \tag{4.12}$$

where W is the bandwidth in MHz and $\text{SNR}_{1.0\,\text{MHz}}$ is the SNR within 1.0 MHz bandwidth with the constant transmission power P. It can be observed that increasing bandwidth does not result in linear increase in capacity if the transmission power is kept constant. The channel capacity as a function of bandwidth for various values of $\text{SNR}_{1.0\,\text{MHz}}$ is shown in Figure 4.4. The capacity gains at 20.0 MHz relative to 1.0 MHz bandwidth for $\text{SNR}_{1.0\,\text{MHz}}$ of 0, 5 and 10 dB are also summarized in Table 4.1. It can be noted that for the low SNR scenario of $\text{SNR}_{1.0\,\text{MHz}} = 0.0$ dB, there is a very small increase in capacity when the bandwidth is increased from 1 to 20 MHz. In order to obtain linear increase in capacity at low SNR, the transmission power also needs to be scaled with bandwidth effectively keeping the same power spectral density. This would, however, require very expensive high power broadband power amplifiers. It is therefore desirable to allow multiple simultaneous transmissions in the uplink to maximize the total transmitted power.

4.2.2   Capacity of WCDMA

Earlier we noted that multiple users transmitting simultaneously using WCDMA interfere with each other due to the asynchronous nature of the received uplink transmissions. The uplink capacity limit of a WCDMA system can be approximated as:

$$C_{\text{WCDMA}} = K \cdot \log_2 \left(1 + \frac{P}{(1+f)KP + (\alpha - 1)P + N_0} \right) \text{ [b/s/Hz]}, \tag{4.13}$$

Table 4.1. Capacity gains with fixed transmit power.

SNR$_{1.0\,MHz}$	C$_{20\,MHz}$/C$_{1MHZ}$
0.0 dB	1.41
5.0 dB	2.06
10.0 dB	3.38

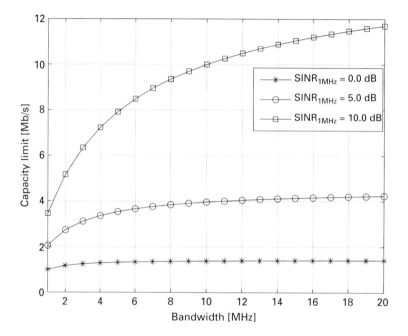

Figure 4.4. Channel capacity with fixed transmit power.

where

    K : number of users transmitting simultaneously
    P : received power for a user
    f : ratio between other-cell and own-cell signal
    α : fraction of the own-user signal considered as interference
    N_0 : background noise.

For the special case where $f = 0$ and $\alpha = 0$, the above equation simplifies to:

$$C_{WCDMA} = K \cdot \log_2 \left(1 + \frac{P}{(K-1)P + N_0} \right) \text{ [b/s/Hz].} \tag{4.14}$$

For large K, i.e. $K \to \infty$

$$C_{WCDMA} \approx \log_2(e) = 1.44 \text{ [b/s/Hz].} \tag{4.15}$$

An example of WCDMA capacity as a function of the number of users in the cell for the case of a single isolated cell is shown in Figure 4.5. The two curves indicate the cases where the single user SNR is 0.0 or 10.0 dB. The single user SNR is defined as P/N_0 i.e. no inter-cell or intra-cell interference present in the system. The 10.0 and 0.0 dB cases represent a good user and a relatively weak user respectively. The capacity at point $K = 1$ represents the situation of a single user transmitting at a time (i.e. a TDMA approach). It can be noted that it is advantageous to schedule good users in a TDM fashion and weak users in a WCDMA fashion in order to maximize the system capacity. Such a hybrid WCDMA/TDMA approach is described in [4] and it is shown that the hybrid WCDMA/TDMA approach provides significant gains over a pure WCDMA scheme in smaller bandwidths such as 1.25 MHz. However, for larger bandwidths a TDMA approach suffers from link budget limitation due to limited UE transmission power i.e. a single user transmitting in a TDM fashion over a large bandwidth such as 5.0, 10.0, 20.0 MHz may not be able to efficiently use the whole bandwidth due to its transmission power limitation.

4.2.3 Capacity of TDMA

In a TDMA (Time Division Multiple Access) system, a single user transmits at a given time slot. The total system resource is shared among multiple users accessing the link on a time slot by time slot basis. The capacity limit of a TDMA system can be approximated by setting $K = 1$ (i.e. single user transmitting at a given time) in the capacity relationship for a WCDMA system as below:

$$C_{\text{TDMA}} = \log_2 \left(1 + \frac{P}{(f + \alpha)P + N_0} \right) \text{ [b/s/Hz]}. \tag{4.16}$$

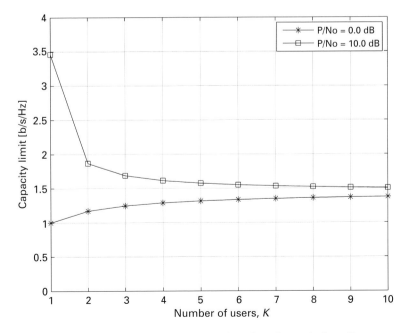

Figure 4.5. WCDMA capacity as a function of number of users in the cell.

It can be noted that there is no intra-cell (multiple access) interference in case of TDMA because a single user transmits in a given slot. However, for larger system bandwidths a TDMA approach suffers from uplink coverage limitation because, for a fixed UE transmission power as the system bandwidth increases, the power spectral density goes down. This is because in a TDMA approach, a user needs to transmit over the full bandwidth, as multiple users cannot share the resources in the frequency or code domains. Let us consider the TDMA coverage issue in more detail. The thermal noise power at a receiver is given as:

$$P = k_B\, TW \quad \text{watts,} \tag{4.17}$$

where $k_B = 1.38 \times 10^{-23}$ joules/K is the Boltzmann constant, T temperature in Kelvin and W bandwidth in Hz. The thermal noise power spectral density N_0 in dBm/Hz at room temperature $T = 300$ can be written as:

$$N_0 = -174 \text{ dBm/Hz.} \tag{4.18}$$

If we assume a transmission power of P_t watts and a required received power of P_r watts, the total path-loss permitted is:

$$PL = \frac{P_t}{P_r}. \tag{4.19}$$

The received SNR (signal to noise ratio) ρ is given as:

$$\rho = \frac{P_r}{N_0 W}. \tag{4.20}$$

The path-loss allowed to guarantee a received SNR of ρ can then be written as:

$$PL = \frac{P_t}{(\rho \times N_0 W)}. \tag{4.21}$$

The pathloss PL in dB supported for a transmission power of P_t dBm and the received SNR of ρ dB can be written as:

$$PL_{dB} = (P_t - N_0 W - \rho) \text{ dB.} \tag{4.22}$$

Assuming COST231Hata urban propagation model with Node-B antenna height of 32 meters, UE antenna height of 1.5 meter and carrier frequency of 1.9 GHz, the supported distance d between transmitter and receiver is given as:

$$d = 10^{\left(\frac{PL_{dB} - 31.5}{35}\right)} \text{ m.} \tag{4.23}$$

The allowed distances between UE and Node-B as a function of system bandwidth for $\rho = 0, 10, 20$ dB are plotted in Figure 4.6. We assumed the transmit power of $P_t = 24$ dBm as this is the maximum UE power permitted in most cellular standards. As we expect, the communication range is adversely affected as the bandwidth increases for a fixed transmission power. For example, for a received SNR threshold of 20 dB, as the bandwidth increases from 1 to 20 MHz, the communication range decreases from about 300 to only 125 m. It should be noted that we didn't consider inter-cell interference in our anlaysis. In the presence of inter-cell interference, the communication range would be further limited. We should also note that loss

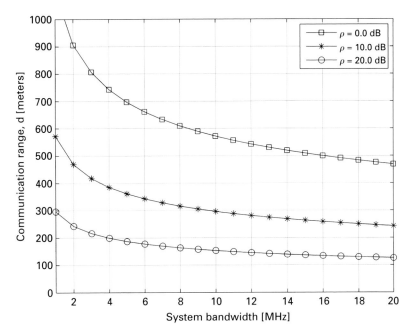

Figure 4.6. Communication range as a function of system bandwidth for $\rho = 0, 10, 20$ dB and $P_t = 24$ dBm.

in the coverage area with reduced range is even more severe as the coverage area varies as the square of the distance. Due to this coverage limitation issue at larger bandwidths, the TDMA approach was ruled out at a very early stage of LTE system design because LTE requirements mandated support for larger system bandwidths of up to 20 MHz.

4.2.4 Capacity of OFDMA

Unlike TDMA, OFDMA allows sharing resources among multiple users accessing the system by allocating to a user only a fraction of the total bandwidth. Therefore, multiple users can transmit simultaneously on orthogonal subcarriers. The transmissions from multiple users are orthogonal as long as the relative delay between the received transmissions is within the cyclic prefix (CP) length. In general, the CP length is several microseconds, to account for the multi-path delay spread, and therefore makes the timing synchronization within the CP length feasible. This is in contrast to synchronous WCDMA where subchip level synchronization (generally a small fraction of a microsecond depending upon the chip rate) is required to guarantee orthogonal transmissions. The uplink capacity limit for an OFDMA system can be written as:

$$C_{\text{OFDMA}} = \sum_{i=1}^{K} \beta_i \cdot \log_2 \left(1 + \frac{P}{fP + \beta_i N_0} \right) \text{ [b/s/Hz]}, \qquad (4.24)$$

where β_i is the fraction of bandwidth allocated to user i. For the case where the bandwidth is equally divided among the K users transmitting simultaneously, the above formula can be simplified as below:

$$C_{OFDMA} = \sum_{i=1}^{K} \frac{1}{K} \cdot \log_2 \left(1 + \frac{P}{fP + \frac{N_0}{K}} \right) \text{ [b/s/Hz]}$$

$$= \log_2 \left(1 + \frac{KP}{fKP + N_0} \right) \text{ [b/s/Hz]}. \qquad (4.25)$$

We note that there is no intra-cell (multiple access) interference or inter-symbol-interference (ISI) due to orthogonal subcarriers used by different users and 1-tap OFDM subcarrier equalization. However, cyclic prefix (guard interval) overhead (typically around 10%) needs to be taken into account for the OFDM case. Therefore, the capacity of an OFDMA system can be scaled-down to account for CP overhead as below:

$$C_{OFDMA} = \left(\frac{T_s}{T_s + \Delta} \right) \times \log_2 \left(1 + \frac{KP}{fKP + N_0} \right) \text{ [b/s/Hz]}, \qquad (4.26)$$

where T_s is the OFDM symbol duration and Δ is the cyclic prefix duration.

4.2.5 Capacity of SC-FDMA

Like OFDMA, SC-FDMA avoids intra-cell interference in the uplink. However, SC-FDMA can also benefit from frequency diversity because a given modulation symbol is transmitted over the whole bandwidth allocated to the UE. However, the downside of this approach is that performance of SC-FDMA suffers in a frequency-selective fading channel due to noise enhancement. This is because the IDFT operation after frequency-domain equalization at the receiver spreads out the noise over all the modulation symbols. It should be noted that noise enhancement results in inter-symbol-interference (ISI) and not the inter-user interference, that is, there is no intra-cell interference among UEs transmitting over orthogonal frequency resources. The losses of SC-FDMA link performance relative to OFDMA have been estimated ranging from no loss or a slight gain due to diversity at low SINR for QPSK modulation to about 1 dB for 16-QAM and 64-QAM modulations typically used at higher SINR [8]. Therefore, the uplink capacity limit for an SC-FDMA system is given as:

$$C_{SC\text{-}FDMA} = \left(\frac{T_s}{T_s + \Delta} \right) \times \log_2 \left(1 + \frac{KP}{fKP + N_0} \times \frac{1}{10^{(L_{SC\text{-}FDMA}/10)}} \right) \text{ [b/s/Hz]}, \qquad (4.27)$$

where $L_{SC\text{-}FDMA}$ represents the SC-FDMA link loss in dBs relative to OFDMA. This loss occurs at higher SINR when frequency-domain linear equalization is used. It should be noted that some or all of this loss can be recovered by using a more advanced receiver [9] at the Node-B at the expense of additional complexity.

4.3 Capacity results

We provide some numerical results for some selected cases in Figures 4.7 and 4.8 for single user SINR, P/N_0 of 0.0 and 10.0 dB respectively. The single user SINR, P/N_0 is defined as the SINR seen when no interference is present. In these results we assumed a flat fading channel

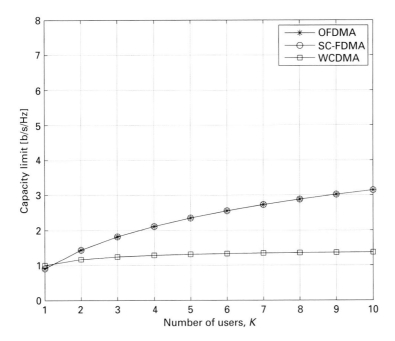

Figure 4.7. Capacity limits for OFDMA, SC-FDMA and WCDMA for a single cell scenario ($f = 0.0$) at $P/N_0 = 0.0$ dB.

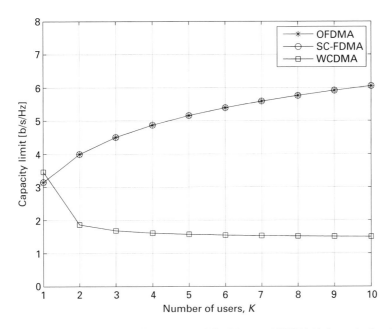

Figure 4.8. Capacity limits for OFDMA, SC-FDMA and WCDMA for a single cell scenario at $P/N_0 = 10.0$ dB.

scenario i.e. no inter-symbol interference, which means $L_{SC\text{-}FDMA} = 0.0\,dB$ i.e. no loss for SC-FDMA relative to OFDMA. In our further discussions on performance comparison, we will collectively refer to SC-FDMA and OFDMA as orthogonal schemes. We also assumed $f = 0.0$, which indicates no inter-cell interference. We are interested in performance impact of intra-cell or multiple access interference (MAI) on uplink performance of different access techniques. We provide performance for OFDMA, SC-FDMA and WCDMA schemes. The TDMA scheme can be seen as a special case when the number of users transmitting is one ($K = 1$). The horizontal axis shows the number of users simultaneously accessing the system.

The $P/N_0 = 0.0\,dB$ case in Figure 4.7 represents a weak user in the system while the $P/N_0 = 10.0\,dB$ case in Figure 4.8 depicts a relatively good user in the system, which is generally the case for users closer to the base station. We note that gains of orthogonal access over non-orthogonal access for the larger SINR user case increase as the number of users increases. This is explained by the fact that the performance of a high SINR user is dominated by intra-cell (or inter-user) interference and an orthogonal access benefits by eliminating the intra-cell interference. However, the performance of a weak user is dominated by the inter-cell interference and the background noise and eliminating intra-cell interference by using orthogonal access only provides a small advantage. It should be noted that performance of a non-orthogonal scheme for the larger SINR case ($P/N_0 = 10.0\,dB$) degrades as the number of users increases. This is explained by the fact that increasing number of users also results in increased intra-cell interference. In fact, for the large SINR case, it is better to let a single user transmit in case of non-orthogonal access, which effectively means a TDMA scheme with $K = 1$. We note that orthogonal schemes provide approximately four times improvement over WCDMA for 10 users case in a single-isolated cell scenario considering a good SINR user.

We also note that performance of OFDMA and SC-FDMA with $K = 1$ is similar to TDMA performance. However, the benefit of OFDMA and SC-FDMA is that multiple users can simultaneously transmit providing overall larger power transmitted in the system. Therefore, a higher capacity than the TDMA case can be achieved for the case of $K > 1$.

Although the numerical results for other-cell to own-cell signal ratio f of greater than zero are not presented, we can note from the analysis in Section 4.2 that the gains of orthogonal access are larger when the other-cell to own-cell signal ratio f is small i.e. the highest gains of orthogonal access are seen in an isolated hotspot cell case. We know that as the other-cell to own-cell signal ratio f increases, the performance is dominated by other-cell interference rather than only by intra-cell interference. Therefore, in case of heavy inter-cell interference, the gains of orthogonal access over non-orthogonal access go down.

In summary, the numerical results show that orthogonal uplink multiple access schemes OFDMA and SC-FDMA can provide significant capacity gains over a non-orthogonal uplink multiple access scheme WCDMA. We expect that the gains of OFDMA and SC-FDMA over WCDMA will be even larger when a frequency selective channel is considered. In this case, WCDMA using Rake receiver will suffer from ISI. There can also be some performance degradation of OFDMA due to frequency selectivity because different subcarriers experience different fading conditions. Also, SC-FDMA performance can be degraded due to loss of orthogonality or noise enhancement in a frequency-selective channel. However, these losses are expected to be smaller relative to the loss experienced by WCDMA due to ISI. We can assume that the gains of OFDMA and SC-FDMA over WCDMA in a frequency-selective channel in the uplink would be similar to the gains of OFDM over WCDMA for downlink transmission. This is because in WCDMA downlink transmissions, there is no MAI interference. We should note that these gains are in addition to the gains that OFDMA and SC-FDMA provide by eliminating multiple-access interference (MAI) in the uplink.

4.4 Hybrid uplink access

We mentioned earlier that SC-FDMA can result in link performance loss relative to OFDMA ranging from no loss at low SINR for QPSK modulation to 1–1.5 dB at higher SINR for 16-QAM and 64-QAM modulations. The link performance comparison between OFDM and SC-FDMA for 64-QAM modulation with 1/2 and 2/3 coding rates in an SCM channel (see Chapter 18) for an MMSE receiver is shown in Figure 4.9. In the next chapter we will learn that SC-FDMA provides a low-peakiness signal relative to OFDMA. The reduced signal peakiness allows to increase the UE transmission power providing larger range and coverage. A hybrid scheme as shown in Figure 4.10 using DFT-precoding for QPSK transmissions at low SINR while by-passing the DFT-precoding for higher order modulations at higher SINR can then be considered. Such a scheme can provide the low-signal peakiness advantage of SC-FDMA for coverage limited UEs using QPSK while avoiding the link performance loss for high SINR UEs using higher order modulations. Also, the low signal-peakiness advantage of SC-FDMA goes down for higher order modulations, as we will discuss in the next chapter. It then seems appropriate not to DFT-precode higher order modulation transmissions.

The receiver for the hybrid scheme is shown in Figure 4.11. It can be seen that the receiver structure is very similar to the SC-FDMA receiver with the only difference that

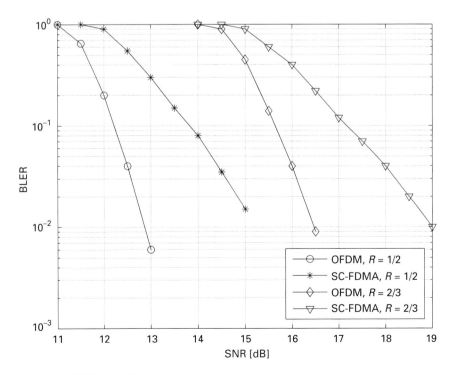

Figure 4.9. OFDM and SC-FDMA link performance for 64-QAM.

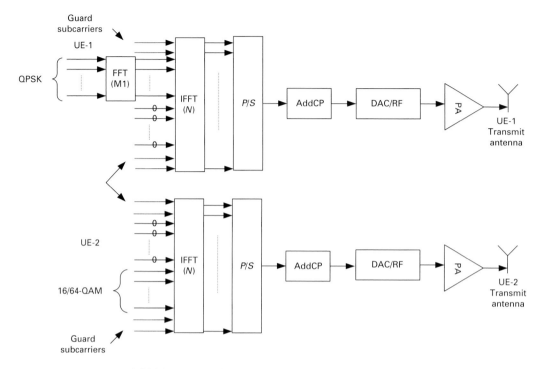

Figure 4.10. Hybrid SC-FDMA and OFDMA transmitter.

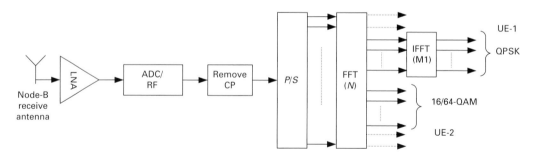

Figure 4.11. Hybrid SC-FDMA and OFDMA receiver.

the transmissions that are not DFT-precoded at the transmitter are available directly at the output of DFT. That is, there is no need to perform additional IDFT operation. However, the transmissions that are DFT-precoded at the transmitter are obtained after performing the IDFT operation.

In the hybrid scheme, we can also consider not to DFT-precode MIMO transmissions. This will simplify the MIMO receiver as 1-tap equalizer can be used. A more complex MIMO receiver processing is required for SC-FDMA as each modulations symbol is spread over multiple subcarriers. The hybrid scheme can further be extended by only using DFT-precoding when a transmission is power-limited irrespective of the modulation or MIMO scheme used.

Table 4.2. FFT-precoding sizes.

Multiple of 12	FFT size M	Prime number						
		2, $i=$	3, $j=$	5, $k=$	7, $l=$	11, $m=$	13, $n=$	17, $p=$
1	12	2	1					
2	24	3	1					
3	36	2	2					
4	48	4	1					
5	60	2	1	1				
6	72	3	2					
7	84	2	1		1			
8	96	5	1					
9	108	2	3					
10	120	3	1	1				
11	132	2	1			1		
12	144	4	2					
13	156	2	1				1	
14	168	3	1		1			
15	180	2	2	1				
16	192	6	1					
17	204	2	1					1

4.5   FFT precoding complexity

SC-FDMA scheme requires an additional DFT operation at the transmitter and an IDFT operation at the receiver relative to an OFDMA scheme. The complexity of these additional operations can be reduced if the DFT/IDFT sizes allowed can be factorized into a small number of (prime) numbers. The smaller the number of prime numbers in the factorization, the simpler the implementation of DFT/IDFT. The size of the DFT precoder M depends on the amount of resources allocated for uplink data transmission and can be written as:

$$M = 2^i \times 3^j \times 5^k \times 7^l \times 11^m \times 13^n \times 17^p \times \cdots . \tag{4.28}$$

Here, we assumed that the FFT sizes are limited to factors of seven prime numbers 2, 3, 5, 7, 11, 13 and 17. If we limit the minimum DFT size to 12, this would mean that a minimum of two prime factors of 2 and 3 would be required. In Table 4.2, we list the FFT sizes up to a maximum size of 204 that are multiples of 12 and the corresponding prime factors. If we just allow two prime factors of 2 and 3, the FFT sizes of 60, 84, 120, 132, 156, 168, 180 and 204 will not be allowed. Note that when we allow a smaller number of prime factors, the flexibility in resource allocation is impacted. However, if we allow a third prime number 5, FFT sizes of 60, 120 and 180 can also be supported. This means that as the number of prime factors is increased, more flexibility in resource allocation becomes available. By looking at Table 4.2, it appears reasonable to allow three prime numbers (2, 3 and 5). This would allow using 12 out of 17 possibilities shown in Table 4.2. The five possibilities that are not permitted are highlighted and are FFT sizes of 84, 132, 156, 168 and 204. By limiting the prime numbers to 2, 3 and 5, we can expect some reduction in FFT implementation complexity.

4.6   Summary

We have discussed some of the problems associated with the WCDMA scheme used in the uplink of 3G systems. The major problem of WCDMA is intra-cell interference also referred to as multiple access interference (MAI), which results from its non-orthogonal nature. The intra-cell interference limits the maximum achievable data rates and limits the capacity of the uplink. In order to resolve the MAI issue in the uplink, two orthogonal schemes, namely OFDMA and SC-FDMA, were proposed for LTE. Both schemes allow orthogonal transmissions from multiple UEs transmitting simultaneously. The orthogonality of transmission is guaranteed by allocating orthogonal frequency bands to different UEs.

A capacity analysis of both orthogonal and non-orthogonal schemes was conducted based on simple models. We were interested in estimating achievable gains by using an orthogonal scheme compared to the currently 3G non-orthogonal WCDMA scheme. Therefore, we assumed the case of a single-path flat fading channel i.e. no ISI was considered. The results confirmed that orthogonal multiple access schemes OFDMA and SC-FDMA can provide large capacity gains over a non-orthogonal WCDMA scheme. We also noted that due to uplink power limitation, the typical situation for uplink would be the case where multiple users transmit simultaneously. We noted that gains of orthogonal schemes over non-orthogonal schemes increase as the number of UEs transmitting simultaneously increases.

Based on the performance comparison, both OFDMA and SC-FDMA emerged as strong candidates for LTE uplink. One advantage of SC-FDMA relative to OFDM, which is discussed in more detail in the next chapter, is low signal-peakiness of the SC-FDMA signal. A low signal-peakiness allows a UE to transmit at a higher power providing greater coverage. However, one drawback of SC-FDMA relative to OFDMA is link performance loss in a frequency-selective channel. This loss can be over one dB at higher SINR. We discussed a hybrid uplink access scheme where DFT-precoding is only used for power limited UEs. This scheme effectively uses OFDMA when power limitation is not an issue and therefore avoids link performance loss for higher order modulation transmissions. Another drawback of SC-FDMA relative to OFDMA is additional DFT and IDFT operations at the transmitter and receiver respectively resulting in increased implementation complexity. However, we noted that the DFT and IDFT operation can be relatively simplified if we limit the DFT-precoding sizes to a few prime factors.

During the LTE proposals evaluation phase, a great deal of emphasis was given to the coverage aspect and hence the low signal-peakiness was considered highly desirable. Based on this simple fact, SC-FDMA was selected as the multiple access scheme for the LTE uplink. However, in the later phase of LTE development, it was realized that SC-FDMA complicates the signaling and control multiplexing and other aspects of uplink design. Only time will prove if SC-FDMA selection over OFDMA was the right choice or not. In the author's view a hybrid scheme that combines the benefits of both SC-FDMA and OFDMA would have been a better approach for LTE uplink.

There was also a big debate over whether both localized and distributed transmission flavors of SC-FDMA should be allowed or a single scheme should be selected. The benefit of distributed transmission is greater frequency diversity because a given transmissions is spread over a larger bandwidth. These frequency-diversity gains from distributed SC-FDMA can be of the order of 1 dB in certain scenarios [8]. On the other hand, localized transmissions allow employing frequency-selective multi-user scheduling as a UE can be scheduled over a frequency band experiencing high signal quality. When the channel can be tracked with fairly high accuracy, which can be the case for low UE speeds, localized transmission outperforms distributed transmission. However, at higher UE speeds when channel quality cannot be tracked with reasonable accuracy, distributed transmission can be beneficial. However, DFDMA requires channel estimates over a wideband to exploit the frequency diversity gains. This requires high pilot or reference signal overhead on the uplink. Due to concerns on reference signal overhead, DFDMA scheme was not adopted in the standard leaving only localized SC-FDMA as the single scheme in LTE uplink. However, a frequency hopping approach was allowed to be used in order to gain frequency-diversity with localized SC-FDMA.

References

[1] Boyer, P., Stojanovic, M. and Proakis, J., "A simple generalization of the CDMA reverse link pole capacity formula," *IEEE Transactions on Communications*, vol. 49, com-I no. 10, pp. 1719–1722, Oct. 2001.

[2] Hyung, M. G., Lim, J. and Goodman, D. J., "Single carrier FDMA for uplink wireless transmission," *IEEE Vehicular Technology Magazine*, pp. 30–38, Sept. 2006.

[3] Hyung, M.G., Lim, J. and Goodman, D. J., "Peak-to-average power ratio of single carrier FDMA signals with pulse shaping," *IEEE 17th International Symposium on Personal, Indoor and Mobile Radio Communications*, pp. 1–5, Sept. 2006.

[4] Khan, F., "A time-orthogonal CDMA high speed uplink data transmission scheme for 3G and beyond," *IEEE Communication Magazine*, pp. 88–94, Feb. 2005.

[5] Sorger, U., De Broeck, I. and Schnell, M., "Interleaved FDMA – A new spread-spectrum multiple-access scheme," *Proc. of ICC'98*, pp. 1013–1017, June 1998.

[6] Tse, D. and Viswanath, P., *Fundamentals of Wireless Communication*, New York: Cambridge University Press, 2005.

[7] Hanzo, L., Münster, M., Choi, B. J. and Keller, T., *OFDM and MC-CDMA for Broadband Multi-User Communications, WLANs and Broadcasting*, Chichester: Wiley, 2003.

[8] Priyanto, B. E., Codina, H., Rene, S., Sorensen, T. B. and Mogensen, P., "Initial performance evaluation of DFT-spread OFDM based SC-FDMA for UTRA LTE uplink," *Proceedings of IEEE 65th Vehicular Technology Conference, VTC2007-Spring,* April 2007, pp. 3175–3179.

[9] Falconer, D., Ariyavisitakul, S. L., Benyamin-Seeyar, A. and Eidson, B., "Frequency domain equalization for single-carrier broadband wireless systems," *IEEE Communications Magazine*, vol. 40, no. 4, pp. 58–66, Apr. 2002.

5 Reducing uplink signal peakiness

In cellular systems, the wireless communication service in a given geographical area is provided by multiple Node-Bs or base stations. The downlink transmissions in cellular systems are one-to-many, while the uplink transmissions are many-to-one. A one-to-many service means that a Node-B transmits simultaneous signals to multiple UEs in its coverage area. This requires that the Node-B has very high transmission power capability because the transmission power is shared for transmissions to multiple UEs. In contrast, in the uplink a single UE has all its transmission power available for its uplink transmissions to the Node-B. Typically, the maximum allowed downlink transmission power in cellular systems is 43 dBm, while the uplink transmission power is limited to around 24 dBm. This means that the total transmit power available in the downlink is approximately 100 times more than the transmission power from a single UE in the uplink. In order for the total uplink power to be the same as the downlink, approximately 100 UEs should be simultaneously transmitting on the uplink.

Most modern cellular systems also support power control, which allows, for example, allocating more power to the cell-edge users than the cell-center users. This way, the cell range in the downlink can be extended because the Node-B can always allocate more power to the coverage-limited UE. However, in the uplink, the maximum transmission power is constrained by the maximum UE transmission power. This limits the achievable coverage range on the uplink, thus the uplink is the limiting link in terms of coverage and range. Due to this imbalance between uplink and downlink, 1G and 2G cellular systems employed receiver diversity on the uplink to improve the uplink range. However, receiver diversity is also becoming common on the downlink in 3G systems. A two receiver antenna diversity in the UEs again makes uplink the limiting link.

In voice communications, many UEs transmit and receive data simultaneously. However, with bursty Internet data traffic, a small number of UEs are generally receiving or transmitting data at a given time. The bursty Internet data traffic can be served more efficiently in the downlink by employing scheduling. The Node-B can allocate a large amount of power and bandwidth resources to a UE resulting in a very high instantaneous data rate. This strategy also results in lower signaling overhead, as scheduling grants need to be carried for a small number of UEs. However, bursty data traffic in the uplink means that only a small number of UEs transmit in the uplink. However, these UEs may not be able to transmit at very high data rates due to peak transmission power limitation. It can be argued that the cell-edge peak data rate in a coverage limited situation in the uplink is a factor of 100 smaller than the downlink peak data rate due to difference in Node-B and UE transmission powers. This assumes that at the cell-edge in a coverage-limited situation, the SINR is low enough so that capacity and data rates scale linearly with power. Also, we assume that Node-B can allocate all of its transmit power to serve a cell-edge user at peak data rate. When the uplink data transmission rates are limited due to transmit power limitations, it takes longer to transmit a given amount of data. This adds delays in the system and also a signaling overhead because of an increase in the number of scheduling grants as a consequence of the increased number of users requiring simultaneous transmissions.

A simple solution to increase the uplink data rates would be to increase the uplink power amplifier size. However, this would come at the expense of increased power amplifier cost and UE battery power consumption. A source of inefficiency in transmit power is the backoff required due to high signal peakiness or peak-to-average power ratio (PAPR) of the transmit signal. By reducing uplink signal peakiness, higher power can be transmitted in the uplink for the same power amplifier size. In this chapter, we will discuss measures of signal peakiness, signal peakiness for different modulations, low PAPR modulations and spectrum shaping techniques for reducing signal peakiness for SC-FDMA.

5.1 Measures of signal peakiness

The signal peakiness is an important consideration for RF amplifier performance particularly for the uplink transmissions. The reduction in uplink signal peakiness can provide increased transmit power enabling higher data rates in the uplink. We discuss two measures of signal peakiness: PAPR and cubic power metric.

5.1.1 Peak-to-average-power ratio

The peak-to-average-power ratio (PAPR) is generally defined as the ratio of the instantaneous signal power to the average signal power as below:

$$\text{PAPR} = \frac{|s(t)|^2}{E\left[|s(t)|^2\right]}. \tag{5.1}$$

In general, a low PAPR is desired for improved power amplifier efficiency i.e. smaller power backoff is needed. In the current 3G systems, WCDMA is used as the multiple access scheme. A large number of parallel codes are used for transmission to various users or a single user (e.g. HSDPA uses up to 15 parallel codes for high-speed data transmission to a single user) on the downlink. A large PAPR results from the large number of parallel code channels transmission. In general, the WCDMA PAPR for a large number of parallel codes transmission can be approximated by the complementary cumulative distribution function (CCDF) of a complex Gaussian random variable. Similarly, in case of OFDM a large number of tones is used for transmission to different UEs or transmission to a single UE in the downlink. This also results in high PAPR that can again be approximated by a complex Gaussian random variable as long as the number of tones used is sufficiently large (16 or greater). Therefore, there is practically no difference in PAPR for WCDMA and OFDM for downlink. It is assumed that the existing 3G Node-Bs are already designed to cope with the PAPR resulting from a complex Gaussian signal.

In order to illustrate why superposition of parallel subchannels (codes or tones) results in large signal variations, Figure 5.1 shows superposition of 4 OFDM tones at frequencies of f, $2f$, $3f$ and $4f$. The bottom-left plot shows simple addition of four tones (all tones modulated by '+1') while the plot on the bottom-right shows the case where tones at frequencies f and $3f$

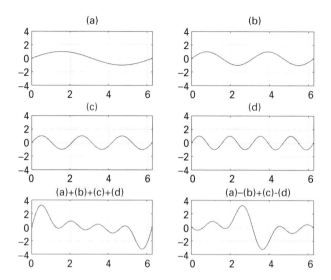

Figure 5.1. Superposition of sinusoids.

are modulated by '+1' while tones at frequencies $2f$ and $4f$ are modulated by '−1'. We can note that superposition of multiple sinusoids can result in large variations of the composite signal.

In Figure 5.1, we assumed superposition of only four sinusoids. In practice, the number of tones used in OFDM is much larger. Also, we noted that in WCDMA downlink, up to 15 parallel code channels can be used for transmission. As pointed out earlier, the PAPR of OFDM and WCDMA for a large number of subchannels (codes or tones) can be approximated by a complex Gaussian signal. According to the central limit theorem, the real and imaginary parts of the transmitted signal in OFDM and WCDMA for a large number of subchannels (subcarriers or codes) are Gaussian distributed. Therefore, the PAPR distribution can be approximated by CCDF of the complex Gaussian signal given as:

$$\text{probability}(\text{PAPR} > x) = e^{-x}. \tag{5.2}$$

Unlike the downlink, fewer parallel code channels (preferably a single code channel) are used for transmissions from a single UE in the WCDMA uplink. This is to provide lower PAPR relative to downlink, as the power amplifier efficiency is more critical for the UE due to its form factor and lower cost requirements. When a multi-carrier multiple access scheme such as OFDMA is employed in the uplink, the PAPR becomes comparable to the downlink PAPR. A single-carrier scheme then becomes attractive for the uplink due to its low PAPR characteristic.

5.1.2   Cubic metric

In the course of LTE and earlier HSPA (high speed packet access) standards performance evaluation, it was realized that PAPR may not be an accurate measure of UE power amplifier backoff or power de-rating. Another metric referred to as cubic metric (CM) was developed as a more effective predictor of the actual reduction in power capability, or power de-rating, of a typical power amplifier in a UE. CM was found to have a higher correlation with the measured power de-ratings and resulted in a tighter distribution of errors. This was found to be true for several different power amplifiers using different technologies and various signals with a range of PAPRs.

The CM is defined as:

$$CM = \frac{20 \log_{10} \left[\frac{(v_n^3)_{rms}}{(v_{ref}^3)_{rms}} \right]}{F}, \tag{5.3}$$

where v_n is the normalized voltage waveform of the input signal and v_{ref} is the normalized voltage of the 12.2 Kb/s AMR (adaptive multi-rate) speech signal in WCDMA. We note that, by definition, CM of the reference signal is 0 dB. The empirical factor F is obtained by linear curve fit of CM to the actual power amplifier power de-rating curve. A value of $F = 1.85$ was used extensively during LTE performance evaluation. The term $20 \times \log_{10} \left(v_{ref}^3 \right)_{rms}$, which represents the cubic power of the WCDMA voice signal, is assumed to be 1.5237.

The CM can then be approximated as:

$$CM = \frac{\left[20 \log_{10} \left(V_n^3 \right)_{rms} - 1.5237 \right]}{1.85}. \tag{5.4}$$

The motivation for using CM was to model the impact of power amplifier non-linearity on the adjacent channel leakage (ACL). The primary cause of adjacent channel leakage (ACL) is the third order non-linearity of the amplifier's gain characteristic. The amplifier voltage gain characteristic in this case can be written as:

$$v_o(t) = G_1 \times v_i(t) + G_3 \times [v_i(t)]^3, \tag{5.5}$$

where $v_i(t)$ is the input voltage to the power amplifier and $v_o(t)$ is the output voltage. Also G_1 and G_3 are the linear and non-linear gains of the amplifier. These values generally depend only on the amplifier design, and are independent of the input signal $v_i(t)$.

The cubic term in Equation (5.5) will generate several types of degradation to the output signal. This includes channel distortion terms that contribute to EVM, as well as signals at the third harmonic of the carrier frequency, and signals in the upper and lower adjacent channel bands. For a given amplifier, the total energy in the cubic term is determined by the input signal $v_i(t)$ only. In addition, the total energy is distributed among the various distortion components in some predefined, signal dependent way.

5.2 PAPR of QAM modulations

In the current 3G systems that use non-orthogonal uplink access based on WCDMA the maximum SINR is limited due to intra-cell interference. Also, since the uplink access is non-orthogonal, uplink bandwidth sharing is not a concern as transmissions from multiple UEs transmitting simultaneously can overlap in frequency. It is for these reasons that higher order modulations are generally not used in the WCDMA uplink. However, in the LTE system employing orthogonal uplink access based on SC-FDMA, there is no intra-cell interference. Total bandwidth also needs to be shared among multiple UEs accessing the uplink data channel simultaneously. Therefore, LTE uses higher order modulations such as 16-QAM and 64-QAM

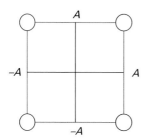

Figure 5.2. QPSK constellation.

to achieve higher data rates. Unfortunately, the higher order modulations result in increased PAPR and larger power amplifier backoffs on the uplink reducing the total available transmission power. Another alternative to achieve higher data rates in the uplink is to use MIMO (Multiple Input Multiple Output) spatial multiplexing techniques. However, MIMO spatial multiplexing schemes require multiple RF chains and multiple power amplifiers and therefore impact the UE complexity and battery power consumption. In this section, we discuss PAPR for QPSK, 16-QAM and 64-QAM modulations.

5.2.1   Quadrature Phase Shift Keying (QPSK)

The QPSK constellation is shown in Figure 5.2. Assuming a unit average power, the real and imaginary components take amplitude values of A and $-A$ where A is given as below:

$$\frac{4(A^2 + A^2)}{4} = 1$$

$$\Rightarrow A = \frac{1}{\sqrt{2}}. \tag{5.6}$$

In QPSK modulation, all the constellation points have the same power and the PAPR can be calculated as:

$$\text{PAPR}_{\text{QPSK}} = \frac{(A^2 + A^2)}{1} = 1.0 = 0.0\,\text{dB}. \tag{5.7}$$

This means that PAPR for QPSK is 0.0 dB i.e. a constant envelope signal.

5.2.2   16 quadrature amplitude modulation (16-QAM)

The 16-QAM constellation is shown in Figure 5.3. Again, assuming a unit average power, the real and imaginary components take amplitude values of A, $3A$, $-A$ and $-3A$ where A is given as below:

$$\frac{4(A^2 + A^2 + A^2 + 9A^2 + A^2 + 9A^2 + 9A^2 + 9A^2)}{16} = 1$$

$$\Rightarrow A = \frac{1}{\sqrt{10}}. \tag{5.8}$$

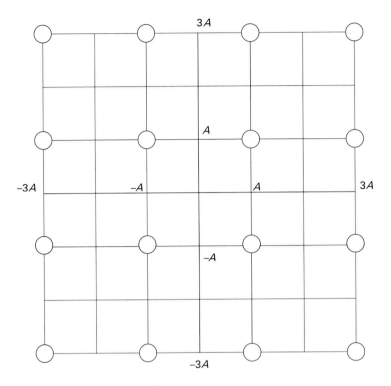

Figure 5.3. 16-QAM constellation.

The maximum PAPR for 16-QAM can be calculated as below:

$$\text{PAPR}_{16\text{-QAM}} = \frac{(9A^2 + 9A^2)}{1} = 1.8 = 2.55 \text{ dB}. \tag{5.9}$$

That is, 16-QAM results in a maximum PAPR of 2.55 dB with 25% probability i.e. four symbols (corner points in the constellation) at PAPR of 2.55 dB out of a total of 16 symbols. Similarly, we can obtain the PAPR for the four inner constellation points as below:

$$\text{PAPR}_{16\text{-QAM}} = \frac{(A^2 + A^2)}{1} = 0.2 = -7 \text{ dB}. \tag{5.10}$$

The PAPR for the remaining eight constellation points is:

$$\text{PAPR}_{16\text{-QAM}} = \frac{(A^2 + 9A^2)}{1} = 1.0 = 0.0 \text{ dB}. \tag{5.11}$$

5.2.3   64 quadrature amplitude modulation (64-QAM)

The 64-QAM constellation is shown in Figure 5.4. Once again, assuming a unit average power, the real and imaginary components take amplitude values of A, $3A$, $5A$, $7A$, $-A$, $-3A$, $-5A$

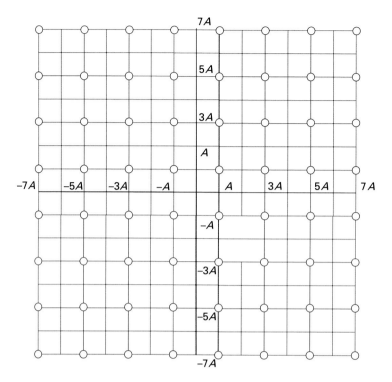

Figure 5.4. 64-QAM constellation.

and $-7A$ where A is given as below:

$$\frac{4\left(\begin{array}{c}2A^2 + 2 \times 10A^2 + 18A^2 + 2 + 26A^2 + 2 \\ \times 34A^2 + 3 \times 50A^2 + 2 + 58A^2 + 2 \times 74A^2 + 98A^2\end{array}\right)}{64} = 1$$

$$\Rightarrow A = \frac{1}{\sqrt{42}}. \tag{5.12}$$

The maximum PAPR for 64-QAM can be calculated as below:

$$\text{PAPR}_{64\text{-QAM}} = \frac{(49A^2 + 49A^2)}{1} = \frac{98}{42} = 3.68 \text{ dB}. \tag{5.13}$$

This shows that 64-QAM results in a PAPR of 3.68 dB with 6.25% (1/16) probability i.e. four symbols (corner symbols in the constellation) at PAPR of 3.68 dB out of a total of 64 symbols.

We noted that for a single subchannel (CDMA code or OFDM subcarrier) transmission, QPSK, 16-QAM and 64-QAM results in maximum PAPR of 0.0, 2.55 and 3.68 dB respectively. Let us now have a look at the PAPR for these modulations for a large number of parallel subchannels transmission. Figure 5.5 compares complex Gaussian CCDF with QPSK, 16-QAM and 64-QAM PAPR with 16 parallel sub channels transmission. As expected for a large number of subchannels, the PAPR can be approximated with the Complex Gaussian CCDF.

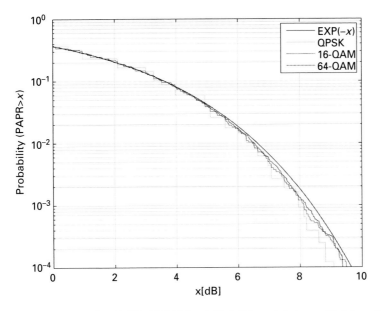

Figure 5.5. PAPR for QPSK, 16-QAM and 64-QAM modulations with 16 parallel channels.

It is also interesting to note that the modulation order does not make much difference when the number of parallel subchannels is large. This indicates that once we design a system to accommodate the worst case PAPR (that of complex Gaussian signal), the number of subchannels used and modulation order do not make any difference.

5.3   Peakiness of SC-FDMA signal

As we noted earlier, SC-FDMA is used in the LTE uplink due to its low signal peakiness characteristic. In Figure 5.6, we plot the PAPR CCDF for SC-FDMA and compare it against the CCDF of a complex Gaussian signal. We assumed both QPSK and 16-QAM modulations, FFT size $M = 64$ and IFFT size $N = 512$. We note that at the 0.1% point on CCDF, the PAPR of SC-FDMA using QPSK is approximately 2.5 dB lower than the PAPR of a complex Gaussian signal. Since the PAPR of an OFDM signal transmitted on a sufficiently large number of tones or subcarriers is similar to the PAPR of a complex Gaussian signal, we can say that SC-FDMA provides 2.5 dB lower PAPR than PAPR of an OFDM signal. As a reference, for QPSK modulation, the cubic metric for SC-FDMA and OFDM is 1.0 and 3.4 dB respectively. This shows a similar difference to the difference in PAPR at the 0.1% point. The signal peakiness benefit of SC-FDMA reduces for higher order modulations such as 16-QAM and 64-QAM. We note that for 16-QAM modulation at the 0.1% point on CCDF, the PAPR of SC-FDMA is approximately 2.0 dB lower than the PAPR of a complex Gaussian signal. Also, the cubic metric of SC-FDMA with 16-QAM is 1.8 dB, which is only 1.6 dB better than the cubic metric for OFDM. As pointed out earlier, a complex Gaussian signal gives the worst-case PAPR and therefore its PAPR or CM is not affected by the modulation. The cubic metric for SC-FDMA using 64-QAM modulation is approximately 2.0 dB, which is only 1.4 dB better than the cubic metric for OFDM.

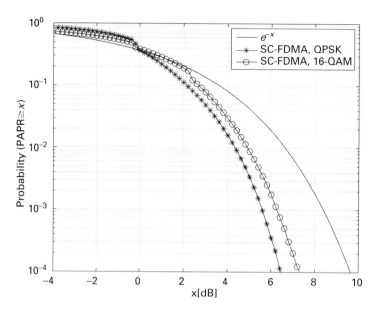

Figure 5.6. PAPR for SC-FDMA for $M = 64, N = 512$.

It can, however, be argued that lower PAPR/CM is more desirable for power-limited UEs using QPSK modulation. In general, QPSK is used at relatively lower SINR where capacity scales approximately linearly with SINR. This means that additional power that becomes available due to lower PAPR/CM directly translates into higher data throughputs. However, higher order modulations such as 16-QAM and 64-QAM are used at relatively higher SINR where channel capacity scales logarithmically with SINR. Therefore, the additional power that becomes available due to lower PAPR/CM translates into only marginal improvements in data rates.

In order to maintain a single-carrier property in localized SC-FDMA, the FFT-precoded data symbols should be mapped to contiguous subcarriers at the input of IFFT as shown for the case of contiguous allocation (CA) in Figure 5.7. The PAPR results in Figure 5.6 are plotted for this contiguous allocation case. When the FFT-precoded data symbols are mapped to non-contiguous groups of subcarriers, the PAPR and CM increases. Figure 5.7 shows two non-contiguous allocations (NCA), a scheme referred to as NCA2 where the FFT precoded symbols are mapped to two groups of contiguous subcarriers and another scheme referred to as NCA4 where the FFT precoded symbols are mapped to four groups of contiguous subcarriers. The signal peakiness increases with the increasing number of groups the transmission is mapped to as shown in Figure 5.8. The increase in cubic metric for NCA2 and NCA4 relative to the CA scheme is 0.65 and 0.83 dB respectively. This means that in order to keep the low signal peakiness benefit of SC-FDMA scheme, certain constraints in resource allocation are introduced.

5.4   Low PAPR modulations

The PAPR of SC-FDMA can possibly be further reduced by using π/M-shifted MPSK modulation, which employs per-symbol phase rotation. The constellations for $\pi/2$-shifted BPSK and

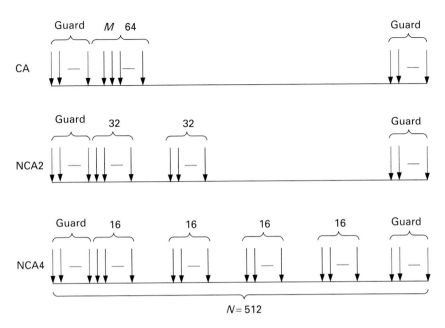

Figure 5.7. SC-FDMA with contiguous allocation (CA) and non-contiguous allocation (NCA) of subcarrier groups.

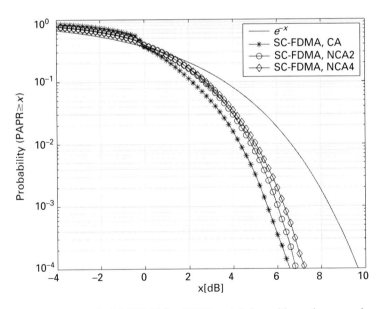

Figure 5.8. PAPR for SC-FDMA for QPSK modulation with contiguous and non-contiguous allocations.

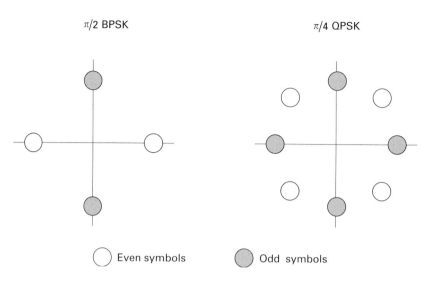

Figure 5.9. $\pi/2$-shifted BPSK and $\pi/4$-shifted QPSK signal constellations.

$\pi/4$-shifted QPSK are given in Figure 5.9. In case of $\pi/2$-shifted BPSK, the even-numbered symbols are transmitted as in a BPSK modulation. However, the odd-numbered symbols are rotated by a phase shift of $\pi/2$. Similarly, for $\pi/4$-shifted QPSK, the even-numbered symbols are transmitted as in a normal QPSK modulation. However, the odd-numbered symbols are rotated by a phase shift of $\pi/4$. Note that phase shifts on the odd-numbered symbols have no impact on the link performance relative to standard BPSK and QPSK modulations.

The PAPR CCDF for $\pi/2$-shifted BPSK and $\pi/4$-shifted QPSK signal constellations is shown in Figure 5.10. We note that $\pi/4$-shifted QPSK results in marginal improvement in PAPR. At the 0.1% point on CCDF, the PAPR for $\pi/4$-shifted QPSK is approximately 0.2 dB lower than the standard QPSK constellation. However, $\pi/2$-shifted BPSK provides approximately 1.2 dB lower PAPR than a standard QPSK constellation. The cubic metric for $\pi/4$-shifted QPSK shows no improvement relative to the standard QPSK constellation. In case of $\pi/2$-shifted BPSK, improvements of 0.8 dB are observed in the cubic metric as well. The cubic metric for $\pi/2$-shifted BPSK is only 0.2 dB relative to 1.0 dB for standard QPSK modulation.

5.5   Spectrum shaping

In the previous section, we noted that π/M-shifted MPSK modulations particularly $\pi/2$-shifted BPSK can further reduce the signal peakiness of a single-carrier transmission. In this section, we discuss pulse-shaping or spectrum-shaping techniques that can even further reduce the signal peakiness of a single-carrier transmission. For time-domain processing of SC-FDMA, pulse shaping can be used on the time-domain signal to reduce signal peakiness. In case of frequency-domain processing of SC-FDMA such as is the case in DFT-Spread OFDM, spectrum shaping can be performed in the frequency-domain as shown in Figure 5.11. The spectrum-shaping function can be obtained by simply taking the Fourier transform of the pulse-shaping function.

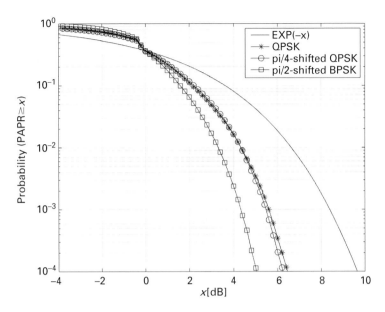

Figure 5.10. PAPR for $\pi/2$-shifted BPSK and $\pi/4$-shifted QPSK signal constellations.

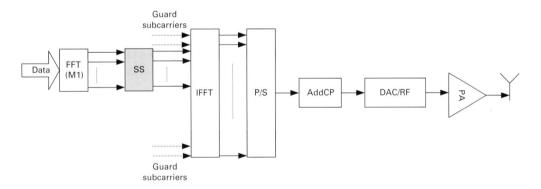

Figure 5.11. SC-FDMA Transmitter using Spectrum shaping.

The spectrum shaping is performed on the DFT-precoded samples $X_l[l = 0, 1, \ldots, (M-1)]$ before mapping to IDFT. The DFT-precoded samples are simply multiplied with the spectrum shaping function as below:

$$\tilde{X}_l = X_l \times S(l) \quad l = 0, 1, \ldots, (M-1), \tag{5.14}$$

where $S(l)$ are the samples of the spectrum-shaping function and M is the number of subcarriers used for transmission. Two classes of spectrum-shaping functions namely Kaiser window [3] and raised-cosine spectrum-shaping were proposed for signal peakiness reduction in the LTE system. We will discuss, in detail, these two spectrum shaping functions in the next sections.

5.5.1   Kaiser window spectrum-shaping

The spectrum-shaping function for the Kaiser window [2] is defined as:

$$S_{\text{Kaiser}}(l) = \frac{I_0\left(\beta\sqrt{1 - \left(\frac{2l}{M-1} - 1\right)^2}\right)}{I_0(\beta)} \qquad l = 0, 1, \ldots, (M-1), \qquad (5.15)$$

where β is an arbitrary real number that determines the shape of the window, and the integer M gives the length of the window. $I_0(.)$ is the zeroth-order modified Bessel function of the first kind which is defined by:

$$I_0(x) = \frac{1}{\pi}\int_0^{\pi} e^{x\cos t}\, dt. \qquad (5.16)$$

By construction, the Kaiser window function in Equation (5.15) peaks at unity at the center of the window, and decays exponentially towards the window edges as shown in Figure 5.12.

5.5.2   Raised-cosine spectrum-shaping

The raised-cosine filters are frequently used in wireless systems to produce a band-limited signal. A main characteristic of these filters is that they produce Nyquist pulses after matched filtering and therefore do not cause any inter-symbol interference. In the frequency-domain,

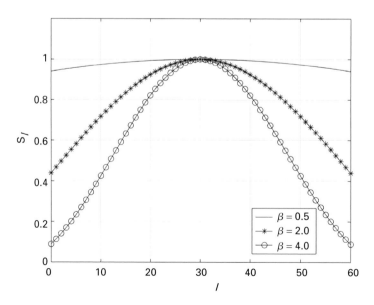

Figure 5.12. Kaiser window spectrum-shaping function for $M = 60$ and $\beta = 0.5, 2.0, 4.0$.

the raised-cosine spectrum-shaping function is defined as:

$$
S_{RC}(l) = \begin{cases} 1 & |l| \le \dfrac{(1-\beta)M}{2} \\ \dfrac{1}{2}\left[1 + \cos\left(\dfrac{\pi}{M\beta}\left[|l| - \dfrac{(1-\beta)M}{2}\right]\right)\right] & \dfrac{(1-\beta)M}{2} < |l| \le \dfrac{(1+\beta)M}{2} \\ 0 & \text{otherwise} \end{cases},
$$

(5.17)

where $0 \le \beta \le 1$ is the roll-off factor. The roll-off factor is a measure of the excess bandwidth. As β approaches zero, the roll-off zone becomes infinitesimally narrow leading to a rectangular spectrum-shaping function. When $\beta = 1$, the non-zero portion of the spectrum is a pure raised cosine, leading to the simplification:

$$
S_{RC}(l)|_{\beta=1} = \begin{cases} \frac{1}{2}\left[1 + \cos\left(\frac{\pi l}{M}\right)\right] & |l| \le M \\ 0 & \text{otherwise}. \end{cases}
$$

(5.18)

When a matched filter is used at the receiver,

$$
S_R(l) = S_T^*(l),
$$

(5.19)

where $S_R(l)$ and $S_T(l)$ represent the spectrum-shaping function at the transmitter and the receiver respectively and x^* denotes the complex conjugate of x. In order to provide a total response of the system as raised-cosine, a root-raised-cosine (RRC) filter is typically used at each end of the communication system. Therefore $S_R(l)$ and $S_T(l)$ are given as:

$$
|S_R(l)| = |S_T(l)| = \sqrt{|S_{RC}(l)|}.
$$

(5.20)

A drawback of the RRC filter is that it results in bandwidth expansion by a factor of $(1 + \beta)$. The bandwidth of a the raised-cosine filter is defined as the width of the non-zero portion of its spectrum. We note from Figure 5.13 that bandwidth occupied by the raised-cosine filter for $\beta = 0.2, 0.5, 1.0$ is respectively 20%, 50% and 100% more than the rectangular spectrum-shaping function. In Figure 5.13, we assumed a DFT-precoding size of $M = 60$ and this would be the number of subcarriers required for transmission if a rectangular spectrum-shaping function was used. However, the number of subcarriers required for transmission for $M = 60$ and $\beta = 0.2, 0.5, 1.0$ is 72, 92 and 120 respectively.

The bandwidth expansion and spectrum shaping using the raised-cosine spectrum-shaping function is illustrated in Figure 5.14. M FFT-precoded data symbols are expanded by copying $\beta M/2$ samples from the beginning and appending them to the end and copying $\beta M/2$ samples from the end and appending them to the beginning. This results in a total of $(1 + \beta) M$ samples required for RRC spectrum shaping. A total of $(1 + \beta) M$ FFT-precoded data symbols are renumbered from $\frac{-(1-\beta)M}{2} + 1$ to $\frac{(1+\beta)M}{2}$ and RRC spectrum shaping is performed as below:

$$
\tilde{X}_l = X_l \times \sqrt{S_{RC}(l)}, \qquad \frac{-(1-\beta)M}{2} < l \le \frac{(1+\beta)M}{2}.
$$

(5.21)

5.5.3   PAPR/CM performance with spectrum shaping

The PAPR/CM performances for QPSK and $\pi/2$-BPSK using Kaiser window and RRC spectrum shaping are given in Tables 5.1 and 5.2 respectively. For QPSK modulation, the best

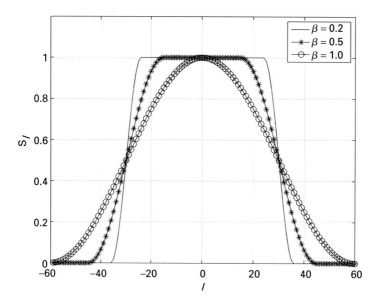

Figure 5.13. Root-raised spectrum-shaping function for $M = 60$ and $\beta = 0.2, 0.5, 1.0$.

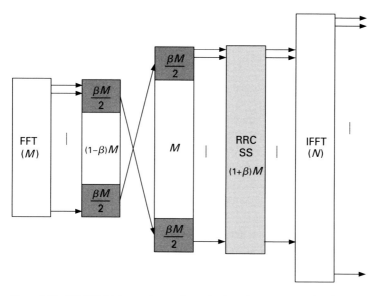

Figure 5.14. SC-FDMA transmitter using raised-cosine spectrum shaping.

achievable PAPR/CM performance for Kaiser window and RRC spectrum shaping is 4.4 dB
(CM = 0.85 dB) and 2.9 dB (CM = 0.11 dB) respectively. This performance is achieved for a
Kaiser shaping parameter of $\beta = 2.5$ and RRC roll-off of $\beta = 0.5$. It should be noted that the
spectral efficiency of QPSK using RRC with roll-off of $\beta = 0.5$ is only 1.33 bits/subcarrier.
In contrast, the spectral efficiency of QPSK using Kaiser window with a shaping parameter of
$\beta = 2.5$ is 2.0 bits/subcarrier. This is because unlike RRC spectrum shaping, Kaiser window
spectrum shaping does not result in bandwidth expansion. The bandwidth efficiency (BE) in

Table 5.1. PAPR/CM performance for QPSK and $\pi/2$-BPSK using Kaiser window spectrum shaping.

Bandwidth efficiency [bits/subcarrier]	Modulation	Shaping parameter β	0.1%PAPR [dB]	Cubic metric [dB]
2	QPSK	2.0	4.8	0.82
2	QPSK	2.5	4.4	0.85
1	$\pi/2$-BPSK	2.0	2.8	0.45
1	$\pi/2$-BPSK	2.5	2.1	−0.6

Table 5.2. PAPR/CM performance for QPSK and $\pi/2$-BPSK using RRC spectrum shaping.

Bandwidth efficiency [bits/subcarrier]	Modulation	RRC roll-off β	0.1 %PAPR [dB]	Cubic metric [dB]
2	QPSK	0.0	5.7	1.0
1.67	QPSK	0.2	4.4	0.45
1.333	QPSK	0.5	2.9	0.11
1	$\pi/2$-BPSK	0.0	4.5	0.2
0.83	$\pi/2$-BPSK	0.2	2.2	−0.32
0.67	$\pi/2$-BPSK	0.5	2.0	−0.4

bits per subcarrier for RRC spectrum shaping with roll-off factor β is written as:

$$BE = \frac{\log_2 (K)}{(1 + \beta)} \text{ bits/subcarrier,} \qquad (5.22)$$

where K is the modulation order, which is, for example, 2, 4 and 16 for BPSK, QPSK and 16-QAM modulations respectively. For $\pi/2$-BPSK modulation, the best achievable PAPR/CM performance is 2.0 dB (CM $= -0.4$ dB) with RRC roll-off of $\beta = 0.5$ and 2.1 dB (CM $= -0.6$ dB) with a Kaiser shaping parameter of $\beta = 2.5$ respectively. Again we note that in this case, Kaiser window spectrum shaping provides bandwidth efficiency of 1.0 bits/subcarrier while RRC spectrum shaping with roll-off of $\beta = 0.5$ provides a bandwidth efficiency of only 0.66 bits/subcarrier.

Another point worth noting is that for QPSK using RRC spectrum shaping with roll-off of $\beta = 0.5$, both PAPR/CM and bandwidth efficiency are better than $\pi/2$-BPSK modulation. QPSK provides a PAPR of 2.9 dB (CM $= 0.11$ dB) with bandwidth efficiency of 1.33 bits/subcarrier. On the other hand simple $\pi/2$-BPSK modulation results in a PAPR of 4.5 dB (CM $= 0.2$ dB) and bandwidth efficiency of 1.0 bits/subcarrier. Therefore, it is fair to say that when spectrum shaping is applied, QPSK can outperform simple $\pi/2$-BPSK modulation. However, if a lower PAPR/CM is the requirement and bandwidth efficiency is not a concern then $\pi/2$-BPSK modulation with spectrum shaping can be employed.

Earlier we noted that RRC spectrum shaping does not cause any inter-symbol interference. On the other hand, Kaiser window spectrum shaping which is not a Nyquist filter would cause inter-symbol interference. It is therefore required to assess the impact of spectrum shaping on link performance. The SNR required for QPSK at 10% packet error rate in an AWGN channel for RRC and Kaiser window spectrum shaping are given in Table 5.3. We assumed a turbo coding rate of 1/3, a UE speed of 3 km/h and FFT-precoding size of $M = 60$. A 10% packet

Table 5.3. CM and SNR for QPSK.

Spectrum-shaping function	Roll-off or shaping parameter β	Cubic metric (dB)	SNR at 10% (dB)	Combined CM+SNR (dB)
RRC	0.0	1.0	−3.8	−2.8
	0.2	0.45	−3.8	−3.35
	0.5	0.11	−3.8	−3.69
Kaiser	2.0	0.82	−3.5	−2.68
	2.5	0.85	−3.1	−2.25

Table 5.4. SNR and CM for $\pi/2$-BPSK.

Spectrum-shaping function	Roll-off or shaping parameter β	Cubic metric (dB)	SNR at 10% (dB)	Combined CM+SNR (dB)
RRC	0.0	0.2	−6.7	−6.5
	0.2	−0.32	−6.7	−7.02
	0.5	−0.4	−6.7	−7.1
Kaiser	2.0	0.45	−6.65	−6.2
	2.5	−0.6	−6.6	−7.2

error rate point is used because this is generally the packet error rate on initial transmission in the presence of hybrid ARQ. As expected, the performance of RRC spectrum shaping for different roll-off factors is the same, that is, the required SNR is −3.8 dB in all cases. However, the performance of Kaiser window spectrum shaping is degraded by 0.3 and 0.7 dB respectively for $\beta = 2$ and $\beta = 2.5$ respectively. A larger shaping parameter $\beta = 2.5$ leads to greater performance degradation as the filter becomes narrower as seen from Figure 5.12.

In terms of coverage, the overall performance is determined by both the cubic metric and the required SNR. A lower cubic metric means a lower back-off for the UE power amplifier and hence a larger effective transmit power improving the transmission range. A lower required SNR means a lower transmit power is require to guarantee a certain packet error rate performance. In the last column, we provide combined cubic metric and SNR performance. We can see that RRC spectrum shaping outperforms Kaiser window spectrum shaping in all the cases considered for QPSK. We note that QPSK without any spectrum shaping, that is RRC roll-off of zero, outperforms Kaiser window spectrum shaping by more than 0.1 dB. It can, therefore, be concluded that Kaiser window spectrum shaping for QPSK should never be used. On the other hand, RRC based spectrum shaping can provide approximately 1 dB improvement in range for QPSK by lowering the cubic metric while causing no inter-symbol interference.

The SNR and CM performance for $\pi/2$-BPSK under the same conditions as considered for QPSK are given in Table 5.4. We note that Kaiser spectrum shaping with a shaping parameter of $\beta = 2.5$ provides the lowest combined cubic metric and SNR metric. The combined cubic metric for Kaiser spectrum shaping with a shaping parameter of $\beta = 2.5$ is 0.1 dB better than RRC spectrum shaping with roll-off of $\beta = 0.5$. Moreover, from Tables 5.1 and 5.2, we know that the bandwidth efficiency of RRC spectrum shaping with roll-off of $\beta = 0.5$ is lower than Kaiser spectrum shaping. It can therefore be concluded that Kaiser window spectrum shaping can be useful for $\pi/2$-BPSK.

5.6   Coverage gain due to low signal peakiness

In the discussions until now we have emphasized the importance of transmit power gains of a few dBs or even gains of a small fraction of a dB. A relevant question to ask here is what these small gains mean in terms of cellular system range and coverage improvements. In order to get some insight into range and coverage extension with lower signal peakiness, we start with propagation analysis in free space. Let P_T be the radiated power of a source (an isotropic radiator), then power flux density at any point on a spherical surface at a point distance d from the source is:

$$P_D(W/m^2) = \frac{P_T}{4\pi d^2},$$ (5.23)

where $4\pi d^2$ is the surface area of a sphere of radius d. If a receiving antenna is placed on a spherical wavefront, then the received power can be written as:

$$P_R = \frac{P_T A_R}{4\pi d^2},$$ (5.24)

where A_R is the effective aperture of the receiving antenna. The gain of the transmitter antenna related to an isotropic radiator is given as:

$$G_T = \frac{4\pi A_T}{\lambda^2},$$ (5.25)

where λ is the wavelength of the radiation and A_T is the effective aperture of the transmitting antenna. The received power then can be written as:

$$P_R = P_T \frac{4\pi A_T}{\lambda^2} \frac{A_R}{4\pi d^2} = P_T \frac{4\pi A_T}{\lambda^2} \frac{4\pi A_R}{\lambda^2} \left(\frac{\lambda}{4\pi d}\right)^2$$

$$= P_T G_T G_R \left(\frac{\lambda}{4\pi d}\right)^2,$$ (5.26)

where G_T is the transmitter antenna gain and G_R is the receiver antenna gain. The total path-loss in free space can then be written as:

$$PL_{FS} = 10\log_{10}\left(\frac{P_T}{P_R}\right)$$

$$= 20\log_{10}\left(\frac{4\pi d}{\lambda}\right) - 10\log_{10} G_T - 10\log_{10} G_R \quad [\text{dB}].$$ (5.27)

The above equation can further be simplified as:

$$PL_{FS} = 32.44 + 20\log_{10}(d) + 20\log_{10}(f) - 10\log_{10} G_T - 10\log_{10} G_R \quad [\text{dB}],$$ (5.28)

where d is in meters and f is in GHz. The fixed term 32.44 dB represents the free space path-loss at a distance of 1 meter from a transmitter radiating at 1 GHz frequency. It can be noted

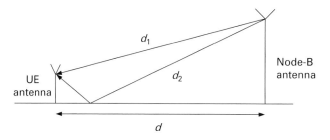

Figure 5.15. Direct and reflected path off a ground plane.

that with every doubling of the distance, the path-loss is increased by $20 \log_{10}(2) = 6\,\text{dB}$. This also means that with every 6.0 dB additional gain available, for example, via increased transmit power the communication range in free space is doubled.

In cellular systems, the received power at the UE (downlink) or the Node-B (uplink) as a function of distance between the Node-B and the UE, d decreases more than in free space. This is particularly true when the distance d is sufficiently large so that the difference between the direct path length d_1 and the reflected path length d_2 as shown in Figure 5.15 becomes comparable to the signal wavelength λ. Under this condition, the direct path wave and the reflected path wave start to cancel each other out because the sign of the reflected wave electric field is reversed [4]. The electric wave power at the receiver is then attenuated as $1/d^4$ rather than $1/d^2$ experienced in free space. The power attenuation factor is generally referred to as the path-loss exponent denoted as α. The received power P_R is then

$$P_R \propto \left(\frac{1}{d} \right)^{\alpha}. \tag{5.29}$$

In cellular systems modeling, the pathloss exponent α is assumed to be between 2 and 4 depending upon the wireless channel environment. For example, in the COST231 Hata urban propagation model [5], which is widely used in cellular systems performance evaluation, the path-loss exponent is assumed as $\alpha = 3.5$.

Let us now turn our attention to the range and coverage improvements due to power gains achieved through low signal-peakiness. The incremental range extension ΔR by a power gain of g_P dBs for a pathloss exponent α can be written as:

$$\Delta R = \left(\frac{d_1 - d_0}{d_0} \right) = \left(10^{\left(\frac{g_P}{10} \right)} \right)^{\frac{1}{\alpha}} - 1, \tag{5.30}$$

where d_0 is the original range and d_1 is the range with power gain of g_P dBs as shown in Figure 5.16. We note from Equation (5.30) that for the same power gain, the incremental range extension is larger for smaller path loss exponent α. Now assuming a circular-shape omni-cell, the gain in coverage area ΔA can be calculated as:

$$\Delta A = \frac{A_1 - A_0}{A_0} = \frac{\pi d_1^2 - \pi d_0^2}{\pi d_0^2} = \left(\left[\frac{d_1}{d_0} \right]^2 - 1 \right). \tag{5.31}$$

We note that the incremental gains in area are much larger as doubling of the range, for example, provides a 3 times gain in coverage area. We note from Figure 5.16 that the

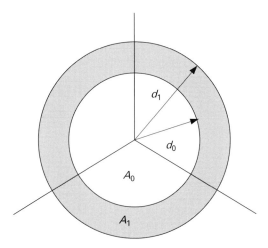

Figure 5.16. Illustration of range and coverage area extension.

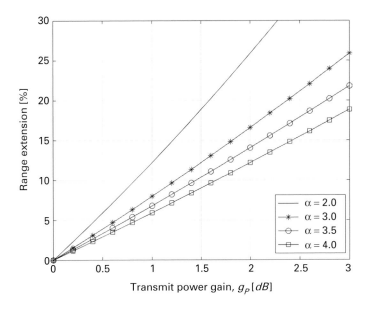

Figure 5.17. Range extension as a function of transmit power gain g_P.

incremental coverage area gains in a sectorized system will be of the same order as the gains in an omni-cell.

We plot range extension and coverage area gains as a function of the transmit power gain g_P in Figures 5.17 and 5.18 respectively. We note that a transmit power gain of, for example, 2.5 dB can extend the communication range by 15 to over 20% for a path-loss exponent α in the range of 3 to 4. The gains in coverage area for a path-loss exponent α in the range of 3–4 are between 30 and 50%. Note that a coverage area gain of 50% means that 50% less number of Node-Bs would be required to cover a geographical area resulting in large savings in system deployment cost. We remark that a single dB power gain can provide coverage area gains of 10–15%.

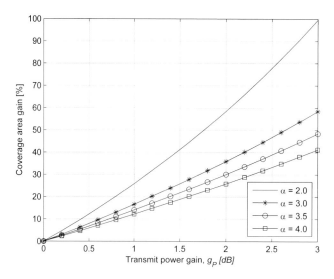

Figure 5.18. Coverage area gain as a function of transmit power gain g_P.

We should, however, note that transmit power gain is important for increasing the coverage for a certain data rate. A larger power gain for a UE can provide higher data rates at a certain distance from the Node-B. If the system is interference limited and the goal is to improve the overall system capacity in a geographical area, a larger number of Node-Bs would be required.

5.7   Summary

We noted that reducing uplink signal peakiness by a few dBs could result in huge improvements in coverage area and range. In particular, a 2.5 dB reduction in signal peakiness due to SC-FDMA relative to OFDM provides approximately 40% increase in the coverage area. This means huge savings in deployment cost as SC-FDMA would require half the number of Node-Bs to cover a geographical area as that required by OFDM. However, this comparison only holds for coverage limited situations in, for example, rural areas. In interference-limited situations, UEs do not need to transmit at peak power and hence the benefit of SC-FDMA would not be apparent. Also, to provide higher system capacity in a geographical area, a larger number of Node-Bs would be required. The signal peakiness benefit of SC-FDMA goes down as higher order modulations such as 16-QAM and 64-QAM are used. In case of 16-QAM, SC-FDMA provides approximately 1.8 dB lower signal peakiness than OFDM.

The signal peakiness of localized SC-FDMA increases when the FFT-precoded data symbols are mapped to non-contiguous subcarrier groups. The larger the number of groups of subcarriers the transmission is mapped to, the larger the increase in signal peakiness. It is therefore required that FFT-precoded symbols are mapped to a contiguous set of subcarriers to keep the low-signal-peakiness single-carrier property of SC-FDMA. This introduces an undesirable restriction in resource allocation limiting the flexibility in uplink scheduling.

We also discussed additional techniques that can further reduce peakiness of single-carrier transmission. This includes π/M-shifted MPSK modulations particularly $\pi/2$-shifted BPSK and frequency-domain spectrum shaping. We remarked that $\pi/2$-shifted BPSK could further reduce signal peakiness by approximately 1 dB. The bandwidth efficiency of $\pi/2$-shifted BPSK, however, is only half the bandwidth efficiency of QPSK. In fact, using spectrum shaping for QPSK can provide lower signal peakiness and higher bandwidth efficiency than $\pi/2$-shifted BPSK. However, in extreme coverage limited situations where bandwidth efficiency is not a concern, $\pi/2$-shifted BPSK with spectrum shaping can be used to further reduce SC-FDMA signal peakiness. The total transmit power gain due to $\pi/2$-shifted BPSK and spectrum shaping relative to SC-FDMA using QPSK is approximately 1.5 dB. We noted that root raised cosine (RRC) spectrum shaping can be beneficial for QPSK while Kaiser window spectrum shaping does not provide any gain for QPSK. The RRC spectrum shaping does not cause any inter-symbol interference. However, a drawback of RRC spectrum shaping is that it results in bandwidth expansion. On the other hand, Kaiser window spectrum-shaping does not result in bandwidth expansion but causes inter-symbol interference degrading link performance. When the bandwidth expansion for RRC and inter-symbol interference for Kaiser window spectrum shaping are taken into account, it turns out that RRC provides overall best performance for QPSK modulation while Kaiser window has a slight advantage over RRC for $\pi/2$-shifted BPSK. However, implementing two spectrum-shaping functions would also mean additional complexity in both the transmitter and the receiver. If a single spectrum-shaping function is desired for both QPSK and $\pi/2$-shifted BPSK then RRC spectrum shaping seems the appropriate candidate.

Another aspect that was not considered until now is regulatory requirements on maximum UE transmit power. It turns out that if SC-FDMA uses a WCDMA HSPA power amplifier, transmission using QPSK without spectrum shaping can achieve the maximum specified transmit power of 24 dBm. This means that the additional signal peakiness reduction due to $\pi/2$-shifted BPSK or spectrum shaping would not allow increasing transmit power beyond 24 dBm. This means that $\pi/2$-shifted BPSK and spectrum shaping would not help improve coverage beyond what is already achievable with SC-FDMA using QPSK. However, it can be argued that signal peakiness reduction using $\pi/2$-shifted BPSK or spectrum shaping can help reduce UE battery power consumption while allowing maximum transmit power of 24 dBm. Also, the LTE system can use a lower rating power amplifier than the WCDMA HSPA system again reducing UE battery power consumption and also cost. These arguments were not convincing enough and hence $\pi/2$-shifted BPSK and spectrum shaping were not included in the LTE standard specifications. However, it was understood that if, in the future, maximum specified transmit power limit can be increased beyond 24 dBm, techniques such as $\pi/2$-shifted BPSK and spectrum shaping can be useful to improve cell range and coverage.

References

[1] 3GPP TR 25.814, Physical Layer Aspects for Evolved UTRA.

[2] Kaiser F. J. and Schafer R. W., "On the use of the Io–Sinh window for spectrum analysis," *IEEE Transactions on Acoustics, Speech and Signal Processing*, vol. ASSP-28, no. 1, pp. 105–107, Feb. 1980.

[3] Mauritz, O. and Papovic, B. "Optimum family of spectrum-shaping functions for PAPR reduction on DFT-spread OFDM signals," *64th IEEE Vehicular Technology Conference, VTC-2006 Fall*, Montreal, Canada, pp. 1–5, Sept. 2006.

[4] Jakes, W. C., *Microwave Mobile Communications*, New York: Wiley, 1974.

[5] COST Action 231, "Digital radio mobile towards future generation systems, final report," Technical Report, European Communities, EUR 18957, 1999.

6 Transmit diversity

The LTE system design goal is optimization for low mobile speeds ranging from stationary users to up to 15 km/h mobile speeds. At these low speeds, eNode-B can exploit multi-user diversity gains by employing channel sensitive scheduling. For downlink transmissions, UEs feed back downlink channel quality information back to the eNode-B. Using a channel quality sensitive scheduler such as proportional fair scheduler, eNode-B can serve a UE on time-frequency resources where it is experiencing the best conditions. It is well known that when multi-user diversity can be exploited, use of other forms of diversity such as transmit diversity degrades performance. This is because multi-user diversity relies on large variations in channel conditions while the transmit diversity tries to average out the channel variations.

The LTE system is also required to support speeds ranging from 15–120 km/h with high performance. Actually, the system requirements state mobility support up to 350 km/h or even up to 500 km/h. At high UE speeds, the channel quality feedback becomes unreliable due to feedback delays. When reliable channel quality estimates are not available at eNode-B, channel-sensitive scheduling becomes infeasible. Under these conditions, it is desired to average out the channel variations by all possible means. Moreover, the channel sensitive scheduler has to wait for the right (good) channel conditions when a UE can be scheduled. This introduces delays in packet transmissions. For delay-sensitive traffic such as VoIP application, channel-sensitive scheduling cannot be used under most conditions. In this case, it is also desired to average out the channel variations even for low speed UEs using a delay sensitive service.

Various sources of diversity can be used in an OFDM system. This includes time-diversity, frequency diversity, receive diversity and transmit diversity. In general, with very short transmission time of 1 ms subframe in the LTE system, there is not much time-diversity available (except for the case of very high UE speeds). For example, the coherence time at a UE speed of 120 km/h ($v = 120$ km/h) and carrier frequency of 2 GHz ($f = 2$ GHz) is $c/(4fv) = 1.1$ ms. The time diversity can, however, be exploited in the LTE system by using hybrid ARQ where retransmissions are spaced approximately 8 ms apart at the expense of additional transmission delay.

The frequency diversity in the LTE system can be exploited by scheduling transmissions over distributed resources. In the uplink slot-level (0.5 ms half subframe), hopping where the transmission is hopped at two frequencies within a 1 ms subframe is used. The LTE system also requires support for receive diversity with at least two receive antennas.

The transmit diversity provides another additional source of diversity for averaging out the channel variation either for operation at higher UE speeds or for delay sensitive services at both low and high UE speeds. In the standardization phase various types of transmit diversity schemes were discussed and evaluated for both two transmit antennas and four transmit antennas cases. In the first part of this chapter, we study the details of these various schemes. In the second part, we will elaborate on the transmit diversity scheme used in the LTE system.

6.1 Transmit diversity schemes

6.1.1 Cyclic delay diversity (CDD)

In the cyclic delay diversity (CDD) scheme when applied to an OFDM system, delayed versions of the same OFDM symbol are transmitted from multiple antennas as shown in Figure 6.1 where we assumed a case of four transmit antennas. Let $x_0, x_1, \ldots, x_{(N-1)}$ be the sequence of modulation symbols at the input of IFFT, then the sequence of samples at the output of the IFFT $z_0, z_1, \ldots, z_{(N-1)}$ is written as:

$$z_n = \frac{1}{N} \sum_{k=0}^{(N-1)} x_k e^{j2\pi \frac{kn}{N}} \quad n = 0, 1, \ldots, (N-1). \tag{6.1}$$

The sequence of samples at the output of the IFFT $z_0, z_1, \ldots, z_{(N-1)}$ is cyclically shifted before transmission from different antennas. In the example of Figure 6.1, we assume a cyclic delay of 0, 1, 2 and 3 samples from transmission on antenna 0, 1, 2 and 3 respectively. It should be noted that cyclic delay is applied before adding the cyclic prefix (CP) and, therefore, there is no impact on multi-path robustness of the transmitted signal.

A cyclic delay diversity scheme can be implemented in the frequency domain with a phase shift of $e^{j\varphi_p k}$ applied to OFDM subcarrier k transmitted from the pth transmit antenna. The angle φ_p for the pth transmit antenna is given as:

$$\varphi_p = \frac{2\pi}{N} D_p, \tag{6.2}$$

where D_p is the cyclic delay in samples applied from the pth transmitting antenna.

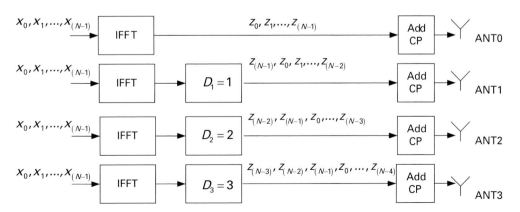

Figure 6.1. Cyclic Delay Diversity (CDD) scheme.

When implemented in the frequency domain, the CDD can be seen as precoding the transmitted modulation symbols with a diagonal matrix as given below for the case of four transmit antennas.

$$
\begin{bmatrix} y_k^0 \\ y_k^1 \\ y_k^2 \\ y_k^3 \end{bmatrix} = \frac{1}{\sqrt{4}} \begin{bmatrix} 1 & 0 & 0 & 0 \\ 0 & e^{j\phi_1 k} & 0 & 0 \\ 0 & 0 & e^{j\phi_2 k} & 0 \\ 0 & 0 & 0 & e^{j\phi_3 k} \end{bmatrix} \times \begin{bmatrix} x_k \\ x_k \\ x_k \\ x_k \end{bmatrix} = \begin{bmatrix} x_k \\ x_k e^{j\phi_1 k} \\ x_k e^{j\phi_2 k} \\ x_k e^{j\phi_3 k} \end{bmatrix}
$$
$$
k = 0, 1, \ldots, (N-1), \tag{6.3}
$$

where x_k represents the modulation symbol to be transmitted at the kth subcarrier and y_k^p represents the phase shifted modulation symbol transmitted from antenna p on the kth subcarrier. The CDD transmission in the frequency-domain using precoding is pictorially depicted in Figure 6.2.

Let $H_p(k)$ represent the channel gain on antenna p and the kth subcarrier, then the composite channel gain $H_c(k)$ experienced by a modulation symbol transmitted on the kth subcarrier is given as:

$$
H_c(k) = H_0(k) + H_1(k) \cdot e^{j\phi_1 k} + \cdots + H_p(k) \cdot e^{j\phi_p k}. \tag{6.4}
$$

The channel gain power $|H_p(k)|^2$ is plotted in Figures 6.3 and 6.4 for the cases of two and four antennas respectively. In Figure 6.3, we assumed delay values of 0 and 4 samples from the second antenna. We note that the larger the delay, the larger the frequency selectivity introduced over the transmitted bandwidth. As can be noted, a delay sample value of one results in one cycle over the total bandwidth while a delay sample of four results in four cycles. We note that two antennas CDD results in peaks of up to 3 dB relative to a flat-fading case when the signals transmitted from the two antennas combine coherently. The results for four antennas CDD in Figure 6.4 show peaks of up to 6 dB when the signals from the four antennas are combined coherently. However, the peaks in the case of 4 antennas are narrower relative to the case of two antennas CDD. This is because when looking at the whole bandwidth, CDD,

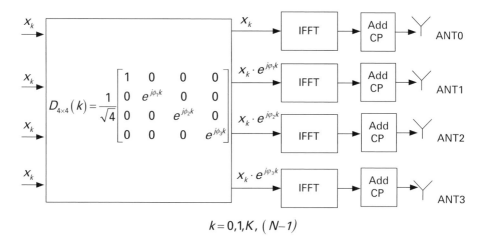

Figure 6.2. Frequency-domain implementation of cyclic delay diversity (CDD) scheme.

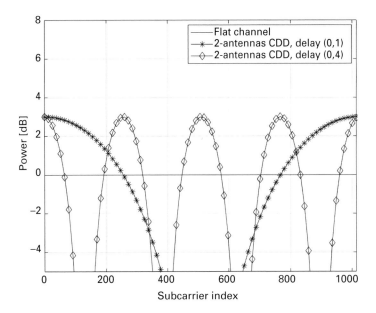

Figure 6.3. CDD power spectral density for 2-antenna case.

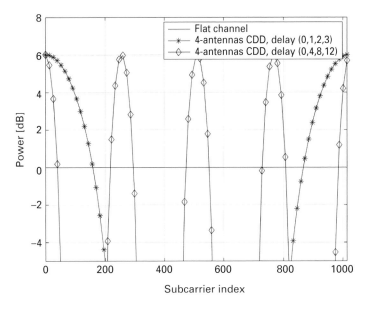

Figure 6.4. CDD power spectral density for 4-antenna case.

does not change the average received signal power, it only provides different channel gains for symbols transmitted on different subcarriers. Note that we assumed that the received power from each of the transmit antennas is equal.

Let us now see how CDD provides diversity gain by considering the case of two transmit antennas where the received power levels from the two antennas are different due to fading.

We plot the CDD composite channel power gains for the case where signal power from the second transmit antenna is 6 dB below the signal power from the first transmit antennas. We also assume that for open-loop transmit diversity the transmitter has no knowledge of the channel gains experienced by the receiver. In the absence of CDD, the transmitter has two choices. The first choice is to transmit the signal from a single antenna. In this case if the signal is transmitted from the first antenna, the received data packet would experience a better SNR than if it was transmitted from the second antenna. However, if the data packet is transmitted from the second antenna, the transmission may not be recovered because the received power gain on the second antenna is -6 dB below the first antenna. The second choice can be to transmit the signal from both the antennas without using CDD, that is transmitting the signal without introducing any phase shifts from the second antenna. In this case, the best scenario could be that the two signals add up in-phase with 0 degrees phase difference between the channel gains from the two antennas. This will result in a received power gain of 0.5 dB relative to the first antenna.

$$|H_c(k)|^2 = |H_0(k) + H_1(k)|^2 = \left(\frac{1}{\sqrt{2}} \left(1 + \frac{1}{\sqrt{4}} \right) \right)^2 = 1.125 = 0.5 \text{ dB}. \tag{6.5}$$

However, if the signals add up out of phase with a π radian phase difference between the channel gains from the two antennas, the received power is only -9 dB.

$$|H_c(k)|^2 = |H_0(k) + H_1(k)|^2 = \left(\frac{1}{\sqrt{2}} \left(1 - \frac{1}{\sqrt{4}} \right) \right)^2 = 0.125 = -9.0 \text{ dB}. \tag{6.6}$$

We note that this scheme increases the received power gain variations relative to the case when a single antenna is used for transmission where worst-case power gain is only -6 dB.

Note that we assumed a frequency-flat fading channel $H_p(k) = H_p$. We also remark that when multiple frequency-flat fading channel gains H_p are added, the resulting channel H_c is also flat. Therefore, transmitting a signal from a single transmit antenna or two transmit antennas (without frequency-specific phase applied) provides no transmit diversity. In fact, we noted that the performance can be worse when the signal is transmitted from multiple antennas compared to the case where the signal is transmitted from a single transmit antenna.

Now let us turn our attention to CDD where frequency-dependent phase shift is applied to the transmitted signal from the second transmit antenna. The resulting composite channel gain $H_c(k)$ experienced by a modulation symbol transmitted on the kth subcarrier is given as

$$H_c(k) = H_0(k) + H_1(k) \cdot e^{j\phi_1 k}. \tag{6.7}$$

We note that for a few subcarriers, the signals add in-phase providing power gain of 0.5 dB while for another few subcarriers, the signals add out-of-phase resulting in power gain of -9.0 dB. The majority of the subcarriers experience a power gain between 0.5 and -9.0 dB. For example, for the subcarrier with index 256 in the case of $(0, 1)$ delay, the phase applied at the second transmit antenna is $\pi/2$ radians:

$$\varphi_1(256) = \frac{2\pi \times 256}{1024} = \frac{\pi}{2} \text{ radians}. \tag{6.8}$$

The received power gain on a subcarrier with index 256 is then given as:

$$|H_c(k)|^2 = |H_0(k) + jH_1(k)|^2 = \left(\frac{1}{\sqrt{2}}\left(1 + j\frac{1}{\sqrt{4}}\right)\right)^2 = 0.625 = -2.04\,\text{dB}. \qquad (6.9)$$

The composite channel power gain $|H_c(k)|^2$ is plotted in Figure 6.5 for the case of $(0, 1)$ and $(0, 4)$ delay. The mean power gain across all the subcarriers can be written as:

$$E\left\{|H_c(k)|^2\right\} = \frac{1}{N}\sum_{k=0}^{N}|H_c(k)|^2 = \left(\frac{1 + 0.25}{2}\right) = 0.625 = -2.04\,\text{dB}. \qquad (6.10)$$

Note that this is simply the average of the power gains from the two transmit antennas. Therefore, we can say that CDD provides an average of the power gains from the transmit antennas and thus provides transmit diversity.

With the assumption of a flat-fading channel, the channel appears as AWGN for a given transmission in a frequency that is the power gains for all the subcarriers and hence the modulation symbols transmitted on these subcarriers are the same. The introduction of CDD creates fading in the frequency-domain, which results in capacity loss as indicated by Jensen's inequality, which states that if f is a strictly concave function then $E\,[f\,(x)] \le f\,(E\,[x])$:

$$E\left[\log_2\left(1 + |H_c(k)|^2\right)\right] \le \log_2\left(1 + E\left[|H_c(k)|^2\right]\right). \qquad (6.11)$$

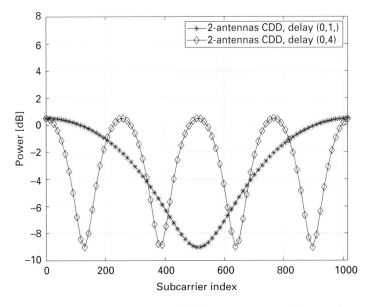

Figure 6.5. CDD power spectral density for 2-antenna case with second antenna power 6 dB below the first antenna power.

In the example we are considering, the two quantities in the above inequality are:

$$E\left[\log_2\left(1 + |H_c(k)|^2\right)\right] = \frac{1}{N}\sum_{k=0}^{(N-1)}\log_2\left(1 + |H_c(k)|^2\right) = 0.67 \text{b/s/Hz} \qquad (6.12)$$

$$\log_2\left(1 + E\left[|H_c(k)|^2\right]\right) = \log_2\left(1 + 0.625\right) = 0.7 \text{b/s/Hz}. \qquad (6.13)$$

In this case, the capacity loss due to CDD of only 0.03b/s/Hz appears rather small. This is because we are considering a low SNR case where average power levels from the first and second antennas are 0.0 dB and -6 dB respectively with average power level across both antennas of -2.0 dB. At low SNR, the capacity scales approximately linearly with SNR as given by:

$$C = E\left[\log_2\left(1 + |H_c(k)|^2 \, SNR\right)\right] \approx E\left[|H_c(k)|^2 \, SNR\right]\log_2 = SNR \times \log_2 e. \qquad (6.14)$$

This means that at low SNR, capacity is not a concave function of SNR and hence there is no capacity penalty due to CDD:

$$C = E\left[\log_2\left(1 + |H_c(k)|^2 \, SNR\right)\right]. \qquad (6.15)$$

At very high SNR, $|H_c(k)|^2 \, SNR >> 1$ and the above equation can be simplified as:

$$C = C \approx E\left[\log_2\left(SNR \times |H_c(k)|^2\right)\right] = \log_2 SNR + E\left[\log_2\left(|H_c(k)|^2\right)\right], \qquad (6.16)$$

where $E\left[\log_2\left(|H_c(k)|^2\right)\right]$ represents the penalty due to CDD at high SNR.

Another issue with CDD-based transmit diversity is symbol puncturing in the correlated antennas case. We demonstrate this effect by considering a case of two perfectly correlated transmit antennas. Also, we assume that the delay applied to the signals transmitted from the second transmit antenna is $N/2$ samples. This results in φ_1 phase of π radians as below:

$$\varphi_1 = \frac{2\pi}{N}D_1 = \frac{2\pi}{N}\left(\frac{N}{2}\right) = \pi. \qquad (6.17)$$

Now for the case of perfectly correlated antennas we assume $H_0(k) = H_1(k)$. The composite channel power gain can then be written as:

$$|H_c(k)|^2 = \left|\frac{1}{\sqrt{2}}\left(H_0(k) + H_1(k)\cdot e^{j\phi_1 k}\right)\right|^2 = \left|\frac{H_0(k)}{\sqrt{2}}\left[1 + (-1)^k\right]\right|^2. \qquad (6.18)$$

We note that the signal is completely erased for odd subcarriers ($k = 1, 2, 3, \ldots$). This will result in puncturing of the transmitted codeword symbols resulting in coding loss. We also note that there is no energy loss because, for even subcarriers, the received signals combine coherently providing a 3 dB power gain.

In order to overcome the puncturing effect, CDD can be precoded using, for example, a unitary matrix such as a DFT-matrix. The precoded-CDD precoder combines CDD-delay-based phase shifts with the Fourier-based precoding as below for the case of two transmit antennas:

$$C_{2\times2} = D_{2\times2} \times W_{2\times2} = \frac{1}{\sqrt{2}} \begin{bmatrix} 1 & 0 \\ 0 & e^{j\varphi_1 k} \end{bmatrix} \times \begin{bmatrix} 1 & 1 \\ 1 & -1 \end{bmatrix}$$

$$= \frac{1}{\sqrt{2}} \begin{bmatrix} 1 & 1 \\ e^{j\phi_1 k} & -e^{j\phi_1 k} \end{bmatrix}. \tag{6.19}$$

Let $x_1(k)$ represent the modulation symbol to be transmitted on the k subcarrier, the signals transmitted from the two physical antennas, $y_1(k)$ and $y_2(k)$ are given as below:

$$\begin{bmatrix} y_0(k) \\ y_1(k) \end{bmatrix} = \frac{1}{\sqrt{2}} \begin{bmatrix} 1 & e^{j\phi_1 k} \\ 1 & -e^{j\phi_1 k} \end{bmatrix} \times \begin{bmatrix} x(k) \\ x(k) \end{bmatrix} = \frac{1}{\sqrt{2}} \begin{bmatrix} x(k)\left(1 + e^{j\phi_1 k}\right) \\ x(k)\left(1 - e^{j\phi_1 k}\right) \end{bmatrix}. \tag{6.20}$$

The composite channel power gain for the precoded-CDD on the kth subcarrier can be written as:

$$|H_c(k)|^2 = \left| \frac{(H_0(k) + H_1(k))}{2} + \frac{(H_0(k) - H_1(k))}{2} e^{j\phi_1 k} \right|^2. \tag{6.21}$$

For the case of $H_0(k) = H_1(k)$, the composite channel gain is simply:

$$|H_c(k)|^2 = |H_0(k)|^2. \tag{6.22}$$

The power gains from the first transmit antenna $\left|H_0(k)\left(1 + e^{j\phi_1 k}\right)\right|^2$ and from the second transmit antenna $\left|H_1(k)\left(1 - e^{j\phi_1 k}\right)\right|^2$ and the composite channel gain $|H_c(k)|^2$ are plotted in Figure 6.6 for the case of $H_0(k) = H_1(k)$. We assumed a frequency-flat channel, which means that $H_0(k)$ and $H_1(k)$ are not functions of the subcarrier index. This results in a constant composite channel gain as indicated by Equation (6.22). It should be noted that the receiver only sees the composite channel gain because the transmitted signals from the two transmit antennas get combined in the air. These three channel gains for the case when the second transmit antenna average power is 6 dB below the first antenna are shown in Figure 6.7. In comparing the plots in Figures 6.6 and 6.7 with the plots in Figures 6.3 and 6.5, we note that precoded CDD results in fewer variations in channel gains than the simple CDD scheme. In fact from Figure 6.6, we note that for the case of correlated antennas exhibiting frequency-flat fading, the precoded CDD composite channel is also a frequency-flat fading channel. In case of perfectly correlated antenna, there is no diversity to be gained. We noted from Figure 6.3 that the simple CDD scheme could create artificial frequency selectivity even in this case resulting in channel capacity loss. However, in the same case, precoded-CDD does not incur any channel capacity penalty.

Now let us again consider the example that we used to show the puncturing effect with ϕ_1 phase of π radians. In this case, the above equation can be written as:

$$\begin{bmatrix} y_0(k) \\ y_1(k) \end{bmatrix} = \begin{bmatrix} x(k)\left(1 + e^{j\phi_1 k}\right) \\ x(k)\left(1 - e^{j\phi_1 k}\right) \end{bmatrix} = \begin{bmatrix} x(k)\left(1 + (-1)^k\right) \\ x(k)\left(1 - (-1)^k\right) \end{bmatrix}. \tag{6.23}$$

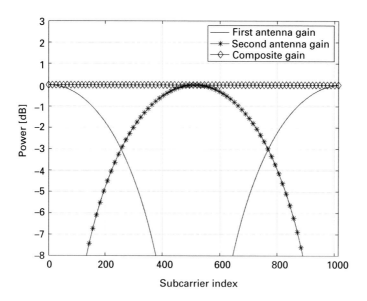

Figure 6.6. Precoded-CDD power spectral density for 2-antennas.

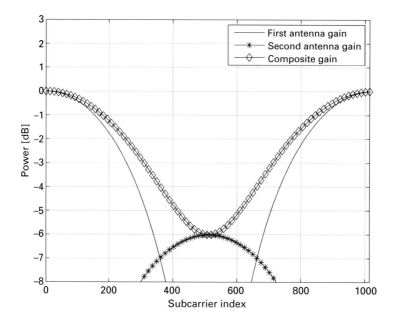

Figure 6.7. Precoded-CDD power spectral density for the 2-antenna case with the second antenna power 6 dB below the first antenna power.

Let us now see what happens for the case of even and odd subcarriers:

$$\begin{bmatrix} y_0(k) \\ y_1(k) \end{bmatrix} = \begin{bmatrix} x(k) \\ 0 \end{bmatrix} \quad k = 0, 2, 4, \ldots$$

$$\begin{bmatrix} y_0(k) \\ y_1(k) \end{bmatrix} = \begin{bmatrix} 0 \\ x(k) \end{bmatrix} \quad k = 1, 3, 5, \ldots$$

(6.24)

We note that for the case of even subcarriers, the modulation symbols are transmitted from the first antenna only while for the case of odd subcarriers, the modulation symbols are transmitted from the second antenna only as shown in Figure 6.8. We also note that there is no puncturing of modulation symbols happening when the two transmit antennas are correlated, that is $H_0(k) = H_1(k)$. In this case, the symbol transmitted over the kth subcarrier, $x(k)$, experiences a channel gain of $H_0(k)$. This scheme described by Equation (6.24) effectively becomes the Frequency Shift Transmit Diversity (FSTD) scheme discussed in the next section.

Let us now consider the precoded-CDD scheme for the case of four transmit antennas. We will use a 4×4 DFT matrix for CDD precoding:

$$
W_{4 \times 4} = \frac{1}{\sqrt{4}} \begin{bmatrix} 1 & 1 & 1 & 1 \\ 1 & e^{j\pi/2} & e^{j\pi} & e^{j3\pi/2} \\ 1 & e^{j\pi} & e^{j2\pi} & e^{j3\pi} \\ 1 & e^{j3\pi/2} & e^{j3\pi} & e^{j9\pi/2} \end{bmatrix} = \frac{1}{\sqrt{4}} \begin{bmatrix} 1 & 1 & 1 & 1 \\ 1 & j & -1 & -j \\ 1 & -1 & 1 & -1 \\ 1 & -j & -1 & j \end{bmatrix}.
\tag{6.25}
$$

The composite CDD precoder can then be written as:

$$
C_{4 \times 4} = W_{4 \times 4} \times D_{4 \times 4} = W_{4 \times 4} \times \begin{bmatrix} 1 & 0 & 0 & 0 \\ 0 & e^{j\phi_1 k} & 0 & 0 \\ 0 & 0 & e^{j\phi_2 k} & 0 \\ 0 & 0 & 0 & e^{j\phi_3 k} \end{bmatrix}
\tag{6.26}
$$

$$
= \frac{1}{\sqrt{4}} \begin{bmatrix} 1 & e^{j\phi_1 k} & e^{j\phi_2 k} & e^{j\phi_3 k} \\ 1 & je^{j\phi_1 k} & -e^{j\phi_2 k} & -je^{j\phi_3 k} \\ 1 & -e^{j\phi_1 k} & e^{j\phi_2 k} & -e^{j\phi_3 k} \\ 1 & -je^{j\phi_1 k} & -e^{j\phi_2 k} & je^{j\phi_3 k} \end{bmatrix}
$$

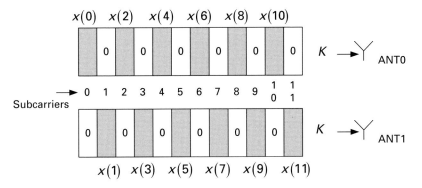

Figure 6.8. Precoded-CDD mapping for the case of $\varphi_1 = \pi$.

$$
\begin{bmatrix} y_0\,(k) \\ y_1\,(k) \\ y_2\,(k) \\ y_3\,(k) \end{bmatrix} = C_{4\times 4} \times \begin{bmatrix} x\,(k) \\ x\,(k) \\ x\,(k) \\ x\,(k) \end{bmatrix} = \begin{bmatrix} \left(1 + e^{j\phi_1 k} + e^{j\phi_2 k} + e^{j\phi_3 k}\right)x\,(k) \\ \left(1 + je^{j\phi_1 k} - e^{j\phi_2 k} - je^{j\phi_3 k}\right)x\,(k) \\ \left(1 - e^{j\phi_1 k} + e^{j\phi_2 k} - e^{j\phi_3 k}\right)x\,(k) \\ \left(1 - je^{j\phi_1 k} - e^{j\phi_2 k} + je^{j\phi_3 k}\right)x\,(k) \end{bmatrix}. \tag{6.27}
$$

Let us assume the following case where:

$$
\varphi_1 = \pi \quad \varphi_2 = 2\varphi_1 \quad \varphi_3 = 3\varphi_1. \tag{6.28}
$$

Then for various values of the subcarrier index k, Equation (6.27) can be expressed as:

$$
[6pt]\quad \begin{matrix} k = 0,4,\ldots & k = 1,5,\ldots & k = 2,6,\ldots & k = 3,7,\ldots \end{matrix}
$$

$$
\begin{bmatrix} y_0\,(k) \\ y_1\,(k) \\ y_2\,(k) \\ y_3\,(k) \end{bmatrix} = \begin{bmatrix} x\,(k) \\ 0 \\ 0 \\ 0 \end{bmatrix} \begin{bmatrix} 0 \\ x\,(k) \\ 0 \\ 0 \end{bmatrix} \begin{bmatrix} 0 \\ 0 \\ x\,(k) \\ 0 \end{bmatrix} \begin{bmatrix} 0 \\ 0 \\ 0 \\ x\,(k) \end{bmatrix}. \tag{6.29}
$$

We note that in a given subcarrier, a single modulation symbol is transmitted from a single transmit antenna. Therefore, we note that in this special case, precoded-CDD scheme reduces to an FSTD scheme.

6.1.2   Frequency shift transmit diversity

In the frequency shift transmit diversity scheme, a given modulation symbol is transmitted on a single antenna in a given subcarrier. Let $y_p\,(k)$ represent the signal transmitted from the pth antenna on the kth subcarrier, then the transmit matrix for FSTD can be written as:

$$
\begin{bmatrix} y_0\,(4i) & y_0\,(4i+1) & y_0\,(4i+2) & y_0\,(4i+3) \\ y_1\,(4i) & y_1\,(4i+1) & y_1\,(4i+2) & y_1\,(4i+3) \\ y_2\,(4i) & y_2\,(4i+1) & y_2\,(4i+2) & y_2\,(4i+3) \\ y_3\,(4i) & y_3\,(4i+1) & y_3\,(4i+2) & y_3\,(4i+3) \end{bmatrix}
$$

$$
= \begin{bmatrix} x\,(4i) & 0 & 0 & 0 \\ 0 & x\,(4i+1) & 0 & 0 \\ 0 & 0 & x\,(4i+2) & 0 \\ 0 & 0 & 0 & x\,(4i+3) \end{bmatrix}, \tag{6.30}
$$

where $i = 0, 1, 2, \ldots$

Let us consider the case for $i = 0$:

$$
\begin{bmatrix} y_0\,(0) & y_0\,(1) & y_0\,(2) & y_0\,(3) \\ y_1\,(0) & y_1\,(1) & y_1\,(2) & y_1\,(3) \\ y_2\,(0) & y_2\,(1) & y_2\,(2) & y_2\,(3) \\ y_3\,(0) & y_3\,(1) & y_3\,(2) & y_3\,(3) \end{bmatrix} = \begin{bmatrix} x\,(0) & 0 & 0 & 0 \\ 0 & x\,(1) & 0 & 0 \\ 0 & 0 & x\,(2) & 0 \\ 0 & 0 & 0 & x\,(3) \end{bmatrix}. \tag{6.31}
$$

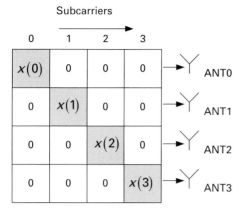

Figure 6.9. Frequency shift transmit diversity scheme for four transmit antennas.

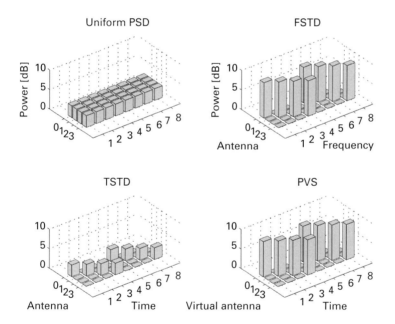

Figure 6.10. Power spectral density for FSTD, TSTD amd PVS schemes.

We note that a single antenna transmits in a given subcarrier as is also shown in Figure 6.9. We also remark that since each antenna transmits on 1/4 of the subcarriers, the power on each transmitted subcarrier is 6 dB higher than the case if the antenna was transmitting over all the subcarriers. This is pictorially shown in Figure 6.10 for the case of four antennas and 8 subcarriers. We assumed a 3 dB uniform power spectral density per antenna when transmitting over the whole bandwidth. The power spectral density per antenna on the transmitted subcarriers in case of FSTD is 6 dB higher than in the uniform PSD case. We also note that each antenna is able to transmit at full power across multiple subcarriers, that is, transmitting at 6 dB higher power on 1/4 of the subcarriers.

6.1.3   Time shift transmit diversity

The time shift transmit diversity (TSTD) scheme is similar to the FSTD scheme with the difference that switching now happens in time. Let $y_p(n)$ represent the signal transmitted from the pth antenna on the nth slot or OFDM symbol, then the transmit matrix for TSTD can be written as:

$$
\begin{bmatrix}
y_0(4i) & y_0(4i+1) & y_0(4i+2) & y_0(4i+3) \\
y_1(4i) & y_1(4i+1) & y_1(4i+2) & y_1(4i+3) \\
y_2(4i) & y_2(4i+1) & y_2(4i+2) & y_2(4i+3) \\
y_3(4i) & y_3(4i+1) & y_3(4i+2) & y_3(4i+3)
\end{bmatrix}
$$

$$
=
\begin{bmatrix}
x(4i) & 0 & 0 & 0 \\
0 & x(4i+1) & 0 & 0 \\
0 & 0 & x(4i+2) & 0 \\
0 & 0 & 0 & x(4i+3)
\end{bmatrix},
\tag{6.32}
$$

where $i = 0, 1, 2, \ldots$

Let us consider the case for $i = 0$:

$$
\begin{bmatrix}
y_0(0) & y_0(1) & y_0(2) & y_0(3) \\
y_1(0) & y_1(1) & y_1(2) & y_1(3) \\
y_2(0) & y_2(1) & y_2(2) & y_2(3) \\
y_3(0) & y_3(1) & y_3(2) & y_3(3)
\end{bmatrix}
=
\begin{bmatrix}
x(0) & 0 & 0 & 0 \\
0 & x(1) & 0 & 0 \\
0 & 0 & x(2) & 0 \\
0 & 0 & 0 & x(3)
\end{bmatrix}.
\tag{6.33}
$$

We note that signal is transmitted from a single transmit antenna at a given time. However, we also note that now the trasmission happens on all the subcarriers in a given time slot. Therefore, the power spectral density on the trammitted subcarriers is the same as the uniform PSD reference case. However, with TSTD, only a single antenna can transmit at a given time. Therefore, the TSTD scheme results in 6 dB power loss compared to the reference case as is illustraed in Figure 6.10. Note that power can be shifted in frequency by increasing PSD on certain parts of the band while reducing PSD on certain other parts of the band. However, assuming a maximum transmit power limitation per antenna, power cannot be shifted in time from one slot to the other.

6.1.4   Precoding vector switching

In order to overcome the power loss problem of TSTD, the switching in time can be performed across virtual antenna created by precoding rather than by the physical antennas. The resulting scheme is referred to as precoding vector switching (PVS). Let us consider a set of four virtual antennas created by using a 4 × 4 DFT precoding matrix. Each column of the matrix forms a virtual antenna. The elements of the columns refer to scaling applied to the signal transmitted from the four physical antennas. For example if symbol $x(1)$ is transmitted from virtual antenna 1 (VA1), then the symbols transmitted from the four physical antennas are $x(1)$, $jx(1)$, $-x(1)$, $jx(1)$ respectively:

$$
\begin{array}{cccc}
\text{VA0} & \text{VA1} & \text{VA2} & \text{VA3}
\end{array}
$$

$$
\begin{bmatrix} 1 \\ 1 \\ 1 \\ 1 \end{bmatrix}
\begin{bmatrix} 1 \\ j \\ -1 \\ j \end{bmatrix}
\begin{bmatrix} 1 \\ -1 \\ 1 \\ -1 \end{bmatrix}
\begin{bmatrix} 1 \\ -j \\ -1 \\ j \end{bmatrix}.
\tag{6.34}
$$

Let $y_p(n)$ represent the signal transmitted from the pth antenna on the nth slot or OFDM symbol, then the transmit matrix for PVS can be written as:

$$
\begin{bmatrix}
y_0(4i) & y_0(4i+1) & y_0(4i+2) & y_0(4i+3) \\
y_1(4i) & y_1(4i+1) & y_1(4i+2) & y_1(4i+3) \\
y_2(4i) & y_2(4i+1) & y_2(4i+2) & y_2(4i+3) \\
y_3(4i) & y_3(4i+1) & y_3(4i+2) & y_3(4i+3)
\end{bmatrix}
$$
$$
=
\begin{bmatrix}
x(4i) & x(4i+1) & x(4i+2) & x(4i+3) \\
x(4i) & jx(4i+1) & -x(4i+2) & jx(4i+3) \\
x(4i) & -x(4i+1) & x(4i+2) & -x(4i+3) \\
x(4i) & jx(4i+1) & -x(4i+2) & jx(4i+3)
\end{bmatrix},
\tag{6.35}
$$

where $i = 0, 1, 2, \ldots$

We note that there are no zero elements in the transmission matrix for PVS. This means that the signal is transmitted from all the antennas in all the time slots. Let us consider the case for $i = 0$:

$$
\begin{bmatrix}
y_0(0) & y_0(1) & y_0(2) & y_0(3) \\
y_1(0) & y_1(1) & y_1(2) & y_1(3) \\
y_2(0) & y_2(1) & y_2(2) & y_2(3) \\
y_3(0) & y_3(1) & y_3(2) & y_3(3)
\end{bmatrix}
=
\begin{bmatrix}
x(0) & x(1) & x(2) & x(3) \\
x(0) & jx(1) & -x(2) & jx(3) \\
x(0) & -x(1) & x(2) & -x(3) \\
x(0) & jx(1) & -x(2) & jx(3)
\end{bmatrix}.
\tag{6.36}
$$

We note that a given symbol is transmitted from a single virtual antenna in a given time slot. However, similarly to the case of FSTD, the power on each virtual antenna can now be 6 dB higher. Figure 6.10 shows power transmitted on each virtual antenna for the case of PVS. In terms of power transmitted from the physical antennas, PVS power spectral density is uniform across space (antenna) and time.

Let $h_p(k)$ represent the channel gain on physical antenna p and kth subcarrier, then the composite channel gain $h_{cv}(k)$ on the vth virtual antenna as experienced by a modulation symbol transmitted on the kth subcarrier is given as:

$$
\begin{aligned}
\text{VA0 } h_{c0}\,(k) &= h_0\,(k) + h_1\,(k) + h_2\,(k) + h_3\,(k) \\
\text{VA1 } h_{c1}\,(k) &= h_0\,(k)\,jh_1\,(k) - h_2\,(k) + jh_3\,(k) \\
\text{VA2 } h_{c2}\,(k) &= h_0\,(k) - h_1\,(k) + h_2\,(k) - h_3\,(k) \\
\text{VA3 } h_{c3}\,(k) &= h_0\,(k) + jh_1\,(k) - h_2\,(k) + jh_3\,(k).
\end{aligned}
\tag{6.37}
$$

We note that the composite channel gains are different for different virtual antennas and switching on these virtual antennas for different modulation symbols can provide transmit diversity. However, one drawback of the PVS scheme is that some of the modulation symbols can be completely punctured if the phases from different physical antennas add destructively. For example, the composite channel gain on VA2 $h_{c2}\,(k)$ can be completely nulled out if the transmissions from the four antennas are correlated and in phase.

6.1.5   Block-codes-based transmit diversity

An example of a two transmit antennas block-code space-time diversity scheme is the Alamouti code [1]. In this approach during any symbol period, two data symbols are transmitted simultaneously from the two transmit antennas. Suppose during the first symbol period t_0, the symbols transmitted from antenna 0 and 1 are denoted as $x\,(0)$ and $x\,(1)$ respectively as shown in Figure 6.11. During the next symbol period t_1, the symbols transmitted from antennas 0 and 1 are $-x\,(1)^*$ and $x\,(0)^*$, where x^* represents the complex conjugate of x. The Alamouti scheme can also be implemented in a space-frequency coded form. In this case, the two symbols are sent on two different frequencies, for example, on different subcarriers in an Orthogonal Frequency Division Multiplexing (OFDM) system as shown in Figure 6.11.

Let $y_p\,(k)$ represent the signal transmitted from the pth antenna on the kth subcarrier, then:

$$
\begin{bmatrix} y_0\,(0) & y_0\,(1) \\ y_1\,(0) & y_1\,(1) \end{bmatrix} = \begin{bmatrix} x\,(0) & -x\,(1)^* \\ x\,(1) & x\,(0)^* \end{bmatrix}.
\tag{6.38}
$$

In case of STBC, the subcarrier indices are replaced by the time indices. In an OFDM system, the time index is the same as the OFDM symbol index. A fundamental requirement for

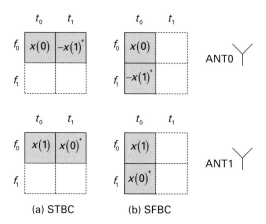

(a) STBC   (b) SFBC

Figure 6.11. STBC and SFBC transmit diversity schemes for 2-Tx antennas.

block-codes-based transmit diversity schemes is that the channel needs to be constant over the pair of modulation symbols transmission. This means a constant channel over two adjacent OFDM symbols or subcarriers for the STBC and SFBC schemes respectively. Depending upon the channel scenario, one scheme can be shown outperforming the other scheme. For example, in a relatively frequency-flat fading channel for a very-high-speed UE, it can be shown that the SFBC scheme outperforms the STBC scheme. On the other hand, over a highly dispersive channel for a slow moving UE, the STBC scheme outperforms the SFBC scheme. Under typical channel conditions experienced in mobile environments, both schemes perform similarly. Another important consideration is that both the STBC and SFBC schemes require an even number of OFDM symbols or subcarriers respectively for pairing of the modulation symbols. In a mobile OFDM system, the number of OFDM symbols within a subframe is generally much smaller than the number of OFDM subcarriers. In the LTE system, the maximum number of OFDM symbols within a subframe is 14 while the number of subcarriers varies from a minimum of 72 to over 1200. It is, therefore, much easier to pair subcarriers together than the OFDM symbols. This is particularly true for the LTE system where the time-multiplexed control information uses a dynamically varying number of OFDM symbols between 1 and 3, thus sometimes leaving an odd number of OFDM symbols for data transmission. This condition favors the SFBC scheme over the STBC scheme. We will consider only the SFBC-based transmit diversity scheme in our further discussions.

Let us now see how the original symbols $x(0)$ and $x(1)$ can be recovered with some processing at the receiver. Let r_0 and r_1 denote the received signal in subcarrier f_0 and subcarrier f_1 respectively for the case of SFBC. These two received signals can be written as:

$$r_0 = h_0 x(0) + h_1 x(1) + n_0$$
$$r_1 = -h_0 x(1)^* + h_1 x(0)^* + n_1, \tag{6.39}$$

where n_0 and n_1 represent the additive white Gaussian noise (AWGN) in subcarrier f_0 and subcarrier f_1 respectively. Also h_0 and h_1 are the channel gains on antenna 0 and antenna 1 respectively. We assume that the channel gains do not change across the two subcarriers.

The estimates of the two transmitted symbols $\widehat{x}(0)$ and $\widehat{x}(1)$, are obtained by applying the following operations:

$$\widehat{x}(0) = h_0^* r_0 + h_1 r_1^*$$
$$= h_0^* (h_0 x(0) + h_1 x(1) + n_0) + h_1 \left(-h_0 x(1)^* + h_1 x(0)^* + n_2\right)^*$$
$$= \left(|h_0|^2 + |h_1|^2\right) x(0) + h_0^* n_0 + h_1 n_1^*$$
$$\widehat{x}(1) = h_1^* r_0 - h_0 r_1^*$$
$$= h_1^* (h_0 x(0) + h_1 x(1) + n_0) - h_0 \left(-h_0 x(1)^* + h_1 x(0)^* + n_1\right)^*$$
$$= \left(|h_0|^2 + |h_1|^2\right) x(1) + h_1^* n_0 - h_0 n_1^*. \tag{6.40}$$

The set of Equations (6.39) can be written as:

$$\begin{bmatrix} r_0 \\ -r_1^* \end{bmatrix} = \begin{bmatrix} h_0 & h_1 \\ -h_1^* & h_0^* \end{bmatrix} \times \begin{bmatrix} x(0) \\ x(1) \end{bmatrix} + \begin{bmatrix} n_0 \\ n_1 \end{bmatrix}, \tag{6.41}$$

where the equivalent channel matrix for the Alamouti code H_2 is:

$$H_2 = \begin{bmatrix} h_0 & h_1 \\ -h_1^* & h_0^* \end{bmatrix}. \tag{6.42}$$

The operations given in Equation (6.40) are based on a simple matched filter receiver. Assuming a matched filter receiver, the resulting channel gains matrix can be written as:

$$H_2^H H_2 = \begin{bmatrix} h_0^* & -h_1 \\ h_1^* & h_0 \end{bmatrix} \begin{bmatrix} h_0 & h_1 \\ -h_1^* & h_0^* \end{bmatrix} = \left(h_0^2 + h_1^2 \right) \begin{bmatrix} 1 & 0 \\ 0 & 1 \end{bmatrix}. \tag{6.43}$$

We note that the Alamouti code is an orthogonal code. We also remark that the instantaneous channel gain estimates h_0 and h_1 on antenna 0 and antenna 1 respectively are required for received symbols processing at the receiver. This requires separate pilot or reference symbols transmitted from both the antennas for channel estimation at the receiver. We also note that the diversity gain achieved by Alamouti coding is the same as that achieved in maximum ratio combining (MRC).

An alternative representation of Alamouti code is obtained by taking the transpose of the 2×2 matrix in Equation (6.38) as below:

$$\begin{bmatrix} y_0(0) & y_0(1) \\ y_1(0) & y_1(1) \end{bmatrix} = \begin{bmatrix} x(0) & x(1) \\ -x(1)^* & x(0)^* \end{bmatrix}. \tag{6.44}$$

We will assume this alternative definition of the Alamouti code from now onwards.

Let us now consider the case of more than two transmit antennas. For more than two transmit antennas, orthogonal full-diversity block codes are not available. An example of quasi-orthogonal block code referred to as ABBA code [2] is described below. Let $y_p(k)$ represent the signal transmitted from the pth antenna on the kth subcarrier, then the transmit matrix for ABBA can be written as:

$$
\begin{bmatrix} y_0(0) & y_0(1) & y_0(2) & y_0(3) \\ y_1(0) & y_1(1) & y_1(2) & y_1(3) \\ y_2(0) & y_2(1) & y_2(2) & y_2(3) \\ y_3(0) & y_3(1) & y_3(2) & y_3(3) \end{bmatrix} = \begin{bmatrix} A & B \\ B & A \end{bmatrix}
$$

$$
= \begin{bmatrix} x(0) & x(1) & x(1) & x(3) \\ -x(1)^* & x(0)^* & -x(3)^* & x(2)^* \\ x(2) & x(3) & x(0) & x(1) \\ -x(3)^* & x(2)^* & -x(1)^* & x(0)^* \end{bmatrix}, \tag{6.45}
$$

where A and B are given as:

$$
A = \begin{bmatrix} x(0) & x(1) \\ -x(1)^* & x(0)^* \end{bmatrix}
$$

$$
B = \begin{bmatrix} x(2) & x(3) \\ -x(3)^* & x(2)^* \end{bmatrix}. \tag{6.46}
$$

The equivalent channel matrix for the ABBA code H_{ABBA} is given as:

$$H_{ABBA} = \begin{bmatrix} h_0 & -h_1^* & h_2 & -h_3^* \\ h_1 & h_0^* & h_3 & h_2^* \\ h_2 & -h_3^* & h_0 & -h_1^* \\ h_3 & h_2^* & h_1 & h_0^* \end{bmatrix}. \tag{6.47}$$

Assuming a matched filter receiver, the resulting channel gains matrix can be written as:

$$H_{ABBA}^H H_{ABBA} = h^2 \begin{bmatrix} 1 & 0 & \frac{2(h_0 h_2^* + h_1 h_3^*)}{h^2} & 0 \\ 0 & 1 & 0 & \frac{2(h_0 h_2^* + h_1 h_3^*)}{h^2} \\ \frac{2(h_0 h_2^* + h_1 h_3^*)}{h^2} & 0 & 1 & 0 \\ 0 & \frac{2(h_0 h_2^* + h_1 h_3^*)}{h^2} & 0 & 1 \end{bmatrix}, \tag{6.48}$$

where $h^2 = h_0^2 + h_1^2 + h_2^2 + h_3^2$. We notice that the resulting channel gains matrix is non-orthogonal with the presence of the interference term $\left[2\left(h_0 h_2^* + h_1 h_3^*\right)/h^2\right]$. More advanced receivers such as the Maximum Likelihood (ML) receiver can be used to recover from loss of orthogonality at the expense of increased receiver complexity.

On the other hand if we want an orthogonal code for the case of four transmit antennas, we have to live with some loss in code rate. An example of rate-3/4 orthogonal block code for four transmit antennas is given below [3]:

$$\begin{bmatrix} y_0(0) & y_0(1) & y_0(2) & y_0(3) \\ y_1(0) & y_1(1) & y_1(2) & y_1(3) \\ y_2(0) & y_2(1) & y_2(2) & y_2(3) \\ y_3(0) & y_3(1) & y_3(2) & y_3(3) \end{bmatrix} = \begin{bmatrix} x(0) & 0 & x(1) & -x(2) \\ 0 & x(0) & x(2)^* & x(1)^* \\ -x(1)^* & -x(2) & x(0)^* & 0 \\ x(2)^* & -x(1) & 0 & x(0)^* \end{bmatrix}. \tag{6.49}$$

We note that three symbols are transmitted over four subcarriers on four transmit antennas. This means that the bandwidth efficiency of this code is only 75% of the full rate code. This low bandwidth efficiency will result in increased coding rate for the same use of subcarriers and therefore resulting in some channel coding loss. The apparent tradeoff here is between transmitting diversity gain versus the channel coding gain. In cases where very low coding rates with repetition are used such as may be the case for the cell-edge users in a cellular system, a lower rate block code does not result in reduced channel coding gain. The simple effect is less repetition. In these special cases, a lower rate block code may still be useful. On the other hand, for high-data rate users closer to the cell center using higher coding rates and higher order modulation, the bandwidth loss due to a lower rate block code generally does not offset the transmit diversity gain. This is particularly true when other forms of diversity sources such as receive diversity and frequency-diversity are available.

We now consider a few schemes that provide orthogonal transmission by combining 2 transmit antennas Alamouti block code with CDD, FSTD, TSTD or PVS schemes.

Let us first consider the combined SFBC-CDD scheme. In this case, four antennas are paired into two groups of two antennas each. The SFBC scheme is applied across two antennas within each pair. The CDD is applied across the two groups of antennas as shown in Figure 6.12. In the example of Figure 6.12, we have shown transmission of two modulation symbols over two subcarriers. The transmission from the first set of antennas (antenna 0 and antenna 1) happens

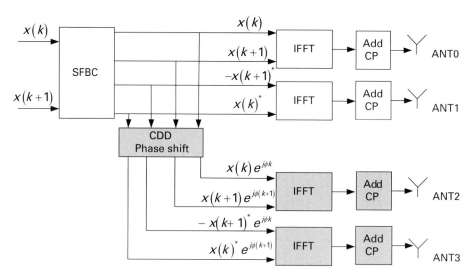

Figure 6.12. Combined SFBC-CDD scheme for 4-Tx antennas.

as a normal SFBC scheme for 2-Tx antennas. A frequency-specific phase shift is applied for SFBC transmission from the second set of antennas (antennas 2 and 3). This scheme is also sometimes referred to as the combined SFBC-PSD (phase shift diversity) scheme. Since the CDD is applied in the frequency-domain, the phase shift across a pair of subcarriers used by SFBC block code can be constant. The phase can change from one pair of subcarriers to the next pair of subcarriers.

Let $y_p(k)$ represent the signal transmitted from the pth antenna on the kth subcarrier, then the transmit matrix for the combined SFBC-CDD scheme can be written as:

$$
\begin{bmatrix}
y_0(k) & y_0(k+1) \\
y_1(k) & y_1(k+1) \\
y_2(k) & y_2(k+1) \\
y_3(k) & y_3(k+1)
\end{bmatrix}
=
\begin{bmatrix}
x(0) & x(1) \\
-x(1)^* & x(0)^* \\
x(0)\,e^{j\phi k} & x(1)\,e^{j\phi(k+1)} \\
-x(1)^*\,e^{j\phi k} & x(0)^*\,e^{j\phi(k+1)}
\end{bmatrix}.
\tag{6.50}
$$

As pointed out the phase shift applied from the second pair of antennas can be constant across the pair of subcarriers resulting in the following transmit matrix:

$$
\begin{bmatrix}
y_0(k) & y_0(k+1) \\
y_1(k) & y_1(k+1) \\
y_2(k) & y_2(k+1) \\
y_3(k) & y_3(k+1)
\end{bmatrix}
=
\begin{bmatrix}
x(0) & x(1) \\
-x(1)^* & x(0)^* \\
x(0)\,e^{j\phi k} & x(1)\,e^{j\phi k} \\
-x(1)^*\,e^{j\phi k} & x(0)^*\,e^{j\phi k}
\end{bmatrix}.
\tag{6.51}
$$

Note that in this case, the phase is dependent upon the first subcarrier in the pair of subcarriers used by SFBC. The phase shift applied to the second subcarrier in the pair is the same as the first subcarrier.

The combined SFBC-CDD scheme can face the same problem of symbol puncturing as does the CDD scheme as discussed in Section 6.1. We demonstrate this effect by considering a case of four perfectly correlated transmit antennas, that is:

$$h_0^2 = h_1^2 = h_2^2 = h_3^2. \tag{6.52}$$

Also, we assume that the delay applied to the signals transmitted from the second pair of antennas (antenna 2 and antenna 3) is $N/2$ samples. This results in φ_1 phase of π radians as below:

$$\varphi_1 = \frac{2\pi}{N} D = \frac{2\pi}{N} \left(\frac{N}{2}\right) = \pi. \tag{6.53}$$

We also assume the scheme of Equation (6.51) where the phase shift on the pair of subcarriers used by SFBC block code is fixed. The composite channel power gain can then be written as:

$$h^2(k) = \left(\frac{h_0^2(k) + h_0^2(k) \cdot e^{j\phi_{1k}}}{2}\right) = \frac{h_0^2(k)}{2}(1 + (-1)^k)^2. \tag{6.54}$$

We note that the signal is completely erased when k is an odd number. This will result in puncturing of the transmitted codeword symbols resulting in coding loss. We also note that there is no energy loss because when k is an even number, the received signals combine coherently providing 3 dB power gain.

Let us now consider the combined SFBC-PVS scheme. In fact, in the case of SFBC combined with precoding switching, a precoding matrix is used rather than a precoding vector. Therefore, in more accurate terms, the scheme should be referred to as SFBC-PMS (precoding matrix switching) scheme. For illustrating the principle of SFBC-PMS, let us assume a set of four virtual antennas created by using a 4×4 DFT precoding matrix as given in Equation (6.34), which is reproduced below for convenience:

$$\begin{array}{cccc} \text{VA0} & \text{VA1} & \text{VA2} & \text{VA3} \\ \begin{bmatrix} 1 \\ 1 \\ 1 \\ 1 \end{bmatrix} & \begin{bmatrix} 1 \\ j \\ -1 \\ j \end{bmatrix} & \begin{bmatrix} 1 \\ -1 \\ 1 \\ -1 \end{bmatrix} & \begin{bmatrix} 1 \\ -j \\ -1 \\ j \end{bmatrix} \end{array}. \tag{6.55}$$

Let us group four virtual antennas into two groups representing precoding matrix 1 (PM1) and PM2:

$$\begin{array}{cc} \text{PM1} & \text{PM2} \\ \begin{bmatrix} 1 & 1 \\ 1 & j \\ 1 & -1 \\ 1 & j \end{bmatrix} & \begin{bmatrix} 1 & 1 \\ -1 & -j \\ 1 & -1 \\ -1 & j \end{bmatrix} \end{array}. \tag{6.56}$$

Let $\{x(i)\}_{i=0}^{i=3}$ be the four modulation symbols that need to be transmitted using the SFBC-PMS scheme. The SFBC code is applied to the first pair of symbols $x(0), x(1)$ using PMS1 and to the second pair of symbols $x(2), x(3)$ using PMS2. Let $y_p(k)$ represent the signal transmitted from the pth antenna on the kth subcarrier, then the transmit vector for $k = 0$ for the SFBC-PMS scheme can be written as:

$$\begin{bmatrix} y_0(0) \\ y_1(0) \\ y_2(0) \\ y_3(0) \end{bmatrix} = \begin{bmatrix} 1 & 1 \\ 1 & j \\ 1 & -1 \\ 1 & j \end{bmatrix} \times \begin{bmatrix} x(0) \\ -x(1)^* \end{bmatrix} = \begin{bmatrix} x(0) - x(1)^* \\ x(0) - jx(1)^* \\ x(0) + x(1)^* \\ x(0) - jx(1)^* \end{bmatrix}. \tag{6.57}$$

Since SFBC is applied on a pair of subcarriers, the transmit vector for $k = 1$ is given as:

$$\begin{bmatrix} y_0(1) \\ y_1(1) \\ y_2(1) \\ y_3(1) \end{bmatrix} = \begin{bmatrix} 1 & 1 \\ 1 & j \\ 1 & -1 \\ 1 & j \end{bmatrix} \times \begin{bmatrix} x(1) \\ x(0)^* \end{bmatrix} = \begin{bmatrix} x(1) + x(0)^* \\ x(1) + jx(0)^* \\ x(1) - x(0)^* \\ x(1) + jx(0)^* \end{bmatrix}. \tag{6.58}$$

For the second pair of modulation symbols, the precoding matrix is switched, that is PMS2 is used for the second set of symbols $x(2), x(3)$. Note that the precoding matrix switching can happen either in time across OFDM symbols or in frequency across subcarriers. Assuming switching in frequency, the transmit vector for $k = 2$ is given as:

$$\begin{bmatrix} y_0(2) \\ y_1(2) \\ y_2(2) \\ y_3(2) \end{bmatrix} = \begin{bmatrix} 1 & 1 \\ -1 & -j \\ 1 & -1 \\ -1 & j \end{bmatrix} \times \begin{bmatrix} x(2) \\ -x(3)^* \end{bmatrix} = \begin{bmatrix} x(2) - x(3)^* \\ -x(2) + jx(3)^* \\ x(2) + x(3)^* \\ -x(2) - jx(3)^* \end{bmatrix}. \tag{6.59}$$

The same precoding matrix PMS2 is applied to subcarriers with $k = 3$ as written below:

$$\begin{bmatrix} y_0(3) \\ y_1(3) \\ y_2(3) \\ y_3(3) \end{bmatrix} = \begin{bmatrix} 1 & 1 \\ -1 & -j \\ 1 & -1 \\ -1 & j \end{bmatrix} \times \begin{bmatrix} x(3) \\ x(2)^* \end{bmatrix} = \begin{bmatrix} x(3) + x(2)^* \\ -x(3) - jx(2)^* \\ x(3) - x(2)^* \\ -x(3) + jx(2)^* \end{bmatrix}. \tag{6.60}$$

Now combining results of Equations (6.57) through (6.60) the transmit matrix for the SFBC-PMS scheme can be written as:

$$\begin{bmatrix} y_0(0) & y_0(1) & y_0(2) & y_0(3) \\ y_1(0) & y_1(1) & y_1(2) & y_1(3) \\ y_2(0) & y_2(1) & y_2(2) & y_2(3) \\ y_3(0) & y_3(1) & y_3(2) & y_3(3) \end{bmatrix} =$$

$$\begin{bmatrix} x(0) - x(1)^* & x(1) + x(0)^* & x(2) - x(3)^* & x(3) + x(2)^* \\ x(0) - jx(1)^* & x(1) + jx(0)^* & -x(2) + jx(3)^* & -x(3) - jx(2)^* \\ x(0) + x(1)^* & x(1) - x(0)^* & x(2) + x(3)^* & x(3) - x(2)^* \\ x(0) - jx(1)^* & x(1) + jx(0)^* & -x(2) - jx(3)^* & -x(3) + jx(2)^* \end{bmatrix}. \tag{6.61}$$

The channel gains on the four virtual antennas expressed by Equation (6.55) are given as below:

$$\begin{aligned} \text{VA0} \quad h_{c0} &= h_0 + h_1 + h_2 + h_3 \\ \text{VA1} \quad h_{c1} &= h_0 + jh_1 - h_2 + jh_3 \\ \text{VA2} \quad h_{c2} &= h_0 - h_1 + h_2 - h_3 \\ \text{VA3} \quad h_{c3} &= h_0 + jh_1 - h_2 + jh_3. \end{aligned} \tag{6.62}$$

Note that we have dropped the subcarrier index for simplicity. We assume that the channel stays constant over two consecutive subcarriers. We note that half of the modulations symbols transmitted on virtual antennas 0 and 1 experience a channel gain of $\left(h_{c0}^2 + h_{c1}^2\right)$ while the remaining half of the symbols transmitted on virtual antennas 2 and 3 experience a channel gain of $\left(h_{c2}^2 + h_{c3}^2\right)$. The SFBC-PMS scheme also faces the symbol puncturing issue when the antennas are correlated. We demonstrate this effect by considering the case of four perfectly correlated in-phase transmission antennas, that is:

$$h_0^2 = h_1^2 = h_2^2 = h_3^2. \tag{6.63}$$

We note from Equation (6.62) that half of the modulation symbols transmitted on virtual antennas 2 and 3 experiencing a channel gain of $\left(h_{c2}^2 + h_{c3}^2\right)$ will be punctured as the channel gains on both virtual antennas 2 and 3 are nulled out, that is $h_{c2}^2 = h_{c3}^2 = 0$.

Let us now consider a few schemes for 4-Tx antennas that do not create the puncturing problem. The first such scheme is the combined SFBC-FSTD scheme. Similar to the combined SFBC-CDD scheme, the four transmit antennas are grouped into two pairs. The SFBC block code is applied within each pair while the FSTD is used across the pair of antennas as shown in Figure 6.13. Let $y_p(k)$ represent the signal transmitted from the pth antenna on the kth subcarrier, then the transmit matrix for the combined SFBC-FSTD scheme can be written as:

$$\begin{bmatrix} y_0(0) & y_0(1) & y_0(2) & y_0(3) \\ y_1(0) & y_1(1) & y_1(2) & y_1(3) \\ y_2(0) & y_2(1) & y_2(2) & y_2(3) \\ y_3(0) & y_3(1) & y_3(2) & y_3(3) \end{bmatrix} = \begin{bmatrix} x(0) & x(1) & 0 & 0 \\ -x(1)^* & x(0)^* & 0 & 0 \\ 0 & 0 & x(2) & x(3) \\ 0 & 0 & -x(3)^* & x(2)^* \end{bmatrix}. \tag{6.64}$$

We note that the SFBC-FSTD scheme is a full rate code with transmission of four modulation symbols over four subcarriers. The equivalent channel matrix for the SFBC-FSTD scheme $H_{\text{4-SFBC-FSTD}}$ can be written as:

$$H_{\text{4-SFBC-FSTD}} = \frac{1}{\sqrt{4}} \begin{bmatrix} h_0 & -h_1^* & 0 & 0 \\ h_1 & h_0^* & 0 & 0 \\ 0 & 0 & h_2 & -h_3^* \\ 0 & 0 & h_3 & h_2^* \end{bmatrix}. \tag{6.65}$$

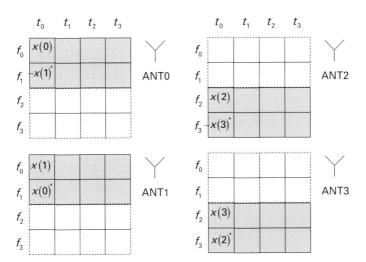

Figure 6.13. SFBC-FSTD transmit diversity schemes for 4-Tx antennas.

Assuming a matched filter receiver, the resulting channel gains matrix can be written as:

$$H^H_{4\text{-SFBC-FSTD}}H_{4\text{-SFBC-FSTD}} = \begin{bmatrix} \dfrac{(h_0^2 + h_1^2)}{2} & 0 & 0 & 0 \\ 0 & \dfrac{(h_0^2 + h_1^2)}{2} & 0 & 0 \\ 0 & 0 & \dfrac{(h_2^2 + h_3^2)}{2} & 0 \\ 0 & 0 & 0 & \dfrac{(h_2^2 + h_3^2)}{2} \end{bmatrix}.$$

(6.66)

We note that half of the modulation symbols experience channel gain of $\frac{(h_0^2 + h_1^2)}{2}$ while the remaining half modulation symbols experience channel gain $\frac{(h_2^2 + h_3^2)}{2}$. The combined STBC-TSTD scheme shown in Figure 6.14 is very similar to the SFBC-FSTD scheme with the subcarrier index in Equation (6.64) replaced with the OFDM symbol (time) index. Another possibility for 4-Tx antennas is to combine the SFBC scheme with the TSTD scheme as shown in Figure 6.15. We note that in the case of the 4-Tx SFBC-FSTD scheme, four modulation symbols are transmitted on four subcarriers in a single OFDM symbol. In the case of the 4-Tx STBC-TSTD scheme, four modulation symbols are transmitted on four OFDM symbols in a single subcarrier. For the SFBC-TSTD scheme, four modulation symbols are transmitted on two subcarriers and two OFDM symbols. Therefore, all the three schemes are full-rate schemes. Assuming that the channel is static on four consecutive subcarriers or four consecutive OFDM symbols, these three schemes also provide similar performance with half of the modulation symbols experiencing channel gain of $\frac{(h_0^2 + h_1^2)}{2}$ while the remaining half modulation symbols experiencing channel gain $\frac{(h_2^2 + h_3^2)}{2}$.

Let us again consider a case of four perfectly correlated transmit antennas, that is:

$$h_0^2 = h_1^2 = h_2^2 = h_3^2. \tag{6.67}$$

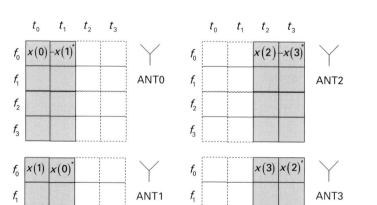

Figure 6.14. STBC-TSTD transmit diversity schemes for 4-Tx antennas.

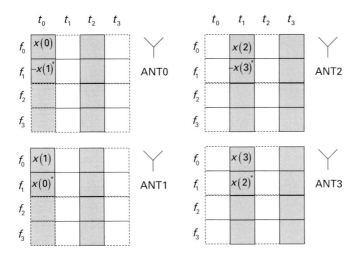

Figure 6.15. SFBC-TSTD transmit diversity schemes for 4-Tx antennas.

From Equation (6.66), we note that there is no symbol-puncturing happening with all modulation symbols experiencing the same channel gain of h_0^2.

The combined SFBC-FSTD, STBC-TSTD and SFBC-TSTD schemes avoid the puncturing problem experienced by the combined SFBC-CDD and SFBC-PMS schemes. However, both STBC-TSTD and SFBC-TSTD schemes result in power inefficiency because only two antennas transmit in a given OFDM symbol as is noted from Figures 6.14 and 6.15. The combined SFBC-FSTD schemes, however, use the full power because the PSD on the transmitted subcarriers can be 3 dB higher. This is because the SFBC-FSTD scheme only transmits on half the subcarriers from a given transmit antenna. The power spectral density for these schemes is shown in Figure 6.16. Note that for the SFBC-CDD and SFBC-PMS schemes, the power spectral density is uniform in frequency and time because all antennas transmit on all subcarriers in all OFDM symbols.

A potential issue with the SFBC-FSTD scheme is that power spectral density in frequency from each transmits antenna is not constant because each antenna transmit over half of the subcarriers. This can generate relatively bursty interference to the neighboring cells. In order to avoid the puncturing problem associated with the SFBC-CDD and SFBC-PMS schemes while being able to transmit from all the antennas in all subcarriers and all OFDM symbols, we can spread the SFBC-FSTD scheme with an orthogonal sequence such as a DFT matrix or a Hadamard matrix.

Let us define A and B as in Equation (6.46):

$$A = \begin{bmatrix} x(0) & x(1) \\ -x(1)^* & x(0)^* \end{bmatrix}$$
$$B = \begin{bmatrix} x(2) & x(3) \\ -x(3)^* & x(2)^* \end{bmatrix}.$$

(6.68)

We will assume a DFT-based spreading of the SFBC block code to create a transmit matrix for four transmit antennas. A DFT matrix is a $N \times N$ square matrix with entries given by:

$$W = e^{j2\pi mn/N} \quad m, n = 0, 1, \ldots, (N-1).$$

(6.69)

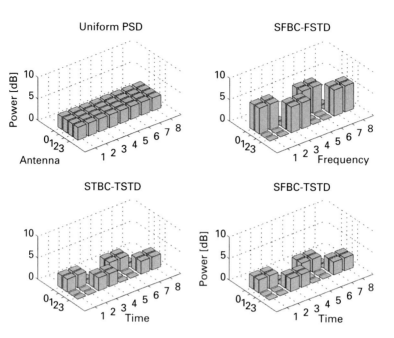

Figure 6.16. Power spectral density for SFBC-FSTD, STBC-TSTD and SFBC-TSTD schemes.

A 2×2 ($N = 2$) DFT matrix can be expressed as:

$$W = \frac{1}{\sqrt{2}} \begin{bmatrix} 1 & 1 \\ 1 & e^{j\pi} \end{bmatrix} = \frac{1}{\sqrt{2}} \begin{bmatrix} 1 & 1 \\ 1 & -1 \end{bmatrix}. \tag{6.70}$$

We can define multiple spreading matrices by introducing a shift parameter (g/G) in the DFT matrix as given by:

$$W_g = e^{j\frac{2\pi m}{N}\left(n + \frac{g}{G}\right)} \quad m, n = 0, 1, \cdots (N - 1). \tag{6.71}$$

We can, for example, define a set of four 2×2 DFT matrices by taking $G = 4$. These four 2×2 matrices with $g = 0, 1, 2, 3$ are given as below:

$$W_0 = \frac{1}{\sqrt{2}} \begin{bmatrix} e^{j0} & e^{j0} \\ e^{j0} & e^{j\pi} \end{bmatrix} = \frac{1}{\sqrt{2}} \begin{bmatrix} 1 & 1 \\ 1 & -1 \end{bmatrix}$$

$$W_1 = \frac{1}{\sqrt{2}} \begin{bmatrix} 1 & 1 \\ e^{j\pi/4} & -e^{j\pi/4} \end{bmatrix} = \frac{1}{\sqrt{2}} \begin{bmatrix} 1 & 1 \\ \frac{1+j}{\sqrt{2}} & \frac{-1-j}{\sqrt{2}} \end{bmatrix}$$

$$W_2 = \frac{1}{\sqrt{2}} \begin{bmatrix} 1 & 1 \\ e^{j\pi/2} & -e^{j\pi/2} \end{bmatrix} = \frac{1}{\sqrt{2}} \begin{bmatrix} 1 & 1 \\ j & -j \end{bmatrix}$$

$$W_3 = \frac{1}{\sqrt{2}} \begin{bmatrix} 1 & 1 \\ e^{j3\pi/4} & -e^{j3\pi/4} \end{bmatrix} = \frac{1}{\sqrt{2}} \begin{bmatrix} 1 & 1 \\ \frac{-1+j}{\sqrt{2}} & \frac{1-j}{\sqrt{2}} \end{bmatrix}. \tag{6.72}$$

We can use any of the above 2×2 DFT matrices to spread the SFBC block code. Let us consider spreading using W_2 as an example. Let $y_p(k)$ represent the signal transmitted from the pth antenna on the kth subcarrier, then the transmit matrix for the spread SFBC-FSTD scheme can be written as:

$$
\begin{bmatrix}
y_0(0) & y_0(1) & y_0(2) & y_0(3) \\
y_1(0) & y_1(1) & y_1(2) & y_1(3) \\
y_2(0) & y_2(1) & y_2(2) & y_2(3) \\
y_3(0) & y_3(1) & y_3(2) & y_3(3)
\end{bmatrix}
= \frac{1}{\sqrt{2}}
\begin{bmatrix}
A & A \\
B & B
\end{bmatrix} \cdot *
\left(W_2 \otimes \begin{bmatrix} 1 & 1 \\ 1 & 1 \end{bmatrix} \right)
$$

$$
= \frac{1}{\sqrt{2}}
\begin{bmatrix}
x(0) & x(1) & x(0) & x(1) \\
-x(1)^* & x(0)^* & -x(1)^* & x(0)^* \\
x(2) & x(3) & x(2) & x(3) \\
-x(3)^* & x(2)^* & -x(3)^* & x(2)^*
\end{bmatrix}
\cdot * \frac{1}{\sqrt{2}}
\begin{bmatrix}
1 & 1 & 1 & 1 \\
1 & 1 & 1 & 1 \\
j & j & -j & -j \\
j & j & -j & -j
\end{bmatrix}
$$

$$
= \frac{1}{\sqrt{4}}
\begin{bmatrix}
x(0) & x(1) & x(0) & x(1) \\
-x(1)^* & x(0)^* & -x(1)^* & x(0)^* \\
jx(2) & jx(3) & -jx(2) & -jx(3) \\
-jx(3)^* & jx(2)^* & jx(3)^* & -jx(2)^*
\end{bmatrix},
\tag{6.73}
$$

where $X \otimes Y$ represents the Kronecker product of matrices X and Y. Also $X \cdot * Y$ represents element-by-element multiplications of matrices X and Y.

The equivalent channel matrix for the spread SFBC-FSTD scheme $H_{\text{4-Spread-SFBC-FSTD}}$ can be written as:

$$
H_{\text{4-Spread-SFBC-FSTD}} =
\begin{bmatrix}
h_0 & -h_1^* & h_0 & -h_1^* \\
h_1 & h_0^* & h_1 & h_0^* \\
jh_2 & -jh_3^* & -jh_2 & jh_3^* \\
jh_3 & jh_2^* & -jh_3 & -jh_2^*
\end{bmatrix}.
\tag{6.74}
$$

Assuming a matched filter receiver, the resulting channel gains matrix can be written as:

$$
H_{\text{4-Spread-SFBC-FSTD}}^{H}
$$
$$
=
\begin{bmatrix}
(h_0^2 + h_1^2) & 0 & 0 & 0 \\
0 & (h_0^2 + h_1^2) & 0 & 0 \\
0 & 0 & (h_2^2 + h_3^2) & 0 \\
0 & 0 & 0 & (h_2^2 + h_3^2)
\end{bmatrix}.
\tag{6.75}
$$

We note that with this spread SFBC-FSTD scheme half of the modulation symbols are transmitted on antennas 0 and 1 and experience a channel gain of $(h_0^2 + h_1^2)$, while the remaining half of the symbols are transmitted on antennas 2 and 3 and experience a channel gain of $(h_2^2 + h_3^2)$. We remark that the diversity performance provided by the spread SFBC-FSTD scheme is the same as the simple SFBC-FSTD scheme. Also unlike SFBC-CDD and SFBC-PMS schemes, the spread scheme does not result in any puncturing of the transmitted modulation symbols when the antennas are correlated. Another benefit of the spread scheme is that transmitted signal power spectral density (PSD) is constant across frequency and time.

However, the spread SFBC-FSTD can suffer from loss of orthogonality when the channel is not constant over the four subcarriers. In order to demonstrate this effect, let us assume that the channel changes between the first pair of subcarriers and the second pair of subcarriers. Furthermore, we assume that the channel is constant over the first pair of subcarriers and also constant over the second pair of subcarriers. Let $\{h_{pi}\}_{i=1}^{2}$ denote the channel on the pth antenna on the ith pair of subcarriers. The equivalent channel matrix for the spread SFBC-FSTD scheme $H_{\text{4-Spread-SFBC-FSTD}}$ can then be written as:

$$
H_{\text{4-Spread-SFBC-FSTD}} =
\begin{bmatrix}
h_{01} & -h_{11}^* & h_{02} & -h_{12}^* \\
h_{11} & h_{01}^* & h_{12} & h_{02}^* \\
jh_{21} & -jh_{31}^* & -jh_{22} & jh_{32}^* \\
jh_{31} & jh_{21}^* & -jh_{32} & -jh_{22}^*
\end{bmatrix}.
$$
(6.76)

Assuming a matched filter receiver, the resulting channel gains matrix can be written as:

$$
=
\begin{bmatrix}
\frac{(h_{01}^2 + h_{11}^2 + h_{02}^2 + h_{12}^2)}{2} & 0 & X & X \\
0 & \frac{(h_{01}^2 + h_{11}^2 + h_{02}^2 + h_{12}^2)}{2} & X & X \\
X & X & \frac{(h_{21}^2 + h_{31}^2 + h_{22}^2 + h_{32}^2)}{2} & 0 \\
X & X & 0 & \frac{(h_{21}^2 + h_{31}^2 + h_{22}^2 + h_{32}^2)}{2}
\end{bmatrix},
$$
(6.77)

where X denotes the non-zero terms due to loss of orthogonality. We note that half of the modulation symbols experience channel gain of $\frac{(h_{01}^2 + h_{11}^2 + h_{02}^2 + h_{12}^2)}{2}$ while the remaining half symbols experience a channel gain of $\frac{(h_{21}^2 + h_{31}^2 + h_{22}^2 + h_{32}^2)}{2}$. The scheme appears to experience a larger diversity than the SFBC-FSTD scheme at the expense of loss of orthogonality.

Now let us see what happens when the channel is not constant over two pairs of subcarriers for the SFBC-FSTD scheme. The equivalent channel matrix in this case is written as:

$$
H_{\text{4-SFBC-FSTD}} =
\begin{bmatrix}
h_{01} & -h_{11}^* & 0 & 0 \\
h_{11} & h_{01}^* & 0 & 0 \\
0 & 0 & h_{22} & -h_{32}^* \\
0 & 0 & h_{32} & h_{22}^*
\end{bmatrix}.
$$
(6.78)

Again, assuming a matched filter receiver, the resulting channel gains matrix can be written as:

$$
H_{\text{4-SFBC-FSTD}}^{H}
=
\begin{bmatrix}
(h_{01}^2 + h_{11}^2) & 0 & 0 & 0 \\
0 & (h_{01}^2 + h_{11}^2) & 0 & 0 \\
0 & 0 & (h_{22}^2 + h_{32}^2) & 0 \\
0 & 0 & 0 & (h_{22}^2 + h_{32}^2)
\end{bmatrix}.
$$
(6.79)

We note that half of the modulation symbols experience channel gain of $(h_{01}^2 + h_{11}^2)$ while the remaining half modulation symbols experience channel gain $(h_{22}^2 + h_{32}^2)$. We note that the diversity performance is unaffected and the SFBC-FSTD scheme stays orthogonal when the channel is not constant over two pairs of subcarriers. However, for the SFBC scheme to be orthogonal, the channel needs to be constant over the two subcarriers within a pair. When the channel is not constant over the two subcarriers within a pair, SFBC schemes also lose orthogonality.

Let $\{h_{pi}\}_{i=1}^2$ denote the channel on the pth antenna on the ith subcarriers. The equivalent channel matrix for the SFBC code H_2 can then be written as:

$$H_2 = \begin{bmatrix} h_{01} & h_{12} \\ -h_{11}^* & h_{02}^* \end{bmatrix}. \tag{6.80}$$

Assuming a matched filter receiver, the resulting channel gains matrix can be written as:

$$H_2^H H_2 = \begin{bmatrix} h_{01}^* & -h_{11} \\ h_{12}^* & h_{02} \end{bmatrix} \begin{bmatrix} h_{01} & h_{12} \\ -h_{11}^* & h_{02}^* \end{bmatrix}$$

$$= \left(h_0^2 + h_1^2\right) \begin{bmatrix} \left(h_{01}^2 + h_{11}^2\right) & \left(h_{01}^* h_{12} - h_{11} h_{02}^*\right) \\ \left(h_{12}^* h_{01} - h_{02} h_{11}^*\right) & \left(h_{02}^2 + h_{12}^2\right) \end{bmatrix}. \tag{6.81}$$

We note that non-diagonal elements show up in the resulting channel gain matrix due to loss of orthogonality. For the SFBC-FSTD scheme, let $\{h_{pi}\}_{i=1}^4$ denote the channel on the pth antenna and the ith subcarriers. The resulting channel gains matrix can then be written as:

$$H_{4\text{-SFBC-FSTD}}^H$$

$$= \begin{bmatrix} \left(h_{01}^2 + h_{11}^2\right) & X & 0 & 0 \\ X & \left(h_{02}^2 + h_{12}^2\right) & 0 & 0 \\ 0 & 0 & \left(h_{23}^2 + h_{33}^2\right) & X \\ 0 & 0 & X & \left(h_{24}^2 + h_{34}^2\right) \end{bmatrix}, \tag{6.82}$$

where X denotes the non-zero terms due to loss of orthogonality. We note that within each pair of SFBC codes, orthogonality is lost. However, the two pairs are still orthogonal because they use orthogonal subcarriers. Now let us see what happens to the spread SFBC-FSTD scheme when the channel is different on the four subcarriers. In this case, we write the resulting channel gain matrix as:

$$= \begin{bmatrix} \dfrac{\left(h_{01}^2 + h_{11}^2 + h_{03}^2 + h_{14}^2\right)}{2} & X & X & X \\ X & \dfrac{\left(h_{11}^2 + h_{02}^2 + h_{13}^2 + h_{04}^2\right)}{2} & X & X \\ X & X & \dfrac{\left(h_{21}^2 + h_{32}^2 + h_{23}^2 + h_{34}^2\right)}{2} & X \\ X & X & X & \dfrac{\left(h_{31}^2 + h_{23}^2 + h_{33}^2 + h_{24}^2\right)}{2} \end{bmatrix}. \tag{6.83}$$

We note that not only does each pair of SFBC codes lose orthogonality but the orthogonality between the two pairs is also lost. This is expected as SFBC code loses orthogonality

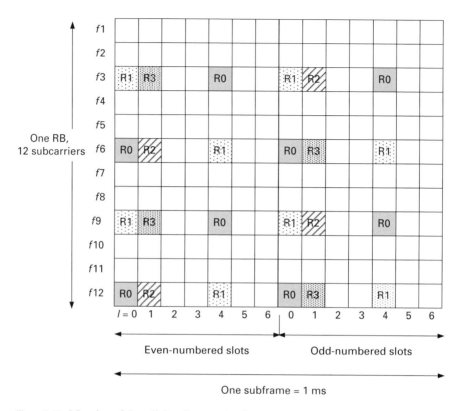

Figure 6.17. Mapping of downlink reference signals.

when the channel changes across two adjacent subcarriers and two pairs of SFBC code lose orthogonality due to the DFT spreading code orthogonality loss. The orthogonality loss problem of the spread SFBC-FSTD scheme is expected to become worse for the modulation symbols transmitted in the OFDM symbol containing the pilot or reference symbols. This is because when four transmit antennas are used such as is the case for the SFBC-FSTD scheme, 6 out of 14 OFDM symbols (out of 12 OFDM symbols for extended cyclic prefix) contain reference symbols as shown in Figure 6.17. The reference symbols use every third subcarrier and therefore two pairs of SFBC codes are further separated by one subcarrier. For the spread SFBC-FSTD scheme, it means that the channel now needs to be constant over five subcarriers which is 75 KHz bandwidth for 15 KHz subcarrier spacing used in the LTE system.

However, another issue we note from Figure 6.17 where R_p represents the reference symbol for antenna p is that time-domain density for R_2 and R_3 is half the density of R_0 and R_1. This results in channel estimates bias with channel estimates on antenna 0 and 1 better than antennas 2 and 3. This motivates the need for channel estimates balancing. A new mapping scheme for SFBC-FSTD where the first pair of modulation symbols $x(0)$, $x(1)$ is mapped to antennas 0 and 2 and a second pair of symbols $x(2)$, $x(3)$ mapped to antennas 1 and 3 as shown in Figure 6.18 can be used. As this scheme balances out the channel estimates effect, we refer to this scheme as the balanced SFBC-FSTD scheme. Let $y_p(k)$ represent the signal transmitted from the pth antenna on the kth subcarrier, then the transmit matrix for the

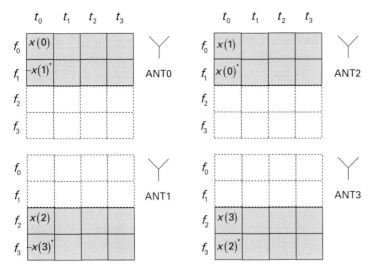

Figure 6.18. Balanced SFBC-FSTD transmit diversity schemes for 4-Tx antennas.

balanced SFBC-FSTD scheme can be written as:

$$
\begin{bmatrix}
y_0(0) & y_0(1) & y_0(2) & y_0(3) \\
y_1(0) & y_1(1) & y_1(2) & y_1(3) \\
y_2(0) & y_2(1) & y_2(2) & y_2(3) \\
y_3(0) & y_3(1) & y_3(2) & y_3(3)
\end{bmatrix}
=
\begin{bmatrix}
x(0) & x(1) & 0 & 0 \\
0 & 0 & x(2) & x(3) \\
-x(1)^* & x(0)^* & 0 & 0 \\
0 & 0 & -x(3)^* & x(2)^*
\end{bmatrix}. \quad (6.84)
$$

The equivalent channel matrix for the balanced SFBC-FSTD scheme $H_{4\text{-Balanced-SFBC-FSTD}}$ can be written as:

$$
H_{4\text{-SFBC-FSTD}} = \frac{1}{\sqrt{4}}
\begin{bmatrix}
h_0 & -h_2^* & 0 & 0 \\
h_2 & h_0^* & 0 & 0 \\
0 & 0 & h_1 & -h_3^* \\
0 & 0 & h_3 & h_1^*
\end{bmatrix}. \quad (6.85)
$$

Assuming a matched filter receiver, the resulting channel gains matrix can be written as:

$$
\begin{bmatrix}
\frac{(h_0^2+h_2^2)}{2} & 0 & 0 & 0 \\
0 & \frac{(h_0^2+h_2^2)}{2} & 0 & 0 \\
0 & 0 & \frac{(h_1^2+h_3^2)}{2} & 0 \\
0 & 0 & 0 & \frac{(h_1^2+h_3^2)}{2}
\end{bmatrix}. \quad (6.86)
$$

We note that half of the modulation symbols experience channel gain of $\frac{(h_0^2+h_2^2)}{2}$ while the remaining half modulation symbols experience channel gain $\frac{(h_1^2+h_3^2)}{2}$. The diversity gain of this scheme is the same as the SFBC-FSTD scheme. The additional benefit is that this scheme balances the effect of imperfect channel estimates as each pair of symbols experiences one good channel estimate and another relatively weak channel estimate. After careful and detailed evaluations of various transmit diversity schemes proposed during the standardization phase, the SFBC and balanced SFBC-FSTD schemes were adopted in the LTE system for two and four transmit antennas respectively.

6.2   Downlink transmission chain

Let us now focus on the details of the 2-Tx antennas SFBC and 4-Tx antennas balanced SFBC-FSTD schemes. In order to be consistent with the structure for MIMO spatial multiplexing, the transmit diversity scheme in the LTE system is defined in terms of layer mapping and transmit diversity precoding. In order to better understand how this structure fits in the bigger scheme of things, we start by looking at the downlink transmission chain shown in Figure 6.19. A turbo coding is first performed on the information codeword. The coded sequence of bits is then scrambled and mapped to complex modulation symbols. In the case of transmission diversity, the complex modulation symbols are mapped to two or four layers for the cases of two-Tx

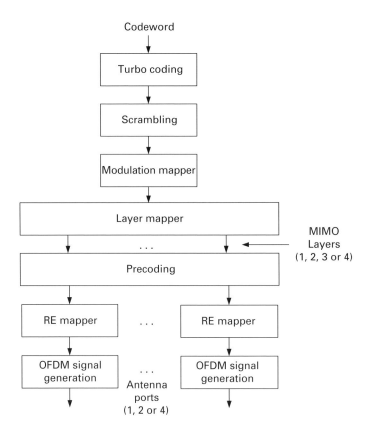

Figure 6.19. Downlink transmission chain.

or four-Tx transmission diversity respectively. We note that the term layer, which generally refers to a stream in MIMO spatial multiplexing, can be confusing when used in the context of transmission diversity. In transmission diversity, a single codeword is transmitted, which is effectively a single rank transmission. After layer mapping, transmission diversity precoding, which is effectively an SFBC block code for 2-Tx antennas and a balanced SFBC-FSTD code for 4-Tx antennas, is applied. The signals after transmission diversity precoding are mapped to time-frequency resources on two or four antennas for the SFBC and balanced SFBC-FSTD cases and OFDM signal generation by use of IFFT takes place. In the following sections, we will only discuss layer mapping and precoding parts that are relevant for transmit diversity discussion.

6.3   Codeword to layer mapping

In the case of transmit diversity transmission, a single codeword is transmitted from two or four antenna ports. The number of layers in the case of transmit diversity is equal to the number of antenna ports. The number of modulation symbols per layer $M_{\text{symb}}^{\text{layer}}$ for 2 and 4 layers is given by:

$$
\begin{aligned}
M_{\text{symb}}^{\text{layer}} &= \frac{M_{\text{symb}}^{(0)}}{2}, \quad \upsilon = 2 \\
M_{\text{symb}}^{\text{layer}} &= \frac{M_{\text{symb}}^{(0)}}{4}, \quad \upsilon = 4,
\end{aligned}
\tag{6.87}
$$

where $M_{\text{symb}}^{(0)}$ represents the total number of modulation symbols within the codeword.

In the case of two antenna ports, the modulation symbols from a single codeword are mapped to 2 ($\upsilon = 2$) layers as below:

$$
\begin{aligned}
x^{(0)}(i) &= d^{(0)}(2i) \\
x^{(1)}(i) &= d^{(0)}(2i+1)
\end{aligned}
\quad i = 0, 1, \ldots \left(M_{\text{symb}}^{\text{layer}} - 1 \right).
\tag{6.88}
$$

In the case of four antenna ports, the modulation symbols from a single codeword are mapped to 4 layers ($\upsilon = 4$) as below:

$$
\begin{aligned}
x^{(0)}(i) &= d^{(0)}(4i) \\
x^{(1)}(i) &= d^{(0)}(4i+1) \\
x^{(2)}(i) &= d^{(0)}(4i+2) \\
x^{(3)}(i) &= d^{(0)}(4i+3)
\end{aligned}
\quad i = 0, 1, \ldots \left(M_{\text{symb}}^{\text{layer}} - 1 \right).
\tag{6.89}
$$

The codeword to layer mapping for two and four antenna ports transmit diversity (TxD) transmissions in the downlink is shown in Figure 6.20. In the case of two antenna ports (two layers), the even-numbered $\left(d^{(0)}(0), d^{(0)}(2), \ldots \right)$ and odd-numbered $\left(d^{(0)}(1), d^{(0)}(3), \ldots \right)$ codeword modulation symbols are mapped to layers 0 and 1 respectively. In the case of four antenna ports 1/4 of the codeword modulation symbols are mapped to a given layer as given by Equation (6.89).

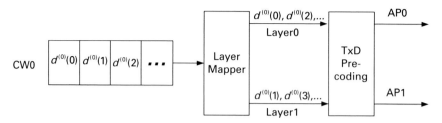

(a) – Two antenna ports TxD layer mapping

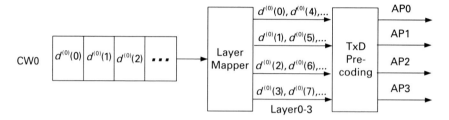

(b) – Four antenna ports TxD layer mapping

Figure 6.20. Codeword to layer mapping for two and four antenna ports transmit diversity (TxD) transmissions in the downlink.

6.4   Transmit diversity precoding

The block of vectors at the output of the layer mapper $x(i) = [x^{(0)}(i) \ldots x^{(\upsilon-1)}(i)]^T$, $i = 0, 1, \ldots, M_{\text{symb}}^{\text{layer}} - 1$ and $(\upsilon = 2, 4)$ is provided as input to the precoding stage as shown in Figure 6.21. The precoding stage then generates another block of vectors $y(i) = [y^{(0)}(i), y^{(1)}(i), \ldots, y^{(P-1)}(i)]^T$, $i = 0, 1, \ldots, M_{\text{symb}}^{\text{layer}} - 1$ and $(P = 2, 4)$. This block of vectors is then mapped onto resources on each of the antenna ports. The symbols at the output of precoding for antenna port p, $y^{(p)}(i)$ are given as:

$$
\begin{bmatrix} y^{(0)}(i) \\ \vdots \\ y^{(P-1)}(i) \end{bmatrix} = W(i) \begin{bmatrix} x^{(0)}(i) \\ \vdots \\ x^{(\upsilon-1)}(i) \end{bmatrix} \qquad P = 2, 4 \quad \upsilon = P. \tag{6.90}
$$

For the case of two antenna ports transmit diversity, the output $y(i) = [y^{(0)}(i) \quad y^{(1)}(i)]^T$ of the precoding operation is written as:

$$
\begin{bmatrix} y^{(0)}(2i) \\ y^{(1)}(2i) \\ y^{(0)}(2i+1) \\ y^{(1)}(2i+1) \end{bmatrix} = \begin{bmatrix} 1 & 0 & j & 0 \\ 0 & -1 & 0 & j \\ 0 & 1 & 0 & j \\ 1 & 0 & -j & 0 \end{bmatrix} \times \begin{bmatrix} x_I^{(0)}(i) \\ x_I^{(1)}(i) \\ x_Q^{(0)}(i) \\ x_Q^{(1)}(i) \end{bmatrix} \qquad i = 0, 1, \ldots, M_{\text{symb}}^{\text{layer}} - 1,
$$

$$
\tag{6.91}
$$

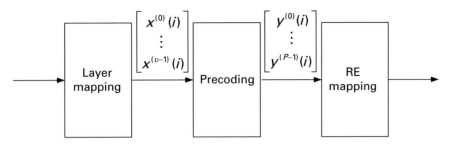

Figure 6.21. Layer mapping, precoding and RE mapping stages.

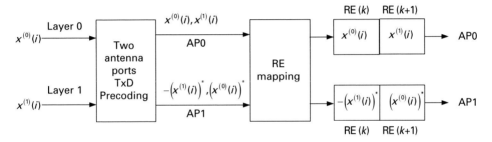

Figure 6.22. Transmit diversity precoding and RE mapping for two antenna ports.

where $x_I^{(0)}(i)$ and $x_Q^{(0)}(i)$ are real and imaginary parts of the moduation symbol on layer 0 and $x_I^{(1)}(i)$ and $x_Q^{(1)}(i)$ are real and imaginary parts of the moduation symbol on layer 1.

$$\begin{bmatrix} y^{(0)}(2i) \\ y^{(1)}(2i) \\ y^{(0)}(2i+1) \\ y^{(1)}(2i+1) \end{bmatrix} = \begin{bmatrix} x_I^{(0)}(i) + jx_Q^{(0)}(i) \\ -x_I^{(1)}(i) + jx_Q^{(1)}(i) \\ x_I^{(1)}(i) + jx_Q^{(1)}(i) \\ x_I^{(0)}(i) - jx_Q^{(0)}(i) \end{bmatrix} = \begin{bmatrix} x^{(0)}(i) \\ -\left(x^{(1)}(i)\right)^* \\ x^{(1)}(i) \\ \left(x^{(0)}(i)\right)^* \end{bmatrix} \quad i = 0, 1, ..., M_{symb}^{layer} - 1.$$

(6.92)

We note that the number of modulation symbols for mapping to resource elements is two times the number of modulation symbols per layer, that is $M_{symb}^{map} = 2 \times M_{symb}^{layer}$. The transmit diversity precoding and RE mapping for two antenna ports is shown in Figure 6.22. We note that the precoding and RE mapping operations result in a space frequency block coding (SFBC) scheme.

For the case of four antenna ports transmit diversity, the output $y(i) = [y^{(0)}(i) \quad y^{(1)}(i) \quad y^{(2)}(i) \quad y^{(3)}(i)]^T$ of the precoding operation is written as:

$$
\begin{bmatrix}
y^{(0)}(4i) \\
y^{(1)}(4i) \\
y^{(2)}(4i) \\
y^{(3)}(4i) \\
y^{(0)}(4i+1) \\
y^{(1)}(4i+1) \\
y^{(2)}(4i+1) \\
y^{(3)}(4i+1) \\
y^{(0)}(4i+2) \\
y^{(1)}(4i+2) \\
y^{(2)}(4i+2) \\
y^{(3)}(4i+2) \\
y^{(0)}(4i+3) \\
y^{(1)}(4i+3) \\
y^{(2)}(4i+3) \\
y^{(3)}(4i+3)
\end{bmatrix}
=
\begin{bmatrix}
1 & 0 & 0 & 0 & j & 0 & 0 & 0 \\
0 & 0 & 0 & 0 & 0 & 0 & 0 & 0 \\
0 & -1 & 0 & 0 & 0 & j & 0 & 0 \\
0 & 0 & 0 & 0 & 0 & 0 & 0 & 0 \\
0 & 1 & 0 & 0 & 0 & j & 0 & 0 \\
0 & 0 & 0 & 0 & 0 & 0 & 0 & 0 \\
1 & 0 & 0 & 0 & -j & 0 & 0 & 0 \\
0 & 0 & 0 & 0 & 0 & 0 & 0 & 0 \\
0 & 0 & 0 & 0 & 0 & 0 & 0 & 0 \\
0 & 0 & 1 & 0 & 0 & 0 & j & 0 \\
0 & 0 & 0 & 0 & 0 & 0 & 0 & 0 \\
0 & 0 & 0 & -1 & 0 & 0 & 0 & j \\
0 & 0 & 0 & 0 & 0 & 0 & 0 & 0 \\
0 & 0 & 0 & 1 & 0 & 0 & 0 & j \\
0 & 0 & 0 & 0 & 0 & 0 & 0 & 0 \\
0 & 0 & 1 & 0 & 0 & 0 & -j & 0
\end{bmatrix}
\times
\begin{bmatrix}
x_I^{(0)}(i) \\
x_I^{(1)}(i) \\
x_I^{(2)}(i) \\
x_I^{(3)}(i) \\
x_Q^{(0)}(i) \\
x_Q^{(1)}(i) \\
x_Q^{(2)}(i) \\
x_Q^{(3)}(i)
\end{bmatrix}
$$

$$
i = 0, 1, ..., M_{\text{symb}}^{\text{layer}} - 1. \tag{6.93}
$$

$$
\begin{bmatrix}
y^{(0)}(4i) \\
y^{(1)}(4i) \\
y^{(2)}(4i) \\
y^{(3)}(4i) \\
y^{(0)}(4i+1) \\
y^{(1)}(4i+1) \\
y^{(2)}(4i+1) \\
y^{(3)}(4i+1) \\
y^{(0)}(4i+2) \\
y^{(1)}(4i+2) \\
y^{(2)}(4i+2) \\
y^{(3)}(4i+2) \\
y^{(0)}(4i+3) \\
y^{(1)}(4i+3) \\
y^{(2)}(4i+3) \\
y^{(3)}(4i+3)
\end{bmatrix}
=
\begin{bmatrix}
x_I^{(0)}(i) + jx_Q^{(0)}(i) \\
0 \\
-x_I^{(1)}(i) + jx_Q^{(1)}(i) \\
0 \\
x_I^{(1)}(i) + jx_Q^{(1)}(i) \\
0 \\
x_I^{(0)}(i) - jx_Q^{(0)}(i) \\
0 \\
0 \\
x_I^{(2)}(i) + jx_Q^{(2)}(i) \\
0 \\
-x_I^{(3)}(i) + jx_Q^{(3)}(i) \\
0 \\
x_I^{(3)}(i) + jx_Q^{(3)}(i) \\
0 \\
x_I^{(2)}(i) - jx_Q^{(2)}(i)
\end{bmatrix}
=
\begin{bmatrix}
x^{(0)}(i) \\
0 \\
-\left(x^{(1)}(i)\right)^* \\
0 \\
x^{(1)}(i) \\
0 \\
\left(x^{(0)}(i)\right)^* \\
0 \\
0 \\
x^{(2)}(i) \\
0 \\
-\left(x^{(3)}(i)\right)^* \\
0 \\
x^{(3)}(i) \\
0 \\
\left(x^{(2)}(i)\right)^*
\end{bmatrix}
\quad i = 0, 1, ..., M_{\text{symb}}^{\text{layer}} - 1.
$$

$$
\tag{6.94}
$$

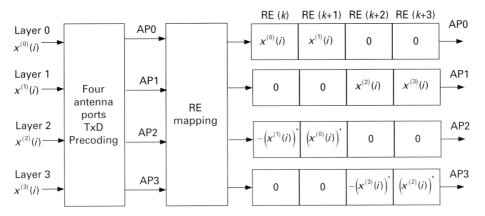

Figure 6.23. Transmit diversity precoding and RE mapping for four antenna ports.

We note that the number of modulation symbols for mapping to resource elements is four times the number of modulation symbols per layer, that is $M_{symb}^{map} = 4 \times M_{symb}^{layer}$. The transmit diversity precoding and RE mapping for four antenna ports is shown in Figure 6.23. We note that the four antenna ports precoding and RE mapping operations results in a balanced SFBC-FSTD scheme as is also illustrated by an alternative representation below:

$$
\begin{bmatrix}
y^{(0)}(4i) & y^{(0)}(4i+1) & y^{(0)}(4i+2) & y^{(0)}(4i+3) \\
y^{(1)}(4i) & y^{(1)}(4i+1) & y^{(1)}(4i+2) & y^{(1)}(4i+3) \\
y^{(2)}(4i) & y^{(2)}(4i+1) & y^{(2)}(4i+2) & y^{(2)}(4i+3) \\
y^{(3)}(4i) & y^{(3)}(4i+1) & y^{(3)}(4i+2) & y^{(3)}(4i+3)
\end{bmatrix}
$$
$$
=
\begin{bmatrix}
x^{(0)}(i) & x^{(1)}(i) & 0 & 0 \\
0 & 0 & x^{(2)}(i) & x^{(3)}(i) \\
-\left(x^{(1)}(i)\right)^* & \left(x^{(0)}(i)\right)^* & 0 & 0 \\
0 & 0 & -\left(x^{(3)}(i)\right)^* & \left(x^{(2)}(i)\right)^*
\end{bmatrix}.
$$
(6.95)

6.5   Summary

Various diversity sources are available in the LTE system to average out the channel variations in a multi-path fading environment. This includes time diversity, frequency diversity, receive diversity and transmit diversity. Various forms of transmit diversity schemes were studied and evaluated in detail during the LTE standardization phase. In the beginning, the cyclic delay diversity scheme was considered to be a strong candidate due to its advantages of simplicity and scalability with the number of transmission antennas. However, one drawback of CDD became apparent in correlated channels. With perfectly correlated antennas, there is no diversity available. However, CDD results in performance degradation because of modulation symbol puncturing due to destructive combining of signals from multiple transmit antennas.

The other strong candidates were schemes based on block codes such as Alamouti block code. A drawback of this approach was that it does not scale with the number of transmission antennas. For more than two antennas, some possibilities are the use of either non-orthogonal block codes or block codes with a rate of less than 1. In either case, the performance is penalized. After long debates over the benefits and drawbacks of CDD and block-code-based schemes, the decision went in favor of a block-code-based transmission diversity. Once the decision was in favor of block codes, it was straightforward to adopt SFBC code for 2-Tx antennas.

Another round of debates took place for selecting the transmit diversity scheme for the case of four transmit antennas. Various schemes such as non-orthogonal block codes and combinations of SFBC with other schemes such as CDD, PVS/PMS, FSTD and TSTD were considered. The non-orthogonal block codes were eliminated due to performance and receiver complexity issues. The problems with combined SFBC-CDD and SFBC-PMS schemes were poor performance in correlated channels due to the modulation symbol puncturing issue. The puncturing issue is less pronounced in the combined SFBC-CDD or SFBC-PMS schemes than in the pure CDD and PVS schemes but it was still a concern. On the other hand schemes that use the TSTD component face the problem of transmit power inefficiency because only a subset of antennas performs transmission in a given OFDM symbol. The schemes with FSTD component do not have the transmit power penalty problem because power can be shifted in frequency from the unused subcarriers on a given antenna to the used subcarriers (note that in case of SFBC-FSTD, each antenna transmits on half the subcarriers only). However, for the TSTD schemes, as the switching happens in time, the power cannot be shifted in time from one OFDM symbol to the other. In the light of these considerations, it was decided to adopt the SFBC-FSTD scheme for the case of four transmit antennas.

Another issue arose for the SFBC-FSTD scheme because each pair of modulations symbols is mapped to two transmit antennas and the reference signals density on antennas 0 and 1 is twice as large as on antennas 2 and 3. This leads to channel estimation biased towards the first pair of symbols in the group. In order to balance out the impact of the imperfect channel estimate, a balanced SFBC-FSTD scheme that maps the first pair of modulation symbols to antenna 0 and 2 and the second pair of modulation symbols to antenna 1 and 3 was adopted.

Another issue that we didn't discuss here was inter-cell interference characteristics of various transmit diversity schemes considered. It was claimed that when inter-cell interference suppression (using, for example, an MMSE receiver) is employed a CDD interferer could be suppressed with a lesser degree of freedom than a block code interferer. This is because a CDD interferer appears as a single rank interferer while an SFBC block code interferer appears as a rank 2 interferer. It was noted that the same receiver degree of freedom is also sufficient to suppress the block code interferer albeit with larger receiver complexity.

References

[1] Alamouti, S. M., "A simple transmit diversity technique for wireless communications," *IEEE Journal on Selected Areas in Communications*, vol. SAC-16, no. 8, pp. 1451–1458, Oct. 1998.

[2] Tirkkonen, O., Boariu, A. and Hottinen, A., "Minimal non-orthogonality rate 1 space-time block code for 3+ Tx antennas," *Proceedings of IEEE Sixth International Symposium on Spread Spectrum Techniques and Applications*, pp. 429–432, Sep. 2000.

[3] Ganesan, G. and Stoica, P., "Space–time block codes: a maximum SNR approach," *IEEE Transactions on Information Theory*, vol. IT-47, no. 4, pp. 1650–1656, May 2001.

7 MIMO spatial multiplexing

In the previous chapter, we discussed how multiple transmission antennas can be used to achieve the diversity gain. The transmission diversity allows us to improve the link performance when the channel quality cannot be tracked at the transmitter which is the case for high mobility UEs. The transmission diversity is also useful for delay-sensitive services that cannot afford the delays introduced by channel-sensitive scheduling. The transmission diversity, however, does not help in improving the peak data rates as a single data stream is always transmitted. The multiple transmission antennas at the eNB in combination with multiple receiver antennas at the UE can be used to achieve higher peak data rates by enabling multiple data stream transmissions between the eNB and the UE by using MIMO (multiple input multiple output) spatial multiplexing. Therefore, in addition to larger bandwidths and high-order modulations, MIMO spatial multiplexing is used in the LTE system to achieve the peak data rate targets. The MIMO spatial multiplexing also provides improvement in cell capacity and throughput as UEs with good channel conditions can benefit from multiple streams transmissions. Similarly, the weak UEs in the system benefit from beam-forming gains provided by precoding signals transmitted from multiple transmission antennas.

7.1 MIMO capacity

A MIMO channel consists of channel gains and phase information for links from each of the transmission antennas to each of the receive antennas as shown in Figure 7.1. Therefore, the channel for the $M \times N$ MIMO system consists of an $N \times M$ matrix $H_{N \times M}$ given as:

$$
H = \begin{bmatrix}
h_{11} & h_{12} & \cdots & h_{1M} \\
h_{21} & h_{22} & \cdots & h_{2M} \\
\vdots & \vdots & \cdots & \vdots \\
h_{N1} & h_{M2} & \cdots & h_{NM}
\end{bmatrix},
\tag{7.1}
$$

where h_{ij} represents the channel gain from transmission antenna j to the receive antenna i. In order to enable the estimations of the elements of the MIMO channel matrix, separate reference signals or pilots are transmitted from each of the transmission antennas.

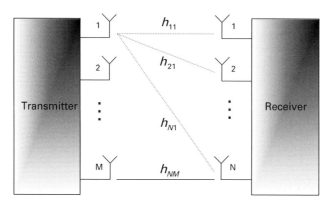

Figure 7.1. An $(M \times N)$ MIMO system.

The capacity of an $M \times N$ MIMO channel can be written as [1]:

$$C_{\text{MIMO}} = E \left[\log_2 \det \left(I_N + \frac{\rho}{M} HH^* \right) \right] \ [\text{b/s/Hz}], \qquad (7.2)$$

where $\rho = \frac{P}{N_0}$ is the received signal-to-noise ratio at each receive antenna.

If $\lambda_1 \geq \lambda_2 \geq \cdots \geq \lambda_{\min}$ are the (random) ordered singular values of the channel matrix H, then we can express (7.2) as:

$$C_{\text{MIMO}} = E \left[\sum_{i=1}^{n_{\min}} \log_2 \left(1 + \frac{\rho}{M} \lambda_i^2 \right) \right]$$

$$= \sum_{i=1}^{n_{\min}} E \left[\log_2 \left(1 + \frac{\rho}{M} \lambda_i^2 \right) \right] \ [\text{b/s/Hz}], \qquad (7.3)$$

where $n_{\min} = \min (M, N)$. Let us first look at the case of low SNR where we can use the approximation $\log_2 (1 + x) \approx x . \log_2 e$

$$C_{\text{MIMO}} \approx \sum_{i=1}^{n_{\min}} \frac{\rho}{M} E \left[(\lambda_i^2) \right] \log_2 e$$

$$= \frac{\rho}{M} E \left[Tr \left[HH^* \right] \right] \log_2 e$$

$$= \frac{\rho}{M} E \left[\sum_{i,j} |h_{ij}|^2 \right] \log_2 e \qquad (7.4)$$

$$= M \times \rho \times \log_2 e \ [\text{b/s/Hz}].$$

It can be noted that at low SNR, an $M \times M$ system yields a power gain of $10 \times \log_{10} (M)$ dBs relative to a single-receiver antenna case. This is because the M receive antennas can coherently combine their received signals to get a power boost.

For the high-SNR case where we can use the approximation $\log_2 (1 + x) \approx \log_2 (x)$, the MIMO capacity formula can be expressed as:

$$
\begin{aligned}
C_{\text{MIMO}} &= \sum_{i=1}^{M} E\left[\log_2 \left(\frac{\rho}{M}\lambda_i^2\right)\right] \\
&= \sum_{i=1}^{M} E\left[\log_2 \left(\frac{\rho}{M}\right) + \log_2 \left(\lambda_i^2\right)\right] \\
&= n_{\min} \times \log_2 \left(\frac{\rho}{M}\right) + \sum_{i=1}^{n_{\min}} E\left[\log_2 \left(\lambda_i^2\right)\right] \text{ [b/s/Hz],}
\end{aligned}
\tag{7.5}
$$

and $E\left[\log_2 \left(\lambda_i^2\right)\right] > -\infty$ for all i. We note that the full $n_{\min}$ degree of freedom can be obtained at high SNR. It can also be noted that maximum capacity is achieved when all the singular values are equal. Therefore, we would expect a high capacity gain when the channel matrix H is sufficiently random and statistically well conditioned. We remark that the number of degrees of freedom is limited by the minimum of the number of transmission and the number of receive antennas. Therefore, a large number of transmission and receive antennas are required in order to get the full benefit from MIMO.

We observed from (7.4) that the capacity of a MIMO system scales linearly with the number of receive antennas at low SNR. We also noted from (7.5) that at high SNR, the capacity scales linearly with $n_{\min}$. Therefore, we can say that at all SNRs, the capacity of an $M \times N$ MIMO system scales linearly with $n_{\min}$. We note, however, that the channel matrix needs to be full rank in order to provide $n_{\min}$ degrees of freedom. In situations where the channel matrix is not full rank due to, for example, correlated antennas or line-of-sight (LOS) propagation, the degrees of freedom can be further limited.

7.2   Codewords and layer mapping

A MIMO transmission chain showing codewords and layers is depicted in Figure 7.2. The maximum number of layers or streams supported υ is equal to the degrees of freedom $n_{\min}$ provided by the MIMO channel. The number of MIMO layers is also referred to as the MIMO rank. For P transmission antenna ports, a rank smaller than $P \geq n_{\min}$ is supported by selecting a subset of the columns of the $P \times P$ precoding matrix. A MIMO codeword is a separately coded and modulated information block that is transmitted on one or more MIMO layers.

7.2.1   Single codeword versus multi-codeword

Both single-codeword (SCW) and multi-codeword (MCW) MIMO schemes were considered for the LTE system. In the case of single-codeword MIMO transmission, a CRC is added to a single information block and then coding and modulation is performed. The coded and modulated symbols are then de-multiplexed for transmission over multiple antennas. In the case of multiple codeword MIMO transmission, the information block is de-multiplexed into smaller information blocks. Individual CRCs are attached to these smaller information blocks and then separate coding and modulation is performed on these smaller blocks. It should be noted that in the case of multi-codeword MIMO transmissions, different modulation and

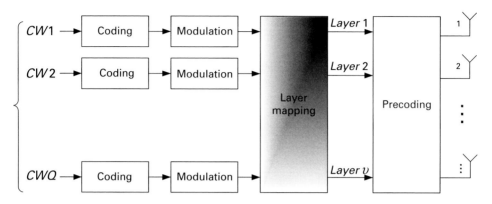

Figure 7.2. MIMO transmit chain showing codewords and layers.

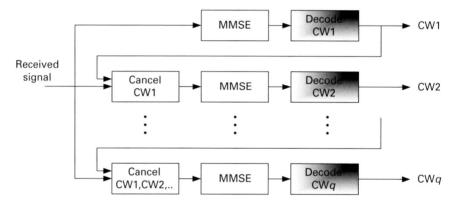

Figure 7.3. MMSE-SIC receiver.

coding can be used on each of the individual streams resulting in a so-called MIMO PARC (per antenna rate control) scheme. In addition, multi-codeword transmission allows for more efficient post-decoding interference cancellation using a MMSE-SIC receiver as depicted in Figure 7.3. This is because a CRC check can be performed on each of the codewords before a codeword is cancelled from the overall received signal. In this way, only correctly received codewords are cancelled avoiding any interference propagation in the cancellation process.

An SCW scheme using an ML (maximum likelihood) receiver can also provide equivalent performance as a MCW scheme with a successive interference cancellation (SIC) receiver. The performance and complexity tradeoffs of SCW and MCW schemes were discussed in detail during the standardization phase. In general, a MMSE-SIC receiver requires a larger buffering because the cancellation process cannot start unless the first codeword is demodulated and decoded correctly. In contrast, the ML receiver demodulates all MIMO layers simultaneously at the expense of additional processing or logic complexity. The buffering required for a MMSE-SIC receiver can be reduced by employing a turbo-codeblock-based successive interference cancellation as discussed in Section 11.3.1.

In terms of signaling overhead, a SCW scheme has an advantage as a single hybrid ARQ ACK/NACK and a single CQI is required to be fed back from the UE. Similarly, only one set of modulation and coding scheme and hybrid ARQ information needs to be signaled on the downlink. In the case of the MCW scheme, hybrid ARQ operation is performed on each codeword requiring larger signaling overhead on both the uplink and the downlink. When different modulation and coding schemes are used as in a MIMO PARC scheme, CQI needs to be provided for each codeword separately. Similarly, the transport block size and MCS needs to be signaled separately on the downlink.

An SCW scheme also has a diversity advantage over an MCW scheme as a single codeword is transmitted over all the MIMO layers. In the case of the MCW scheme, some of the codewords may be successful while others may be in error due to channel quality fluctuations across the MIMO layers. In order to provide diversity in an MCW scheme, large-delay CDD is introduced which enables transmission of each codeword over all the available MIMO layers as discussed in Section 7.5.2.

After carefully evaluating the receiver complexity and signaling overhead aspects of SCW and MCW schemes, the conclusion was to employ a sort of hybrid of single-codeword and multi-codeword schemes where the maximum number of codewords is limited to two.

7.2.2   Codewords to layer mapping

In the case of an SCW scheme, the modulation symbols from a single codeword are mapped to all the MIMO layers and hence the issue of codeword to layer mapping does not arise. In the case of a pure MCW scheme where the number of codewords is equal to the number of MIMO layers, a one-to-one mapping between codewords and layers can be used in a straightforward manner. However, when the number of codewords is smaller than the MIMO layers as is the case in the LTE system for four antenna ports, the codewords to layer mapping needs to be considered carefully.

A maximum of two codewords is supported in the LTE system for rank-2 and greater transmissions. In the case of rank-2, the codeword to layer mapping is straightforward as the number of codewords is equal to the number of MIMO layers. Therefore, CW1 is mapped to layer 1, while CW2 is mapped to layer 2. In the case of rank-3, the number of layers is one more than the number of codewords. Therefore, one codeword needs to be transmitted on one layer while the other codeword needs to be transmitted on two layers. The two possibilities in this case are 1–2 and 2–1 mappings. In the 1–2 mapping, CW1 is transmitted on layer 1, while CW2 is transmitted on layers 2 and 3. In the 2–1 mapping, CW1 is transmitted on layers 1 and 2, while CW2 is transmitted on layer 3. Under the assumption that the three layers are statistically equivalent (no layer ordering, etc.) and that there is no predetermined interference cancellation order the 1–2 and 2–1 mapping schemes are equivalent.

The issue of codeword to layer mapping is more involved for the case of four MIMO layers ($v = 4$). The two schemes that were extensively debated and evaluated are 1–3 mapping and 2–2 mapping. In the 1–3 mapping, CW1 is transmitted on layer 1 while CW2 is transmitted on layers 2, 3 and 4. In the symmetric 2–2 mapping, CW1 is transmitted on layers 1 and 2, while CW2 is transmitted on layers 3 and 4. The SIC receiver structures for these codewords to layer mapping schemes are shown in Figure 7.4. We note that the 1–3 receiver will need a smaller buffer size because it can discard the received signal sooner compared to the 2–2 receiver. This is because it takes twice as long to decode two layers compared to decoding a single layer and therefore the complex received symbols can be discarded quicker, leading to a smaller required buffer memory. If a turbo decoding were completely parallel then there would be no difference in complexity. However, given that all the layers can carry the maximum data rate and that the turbo decoder has a limited amount of parallel processing determined by

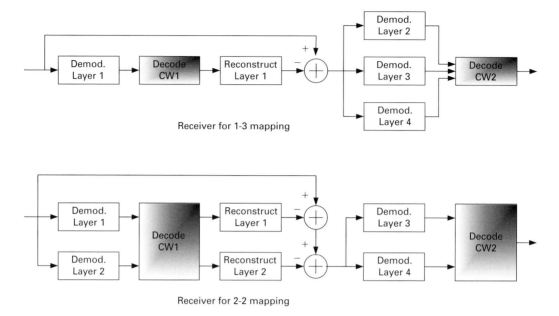

Figure 7.4. Codewords to layer mapping schemes for four MIMO layers ($v = 4$).

the number of parallel decoder circuits, the time required to decode a single layer is smaller than the time required to decode two layers. Moreover, we note that 2–2 receiver requires an additional operation of layer 2 construction as well as two SIC subtractions instead of one in the case of 1–3 mapping. In the 1–3 mapping, UE always cancels the first layer just like in the rank-3 (1–2 mapping) and rank-2 (1–1 mapping) case and thus result in a simpler UE operation irrespective of the MMO rank or configuration.

Another benefit of 1–3 mapping is that a correct channel quality is always available for layer 1. This is because channel quality information (CQI) is provided per codeword. Therefore, in the case of rank override, eNB can always transmit a single codeword on a single Layer 1 for which accurate channel quality is available. In case of 2–2 mapping, CQI is available for each pair of layers (averaged across the two layers) and therefore if eNB decides to perform a single layer transmission, the performance may be affected due to inaccurate CQI. The rank override happens, for example, when eNB does not have sufficient data in its buffers to perform full rank transmission.

An advantage of 2–2 mapping over 1–3 mapping is that each codeword experiences two layers of diversity making the codeword transmission more robust to layer channel quality fluctuations. The performance of 1–3 and 2–2 mapping schemes was evaluated extensively. Both 1–3 and 2–2 mapping schemes provide equivalent performance. In the 1–3 mapping scheme, a single layer is cancelled by the SIC receiver and the remaining three layers benefit from the cancelled interference. Therefore, the amount of interference cancelled is smaller but more layers benefit from the cancellation. In the 2–2 mapping scheme, two layers' worth of interference is cancelled but only two layers benefit from the cancelled interference. Therefore, the amount of cancelled interference is more in this case but the number of layers that benefit from the cancelled interference is smaller than the 1–3 mapping case. These two effects compensate each other leading to similar performance between the 1–3 and 2–2 mapping schemes.

Based on the complexity and performance comparison, it was generally agreed that both schemes could be equally good for the LTE system. However, the final decision was to employ the 2–2 mapping scheme for two codewords mapping in the case of four layers transmission.

7.3   Downlink MIMO transmission chain

The LTE system supports transmission of a maximum of two codewords in the downlink. Each codeword is separately coded using turbo coding and the coded bits from each codeword are scrambled separately as shown in Figure 7.5.

The complex-valued modulation symbols for each of the codewords to be transmitted are mapped onto one or multiple layers. The complex-valued modulation symbols $d^{(q)}(0), ..., d^{(q)}(M_{\text{symb}}^{(q)} - 1)$ for code word q are mapped onto the layers $x(i) = \begin{bmatrix} x^{(0)}(i) & ... & x^{(\upsilon-1)}(i) \end{bmatrix}^T$, $i = 0, 1, ..., M_{\text{symb}}^{\text{layer}} - 1$, where υ is the number of layers and $M_{\text{symb}}^{\text{layer}}$ is the number of modulation symbols per layer. The codeword to layer mapping

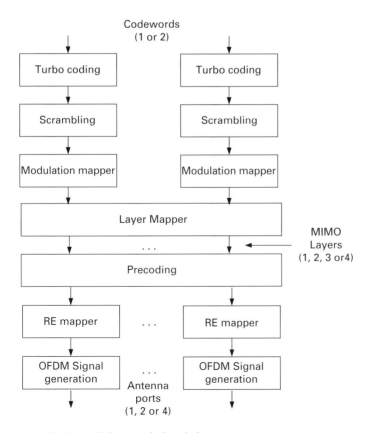

Figure 7.5. Downlink transmission chain.

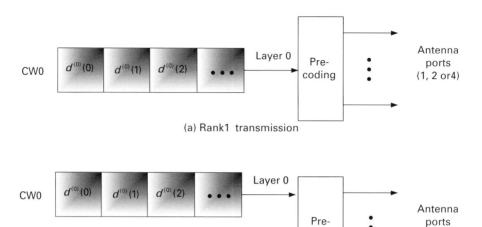

Figure 7.6. Codeword to layer mapping for rank-1 and rank-2 transmissions in the downlink.

for rank-1 and rank-2 transmissions is shown in Figure 7.6. A rank-1 transmission can happen for the case of one, two or four antenna ports while for rank-2 transmission, the number of antenna ports needs to be at least 2. In the case of rank-1 transmission, the complex-valued modulation symbols $d^{(q)}(0), ..., d^{(q)}(M_{\text{symb}}^{(q)} - 1)$ from a single codeword ($q = 0$) are mapped to a single layer ($\upsilon = 0$) as below:

$$x^{(0)}(i) = d^{(0)}(i) \quad M_{\text{symb}}^{\text{layer}} = M_{\text{symb}}^{(0)}. \tag{7.6}$$

Also the number of modulation symbols per layer $M_{\text{symb}}^{\text{layer}}$ is equal to the number of modulation symbols per codeword $M_{\text{symb}}^{(0)}$. It can be noted that for rank-1 transmission, the layer mapping operation is transparent with codeword modulation symbols simply mapped to a single layer.

In the case of rank-2 transmissions, which can happen for both two and four antenna ports, the modulation symbols from the two codewords with ($q = 0, 1$) are mapped to 2 layers ($\upsilon = 0, 1$) as below:

$$\begin{aligned} x^{(0)}(i) &= d^{(0)}(i) \\ x^{(1)}(i) &= d^{(1)}(i) \end{aligned} \quad M_{\text{symb}}^{\text{layer}} = M_{\text{symb}}^{(0)} = M_{\text{symb}}^{(1)}. \tag{7.7}$$

We note that for rank-2 transmission, the codeword to layer mapping is an MCW scheme with two codewords mapped to two layers separately.

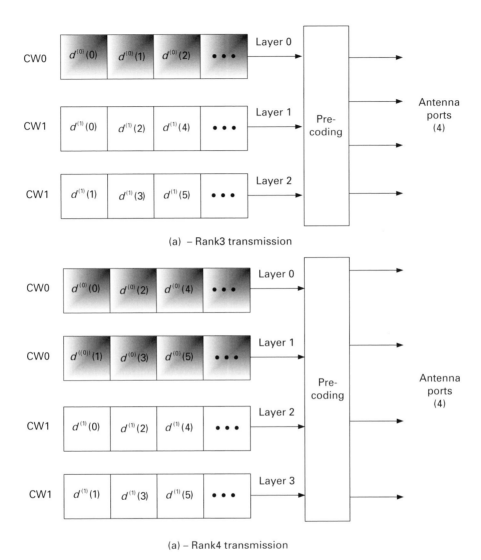

Figure 7.7. Codeword to layer mapping for rank-3 and rank-4 transmissions in the downlink.

The codeword to layer mapping for rank-3 and rank-4 transmissions is shown in Figure 7.7. The rank-3 and rank-4 transmissions can happen for 4 antenna ports. In the case of rank-3 transmission, the modulation symbols from the 2 codewords with $(q = 0, 1)$ are mapped to three layers $(\upsilon = 0, 1, 2)$ as below:

$$x^{(0)}(i) = d^{(0)}(i)$$

$$x^{(1)}(i) = d^{(1)}(2i) \qquad M_{\text{symb}}^{\text{layer}} = M_{\text{symb}}^{(0)} = M_{\text{symb}}^{(1)}\big/2. \qquad (7.8)$$

$$x^{(2)}(i) = d^{(1)}(2i + 1)$$

In the case of rank-4 transmission, the modulation symbols from the 2 codewords with $(q = 0, 1)$ are mapped to 4 layers ($v = 0, 1, 2, 3$) as below:

$$x^{(0)}(i) = d^{(0)}(2i)$$

$$x^{(1)}(i) = d^{(0)}(2i + 1) \qquad M_{symb}^{layer} = M_{symb}^{(0)}/2 = M_{symb}^{(1)}/2. \qquad (7.9)$$

$$x^{(2)}(i) = d^{(1)}(2i)$$

$$x^{(3)}(i) = d^{(1)}(2i + 1)$$

We remark that for rank-3 and rank-4 transmissions, the codeword to layer mapping results in a hybrid of SCW and MCW schemes. In the case of rank-3, the first codeword is mapped to the first layer while the second codeword is mapped to the second and third layers. In the case of rank-4, both the first and the second codewords are mapped to two layers with the first codeword mapped to the first and second layers while the second codeword is mapped to the third and the fourth layers as shown in Figure 7.7.

7.4   MIMO precoding

It is well known that the performance of a MIMO system can be improved with channel knowledge at the transmitter. The channel knowledge at the transmitter does not help to improve the degrees of freedom but power or beam-forming gain is possible [1]. In a TDD system, the channel knowledge can be obtained at the eNB by uplink transmissions thanks to channel reciprocity. However, the sounding signals needs to be transmitted on the uplink, which represents an additional overhead. In an FDD system, the channel state information needs to be fed back from the UE to the eNB. The complete channel state feedback can lead to excessive feedback overhead. For example in a 4×4 MIMO channel, a total of 16 complex channel gains from each of the transmission antennas to each of the receive antennas need to be signaled. An approach to reduce the channel state information feedback overhead is to use a codebook.

In a closed-loop MIMO precoding system, for each transmission antenna configuration, we can construct a set of precoding matrices and let this set be known at both the eNB and the UE. This set of matrices is referred to as MIMO codebook and denoted by $P = \{P_1, P_2, \ldots, P_L\}$. Here $L = 2^r$ denotes the size of the codebook and r is the number of (feedback) bits needed to index the codebook. Once the codebook is specified for a MIMO system, the receiver observes a channel realization, selects the best precoding matrix to be used at the moment, and feeds back the precoding matrix index (PMI) to the transmitter as depicted in Figure 7.8.

7.4.1   Precoding for two antenna ports

The two antenna ports precoding consists of a combination of a 2×2 identity matrix and a discrete fourier transform (DFT) based precoding. A Fourier matrix is an $N \times N$ square matrix with entries given by:

$$W = e^{j2\pi mn/N} \quad m, n = 0, 1, \ldots, (N - 1). \qquad (7.10)$$

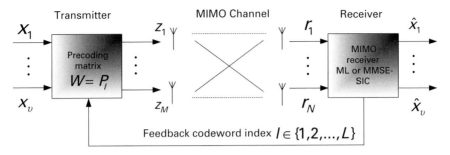

Figure 7.8. Illustration of feedback-based MIMO precoding.

A 2×2 ($N = 2$) Fourier matrix can be expressed as:

$$W = \frac{1}{\sqrt{2}} \begin{bmatrix} 1 & 1 \\ 1 & e^{j\pi} \end{bmatrix} = \frac{1}{\sqrt{2}} \begin{bmatrix} 1 & 1 \\ 1 & -1 \end{bmatrix}. \tag{7.11}$$

We can define multiple precoder matrices by introducing a shift parameter (g/G) in the Fourier matrix as given by:

$$W_g = e^{j \frac{2\pi m}{N}\left(n + \frac{g}{G}\right)} \quad m, n = 0, 1, \cdots (N-1). \tag{7.12}$$

We can, for example, define a set of four 2×2 Fourier matrices by taking $G = 4$. These four 2×2 matrices with $g = 0, 1, 2, 3$ are given as below:

$$
\begin{aligned}
W_0 &= \frac{1}{\sqrt{2}} \begin{bmatrix} e^{j0} & e^{j0} \\ e^{j0} & e^{j\pi} \end{bmatrix} = \frac{1}{\sqrt{2}} \begin{bmatrix} 1 & 1 \\ 1 & -1 \end{bmatrix} \\
W_1 &= \frac{1}{\sqrt{2}} \begin{bmatrix} 1 & 1 \\ e^{j\pi/4} & -e^{j\pi/4} \end{bmatrix} = \frac{1}{\sqrt{2}} \begin{bmatrix} 1 & 1 \\ \frac{1+j}{\sqrt{2}} & \frac{-1-j}{\sqrt{2}} \end{bmatrix} \\
W_2 &= \frac{1}{\sqrt{2}} \begin{bmatrix} 1 & 1 \\ e^{j\pi/2} & -e^{j\pi/2} \end{bmatrix} = \frac{1}{\sqrt{2}} \begin{bmatrix} 1 & 1 \\ j & -j \end{bmatrix} \\
W_3 &= \frac{1}{\sqrt{2}} \begin{bmatrix} 1 & 1 \\ e^{j3\pi/4} & -e^{j3\pi/4} \end{bmatrix} = \frac{1}{\sqrt{2}} \begin{bmatrix} 1 & 1 \\ \frac{-1+j}{\sqrt{2}} & \frac{1-j}{\sqrt{2}} \end{bmatrix}.
\end{aligned} \tag{7.13}
$$

The LTE codebook for two antenna ports consists of four precoders for rank-1 and three precoders for rank-2 as given in Table 7.1. The four rank-1 precoders are simply columns of second and third rank-2 precoder matrices. The second and third rank-2 precoders are DFT matrices W_0 and W_2 respectively. An important criterion for selection of the codebook for two antenna ports was limiting to the precoders that use QPSK alphabets $\{\pm 1, \pm j\}$ and the codebook of Table 7.1 meets this criterion. The rationale for limiting the codebook alphabet to $\{\pm 1, \pm j\}$ was reduction in UE complexity in calculating channel quality information (CQI) by avoiding the need for computing matrix/vector multiplication. The codebook of Table 7.2 also exhibits a nested property, that is, lower rank precoders are a subset of the higher rank precoder. The first rank-2 precoder is a 2×2 identity matrix that enables a simple transmission of two layers from the two antenna ports. This precoder is only used for open-loop MIMO spatial multiplexing without precoding feedback. Also, there are no rank-1 precoder vectors corresponding to the 2×2 identity matrix. This is because each column of the 2×2

Table 7.1. Codebook for transmission on antenna ports $p = 0, 1$.

Codebook index	Number of layers	
	$v = 1$	$v = 2$
0	$\dfrac{W_0^{\{1\}}}{\sqrt{2}} = \dfrac{1}{\sqrt{2}}\begin{bmatrix} 1 \\ 1 \end{bmatrix}$	$\dfrac{I^{\{12\}}}{\sqrt{2}} = \dfrac{1}{\sqrt{2}}\begin{bmatrix} 1 & 0 \\ 0 & 1 \end{bmatrix}$
1	$\dfrac{W_0^{\{2\}}}{\sqrt{2}}\dfrac{1}{\sqrt{2}}\begin{bmatrix} 1 \\ -1 \end{bmatrix}$	$\dfrac{W_0^{\{12\}}}{2} = \dfrac{1}{2}\begin{bmatrix} 1 & 1 \\ 1 & -1 \end{bmatrix}$
2	$\dfrac{W_2^{\{1\}}}{\sqrt{2}} = \dfrac{1}{\sqrt{2}}\begin{bmatrix} 1 \\ j \end{bmatrix}$	$\dfrac{W_2^{\{12\}}}{2} = \dfrac{1}{2}\begin{bmatrix} 1 & 1 \\ j & -j \end{bmatrix}$
3	$\dfrac{W_2^{\{2\}}}{\sqrt{2}} = \dfrac{1}{\sqrt{2}}\begin{bmatrix} 1 \\ -j \end{bmatrix}$	-

identity matrix means physical antenna port selection that leads to a non-constant modulus transmissions scheme because the power from the two antenna ports cannot be used for rank-1 transmission.

7.4.2   Precoding for four antenna ports

For four antenna ports precoding, both DFT-based and householder-transform-based codebooks were extensively discussed and evaluated. A DFT codebook can be obtained by setting $N = 4$ in Equation (7.12). An example of a DFT precoder for the case of $g = 0$ is given below:

$$W_0 = \frac{1}{\sqrt{4}}\begin{bmatrix} 1 & 1 & 1 & 1 \\ 1 & e^{j\pi/2} & e^{j\pi} & e^{j3\pi/2} \\ 1 & e^{j\pi} & e^{j2\pi} & e^{j3\pi} \\ 1 & e^{j3\pi/2} & e^{j3\pi} & e^{j9\pi/2} \end{bmatrix} = \frac{1}{\sqrt{4}}\begin{bmatrix} 1 & 1 & 1 & 1 \\ 1 & j & -1 & -j \\ 1 & -1 & 1 & -1 \\ 1 & -j & -1 & j \end{bmatrix}. \qquad (7.14)$$

Other DFT precoders can be obtained by setting G equal to the number of 4×4 matrices desired and $g = 0, 1, 2, \ldots, (G - 1)$. In the performance analysis of DFT-based and Householder codebooks, it was observed that both codebooks can provide similar performance under similar complexity constraints. Therefore, the choice of codebook was more a matter of taste and the decision was in favor of using a codebook based on the Householder principle.

An $N \times N$ Householder matrix is defined as follows [2]:

$$W = \mathbf{I}_N - 2\mathbf{u}\mathbf{u}^H, \quad \|\mathbf{u}\| = 1. \qquad (7.15)$$

This represents a reflection on the unit vector u in N-dimensional complex space, which is a unitary operation. u is also referred to as the generating vector. Assuming a generating vector $\mathbf{u}_0^T = \begin{bmatrix} 1 & -1 & -1 & -1 \end{bmatrix}$, the 4×4 Householder matrix is given as below:

$$W_0 = \mathbf{I}_4 - 2\mathbf{u}_0\mathbf{u}_0^H / \|\mathbf{u}_0\|^2 = \frac{1}{2}\begin{bmatrix} 1 & 1 & 1 & 1 \\ 1 & 1 & -1 & -1 \\ 1 & -1 & 1 & -1 \\ 1 & -1 & -1 & 1 \end{bmatrix}. \qquad (7.16)$$

Table 7.2. Codebook for transmission on antenna ports $p = 0, 1, 2, 3$.

CB index	u_n	Number of layers v			
		1	2	3	4
0	$u_0 = \begin{bmatrix} 1 & -1 & -1 & -1 \end{bmatrix}^T$	$W_0^{(1)}$	$W_0^{(14)}/\sqrt{2}$	$W_0^{(124)}/\sqrt{3}$	$W_0^{(1234)}/2$
1	$u_1 = \begin{bmatrix} 1 & -j & 1 & j \end{bmatrix}^T$	$W_1^{(1)}$	$W_1^{(12)}/\sqrt{2}$	$W_1^{(123)}/\sqrt{3}$	$W_1^{(1234)}/2$
2	$u_2 = \begin{bmatrix} 1 & 1 & -1 & 1 \end{bmatrix}^T$	$W_2^{(1)}$	$W_2^{(12)}/\sqrt{2}$	$W_2^{(123)}/\sqrt{3}$	$W_2^{(3214)}/2$
3	$u_3 = \begin{bmatrix} 1 & j & 1 & -j \end{bmatrix}^T$	$W_3^{(1)}$	$W_3^{(12)}/\sqrt{2}$	$W_3^{(123)}/\sqrt{3}$	$W_3^{(3214)}/2$
4	$u_4 = \begin{bmatrix} 1 & (-1-j)/\sqrt{2} & -j & (1-j)/\sqrt{2} \end{bmatrix}^T$	$W_4^{(1)}$	$W_4^{(14)}/\sqrt{2}$	$W_4^{(124)}/\sqrt{3}$	$W_4^{(1234)}/2$
5	$u_5 = \begin{bmatrix} 1 & (1-j)/\sqrt{2} & j & (-1-j)/\sqrt{2} \end{bmatrix}^T$	$W_5^{(1)}$	$W_5^{(14)}/\sqrt{2}$	$W_5^{(124)}/\sqrt{3}$	$W_5^{(1234)}/2$
6	$u_6 = \begin{bmatrix} 1 & (1+j)/\sqrt{2} & -j & (-1+j)/\sqrt{2} \end{bmatrix}^T$	$W_6^{(1)}$	$W_6^{(13)}/\sqrt{2}$	$W_6^{(134)}/\sqrt{3}$	$W_6^{(1324)}/2$
7	$u_7 = \begin{bmatrix} 1 & (-1+j)/\sqrt{2} & j & (1+j)/\sqrt{2} \end{bmatrix}^T$	$W_7^{(1)}$	$W_7^{(13)}/\sqrt{2}$	$W_7^{(134)}/\sqrt{3}$	$W_7^{(1324)}/2$
8	$u_8 = \begin{bmatrix} 1 & -1 & 1 & 1 \end{bmatrix}^T$	$W_8^{(1)}$	$W_8^{(12)}/\sqrt{2}$	$W_8^{(124)}/\sqrt{3}$	$W_8^{(1234)}/2$
9	$u_9 = \begin{bmatrix} 1 & -j & -1 & -j \end{bmatrix}^T$	$W_9^{(1)}$	$W_9^{(14)}/\sqrt{2}$	$W_9^{(134)}/\sqrt{3}$	$W_9^{(1234)}/2$
10	$u_{10} = \begin{bmatrix} 1 & 1 & 1 & -1 \end{bmatrix}^T$	$W_{10}^{(1)}$	$W_{10}^{(13)}/\sqrt{2}$	$W_{10}^{(123)}/\sqrt{3}$	$W_{10}^{(1324)}/2$
11	$u_{11} = \begin{bmatrix} 1 & j & -1 & j \end{bmatrix}^T$	$W_{11}^{(1)}$	$W_{11}^{(13)}/\sqrt{2}$	$W_{11}^{(134)}/\sqrt{3}$	$W_{11}^{(1324)}/2$
12	$u_{12} = \begin{bmatrix} 1 & -1 & -1 & 1 \end{bmatrix}^T$	$W_{12}^{(1)}$	$W_{12}^{(12)}/\sqrt{2}$	$W_{12}^{(123)}/\sqrt{3}$	$W_{12}^{(1234)}/2$
13	$u_{13} = \begin{bmatrix} 1 & -1 & 1 & -1 \end{bmatrix}^T$	$W_{13}^{(1)}$	$W_{13}^{(13)}/\sqrt{2}$	$W_{13}^{(123)}/\sqrt{3}$	$W_{13}^{(1324)}/2$
14	$u_{14} = \begin{bmatrix} 1 & 1 & -1 & -1 \end{bmatrix}^T$	$W_{14}^{(1)}$	$W_{14}^{(12)}/\sqrt{2}$	$W_{14}^{(123)}/\sqrt{3}$	$W_{14}^{(3214)}/2$
15	$u_{15} = \begin{bmatrix} 1 & 1 & 1 & 1 \end{bmatrix}^T$	$W_{15}^{(1)}$	$W_{15}^{(12)}/\sqrt{2}$	$W_{15}^{(123)}/\sqrt{3}$	$W_{15}^{(1234)}/2$

The four-antenna ports codebook given in Table 7.2 uses a total of 16 generating vectors $\{u_0, u_1, \ldots u_{15}\}$. These 16 generating vectors result in sixteen 4×4 Householder matrices, which form the precoders for rank-4 transmissions. The precoders for lower ranks are obtained by column subset selection from the rank-4 precoders. The rank-1 precoders always consist of the first column of the matrix. The codebook of Table 7.2 also exhibits a nested property, that is, lower rank precoders are a subset of the higher rank precoder for the same generating vector. For example, for the first rank-4 precoder $W_0^{\{1234\}}/2$ consisting of W_0, the rank-1, 2 and 3 precoders $W_0^{\{1\}}$, $W_0^{\{14\}}/\sqrt{2}$, $W_0^{\{124\}}/\sqrt{3}$ consist of column 1, columns (1,4) and columns (1,2,4) respectively.

As for the case of two-antenna ports codebook, an important criterion for codebook design was the constant modulus property, which is beneficial for avoiding an unnecessary increase in PAPR thereby ensuring power amplifier balance. Similarly, to reduce UE complexity, the alphabet size is limited to QPSK $\{\pm1, \pm j\}$ and 8-PSK $\{\pm1, \pm j, \pm1 \pm j\}$ alphabets. This constraint avoids the need for computing matrix/vector multiplication. For each of the matrices in the Householder codebook, most of the elements take values from the QPSK alphabet set $\{\pm1, \pm j\}$. For 4 out of 16 matrices, some of the elements are taken from the 8PSK alphabet set $\{\pm1, \pm j, \pm1 \pm j\}$. The codebook design achieves chordal distances [3, 4] of 0.7071, 1.0 and 0.7071 for rank-1, rank-2 and rank-3 respectively.

7.4.3 Precoding operation

The symbols at the output of precoding for antenna port p, $y^{(p)}(i)$ are given as:

$$\begin{bmatrix} y^{(0)}(i) \\ \vdots \\ y^{(P-1)}(i) \end{bmatrix} = W(i) \begin{bmatrix} x^{(0)}(i) \\ \vdots \\ x^{(\upsilon-1)}(i) \end{bmatrix} \qquad P = 1,2,4 \quad P \geq \upsilon = 1,2,3,4, \qquad (7.17)$$

where $W(i)$ is size $P \times \upsilon$ precoding matrix, P is number of ports and $\upsilon (\leq P)$ is number of layers transmitted.

An example of rank-2 precoding for two and four antenna ports transmissions is shown in Figure 7.9. We assumed the precoders $W_1^{\{12\}}/2$ and $W_0^{\{14\}}/\sqrt{2}$ for two and four antenna ports respectively. For the case of two antenna ports, the symbols at the output of precoding for antenna ports 0 and antenna port 1 $y^{(0)}(i)$ and $y^{(1)}(i)$ are written as:

$$\begin{bmatrix} y^{(0)}(i) \\ y^{(1)}(i) \end{bmatrix} = \frac{1}{2} \begin{bmatrix} 1 & 1 \\ j & -j \end{bmatrix} \begin{bmatrix} x^{(0)}(i) \\ x^{(1)}(i) \end{bmatrix} = \frac{1}{2} \begin{bmatrix} x^{(0)}(i) + x^{(1)}(i) \\ jx^{(0)}(i) - jx^{(1)}(i) \end{bmatrix} \quad i = 0, 1, \ldots, M_{symb}^{layer}-1,$$
$$(7.18)$$

where $x^{(0)}(i)$ and $x^{(1)}(i)$ represent modulation symbols from codewords 1 and 2 respectively. In the case of four antenna ports, the symbols for antenna ports 0, 1, 2 and 3, $y^{(0)}(i)$, $y^{(1)}(i)$, $y^{(2)}(i)$ and $y^{(3)}(i)$ are given as:

$$\begin{bmatrix} y^{(0)}(i) \\ y^{(1)}(i) \\ y^{(2)}(i) \\ y^{(3)}(i) \end{bmatrix} = \frac{1}{\sqrt{2}} \begin{bmatrix} 1 & 1 \\ 1 & -1 \\ 1 & -1 \\ 1 & 1 \end{bmatrix} \begin{bmatrix} x^{(0)}(i) \\ x^{(0)}(i) \end{bmatrix} = \frac{1}{2} \begin{bmatrix} x^{(0)}(i) + x^{(1)}(i) \\ x^{(0)}(i) - x^{(1)}(i) \\ x^{(0)}(i) - x^{(1)}(i) \\ x^{(0)}(i) + x^{(1)}(i) \end{bmatrix}$$

$$i = 0, 1, \ldots, M_{symb}^{layer} - 1. \qquad (7.19)$$

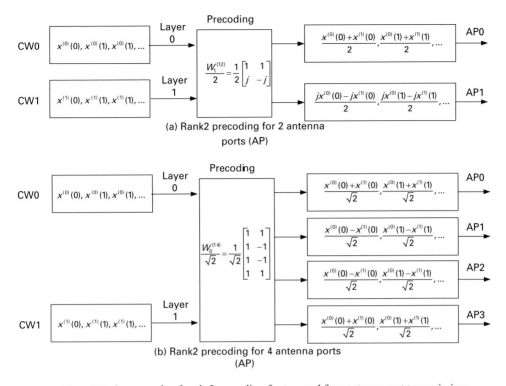

Figure 7.9. An example of rank-2 precoding for two and four antenna port transmissions.

7.5   CDD-based precoding

The LTE system also supports a composite precoding by introducing a cyclic delay diversity (CDD) precoder on top of the precoders described in the previous section. Two flavors of CDD precoding namely small-delay CDD and large-delay CDD precoding were considered. The goal of small-delay precoding is to introduce artificial frequency selectivity for opportunistic scheduling gains [5] with low feedback overhead while the large-delay CDD achieves diversity by making sure that each MIMO codeword is transmitted on all the available MIMO layers.

Both the small-delay and large-delay CDD schemes were incorporated in the LTE standard. However, the small-delay CDD was removed from the specification at the later stages because the scheduling gains promised were small, particularly when feedback-based precoding can be employed for closed-loop MIMO operation.

7.5.1   Small-delay CDD precoding

For small-delay cyclic delay diversity (CDD), the precoding is a composite precoding of CDD-based precoding defined by matrix $D(i)$ and precoding matrix $W(i)$ as given by the relationship below:

$$\begin{bmatrix} y^{(0)}(i) \\ \vdots \\ y^{(P-1)}(i) \end{bmatrix} = D(i) \times W(i) \times \begin{bmatrix} x^{(0)}(i) \\ \vdots \\ x^{(\upsilon-1)}(i) \end{bmatrix} \qquad P = 1, 2, 4 \quad P \geq \upsilon = 1, 2, 3, 4,$$

$$(7.20)$$

where $W(i)$ is size $P \times \upsilon$ precoding matrix, P is number of ports, $\upsilon (\le P)$ is number of layers transmitted and $D(i)$ is a diagonal matrix for support of cyclic delay diversity. In the case of two antenna ports, the CDD diagonal matrix $D(i)$ is given as:

$$D(i) = \begin{bmatrix} 1 & 0 \\ 0 & e^{-j\frac{4\pi \cdot i}{\eta}} \end{bmatrix}. \tag{7.21}$$

In the case of four antenna ports, the CDD diagonal matrix $D(i)$ is given as:

$$D(i) = \begin{bmatrix} 1 & 0 & 0 & 0 \\ 0 & e^{-j\frac{2\pi \cdot i}{\eta}} & 0 & 0 \\ 0 & 0 & e^{-j\frac{4\pi \cdot i}{\eta}} & 0 \\ 0 & 0 & 0 & e^{-j\frac{6\pi \cdot i}{\eta}} \end{bmatrix}, \tag{7.22}$$

where η is the smallest number from the set $\{128, 256, 512, 1024, 2048\}$ such that $\eta \ge (N_{RB}^{DL} \times N_{sc}^{RB})$. N_{RB}^{DL} is the number of resource blocks within the downlink system bandwidth and N_{sc}^{RB} is the number of subcarriers within a resource block. A resource element is uniquely defined by the index pair (k, l) in a slot with $k = 0, ..., (N_{RB}^{DL}N_{sc}^{RB} - 1)$ and $l = 0, ..., (N_{symb}^{DL} - 1)$, where k and l represent the indices in the frequency and time domains, respectively. The CDD precoding is applied in the frequency-domain and does not change across time (OFDM symbols) within a subframe.

The power spectral density for two and four antenna ports small-delay CDD as a function of resource element frequency-domain index is plotted in Figure 7.10. We assumed the case of 10 MHz system bandwidth with $N_{RB}^{DL} \times N_{sc}^{RB} = 600$, which results in $\eta = 1024$. This means

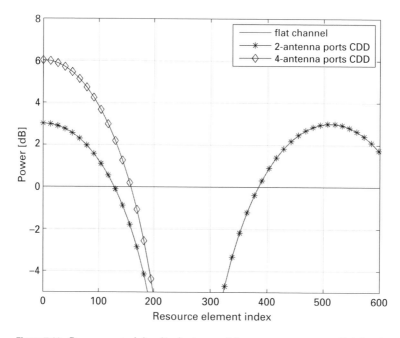

Figure 7.10. Power spectral density for two and four antenna ports small-delay CDD.

Table 7.3. Small-delay CDD phase shifts per RE for 2 and 4 antenna ports.

$N_{RB}^{DL} \times N_{sc}^{RB}$	η	Two antenna ports phase shift per RE [degrees]		Four antenna ports phase shift per RE [degrees]			
		Port 0	Port 1	Port 0	Port 1	Port 2	Port 3
300	512	0	1.40	0	0.70	1.40	2.10
600	1024	0	0.70	0	0.35	0.70	1.05
1200	2048	0	0.35	0	0.175	0.35	0.53

that for two antenna ports, small delay CDD results in a phase shift of 4π radians (720°) from the second antenna port across 1024 resource elements (subcarriers). This represents a phase shift of approximately 0.7° from one resource element to the next. The total phase shift across 600 resource elements is approximately 422 degrees. Therefore, small-delay CDD for two antenna ports results in at least one cycle that is 2π radians (360°) phase shift across the system bandwidth.

In the case of four antenna ports, small-delay CDD results in a total phase shift of 2π radians (360°), 4π radians (720°) and 6π radians (1080°) from the second, third and fourth antenna ports respectively across 1024 resource elements. The means that the composite CDD effect across four antenna ports results in a complete cycle over 1024 resource elements. This is why we do not see a complete cycle of CDD across 600 resource elements in Figure 7.10 for the four-antenna-ports case. The selection of small-delay CDD parameters is a tradeoff between assuring sufficient frequency-selectivity within the system bandwidth while minimizing the phase shift from one subcarrier (RE) to the next. We give phase shifts per RE for two and four antenna ports in Table 7.3. In general, a user is allocated one or more resource blocks with a resource block consisting of 12 contiguous subcarriers. It is then desirable to keep the frequency selectivity to a minimum within the allocated bandwidth. This requires that phase shift per RE is very small. On the other hand we want to introduce sufficient frequency selectivity across the total bandwidth so that a user can see updates on parts of the bandwidth. This is why at least one cycle of CDD across the system bandwidth (600 resource elements, for example) guarantees that a given user experiences upfades at some parts of the bandwidth. However, when a complete cycle of CDD is not present within the system bandwidth, a user can still experience upfades at some parts of the bandwidth. However, it is not guaranteed that these upfades will appear in each subframe. The user will, however, experience upfades in at least some subframe due to the time-varying nature of the channel. The CDD parameters selection for the case of four antenna ports was based on the tradeoff of introducing frequency selectivity over the system bandwidth while minimizing selectivity within the bandwidth allocated to a UE.

The goal of small-delay CDD precoding is to provide gains by exploiting frequency selectivity introduced via multi-user scheduling. We show power spectral density for two-antenna-ports small-delay CDD for four different users in Figure 7.11. We assumed that the phase shift between antenna port 0 and antenna port 1 due to wireless channel for users 1, 2, 3 and 4 is 0, 90, 180 and 270 degrees. When small-delay CDD is applied, these four users see upfades in different parts in frequency. These four users can then be scheduled in the regions where they see upfades, resulting in system throughput improvement.

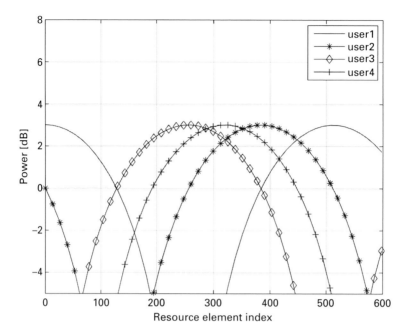

Figure 7.11. Power spectral density for two-antenna-ports small-delay CDD for four different users.

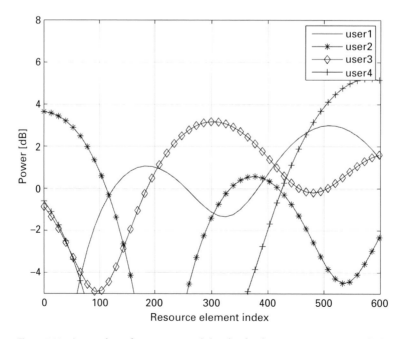

Figure 7.12. A snapshot of power spectral density for four-antenna-ports small-delay CDD for four different users.

As an example, we show power spectral density for four-antenna-ports small-delay CDD for four different users in Figure 7.12. In this case, we assumed random phase shifts for different antennas of different users. We note that when small-delay CDD is applied, these four users see upfades in different parts in frequency due to the random nature of the phases applied from different transmit antennas. Note that CDD phase shifts are applied on top of the random phases due to the wireless channel. The goal with CDD phase shifts is to create frequency selectivity when the wireless channel happens to be flat. In the example of Figure 7.12, the CDD phase shifts are frequency dependent while the random phase shifts for different users on different antennas are frequency independent. We note that these four users can be scheduled in the regions where they see upfades. We also remark that the upfades introduced due to four-antenna-ports small-delay CDD are larger than the two-antenna ports case. This is explained by the fact that with signals from four antenna ports combining coherently, the upfades can be as high as 6 dB relative to the flat-fading reference while with two antenna ports, the maximum upfade is limited to 3 dB.

7.5.2 Large delay CDD precoding

For large-delay cyclic delay diversity (CDD), the precoding is a composite precoding of CDD-based precoding defined by matrix $D(i)$ and precoding matrix $W(i)$ as given by the relationship below:

$$\begin{bmatrix} y^{(0)}(i) \\ \vdots \\ y^{(P-1)}(i) \end{bmatrix} = W(i) \times D(i) \times U \times \begin{bmatrix} x^{(0)}(i) \\ \vdots \\ x^{(\upsilon-1)}(i) \end{bmatrix} \qquad P = 1, 2, 4 \quad P \geq \upsilon = 1, 2, 3, 4,$$

(7.23)

where $W(i)$ is size $P \times \upsilon$ precoding matrix, P is number of ports and $\upsilon \ (\leq P)$ is number of layers transmitted, $D(i)$ is $\upsilon \times \upsilon$ diagonal matrix, U is a $\upsilon \times \upsilon$ DFT matrix for each number of layers transmitted and i represents modulation symbol index within each of the layers with $i = 0, 1, ..., \left(M_{symb}^{layer} - 1 \right)$.

In the case of two layers, the large-delay CDD diagonal matrix $D(i)$ and fixed DFT matrix U are given as:

$$D_{2 \times 2}(i) = \begin{bmatrix} 1 & 0 \\ 0 & e^{-j2\pi i/2} \end{bmatrix}, \quad U_{2 \times 2} = \frac{1}{\sqrt{2}} \begin{bmatrix} 1 & 1 \\ 1 & e^{-j2\pi/2} \end{bmatrix} = \frac{1}{\sqrt{2}} \begin{bmatrix} 1 & 1 \\ 1 & -1 \end{bmatrix}.$$

(7.24)

The CDD diagonal matrix $D(i)$ for odd and even i is written as:

$$D_{2 \times 2}(i) = \begin{cases} \begin{bmatrix} 1 & 0 \\ 0 & -1 \end{bmatrix} & i = 1, 3, \ldots \\ \begin{bmatrix} 1 & 0 \\ 0 & 1 \end{bmatrix} & i = 0, 2, \ldots \end{cases}$$

(7.25)

We note that CDD diagonal matrix $D(i)$ represents a relative phase shift between two antenna ports of either zero or 180 degrees for even and odd i respectively. The composite effect of large-delay CDD precoding and fixed DFT matrix U can be expressed as:

$$D_{2\times2}(i) \times U_{2\times2} = \begin{cases} \begin{bmatrix} 1 & 0 \\ 0 & -1 \end{bmatrix} \times \dfrac{1}{\sqrt{2}} \begin{bmatrix} 1 & 1 \\ 1 & -1 \end{bmatrix} = \dfrac{1}{\sqrt{2}} \begin{bmatrix} 1 & 1 \\ -1 & 1 \end{bmatrix} & i = 1, 3, \ldots \\[4mm] \begin{bmatrix} 1 & 0 \\ 0 & 1 \end{bmatrix} \times \dfrac{1}{\sqrt{2}} \begin{bmatrix} 1 & 1 \\ 1 & -1 \end{bmatrix} = \dfrac{1}{\sqrt{2}} \begin{bmatrix} 1 & 1 \\ 1 & -1 \end{bmatrix} & i = 0, 2, \ldots \end{cases}$$

$$(7.26)$$

We remark that the composite effect of large-delay CDD precoding and fixed DFT matrix U is simply cyclic permutation of columns of the DFT matrix U.

For odd i, the symbols transmitted from antenna ports 0 and 1, $y^{(0)}(i), y^{(1)}(i)$, are given as:

$$\begin{bmatrix} y^{(0)}(i) \\ y^{(1)}(i) \end{bmatrix} = D_{2\times2}(i) \times U_{2\times2} \times \begin{bmatrix} x^{(0)}(i) \\ x^{(1)}(i) \end{bmatrix} = \dfrac{1}{\sqrt{2}} \begin{bmatrix} 1 & 1 \\ -1 & 1 \end{bmatrix} \times \begin{bmatrix} x^{(0)}(i) \\ x^{(1)}(i) \end{bmatrix}$$

$$= \dfrac{1}{\sqrt{2}} \begin{bmatrix} x^{(0)}(i) + x^{(1)}(i) \\ -x^{(0)}(i) + x^{(1)}(i) \end{bmatrix}.$$

$$(7.27)$$

Similarly for even i, the symbols transmitted from antenna ports 0 and 1, $y^{(0)}(i), y^{(1)}(i)$, are given as:

$$\begin{bmatrix} y^{(0)}(i) \\ y^{(1)}(i) \end{bmatrix} = D_{2\times2}(i) \times U_{2\times2} \times \begin{bmatrix} x^{(0)}(i) \\ x^{(1)}(i) \end{bmatrix} = \dfrac{1}{\sqrt{2}} \begin{bmatrix} 1 & 1 \\ 1 & -1 \end{bmatrix} \times \begin{bmatrix} x^{(0)}(i) \\ x^{(1)}(i) \end{bmatrix}$$

$$= \dfrac{1}{\sqrt{2}} \begin{bmatrix} x^{(0)}(i) + x^{(1)}(i) \\ x^{(0)}(i) - x^{(1)}(i) \end{bmatrix}.$$

$$(7.28)$$

The above operations are shown schematically in Figure 7.13. It can be noted that for odd i, the modulation symbols from codeword 0 are transmitted such that symbol $x^{(0)}(i)$ is transmitted from antenna port 0 while $-x^{(0)}(i)$ is transmitted from antenna port 1. On the other hand, the modulation symbols from codeword 1 are transmitted such that the same symbol $x^{(1)}(i)$ is transmitted from both antenna port zero and antenna port 1. For even i, the layer 0 and layer 1 are switched and symbols from codeword 0 are transmitted such that the same symbol $x^{(0)}(i)$ is transmitted from both antenna port 0 and antenna port 1. The symbols from codeword 1 are transmitted such that symbol $x^{(1)}(i)$ is transmitted from antenna port 0 while symbol $-x^{(1)}(i)$ is transmitted from antenna port 1. This operation of switching layers results in transmission of symbols from codeword 1 and codeword 2 on both the layers. This way both the codewords experience the average SINR of the two layers.

In the case of three layers transmission, the large-delay CDD diagonal matrix $D(i)$ and fixed DFT matrix U are given as:

$$D_{3\times3}(i) = \begin{bmatrix} 1 & 0 & 0 \\ 0 & e^{-j2\pi i/3} & 0 \\ 0 & 0 & e^{-j4\pi i/3} \end{bmatrix}, \quad U_{3\times3} = \dfrac{1}{\sqrt{3}} \begin{bmatrix} 1 & 1 & 1 \\ 1 & e^{-j2\pi/3} & e^{-j4\pi/3} \\ 1 & e^{-j4\pi/3} & e^{-j8\pi/3} \end{bmatrix}.$$

$$(7.29)$$

The diagonal elements in the CDD matrix $D(i)$ denote phase shifts by 0, 120 and 240 degrees. For $i = 0$, the CDD diagonal matrix $D(i)$ is simply an identity matrix. Therefore,

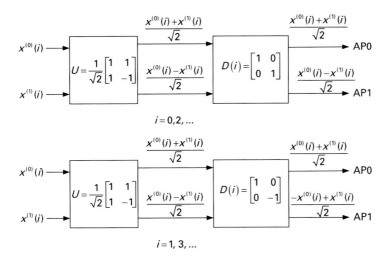

Figure 7.13. Large-delay CDD precoding for two-antenna-ports transmission.

the composite effect of the CDD diagonal matrix $D(i)$ and U results in the matrix U itself. For $i = 1, 2$, the composite effect of the CDD diagonal matrix $D(i)$ and DFT matrix U are given as:

$$D(1)_{3\times3} \times U_{3\times3} = \frac{1}{\sqrt{3}} \begin{bmatrix} 1 & 1 & 1 \\ e^{-j2\pi/3} & e^{-j4\pi/3} & 1 \\ e^{-j4\pi/3} & e^{-j8\pi/3} & 1 \end{bmatrix},$$

$$D_{3\times3}(2) \times U_{3\times3} = \frac{1}{\sqrt{3}} \begin{bmatrix} 1 & 1 & 1 \\ e^{-j4\pi/3} & 1 & e^{-j2\pi/3} \\ e^{-j8\pi/3} & 1 & e^{-j4\pi/3} \end{bmatrix}. \quad (7.30)$$

It can be noted that $i = 1, 2$ results in cyclic shifts of the matrix U by i times. This means that the modulation symbols from codewords 1 and 2 will be transmitted from each of the layers. The codewords SINR is the average of the SINR across the three layers.

In the case of four layers transmission, the large delay CDD diagonal matrix $D(i)$ and fixed DFT matrix U are given as:

$$D_{4\times4}(i) = \begin{bmatrix} 1 & 0 & 0 & 0 \\ 0 & e^{-j2\pi i/4} & 0 & 0 \\ 0 & 0 & e^{-j4\pi i/4} & 0 \\ 0 & 0 & 0 & e^{-j6\pi i/4} \end{bmatrix} \quad (7.31)$$

$$U_{4\times4} = \frac{1}{\sqrt{4}} \begin{bmatrix} 1 & 1 & 1 & 1 \\ 1 & e^{-j2\pi/4} & e^{-j4\pi/4} & e^{-j6\pi/4} \\ 1 & e^{-j4\pi/4} & e^{-j8\pi/4} & e^{-j12\pi/4} \\ 1 & e^{-j6\pi/4} & e^{-j12\pi/4} & e^{-j18\pi/4} \end{bmatrix} = \frac{1}{\sqrt{4}} \begin{bmatrix} 1 & 1 & 1 & 1 \\ 1 & -j & -1 & j \\ 1 & -1 & 1 & -1 \\ 1 & j & -1 & -j \end{bmatrix}. \quad (7.32)$$

The diagonal elements in the CDD matrix $D(i)$ denote phase shifts by 0, 90, 180 and 270 degrees. Let us represent the four columns of the matrix U as four vectors as below:

$$u_0 = \begin{bmatrix} 1 \\ 1 \\ 1 \\ 1 \end{bmatrix}, \quad u_1 = \begin{bmatrix} 1 \\ -j \\ -1 \\ j \end{bmatrix}, \quad u_2 = \begin{bmatrix} 1 \\ -1 \\ 1 \\ -1 \end{bmatrix}, \quad u_3 = \begin{bmatrix} 1 \\ j \\ -1 \\ -j \end{bmatrix}. \tag{7.33}$$

Also, let us now see the composite effect of CDD diagonal matrix $D(i)$ and DFT matrix U for $i = 0, 1, 2, 3, \ldots$ as below:

$$\begin{aligned} D_{4\times4}(0) \times U_{4\times4} &= [u_0, u_1, u_2, u_3] \\ D_{4\times4}(1) \times U_{4\times4} &= [u_1, u_2, u_3, u_0] \\ D_{4\times4}(2) \times U_{4\times4} &= [u_2, u_3, u_0, u_1] \\ D_{4\times4}(3) \times U_{4\times4} &= [u_3, u_0, u_1, u_2] \\ D_{4\times4}(4) \times U_{4\times4} &= D_{4\times4}(0) \times U_{4\times4} \end{aligned} \tag{7.34}$$

$$\vdots$$

Again, we note that $i = 0, 1, 2, 3, \ldots$ results in cyclic shifts of the matrix U by i times. Since there are four columns, the matrix repeats after cyclic shift by four times. This means that the modulation symbols from codeword 1 and codeword 2 will be transmitted from each of the four layers. The codewords SINR is therefore the average of the SINR across the four layers.

7.6   Open-loop spatial multiplexing

A transmission diversity scheme is used for rank-1 open-loop transmissions. However, for rank greater than one, the open-loop transmission scheme uses large-delay CDD along with a fixed precoder matrix for the two-antenna-ports $P = 2$ case, while precoder cycling is used for the four-antenna-ports $P = 4$ case. The fixed precoder used for the case of two antenna ports is the identity matrix. Therefore, the precoder for data resource element index i, denoted by $W(i)$, is simply given as:

$$W(i) = \frac{1}{\sqrt{2}} \begin{bmatrix} 1 & 0 \\ 0 & 1 \end{bmatrix}. \tag{7.35}$$

We note that this fixed precoder is the first rank-2 precoder from Table 7.1. This 2×2 identity matrix is only used for open-loop spatial multiplexing for the case of two antenna ports. The other two rank-2 precoders could also be considered for open-loop spatial multiplexing transmission. However, a concern was fixed beam-forming patterns introduced by the second and third rank-2 precoder from Table 7.1 when used for open-loop spatial multiplexing transmission without precoding feedback.

7.6.1   Precoder cycling for four antenna ports

We noted that large-delay CDD enables each MIMO codeword to experience average SINR across layers. This makes the codeword transmissions robust to layer SINR fluctuations due to CQI feedback delays and inter-cell interference variations, etc. The large-delay CDD allows

Table 7.4. Precoding matrices used for precoder cycling.

	Number of layers υ		
Precoding matrix	2	3	4
C_1	$W_{12}^{\{1\}}$	$W_{12}^{\{12\}}/\sqrt{2}$	$W_{12}^{\{123\}}/\sqrt{3}$
C_2	$W_{13}^{\{1\}}$	$W_{13}^{\{13\}}/\sqrt{2}$	$W_{13}^{\{123\}}/\sqrt{3}$
C_3	$W_{14}^{\{1\}}$	$W_{14}^{\{13\}}/\sqrt{2}$	$W_{14}^{\{123\}}/\sqrt{3}$
C_4	$W_{15}^{\{1\}}$	$W_{15}^{\{12\}}/\sqrt{2}$	$W_{15}^{\{123\}}/\sqrt{3}$

SINR averaging for a given channel realization. However, for open-loop spatial multiplexing without precoding feedback, further diversity can be introduced by precoder cycling with each precoder providing a different realization of layers SINR. A precoder cycling approach is used for open-loop spatial multiplexing transmission mode for the case of four antenna ports $P = 4$ only. A different precoder is used every υ data resource elements, where υ denotes the number of transmission layers in the case of spatial multiplexing. In particular, the precoder for data resource element index i, denoted by $W(i)$, is selected according to $W(i) = C_k$, where k is the precoder index given by:

$$k = \mathrm{mod}\left(\left\lceil\frac{i+1}{\upsilon}\right\rceil - 1, 4\right) + 1 = \mathrm{mod}\left(\left\lfloor\frac{i}{\upsilon}\right\rfloor, 4\right) + 1 \quad k = 1, 2, 3, 4, \qquad (7.36)$$

where the precoder matrices C_1, C_2, C_3, C_4 are given in Table 7.4. We note that the precoder matrices C_1, C_2, C_3, C_4 corresponds to precoder indices 12, 13, 14 and 15 respectively in Table 7.2. The precoder cycling for four antenna ports $P = 4$ and $\upsilon = 2, 3, 4$ is schematically shown in Figure 7.14. The precoder is changed every $\upsilon = 2, 3, 4$ layers in order to be consistent with the large delay CDD operation, which requires υ resource elements to enable transmission of modulation symbols from each codeword across $\upsilon = 2, 3, 4$ layers. The overall effect of precoder cycling and large-delay CDD is that precoder cycling enables different layer SINR realizations while large-delay CDD makes sure that each codeword is transmitted across all the υ layers.

7.6.2 Downlink multi-user MIMO

In single-user MIMO, all the spatial layers within allocated resource blocks are addressed to the same UE. In the case of multi-user MIMO [6], different spatial layers can be addressed to different UEs. Therefore, single-user MIMO can improve both UE peak data rates as well as cell capacity. On the other hand the basic form of multi-user MIMO, where at most one layer is addressed to a UE, does not improve the UE peak data rate. However, under correlated antenna scenarios, multi-user MIMO can improve cell capacity as orthogonal spatial beams can be created for UEs at different spatial locations in the cell. Under these correlated situations, single-user MIMO performance degrades due to channel rank limitation.

The LTE system supports multi-user MIMO for the correlated channel conditions with single layer transmission to a UE. The standard does not limit the number of UEs that can be scheduled using the same resource blocks. The single-user MIMO codebooks for two and four antenna ports are reused for multi-user MIMO. Only rank-1 precoders are needed for

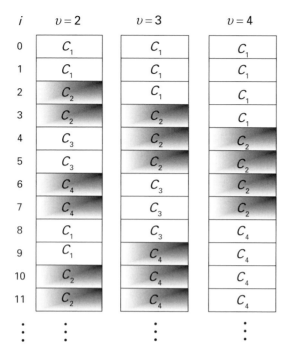

Figure 7.14. Precoder cycling for four antenna ports $P = 4$ and $v = 2, 3, 4$.

multi-user MIMO because from a UE perspective there is always a single layer transmission. A downlink power offset is defined to support multi-user MIMO in the downlink. The power offset information is required for multi-user MIMO scheduling in the downlink because when more than one UE is scheduled on the same resource blocks, the total power is shared among the UEs. The two power offset values represent a power offset of 0 or -3 dB relative to the single user transmission power offset signaled by higher layers.

7.6.3 Summary

The LTE system supports up to 4×4 MIMO in the downlink to achieve peak data rates in excess of 200 Mb/s. For closed-loop MIMO operation based on precoding feedback from the UE, a DFT codebook and a Householder codebook are specified for two and four transmission antenna ports respectively. Both the codebooks exhibit a nested property, that is, lower rank precoders are a subset of the higher rank precoders. This allows eNB to override rank and use a lower rank precoder when the amount of data to be transmitted is less than that which can be transmitted using all the available MIMO layers. Moreover, both the DFT and Householder codebooks are constant-modulus enabling full utilization of transmission power from all the antenna ports.

The LTE system also supports a composite precoding by introducing cyclic delay diversity (CDD) precoder in conjunction with a DFT or Householder precoder for two and four transmission antenna ports respectively. Two flavors of CDD precoding namely small-delay CDD and large-delay CDD precoding were considered. The goal of small-delay precoding is to introduce artificial frequency selectivity for scheduling gains while the large-delay CDD achieves diversity by making sure that each MIMO codeword is transmitted on all the available MIMO layers. The small-delay CDD was removed from the specification at the later stages because the scheduling gains promised by small-delay CDD are small particularly when feedback based precoding can be employed for closed-loop MIMO operation.

A 2×2 identity matrix precoder is used for open-loop MIMO transmission for the case of two antenna ports. A precoder cycling approach on top of large-delay CDD is used for open-loop spatial multiplexing transmission mode for the case of four antenna ports.

The first release of the LTE system does not support single-user MIMO spatial multiplexing in the uplink. However, multi-user MIMO operation where two UEs are scheduled on the same resource blocks in the same subframe is permitted. In this case, the eNB utilizes degrees of freedom provided by multiple receive antennas by using, for example, MMSE processing to separate signals from the two UEs. The multi-user MIMO operation in the uplink can improve the cell capacity but does not help to improve the UE peak data rates.

References

[1] Tse, D. and Viswanath, P., *Fundamentals of Wireless Communication*, New York: Cambridge University Press, 2005.

[2] Golub, G. and Van Loan, C., *Matrix Computations*, Baltimore, MD: The Johns Hopkins University Press, 1996.

[3] Barg, A. and Nogin, D. Y., "Bounds on packings of spheres in the Grassmann manifold," *IEEE Transactions on Information Theory*, vol. IT-48, no. 9, Sept. 2002.

[4] Love, D. and Heath, R., "Limited feedback unitary precoding for spatial multiplexing systems," *IEEE Transactions on Information Theory*, vol. IT-51, no. 8, Aug. 2005.

[5] Khan, F. and Rensburg, C., "An adaptive cyclic delay diversity technique for beyond 3G/4G wireless systems," 64th *IEEE Vehicular Technology Conference, VTC-2006* Fall, Sept. 2006.

[6] Paulraj, A., Nabar, R. and Gore, D., *Introduction to Space-Time Wireless Communications*, Cambridge: Cambridge University Press, 2005.

8 Channel structure and bandwidths

A major design goal for the LTE system is flexible bandwidth support for deployments in diverse spectrum arrangements. With this objective in mind, the physical layer of LTE is designed to support bandwidths in increments of 180 kHz starting from a minimum bandwidth of 1.08 MHz. In order to support channel sensitive scheduling and to achieve low packet transmission latency, the scheduling and transmission interval is defined as a 1 ms subframe. Two cyclic prefix lengths namely normal cyclic prefix and extended cyclic prefix are defined to support small and large cells deployments respectively. A subcarrier spacing of 15 kHz is chosen to strike a balance between cyclic prefix overhead and robustness to Doppler spread. An additional smaller 7.5 kHz subcarrier spacing is defined for MBSFN to support large delay spreads with reasonable cyclic prefix overhead. The uplink supports localized transmissions with contiguous resource block allocation due to single-carrier FDMA. In order to achieve frequency diversity, inter-subframe and intra-subframe hopping is supported. In the downlink, a distributed transmission allocation structure in addition to localized transmission allocation is defined to achieve frequency diversity with small signaling overhead.

8.1 Channel bandwidths

The LTE system supports a set of six channel bandwidths as given in Table 8.1. We note that the transmission bandwidth configuration BW_{config} is 90% of the channel bandwidth BW_{channel} for 3–20 MHz. For 1.4 MHz channel bandwidth, the transmission bandwidth is only 77% of the channel bandwidth. Therefore, LTE deployment in the small 1.4 MHz channel is less spectrally efficient than the 3 MHz and larger channel bandwidths. The relationship between the channel bandwidth BW_{channel} and the transmission bandwidth configuration N_{RB} is shown in Figure 8.1. The transmission bandwidth configuration in MHz is given as:

$$BW_{\text{config}} = \left(\frac{N_{\text{RB}} \times N_{\text{SC}}^{\text{RB}} \times \Delta f}{1000} \right), \tag{8.1}$$

where $\Delta f = 15$ kHz is the subcarrier spacing. The channel edges are defined as the lowest and highest frequencies of the carrier separated by the channel bandwidth, that is at:

$$F_c \pm BW_{\text{channel}}, \tag{8.2}$$

where F_c is the carrier center frequency.

The spacing between carriers depends on the deployment scenario, the size of the frequency block available and the channel bandwidths. The nominal channel spacing between

Table 8.1. Transmission bandwidth configuration BW_{config} in LTE channel bandwidths.

Channel bandwidth $BW_{channel}$ [MHz]	Transmission bandwidth configuration N_{RB}	Transmission bandwidth configuration BW_{config} [MHz]
1.4	6	1.08
3	15	2.7
5	25	4.5
10	50	9
15	75	13.5
20	100	18.0

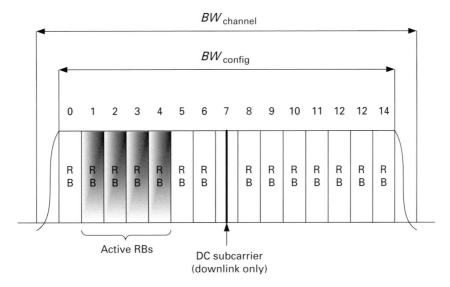

Figure 8.1. Definition of channel bandwidth and transmission bandwidth configuration for one LTE carrier.

two adjacent LTE carriers is defined as the following:

$$\Delta f_{nominal} = \frac{\left(BW_{channel(1)} + BW_{channel(2)}\right)}{2}, \tag{8.3}$$

where $BW_{channel(1)}$ and $BW_{channel(2)}$ are the channel bandwidths of the two respective LTE carriers. The channel spacing can be adjusted to optimize performance in a particular deployment scenario. The channel raster is 100 kHz, which means that the carrier center frequency is always an integer multiple of 100 kHz.

8.2   UE radio access capabilities

The LTE system supports five UE categories with different radio access capabilities as summarized in Table 8.2. The total layer 2 buffer size is defined as the sum of the number of bytes that the UE is capable of storing in the radio link control (RLC) transmission windows and RLC reception and reordering windows for all radio bearers. At the time of writing, it is still being discussed whether it should be possible for the UE to report the uplink and downlink

Table 8.2. Parameter values set by the UE category.

		Downlink				Uplink
UE Category	Total layer 2 buffer size [KBytes]	Maximum number of bits per TTI	Maximum number of bits per transport block per TTI	Total number of soft channel bits	Maximum number of MIMO layers	Maximum number of bits per TTI
Category 1	138	10 296	10 296	250 368	1	5160
Category 2	687	51 024	51 024	1 237 248	2	25 456
Category 3	1373	102 048	75 376	1 237 248	2	51 024
Category 4	1832	150 752	75 376	1 827 072	2	51 024
Category 5	3434	302 752	151 376	3 667 200	4	75 376 (64-QAM)

buffer size separately, which means that buffers are not shared between the uplink and the downlink.

The four parameters defined for the downlink include maximum number of bits per TTI, maximum number of bits per transport block per TTI, total number of soft channel bits and maximum number of MIMO layers. The maximum number of bits per TTI is the maximum number of DL-SCH transport blocks bits that the UE is capable of receiving within a DL-SCH TTI. In the case of spatial multiplexing, this is the sum of the number of bits delivered in each of the two transport blocks. With TTI of 1 ms, the downlink peak data rates, for example, for category 4 and 5 UEs are 150 and 302 Mb/s respectively. The maximum number of bits per transport block per TTI is the maximum number of DL-SCH transport block bits that the UE is capable of receiving in a single transport block within a DL-SCH TTI. A single transport block is used for a single MIMO layer transmission. We note that the minimum downlink data rate supported is approximately 10 Mb/s for category 1 UEs. The parameter soft channel bits indicates the total number of soft channel bits available for hybrid ARQ processing. This parameter is used for deciding whether a hybrid ARQ retransmission is in Chase combining mode (retransmitting the same redundancy version) or in incremental redundancy mode (retransmitting a different redundancy version) as discussed in Chapter 11, Section 11.7.3.

The uplink peak data rates, for category 4 and 5 are 51 and 75 Mb/s respectively. In the uplink, only category 5 UEs support 64-QAM modulation. It should also be noted that in the first release of the LTE system, MIMO spatial multiplexing is not supported in the uplink.

8.3   Frame and slot structure

In the LTE system, uplink and downlink data transmissions are scheduled on a 1 ms subframe basis. A subframe consists of two equal duration (0.5 ms) consecutive time slots with subframe number i consisting of slots $2i$ and $(2i + 1)$. All the time durations are defined in terms of the sample period $T_s = 1/f_s$, where $f_s = 30.72$ Msamples/sec. Some of the control signals such as synchronization and broadcast control in the downlink are carried on a 10 ms radio frame basis, where a radio frame is defined to consist of 10 subframes as shown in Figure 8.2. The transmission of the uplink radio frame number i from a UE starts $N_{TA} \times T_s$ seconds before the start of the corresponding downlink radio frame at the UE, where N_{TA} represents the timing offset between uplink and downlink radio frames at the UE in units of T_s. This timing offset N_{TA} is adjusted for each UE in order to make sure that signals from multiple UEs transmitting on the uplink arrive at the eNode-B at the same time. Each slot is further divided into $N_{\text{symb}}^{\text{UL}}$ SC-FDMA symbols or $N_{\text{symb}}^{\text{DL}}$ OFDM symbols for the uplink and downlink respectively. A resource element is one subcarrier in a single OFDM or SC-FDMA symbol as shown in Figure 8.2. A resource element is defined by the index pair (k, l) in a slot, where k and l are the subcarrier and OFDM/SC-FDMA symbol index respectively.

8.3.1 Physical resource block

A physical resource block (PRB) is defined as $N_{\text{SC}}^{\text{RB}}$ consecutive subcarriers in the frequency-domain and $N_{\text{symb}}^{\text{UL}}$ SC-FDMA symbols in the uplink or $N_{\text{symb}}^{\text{DL}}$ OFDM symbols in the downlink. A physical resource block therefore consists of $\left(N_{\text{symb}}^{\text{UL}} \times N_{\text{SC}}^{\text{RB}}\right)$ resource elements in the

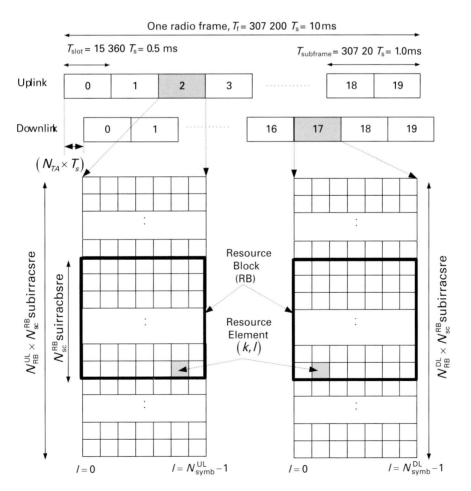

Figure 8.2. Downlink and uplink frame structure.

Table 8.3. Physical resource block parameters.

Configuration		N_{sc}^{RB}	N_{symb}^{DL}	N_{symb}^{UL}
Normal cyclic prefix	$\Delta f = 15\,\text{kHz}$	12	7	7
Extended cyclic prefix	$\Delta f = 15\,\text{kHz}$	12	6	6
	$\Delta f = 7.5\,\text{kHz}$	24	3	NA

uplink and $\left(N_{symb}^{DL} \times N_{SC}^{RB}\right)$ resource elements in the downlink. This corresponds to 180 KHz bandwidth in the frequency-domain and one slot in the time-domain. The number of subcarriers within a resource block N_{SC}^{RB} is 12 or 24 for the case of 15 or 7.5 kHz subcarrier spacing respectively as given in Table 8.3. The subcarrier spacing of 7.5 kHz is only used for the case of MBSFN transmissions with extended cyclic prefix. The minimum number of resource blocks within a slot is six (1.08 MHz transmission bandwidth) while the maximum number of resource blocks within a slot is 110 (19.8 MHz transmission bandwidth).

The physical resource blocks are numbered from 0 to ($N_{RB} - 1$) in the frequency domain. The relation between the physical resource block number n_{PRB} in the frequency domain and resource elements (k, l) in a slot is given by:

$$n_{PRB} = \left\lfloor \frac{k}{N_{sc}^{RB}} \right\rfloor. \tag{8.4}$$

8.3.2 Slot structure

The detailed slot structure is shown in Figure 8.3. A slot consists of 7 or 6 SC-FDMA or OFDM symbols for the case of normal and extended cyclic prefix respectively. The normal cyclic prefix length is 5.2 μs ($160 \times T_s$) in the first SC-FDMA or OFDM symbol and 4.7 μs ($144 \times T_s$) in the remaining six symbols. For the case of unicast traffic, the subcarrier spacing is ($\Delta f = 15$ KHz), which results in SC-FDMA or OFDM symbol duration of $\left(1/\Delta f = 66.6\ \mu s\right)$. The overhead for the case of the normal cyclic prefix can be calculated as below:

$$CP - OH_{normal} = \frac{160 \times T_s + 6 \times 144 \times T_s}{7 \times 2048 \times T_s} = 7.14\%. \tag{8.5}$$

The cyclic prefix lengths of ($160 \times T_s$) and ($144 \times T_s$) were selected so that the result is an integer number of samples for IFFT sizes of 128, 256, 512, 1024 and 2048. The number of samples for the CP for IFFT size of N_{IFFT} can be written as:

$$CP_{normal\text{-}samples} = \begin{cases} \dfrac{166 \times N_{IFFT}}{2048} & l = 0 \\[2mm] \dfrac{144 \times N_{IFFT}}{2048} & l = 1, 2, \ldots, 6 \end{cases}. \tag{8.6}$$

In the case of extended cyclic prefix, the CP length of 16.6 μs ($512 \times T_s$) is the same in all the six symbols. This leads to a CP overhead of 25%. An extended CP length of 33.3μs ($1024 \times T_s$) for subcarrier spacing of ($\Delta f = 15$ kHz) is defined for MBSFN traffic only. The OFDM symbol duration of $\left(1/\Delta f = 133.3\ \mu s\right)$ and CP duration of 33.3 μs ($1024 \times T_s$) results in 25% overhead as well. In fact, the reason for defining a narrower subcarrier spacing

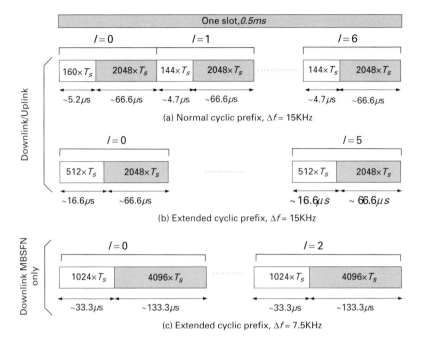

Figure 8.3. Slot structure for normal and extended cyclic prefix.

of ($\Delta f = 15$ kHz) was to keep the overhead relatively low. If the CP duration is increased to 33.3 μs ($1024 \times T_s$) while keeping the subcarrier spacing at ($\Delta f = 15$ kHz), the CP overhead would be 50%. Due to larger OFDM symbol duration for 7.5 kHz subcarrier spacing, only three symbols are carried within a 0.5 ms slot.

8.4   Frame structure type 2

Frame structure type 2 is defined to support unpaired or time-division-duplex (TDD) operation. In this case, each radio frame of length $T_f = 307\,200.T_s = 10$ ms is divided into two half-frames of length $T_f = 153\,600 \cdot T_s = 5$ms each. Each half-frame consists of five subframes of duration 1 ms each. The supported uplink–downlink configurations are shown in Figure 8.4. A special subframe with the three fields DwPTS, GP and UpPTS is defined to allow switching from downlink to uplink and vice versa. The length of DwPTS and UpPTS is given by Table 8.4. The total length of DwPTS, GP and UpPTS is equal to $30720 \cdot T_s = 1$ ms. As for the case of paired or FDD frame structure of Figure 8.2, all subframes which are not special subframes in TDD are defined as two slots each of length $T_{\text{slot}} = 15360 \cdot T_s = 0.5$ ms.

As shown in Figure 8.4, uplink–downlink configurations with both 5 and 10 ms downlink-to-uplink switch-point periodicity are supported. We also note that in the case of 5 ms downlink-to-uplink switch-point periodicity, the special subframe exists in both half-frames. A smaller switch-point periodicity of 5 ms provides lower latency relative to 10 ms periodicity at the expense of low efficiency due to an additional special subframe. This is because in the case of 10 ms downlink-to-uplink switch-point periodicity, the special subframe exists in the first half-frame only.

Table 8.4. Configuration of special subframe in frame structure type 2.

Special subframe configuration	Normal cyclic prefix			Extended cyclic prefix		
	DwPTS	GP	UpPTS	DwPTS	GP	UpPTS
0	$6592 \cdot T_S$	$21\,936 \cdot T_S$		$7680 \cdot T_S$	$20\,480 \cdot T_S$	
1	$19\,760 \cdot T_S$	$8768 \cdot T_S$		$20\,480 \cdot T_S$	$7680 \cdot T_S$	$2560 \cdot T_S$
2	$21\,952 \cdot T_S$	$6576 \cdot T_S$	$2192 \cdot T_S$	$23\,040 \cdot T_S$	$5120 \cdot T_S$	
3	$24\,144 \cdot T_S$	$4384 \cdot T_S$		$25\,600 \cdot T_S$	$2560 \cdot T_S$	
4	$26\,336 \cdot T_S$	$2192 \cdot T_S$		$7680 \cdot T_S$	$17\,920 \cdot T_S$	
5	$6592 \cdot T_S$	$19\,744 \cdot T_S$		$20\,480 \cdot T_S$	$5120 \cdot T_S$	$5120 \cdot T_S$
6	$19\,760 \cdot T_S$	$6576 \cdot T_S$	$4384 \cdot T_S$	$23\,040 \cdot T_S$	$2560 \cdot T_S$	
7	$21\,952 \cdot T_S$	$4384 \cdot T_S$		-	-	-
8	$24\,144 \cdot T_S$	$2192 \cdot T_S$		-	-	-

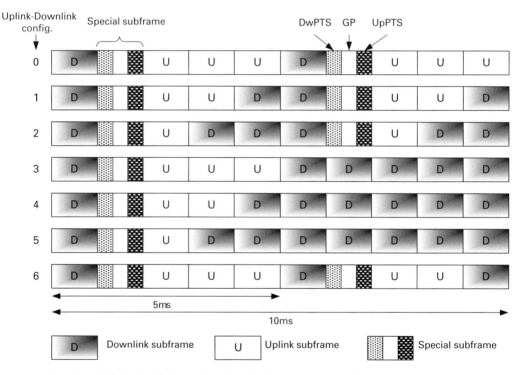

Figure 8.4. Uplink–downlink configurations in frame structure type 2.

The subframes 0 and 5 and DwPTS are always reserved for downlink transmission. UpPTS and the subframe immediately following the special subframe are always reserved for uplink transmission.

8.5   Downlink distributed transmission

In the LTE downlink, the virtual resource block (VRB) concept is defined to enable distributed transmissions. A virtual resource block is of the same size as a physical resource block.

Table 8.5. RB gap values.

System BW (N_{RB}^{DL})	Gap (N_{gap})	
	1st Gap ($N_{gap,1}$)	2nd Gap ($N_{gap,2}$)
6–10	$\left\lceil N_{RB}^{DL}/2 \right\rceil$	N/A
11	4	N/A
12–19	8	N/A
20–26	12	N/A
27–44	18	N/A
45–49	27	N/A
50–63	27	9
64–79	32	16
80–110	48	16

We can differentiate two types of virtual resource blocks namely virtual resource blocks of localized type and virtual resource blocks of distributed type. For each type of virtual resource blocks, a pair of virtual resource blocks over two slots in a subframe is assigned together by a single virtual resource block number, n_{VRB}. The virtual resource blocks of localized type are mapped directly to physical resource blocks such that virtual resource block n_{VRB} corresponds to physical resource block $n_{PRB} = n_{VRB}$. The virtual resource blocks of localized type are numbered from 0 to $\left(N_{VRB}^{DL} - 1\right)$, where $N_{VRB}^{DL} = N_{RB}^{DL}$. The virtual resource blocks of distributed type are numbered from 0 to $\left(N_{VRB}^{DL} - 1\right)$, where N_{VRB}^{DL} is given as:

$$N_{VRB}^{DL} = \begin{cases} N_{VRB,gap1}^{DL} = 2 \cdot \min\left[N_{gap}, \left(N_{RB}^{DL} - N_{gap}\right)\right], & N_{gap} = N_{gap,1} \\ N_{VRB,gap1}^{DL} = \left\lfloor \dfrac{N_{RB}^{DL}}{2N_{gap}} \right\rfloor \cdot 2N_{gap}, & N_{gap} = N_{gap,2}, \end{cases} \tag{8.7}$$

where values of N_{gap} are given in Table 8.5. A single gap value $N_{gap,1}$ is defined for $N_{RB}^{DL} < 50$ while two gap values $N_{gap,1}$ and $N_{gap,2}$ are defined for $N_{RB}^{DL} > 50$. In the case of two gap values, the gap value used is indicated by a 1-bit flag as part of the downlink scheduling assignment.

Consecutive $\tilde{N}_{RB}^{DL}$ VRB numbers compose a unit of VRB number interleaving where $\tilde{N}_{VRB}^{DL}$ is defined as:

$$\tilde{N}_{VRB}^{DL} = \begin{cases} N_{VRB}^{DL}, & N_{gap} = N_{gap,1} \\ 2N_{gap} & N_{gap} = N_{gap,2}. \end{cases} \tag{8.8}$$

Interleaving of VRB numbers of each interleaving unit is performed with four columns and N_{row} rows:

$$N_{row} = \left\lceil \tilde{N}_{VRB}^{DL}/(4P) \right\rceil \times P, \tag{8.9}$$

where P is resource block group (RBG) size, which is a function of the system bandwidth as given in Table 8.6. VRB numbers are written row-by-row in the rectangular matrix, and read out column-by-column. N_{null} nulls are inserted in the last $N_{null}/2$ rows of the second and fourth columns, where

$$N_{null} = 4N_{row} - \tilde{N}_{VRB}^{DL}. \tag{8.10}$$

Table 8.6. RBG sizes.

System bandwidth N_{RB}^{DL}	RBG size (P)
≤ 10	1
11–26	2
27–63	3
64–110	4

The nulls are ignored when reading out. The VRB numbers mapping to PRB numbers including interleaving are derived as follows:

$$
n_{PRB(n_s)} = \begin{cases} \tilde{n}_{PRB}(n_s), & \tilde{n}_{PRB}(n_s) < \dfrac{\tilde{N}_{VRB}^{DL}}{2} \\[3mm] \tilde{n}_{PRB}(n_s) + N_{gap} - \dfrac{\tilde{N}_{VRB}^{DL}}{2} & \tilde{n}_{PRB}(n_s) \geq \dfrac{\tilde{N}_{VRB}^{DL}}{2}, \end{cases} \tag{8.11}
$$

where $\tilde{n}_{PRB}(n_s)$ for even slot number n_s is given as:

$$
\tilde{n}_{PRB}(n_s) = \begin{cases} \tilde{n}'_{PRB} - N_{row} & N_{null} \neq 0, \tilde{n}_{VRB} \geq \left(\tilde{N}_{VRB}^{DL} - N_{null}\right), \tilde{n}_{VRB} \bmod 2 = 1 \\ \tilde{n}'_{PRB} - N_{row} + N_{null}/2 & N_{null} \neq 0, \tilde{n}_{VRB} \geq \left(\tilde{N}_{VRB}^{DL} - N_{null}\right), \tilde{n}_{VRB} \bmod 2 = 0 \\ \tilde{n}''_{PRB} - N_{null}/2 & N_{null} \neq 0, \tilde{n}_{VRB} < \left(\tilde{N}_{VRB}^{DL} - N_{null}\right), \tilde{n}_{VRB} \bmod 4 \geq 2, \\ \tilde{n}''_{PRB} & \text{otherwise,} \end{cases} \tag{8.12}
$$

where

$$
\tilde{n}'_{PRB} = 2N_{row} \cdot (\tilde{n}_{VRB} \bmod 2) + \left\lfloor \frac{\tilde{n}_{VRB}}{2} \right\rfloor + \tilde{N}_{VRB}^{DL} \cdot \left\lfloor \frac{n_{VRB}}{\tilde{N}_{VRB}^{DL}} \right\rfloor
$$

$$
\tilde{n}''_{PRB} = N_{row} \cdot (\tilde{n}_{VRB} \bmod 4) + \left\lfloor \frac{\tilde{n}_{VRB}}{4} \right\rfloor + \tilde{N}_{VRB}^{DL} \cdot \left\lfloor \frac{n_{VRB}}{\tilde{N}_{VRB}^{DL}} \right\rfloor, \tag{8.13}
$$

where

$$
\tilde{n}_{VRB} = n_{VRB} \bmod \tilde{N}_{VRB}^{DL}. \tag{8.14}
$$

The term

$$
\tilde{n}''_{PRB} = N_{row} \cdot (\tilde{n}_{VRB} \bmod 4) + \left\lfloor \frac{\tilde{n}_{VRB}}{4} \right\rfloor \tag{8.15}
$$

in (8.13), for example, implements the interleaver where VRB numbers are written row-by-row in the rectangular matrix with four columns, and read out column-by-column.

For an odd slot number $\tilde{n}_{PRB}(n_s)$ is derived based on the previous even slot number $\tilde{n}_{PRB}(n_s - 1)$ and is given as:

$$
\tilde{n}_{PRB}(n_s) = \left(\tilde{n}_{PRB}(n_s - 1) + \frac{\tilde{N}_{VRB}^{DL}}{2}\right) \bmod \tilde{N}_{VRB}^{DL} + \tilde{N}_{VRB}^{DL} \cdot \left\lfloor \frac{n_{VRB}}{\tilde{N}_{VRB}^{DL}} \right\rfloor. \tag{8.16}
$$

Let us consider a case of 3.0 MHz downlink bandwidth with $\left(N_{RB}^{DL} = 15\right)$. From Table 8.5, we obtain $N_{gap} = N_{gap,1} = 8$. Further by using (8.7) and (8.8), we obtain the number of VRBs $N_{\text{VRB}}^{\text{DL}} = 14$ and also $\tilde{N}_{\text{VRB}}^{\text{DL}} = 14$. The number of rows N_{row} in the interleaving matrix are then determined as:

$$N_{\text{row}} = \left\lceil \frac{\tilde{N}_{\text{VRB}}^{\text{DL}}}{(4P)} \right\rceil \times P = \left\lceil \frac{14}{(4 \times 2)} \right\rceil \times 2 = 4. \tag{8.17}$$

We note that the RBG size P for $\left(N_{\text{RB}}^{\text{DL}} = 15\right)$ equals 2 from Table 8.6. We show the interleaving matrix in Figure 8.5 with $N_{\text{null}} = 2$ as determined by (8.10). Let us now consider a case where eNB allocates $n_{\text{VRB}} = 1, 2$ in the downlink scheduling assignment. By using (8.14) we have $\tilde{n}_{\text{VRB}} = 1, 2$. We obtain $\tilde{n}_{\text{PRB}}'' = 4, 8$ corresponding to $\tilde{n}_{\text{VRB}} = 1, 2$ by using (8.13). Finally, by using (8.11) and (8.12), we get $n_{\text{PRB}}(0) = 4, 8$ and $n_{\text{PRB}}(1) = 1, 11$ for even and odd slots respectively. The mapping to PRBs for this example is shown in Figure 8.6. We note that the transmitted PRBs are distributed across the system bandwidth in the two slots to capture frequency diversity.

The localized or distributed VRB assignment is indicated by a 1-bit flag in downlink control information (DCI) formats 1A and 1B. The formats 1A and 1B are used for low overhead compact scheduling of one PDSCH codeword as described in Chapter 15, Section 15.5.1. In general, a distributed transmission could be signaled by bitmap-based allocation of resource blocks at the expense of additional overhead. It should be noted that distributed VRB assignment not only distributes the allocated PRBs in a slot but also hops the PRBs between the two slots as shown in Figure 8.6. The overhead for the bitmap scheme would be even more to indicate different PRB locations in the two slots. For the example in Figure 8.6, a slot-level PRB bitmap would require 30-bits resource allocation overhead for $\left(N_{\text{RB}}^{\text{DL}} = 15\right)$.

The benefit of distributed VRB assignment is that it allows achieving distributed PRBs allocation in a slot and across the slots with low signaling overhead. In this case, the signaling bits required for compact scheduling grant is simply (see Chapter 15, Section 15.4):

$$\left\lceil \log_2 \left(N_{\text{RB}}^{\text{DL}} (N_{\text{RB}}^{\text{DL}} + 1)/2 \right) \right\rceil = \left\lceil \log_2 \left(15(15 + 1)/2 \right) \right\rceil = 7 \text{ bits}. \tag{8.18}$$

After adding a 1-bit flag to differentiate the localized or distributed VRB assignment, the total overhead is only 8-bits. The difference in overhead between bitmap and compact assignments would be even larger for system bandwidths with larger numbers of resource blocks.

0	1	2	3
4	5	6	7
8	9	10	11
12	NULL	13	NULL

Figure 8.5. Interleaving matrix for $\left(N_{\text{RB}}^{\text{DL}} = 15\right)$.

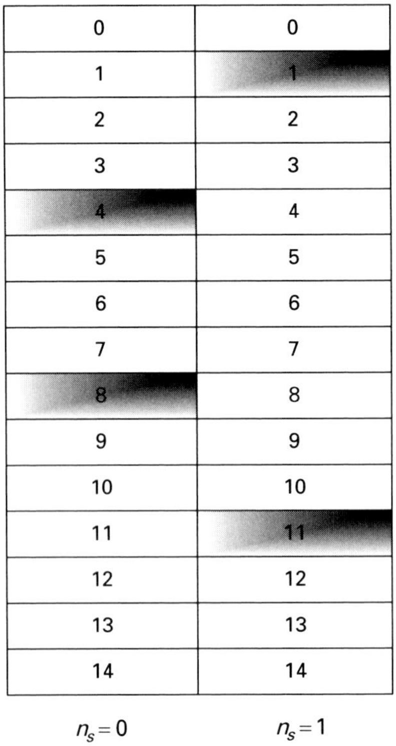

0	0
1	1
2	2
3	3
4	4
5	5
6	6
7	7
8	8
9	9
10	10
11	11
12	12
13	13
14	14
$n_s = 0$	$n_s = 1$

Figure 8.6. An example of PRBs for $\left(N_{RB}^{DL} = 15 \right)$ and $n_{VRB} = 1, 2$ allocation.

8.6   Uplink hopping

The uplink allocations in SC-FDMA are always contiguous to maintain the single-carrier property as discussed in Chapter 5. Therefore, distributed resource allocation cannot be employed to recuperate frequency diversity. However, resource hopping can be used to provide frequency diversity while keeping the resource allocations contiguous. The LTE system allows the configuration of either inter-subframe hopping or both inter-subframe and intra-subframe hopping. In the case of intra-subframe hopping, resources are hopped across the two slots within a subframe. It should be noted that hopping at SC-FDMA symbol level is not permitted, as there is a single reference signal symbol per slot. Moreover, no hopping transmissions mode is supported to enable uplink frequency-selective scheduling where diversity can degrade performance.

The uplink bandwidth is divided into N_{sb} subbands for uplink hopping purposes. The hopping is then performed at the subband level. The size of each subband defined for hopping purposes N_{RB}^{sb} is given by:

$$N_{RB}^{sb} = \left\lfloor \frac{\left(N_{RB}^{UL} - \tilde{N}_{RB}^{PUCCH} \right)}{N_{sb}} \right\rfloor . \tag{8.19}$$

The resources allocated for PUCCH transmission are excluded to determine the bandwidth over which hopping is performed. $\tilde{N}_{RB}^{PUCCH}$ is given as:

$$\tilde{N}_{RB}^{PUCCH} = \begin{cases} N_{RB}^{PUCCH} + 1 & N_{RB}^{PUCCH} \text{ odd} \\ N_{RB}^{PUCCH} & \text{otherwise,} \end{cases} \qquad (8.20)$$

where N_{RB}^{PUCCH} represents the number of resource blocks allocated for PUCCH transmission in a slot. It should be noted that the number of resource blocks excluded due to PUCCH transmission are always even. This is due to the PUCCH mapping at the two edges of the bandwidth with slot-level hopping as described in Chapter 14.

In cases where uplink frequency hopping is disabled, the set of physical resource blocks to be used for transmission are simply given by:

$$n_{PRB} = n_{VRB}, \qquad (8.21)$$

where n_{VRB} is obtained from the uplink scheduling grant.

If uplink frequency hopping with a predefined hopping pattern is enabled, the set of physical resource blocks to be used for transmission in slot n_s is given by the scheduling grant together with a predefined pattern according to:

$$n_{PRB}(n_s) = \left[\frac{n_{VRB} - \lceil N_{RB}^{PUCCH}/2 \rceil + \cdots}{\left(f_{hop}(i) \cdot N_{RB}^{sb} \right)} \right] \bmod \left(N_{RB}^{sb} \cdot N_{sb} \right) + \lceil N_{RB}^{PUCCH}/2 \rceil, \quad (8.22)$$

where n_{VRB} is obtained from the scheduling grant. The index i is incremented every subframe for inter-subframe hopping and every slot for intra-subframe and inter-subframe hopping as:

$$i = \begin{cases} \lfloor n_s/2 \rfloor & \text{inter-subframe hopping} \\ n_s & \text{intra- and inter-subframe hopping.} \end{cases} \qquad (8.23)$$

The hopping function $f_{hop}(i)$ is given by:

$$f_{hop}(i) = \begin{cases} 0, & N_{sb} = 1 \\ \left[f_{hop}(i-1) + 1 \right] \bmod N_{sb}, & N_{sb} = 2 \\ \left[f_{hop}(i-1) + \Delta f_{hop}(i) + 1 \right] \bmod N_{sb}, & N_{sb} > 2, \end{cases} \qquad (8.24)$$

where the hopping offset Δf_{hop} is given as:

$$\Delta f_{hop}(i) = \left(\sum_{k=i\cdot 10+1}^{i\cdot 10+9} c(k) \times 2^{k-(i\cdot 10+1)} \right) \bmod (N_{sb} - 1). \qquad (8.25)$$

Also $f_{hop}(-1) = 0$ and the *PN*-sequence $c(\cdot)$ generator is initialized at the start of each frame in a cell-specific manner as:

$$c_{init} = N_{ID}^{cell}. \qquad (8.26)$$

In this way, hopping happens independently in each cell while avoiding resource collisions due to hopping in the same cell. We note from (8.24) that for the case of a single subband $N_{sb} = 1$, no subband hopping is possible. For the case of two subbands $N_{sb} = 2$, the transmission

Table 8.7. An example of hopping function $f_{hop}(i)$ for $N_{sb} = 4$.

	$f_{hop}(i)$			
$f_{hop}(i-1)$	$\Delta f_{hop} = 0$	$\Delta f_{hop} = 1$	$\Delta f_{hop} = 2$	$\Delta f_{hop} = 3$
0	1	2	3	0
1	2	3	0	1
2	3	0	1	2
3	0	1	2	3

is hopped from one subband to the other with every increment of i. An example of hopping function $f_{hop}(i)$ for the case of four subbands $N_{sb} = 4$ is given in Table 8.7. We note that as hopping offset Δf_{hop} takes random values based on the PN-sequence, the transmissions are hopped from subband to subband in a random fashion. As the new subband information is derived based on the previous subband and the random offset, there are no collisions due to hopping. We note from Table 8.7 that $\Delta f_{hop} = 0, 1, 2, 3$ introduces cyclic shift of $1, 2, 3$ and 0 subband units.

In addition to hopping, mirroring can also be applied to introduce further randomization. A function $f_m(i) \in \{0, 1\}$ determines whether mirroring is used or not. The value of the mirroring function $f_m(i)$ is derived based on the PN-sequence $c(\cdot)$.

$$f_m(i) = \begin{cases} i \bmod 2 & N_{sb} = 1 \\ c(i \cdot 10) & N_{sb} > 1. \end{cases} \tag{8.27}$$

With mirroring, the set of physical resource blocks to be used for transmission in slot n_s is given by:

$$n_{PRB}(n_s) = \left[\begin{array}{c} n_{VRB} - \lceil N_{RB}^{PUCCH}/2 \rceil + \cdots \\ (f_{hop}(i) \cdot N_{RB}^{sb}) + \Delta f_m(i) \end{array} \right] \bmod \left(N_{RB}^{sb} \cdot N_{sb} \right) + \lceil N_{RB}^{PUCCH}/2 \rceil \tag{8.28}$$

$$\Delta f_m(i) = \left[\left(N_{RB}^{sb} - 1 \right) - 2 \left(\left(n_{VRB} - \lceil N_{RB}^{PUCCH}/2 \rceil \right) \bmod N_{RB}^{sb} \right) \right] \cdot f_m(i). \tag{8.29}$$

We give values of Δf_m with mirroring enabled, $f_m(i) = 1$ for $N_{RB}^{UL} = 12$, $N_{sb} = 4$ and $N_{RB}^{sb} = 3$ in Table 8.8. For simplicity, we assumed that $N_{RB}^{PUCCH} = 0$. We note that while hopping function enables subband level hopping, the mirroring function swaps the resource block within the subband as is also shown in Figure 8.7. In Figure 8.7, we assumed $\Delta f_{hop} = 1$ and therefore subbands hop according to the third column in Table 8.7. With three resource blocks per subband $N_{RB}^{sb} = 3$, $\Delta f_m = 0$ means that the position within the subband is not changed, which is the case for the resource blocks in the middle of the subband. However, the resource blocks at the two edges of the subband swap positions as $\Delta f_m = 2, -2$ when $f_m(i) = 1$.

8.7   Summary

The physical layer of LTE supports bandwidths in increments of a resource block starting from a minimum bandwidth of six resource blocks. However, to minimize the number of test

Table 8.8. Values of Δf_m with mirroring enabled, $f_m(i) = 1$ for $N_{RB}^{UL} = 12$, $N_{sb} = 4$ and $N_{RB}^{sb} = 3$.

n_{VRB}	0	1	2	3	4	5	6	7	8	9	10	11
Δf_m	2	0	−2	2	0	−2	2	0	−2	2	0	−2

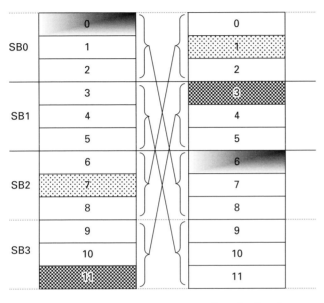

$$\Delta f_{hop} = 1, \Delta f_m = 1$$

Figure 8.7. An example of uplink hopping for $N_{RB}^{UL} = 12$, $N_{sb} = 4$ and $N_{RB}^{sb} = 3$.

scenarios a set of six channel bandwidths comprising 1.4, 3, 5, 10, 15 and 20 MHz is defined. The actual transmission bandwidth is generally smaller than the channel bandwidth to provide the guard bands. For channel bandwidths of 3–20 MHz, the transmission bandwidth is 90% of the channel bandwidth leaving 10% of the channel bandwidth for guard bands on the two edges.

In order to support various cell sizes with reasonable cyclic prefix overhead, two sets of cyclic prefix lengths are defined. The overhead for the normal and extended cyclic prefix is 7.14% and 25% respectively. A subcarrier spacing of 15 kHz is chosen to strike a balance between cyclic prefix overhead and robustness to Doppler spread. An additional smaller 7.5 kHz subcarrier spacing is defined for MBSFN to support large delay spreads with reasonable cyclic prefix overhead.

The uplink supports only localized transmissions with contiguous resource block allocation due to single-carrier FDMA. However, inter-subframe and intra-subframe hopping can be applied to capture frequency diversity. Similarly, virtual resource blocks of the distributed type can be used for frequency-diversity transmission in the downlink.

References

[1] 3GPP TSG RAN v8.2.0, Evolved Universal Terrestrial Radio Access: Base Station Transmission and Reception (3GPP TS 36.104).

[2] 3GPP TSG RAN v8.3.0, Evolved Universal Terrestrial Radio Access: Physical Channels and Modulation (3GPP TS 36.211).

[3] 3GPP TSG RAN v8.3.0, Evolved Universal Terrestrial Radio Access: Multiplexing and Channel Coding (3GPP TS 36.212).

[4] 3GPP TSG RAN v8.3.0, Evolved Universal Terrestrial Radio Access: Physical Layer Procedures (3GPP TS 36.213).

9 Cell search and reference signals

A cell search procedure is used by the UEs to acquire time and frequency synchronization within a cell and detect the cell identity. In the LTE system, cell search supports a scalable transmission bandwidth from 1.08 to 19.8 MHz. The cell search is assumed to be based on two signals transmitted in the downlink, the synchronization signals and broadcast control channel (BCH).

The primary purpose of the synchronization signals is to enable the acquisition of the received symbol timing and frequency of the downlink signal. The cell identity information is also carried on the synchronization signals. The UE can obtain the remaining cell/system-specific information from the BCH. The primary purpose of the BCH is to broadcast a certain set of cell and/or system-specific information. After receiving synchronization signals and BCH, the UE generally acquires information that includes the overall transmission bandwidth of the cell, cell ID, number of transmit antenna ports and cyclic prefix length, etc.

The synchronization signals and BCH are transmitted using the same minimum bandwidth of 1.08 MHz in the central part of the overall transmission band of the cell. This is because, regardless of the total transmission bandwidth capability of an eNB, a UE should be able to determine the cell ID using only the central portion of the bandwidth in order to achieve a fast cell search.

The reference signals are used for channel quality measurements for scheduling, link adaptation and handoff, etc. as well as for data demodulation. In the downlink, three types of reference signals are defined. For non-MBSFN transmissions in the downlink, two types of reference signals are specified: cell-specific reference signals and UE-specific reference signals. For MBSFN transmissions, MBSFN area specific reference signals are defined. In the uplink, there are no common signals due to the many-to-one nature of uplink transmissions. For data demodulation, UE-specific dedicated reference signals are transmitted. For uplink channel quality measurements and power control, sounding reference signals can be transmitted by the UE.

9.1 PN sequence

A *pseudo-noise* (PN) sequence is extensively used in the LTE system for various purposes such as scrambling of reference signals, scrambling of downlink and uplink data transmissions as well as in generating various hopping sequences.

9.1.1 Maximal length sequence

A PN sequence can be generated by using linear feedback shift registers (LFSR). The shift-register sequences with the maximum possible period for an *l*-stage shift register are called *maximal length sequences* or *m-sequences*. A necessary and sufficient condition for a sequence generated by an LFSR to be of maximal length (*m*-sequence) is that its corresponding polynomial be primitive. The periodic autocorrelation function for an *m*-sequence $x(n)$ is defined as:

$$R(k) \triangleq \frac{1}{M} \sum_{n=0}^{(M-1)} x(n) x(n+k). \tag{9.1}$$

The periodic autocorrelation function $R(k)$ equals:

$$R(k) = \begin{cases} 1.0 & k = pM \\ -\frac{1}{M} & k \neq pM. \end{cases} \tag{9.2}$$

We note that the autocorrelation of an *m*-sequence is 1 for zero-lag, and nearly zero $(-1/M,$ where M is the sequence length) for all other lags. In other words, the autocorrelation of the *m*-sequence can be said to approach unit impulse function as *m*-sequence length increases.

9.1.2   Gold sequence

The Gold sequences have been proposed by Gold in 1967 [1]. These sequences are constructed by EXOR-ing two *m*-sequences of the same length as shown in Figure 9.1. Thus, for a Gold sequence of length $n = 2^l - 1$ we need to use two LFSR sequences, each of length $n = 2^l - 1$. If the LSFRs are chosen appropriately, Gold sequences have better cross-correlation properties than *m*-sequences. Gold (and Kasami) showed that for certain well-chosen *m*-sequences, the cross correlation only takes on three possible values, namely $\{-1, -t, (t - 2)\}$. Here t depends solely on the length of the LFSR used. In fact, for an LFSR with l memory elements:

$$t = \begin{cases} 2^{(l+1)/2} + 1, & l, \quad \text{odd} \\ 2^{(l+2)/2} + 1, & l, \quad \text{even}. \end{cases} \tag{9.3}$$

9.1.3   PN-sequence generation in LTE

In LTE, the PN-sequences are defined by a length-31 Gold sequence as illustrated in Figure 9.2. The output sequence $c(n)$, where $n = 0, 1, \ldots, (M_{PN} - 1)$, is defined by:

$$c(n) = (x_1(n + N_C) + x_2(n + N_C)) \bmod 2. \tag{9.4}$$

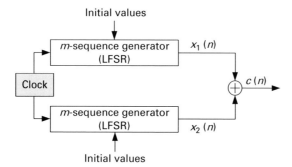

Figure 9.1. A Gold sequence generator.

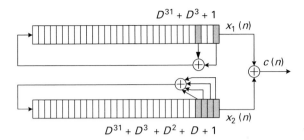

Figure 9.2. PN scrambling code generation in the LTE system.

The first 1600 samples are discarded by setting $N_C = 1600$. The two m-sequences $x_1(n)$ and $x_2(n)$ are respectively generated by feedback polynomials $D^{31} + D^3 + 1$ and $D^{31} + D^3 + D^2 + D + 1$ as below:

$$x_1(n + 31) = (x_1(n + 3) + x_1(n)) \bmod 2$$
$$x_2(n + 31) = (x_2(n + 3) + x_2(n + 2) + x_2(n + 1) + x_2(n)) \bmod 2. \tag{9.5}$$

The first m-sequence is initialized as:

$$x_1(n) = \begin{cases} 1 & n = 0 \\ 0 & 0 < n < 31. \end{cases} \tag{9.6}$$

The initialization of the second m-sequence is given by:

$$c_{\text{init}} = \sum_{i=0}^{30} x_2(i) \cdot 2^i, \tag{9.7}$$

with the value depending on the application of the sequence. For example, when c_{init} is determined based on a cell ID of 255, that is $c_{\text{init}} = N_{\text{ID}}^{\text{cell}} = 255$, the second m-sequence $x_2(n)$ is initialized as:

$$x_2(n) = \begin{cases} 1 & n \leq 8 \\ 0 & 8 < n < 31. \end{cases} \tag{9.8}$$

9.2   Zadoff–Chu (ZC) sequences

Zadoff–Chu (ZC) sequences [2] are also used extensively in the LTE system, for example, for primary synchronization signals, uplink reference signals, uplink physical control channel (PUCCH) and random access channel. A ZC sequence of length N_{ZC} is defined as:

$$x_u(m) = \begin{cases} e^{-j\frac{\pi u m^2}{N_{ZC}}} & \text{when } N_{ZC} \text{ is even} \\ e^{-j\frac{\pi u m(m+1)}{N_{ZC}}} & \text{when } N_{ZC} \text{ is odd} \end{cases} \qquad m = 0, 1, \ldots, (N_{ZC} - 1), \tag{9.9}$$

where u, the sequence index, is relatively prime to N_{ZC} (that is, the only common divisor for u and N_{ZC} is 1). For a fixed u, the ZC sequence has an ideal periodic auto-correlation property (i.e. the periodic auto-correlation is zero for all time shifts other than zero). For different u, ZC sequences are not orthogonal, but exhibit low cross-correlation. If the sequence length N_{ZC} is selected as a prime number, there are $(N_{ZC} - 1)$ different sequences with periodic cross-correlation of $1/\sqrt{N_{ZC}}$ between any two sequences regardless of time shift.

9.3   Downlink frame structure

The downlink transmissions in the LTE system are scheduled on a subframe basis. A subframe of duration 1 ms consists of two slots each of duration 0.5 ms. A group of 20 slots (10 subframes) forms a radio frame of duration 10 ms as shown in Figure 9.3. A slot further consists of seven or six OFDM symbols for the normal cyclic prefix or extended cyclic prefix cases respectively. In the case of 7.5 kHz subcarrier spacing defined for MBSFN and extended cyclic prefix only, a slot consists of three OFDM symbols. For simplicity, we assume the case of normal cyclic prefix with seven OFDM symbols per slot in Figure 9.3.

The cell-specific reference signals are carried in all downlink subframes. The downlink unicast transmissions can be configured to use one, two or four antenna ports. The reference

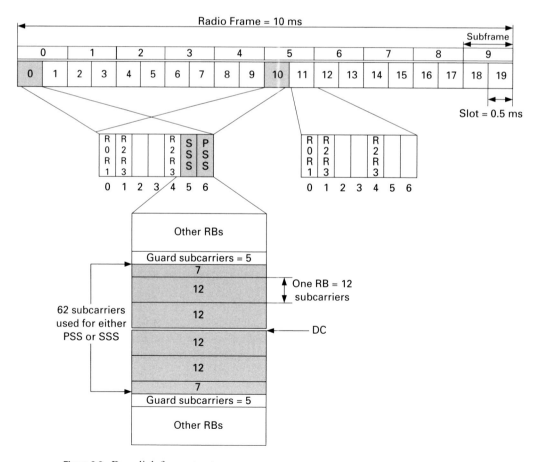

Figure 9.3. Downlink frame structure.

signals for antenna ports 0 and 1 (R_0, R_1) are carried in the first and third last OFDM symbol within a slot. The reference signals for antenna ports 2 and 3 (R_2, R_3) are carried in the second OFDM symbol within a slot. For subframes carrying MBSFN, the reference signals for unicast are present only in the first and second OFDM symbols of the first slot within that subframe. The primary synchronization signal (PSS) and secondary synchronization signal (SSS) are carried in the last and second last OFDM symbols respectively in slot number 0 and slot number 10.

The PSS and SSS are carried in the frequency domain over the middle six resource blocks using 62 subcarriers out of a total of 72 subcarriers (1.08 MHz). Note that the DC subcarrier is not used for any transmission. Moreover, five subcarriers on each side are left as guard subcarriers as shown in Figure 9.3.

9.4   Synchronization signals

A large number of physical layer cell identities (IDs) simplify the task of network planning. This is because neighboring cells are generally required to use different cell IDs. In the LTE system, a total of 504 unique physical layer cell identities are provided. A cell identity is derived from a physical layer cell identity group, $N_{ID}^{(1)}$ in the range of 0–167 and another physical layer identity (within the cell-identity group) $N_{ID}^{(2)}$ in the range 0–2 as shown in Table 9.1.

9.4.1   Primary synchronization signal (PSS)

The primary synchronization signal sequence denoted as $d(n)$ is transmitted in the frequency-domain and is given as below:

$$d_u(n) = \begin{cases} e^{-j\frac{\pi u n(n+1)}{63}} & n = 0, 1, \ldots, 30 \\ e^{-j\frac{\pi u(n+1)(n+2)}{63}} & n = 31, 32, \ldots, 61 \end{cases}, \qquad (9.10)$$

Table 9.1. Physical layer cell identities.

PHY layer cell identity group $N_{ID}^{(1)}$	PHY layer identity $N_{ID}^{(2)}$	PHY layer cell identity $N_{ID}^{\text{cell}} = 3N_{ID}^{(1)} + N_{ID}^{(2)}$
	0	0
0	1	1
	2	2
	0	3
1	1	4
	2	5
$\vdots$	$\vdots$	$\vdots$
	0	501
167	1	502
	2	503

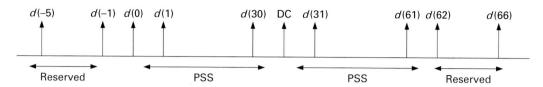

Figure 9.4. Primary Synchronization Signal (PSS) mapping in the frequency-domain. The PSS transmission happens on the last OFDM symbol, $l = N_{\text{symb}}^{\text{DL}} - 1$, in slot numbers 0 and 10.

where the ZC root sequence index u is 25, 29, 34 for $N_{ID}^{(2)} = 0, 1, 2$ respectively. The transmission of primary synchronization signal can use any of the four unicast antenna ports, that is, antenna port 0, 1, 2 or 3. It is also possible that a completely different antenna port from the antenna port 0, 1, 2 or 3 is used for PSS transmission. A cell can create a new antenna port by using one or more of the antenna ports among the available ports, that is, port 0, 1, 2 or 3, to create this new antenna port for the PSS. This allows for use of the proprietary transmit diversity scheme for the PSS. This can be achieved by using some form of multiple antenna precoding.

The mapping of the PSS sequence $d(n)$ to the resource elements is shown in Figure 9.4. The mapping for the non-reserved resource elements (subcarriers other than the guard subcarriers) is performed according to the following relationship:

$$a_{k,l} = d(n), \ k = n - 31 + \left\lfloor \frac{N_{\text{RB}}^{\text{DL}} N_{\text{sc}}^{\text{RB}}}{2} \right\rfloor, \ l = N_{\text{symb}}^{\text{DL}} - 1, \ n = 0, \ldots, 61. \quad (9.11)$$

The resource elements defined by the following relationship are reserved and therefore no transmission happens on these resource elements.

$$k = n - 31 + \left\lfloor \frac{N_{\text{RB}}^{\text{DL}} N_{\text{sc}}^{\text{RB}}}{2} \right\rfloor, \ l = N_{\text{symb}}^{\text{DL}} - 1, \ n = -5, -4, \ldots, -1, 62, 62, \ldots, 66. \quad (9.12)$$

9.4.2 Secondary synchronization signal (SSS)

The antenna port used for transmission of the SSS is the same as the PSS. This allows for the use of the PSS transmission as a phase reference for the detection of the SSS, if so desired. Similar to the PSS, a proprietary transmit diversity scheme can be used for the SSS as well.

An interleaved concatenation of two length-31 binary sequences is used to generate the sequence $d(0), \ldots, d(61)$ for the secondary synchronization signal. Furthermore, the concatenated sequence is scrambled with a scrambling sequence, which is derived from the primary synchronization signal. The combination of two length-31 sequences defining the secondary synchronization signal differs between slot 0 and slot 10. The even-numbered $d(2n)$ and odd-numbered $d(2n+1)$ sequence samples are given as:

$$d(2n) = \begin{cases} s_0^{(m_0)}(n)c_0(n) & \text{in slot 0} \\ s_1^{(m_1)}(n)c_0(n) & \text{in slot 10} \end{cases}$$

$$d(2n+1) = \begin{cases} s_1^{(m_1)}(n)c_1(n)z_1^{(m_0)}(n) & \text{in slot 0} \\ s_0^{(m_0)}(n)c_1(n)z_1^{(m_1)}(n) & \text{in slot 10} \end{cases}, \tag{9.13}$$

where $0 \le n \le 30$. The indices m_0 and m_1 are derived from the physical layer cell ID group $N_{ID}^{(1)}$ according to:

$$\begin{aligned} m_0 &= m' \bmod 31 \\ m_1 &= \left(m_0 + \lfloor m'/31 \rfloor + 1\right) \bmod 31 \\ m' &= N_{ID}^{(1)} + q(q+1)/2, \quad q = \left\lfloor \frac{N_{ID}^{(1)} + q'(q'+1)/2}{30} \right\rfloor, \quad q' = \left\lfloor N_{ID}^{(1)}/30 \right\rfloor . \end{aligned} \tag{9.14}$$

The mapping between PHY layer cell ID group $N_{ID}^{(1)}$ and the indices m_0 and m_1 from the above equation is given in Table 9.2.

Table 9.2. Mapping between PHY layer cell ID group $N_{ID}^{(1)}$ and the indices m_0 and m_1.

PHY layer cell identity group $N_{ID}^{(1)}$	m_0	m_1
0	0	1
1	1	2
⋮	⋮	⋮
29	29	30
30	0	2
31	1	3
⋮	⋮	⋮
58	28	30
59	0	3
60	1	4
⋮	⋮	⋮
86	27	30
87	0	4
88	1	5
⋮	⋮	⋮
113	26	30
114	0	5
115	1	6
⋮	⋮	⋮
139	25	30
140	0	6
141	1	7
⋮	⋮	⋮
164	24	30
165	0	7
166	1	8
167	2	9

Table 9.3. m-sequences $\tilde{s}(n)$, $\tilde{c}(n)$ and $\tilde{z}(n)$.

$\tilde{s}(n)$	$1\ 1\ 1\ 1\ -1\ 1\ 1\ -1\ 1\ -1\ -1\ 1\ 1\ -1\ -1\ -1\ -1\ -1$
	$-1\ 1\ 1\ 1\ -1\ -1\ 1\ 1\ -1\ -1\ -1\ 1\ 1\ -1\ 1\ -1$
$\tilde{c}(n)$	$1\ 1\ 1\ 1\ -1\ 1\ -1\ 1\ 1\ -1\ -1\ -1\ -1\ 1\ 1\ -1\ -1\ 1\ 1\ 1\ 1$
	$-1\ -1\ -1\ -1\ -1\ 1\ 1\ 1\ -1\ -1\ 1\ 1\ -1\ 1\ 1\ 1\ -1$
$\tilde{z}(n)$	$1\ 1\ 1\ 1\ -1\ -1\ -1\ 1\ 1\ 1\ -1\ -1\ 1\ 1\ -1\ -1\ -1\ -1\ -1\ -1$
	$1\ -1\ 1\ 1\ 1\ 1\ -1\ 1\ 1\ 1\ -1\ 1\ 1\ -1\ 1\ -1\ -1$

The two length-31 binary sequences $s_0^{(m_0)}(n)$ and $s_1^{(m_1)}(n)$ are defined as two different cyclic shifts of the m-sequence $\tilde{s}(n)$:

$$s_0^{(m_0)}(n) = \tilde{s}((n + m_0) \bmod 31)$$
$$s_1^{(m_1)}(n) = \tilde{s}((n + m_1) \bmod 31),$$

(9.15)

where $\tilde{s}(i) = 1 - 2x(i)$, $0 \le i \le 30$ with $x(i)$ given as:

$$x(j + 5) = (x(j + 2) + x(j)) \bmod 2, \quad 0 \le j \le 25,$$

(9.16)

with initial conditions $x(0) = 0$, $x(1) = 0$, $x(2) = 0$, $x(3) = 0$, $x(4) = 1$. The expression $(1 - 2x(i))$ converts binary 0 and 1 in $x(i)$ to $+1$ and -1 respectively.

The two length-31 scrambling sequences $c_0(n)$ and $c_1(n)$ which are dependent on the primary synchronization signal are defined by two different cyclic shifts of the m-sequence $\tilde{c}(n)$:

$$c_0(n) = \tilde{c}\left(\left(n + N_{\mathrm{ID}}^{(2)}\right) \bmod 31\right)$$
$$c_1(n) = \tilde{c}\left(\left(n + N_{\mathrm{ID}}^{(2)} + 3\right) \bmod 31\right),$$

(9.17)

where $N_{\mathrm{ID}}^{(2)} \in \{0, 1, 2\}$ is the PHY layer identity within the PHY layer cell ID group $N_{\mathrm{ID}}^{(1)}$ and $\tilde{c}(i) = 1 - 2x(i)$, $0 \le i \le 30$, with $x(i)$ given as:

$$x(j + 5) = (x(j + 3) + x(j)) \bmod 2, \quad 0 \le j \le 25,$$

(9.18)

with initial conditions $x(0) = 0$, $x(1) = 0$, $x(2) = 0$, $x(3) = 0$, $x(4) = 1$.

The two length-31 scrambling sequences $z_1^{(m_0)}(n)$ and $z_1^{(m_0)}(n)$ are defined by a cyclic shift of the m-sequence $\tilde{z}(n)$ (see Table 9.3) according to:

$$z_1^{(m_0)}(n) = \tilde{z}((n + (m_0 \bmod 8)) \bmod 31)$$
$$z_1^{(m_1)}(n) = \tilde{z}((n + (m_1 \bmod 8)) \bmod 31),$$

(9.19)

where m_0 and m_1 are obtained from Table 9.2 and $\tilde{z}(i) = 1 - 2x(i)$, $0 \le i \le 30$, with $x(i)$ given as:

$$x(j + 5) = (x(j + 4) + x(j + 2) + x(j + 1) + x(j)) \bmod 2, \, 0 \le j \le 25,$$

(9.20)

with initial conditions $x(0) = 0$, $x(1) = 0$, $x(2) = 0$, $x(3) = 0$, $x(4) = 1$.

The mapping of the SSS sequence $d(n)$ to the resource elements is similar to PSS mapping as shown in Figure 9.5. The only difference is that the second last OFDM symbol,

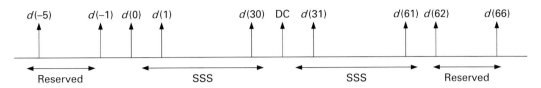

Figure 9.5. Secondary Synchronization Signal (SSS) mapping in the frequency-domain. The SSS transmission happens on the second last OFDM symbol, $l = N_{\text{symb}}^{\text{DL}} - 2$, in slot numbers 0 and 10.

$l = N_{\text{symb}}^{\text{DL}} - 2$, in slot numbers 0 and 10 is used for SSS transmission while the last OFDM symbol, $l = N_{\text{symb}}^{\text{DL}} - 1$, in slot numbers 0 and 10 is used for PSS transmission. Therefore, mapping for the non-reserved resource elements is performed according to the following relationship:

$$a_{k,l} = d\,(n)\,,\ k = n - 31 + \left\lfloor \frac{N_{\text{RB}}^{\text{DL}} N_{\text{sc}}^{\text{RB}}}{2} \right\rfloor,\ l = N_{\text{symb}}^{\text{DL}} - 2,\ n = 0,\ldots,61. \qquad (9.21)$$

The resource elements defined by the following relationship are reserved and therefore no transmission happens on these resource elements.

$$k = n - 31 + \left\lfloor \frac{N_{\text{RB}}^{\text{DL}} N_{\text{sc}}^{\text{RB}}}{2} \right\rfloor,\ l = N_{\text{symb}}^{\text{DL}} - 2, n = -5, -4, \ldots, -1, 62, 62, \ldots, 66. \qquad (9.22)$$

9.4.3 PSS and SSS overhead estimates

The PSS and SSS overheads for some example bandwidths are given in Table 9.4. The overhead is largest for the case of the smallest bandwidth of 1.08 MHz using six resource blocks. This is because PSS and SSS, which are time-multiplexed within a slot, are transmitted using the whole bandwidth in this case. Also, the overhead is larger for an extended cyclic prefix due to the longer duration of the OFDM symbol, which results in fewer OFDM symbols (six per slot) available within a slot relative to the case of normal cyclic prefix (seven OFDM symbols per slot). In the case of a larger bandwidth using, for example, 50 or 100 resource blocks, the PSS/SSS overhead is much smaller because the resource blocks other than the six PSS/SSS resource blocks can be used for data transmission. This provides another reason for deploying the system using the highest available bandwidth where available.

9.5 Broadcast channel

The physical broadcast channel (PBCH) is transmitted over four subframes (one subframe in each frame of 10 ms) with a 40 ms timing interval. The 40 ms timing is detected blindly without requiring any explicit signaling. Also, each subframe transmission of BCH is self-decodable and UEs with good channel conditions may not need to wait for reception of all the four subframes for PBCH decoding.

Table 9.4. PSS/SSS overhead estimates.

		Synchronization overhead [%]		
	Number of RBs	PSS	SSS	Total
Normal cyclic prefix	6	1.43	1.43	2.86
	50	0.17	0.17	0.34
	100	0.085	0.085	0.17
Extended cyclic prefix	6	1.67	1.67	3.33
	50	0.2	0.2	0.4
	100	0.1	0.1	0.2

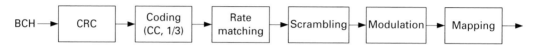

Figure 9.6. BCH transmission chain processing.

The transmission chain processing for the broadcast control channel is depicted in Figure 9.6. The BCH data arrives to the coding unit in the form of a maximum of one transport block every transmission time interval (TTI) of 40 ms.

Error detection is provided on BCH transport blocks through a cyclic redundancy check (CRC). The entire transport block is used to calculate the CRC parity bits. Let us denote the bits in a transport block delivered to layer 1 by

$$a_0, a_1, \ldots, a_{(A-1)}, \tag{9.23}$$

where A is the size of the transport block. The parity bits are given as:

$$p_0, p_1, p_2, p_3, \ldots, p_{L-1}, \tag{9.24}$$

where $L = 16$ is the number of parity bits. The CRC bits are scrambled according to the eNode-B transmit antenna configuration with the sequence $x_{\text{ant},k}$ as indicated in Table 9.5 to form the sequence of bits c_k as below:

$$c_k = \begin{cases} a_k & 0 \leq k < A \\ \left(p_{k-A} + x_{\text{ant},k-A}\right) \bmod 2 & A \leq k < (A+L). \end{cases} \tag{9.25}$$

Therefore, the number of antenna ports supported in the cell is determined blindly by multiple hypothesis testing on PBCH.

After CRC attachment, the bit sequence c_k is coded using a rate 1/3 tail-biting convolutional coding described in Chapter 11. The coded bits are then rate matched using the circular buffer approach to obtain the rate matched sequence $b(0), b(1), \ldots, b(M_{\text{bit}} - 1)$, where M_{bit} is the number of bits transmitted on the physical broadcast channel and depends upon the cyclic prefix length as explained later. The sequence $b(0), b(1), \ldots, b(M_{\text{bit}} - 1)$ is scrambled with a cell-specific sequence prior to modulation, resulting in a block of scrambled bits $\tilde{b}(0), \ldots, \tilde{b}(M_{\text{bit}} - 1)$ according to

$$\tilde{b}(i) = (b(i) + c(i)) \bmod 2. \tag{9.26}$$

Table 9.5. CRC mask for PBCH.

Number of transmit antenna ports at eNode-B	PBCH CRC mask $\langle x_{\text{ant},0}, x_{\text{ant},1}, \ldots, x_{\text{ant},15}\rangle$
1	$\langle 0,0,0,0,0,0,0,0,0,0,0,0,0,0,0,0\rangle$
2	$\langle 1,1,1,1,1,1,1,1,1,1,1,1,1,1,1,1\rangle$
4	$\langle 0,1,0,1,0,1,0,1,0,1,0,1,0,1,0,1\rangle$

The scrambling sequence is initialized with $c_{\text{init}} = N_{\text{ID}}^{\text{cell}}$ in each radio frame number n_f fulfilling the following condition:

$$n_f \bmod 4 = 0. \tag{9.27}$$

With the frame duration of 10 ms, the scrambling sequence is initialized every 40 ms. The block of scrambled bits $\tilde{b}(0), \ldots, \tilde{b}(M_{\text{bit}} - 1)$ is modulated using QPSK modulation, resulting in a block of complex-valued modulation symbols $d(0), \ldots, d(M_{\text{symb}} - 1)$, where

$$M_{\text{symb}} = \frac{M_{\text{bit}}}{Q_m} = \frac{M_{\text{bit}}}{2}, \tag{9.28}$$

where Q_m is the number of bits per modulation symbol and is 2 for QPSK modulation used on the physical broadcast channel. A single antenna, two-antenna SFBC and four-antenna SFBC-FSTD transmit diversity schemes are supported on the physical broadcast channel. The multi-antenna transmission scheme used on the physical broadcast channel is determined blindly based on the CRC masking according to Table 9.5.

The PBCH is transmitted during four consecutive radio frames with 40 ms timing as shown in Figure 9.7. No channel interleaving is used on PBCH and a frequency-first mapping approach is used for mapping of complex-valued symbols $y^{(p)}(0), \ldots, y^{(p)}(M_{\text{symb}} - 1)$ on each antenna port where $p = 0, 1, \ldots, (P - 1)$ and $P = 1, 2, 4$ for single antenna transmission, SFBC and SFBC-FSTD schemes respectively. The transmission of PBCH is performed in the first four OFDM symbols in the second slot within the first subframe of a radio frame. The transmission in the second slot allows one to avoid conflict with downlink control information transmission, which may use up to three OFDM symbols in the first slot in each subframe. The transmission in frequency is centered on the DC subcarrier. This is achieved by setting the starting subcarrier index k as:

$$k = \frac{N_{\text{RB}}^{\text{DL}} N_{\text{sc}}^{\text{RB}}}{2} - 36 + k', \quad k' = 0, 1, \ldots, 71, \quad l = 0, 1, \ldots, 3. \tag{9.29}$$

The transmission of PBCH is centered on the DC subcarrier because when a UE accesses the system and tries to receive PBCH, it is unaware of the system bandwidth used. The total number of subcarriers used for PBCH is 72 in the third and fourth OFDM symbols in the slot that contain no reference signals. The PBCH does not use subcarriers reserved for reference signals of the four antenna ports irrespective of how many antennas are used for PBCH transmission. This is for simplicity reasons because when a UE is receiving PBCH it is not aware of the number of antennas used for transmission. The UE actually performs blind detection of the number of antennas used for PBCH with hypothesis of the single antenna, two antennas SFBC and four antennas SFBC-FSTD schemes. The number of subcarriers available

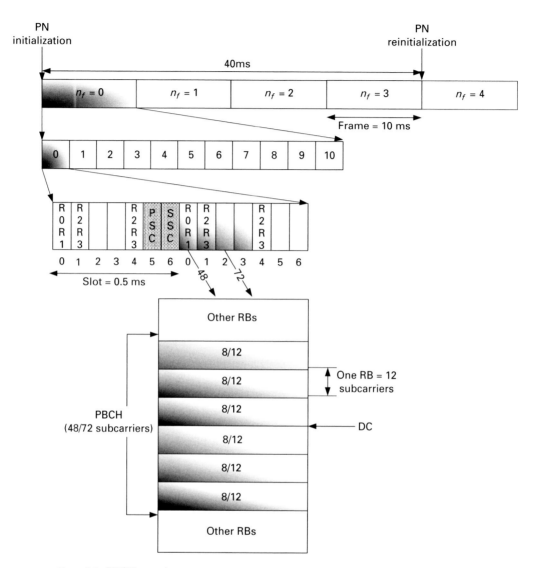

Figure 9.7. PBCH mapping.

for PBCH in the first and second OFDM symbols within the slot containing reference signals is only 48.

We can now determine the total number of bits transmitted on PBCH M_{bit} as:

$$M_{\text{bit}} = 4 \times (2 \times 48 + 2 \times 72) \times Q_m$$
$$= 4 \times (2 \times 48 + 2 \times 72) \times 2 = 1920. \qquad (9.30)$$

For the extended cyclic prefix, there are only six OFDM symbols per slot and hence the fourth OFDM symbol within the slot also contains reference signals and therefore the number of subcarriers available for PBCH is 48 in this OFDM symbol as well. In this case, the total number of bits transmitted on PBCH M_{bit} is given as:

$$M_{bit} = 4 \times (3 \times 48 + 1 \times 72) \times Q_m$$
$$= 4 \times (3 \times 48 + 1 \times 72) \times 2 = 1728. \tag{9.31}$$

The mapping of PBCH to resource elements (k, l) is in increasing order of first the index k, which means frequency-first mapping, then the index l in slot 1 in subframe 0 and finally the radio frame number. The total number of code bits in a single subframe for short cyclic prefix is 480 bits. Given the small size of the messages on BCH, this means that all the coded bits with code rate 1/3 are transmitted in a single subframe at least once. This allows for the decoding of PBCH in a single subframe for UEs with good channel conditions. We also like to point out that the PN sequence used for scrambling of PBCH is initialized every four frames or 40 ms. Therefore, once the UE has decoded the PBCH, it also knows 40 ms system timing.

9.6 Downlink reference signals

In the LTE system design phase, both time-division multiplexing (TDM) and frequency-division multiplexing (FDM) approaches were considered for reference signal multiplexing. These approaches are illustrated in Figure 9.8. A major benefit cited for the TDM approach was the possibility of employing micro-sleep mode at the UE. The idea was that scheduling control information could also be time-multiplexed, for example, in the OFDM symbols just following the TDM pilot. This way UE can perform channel estimation from the TDM pilot and can immediately decode control information. After decoding the TDM control information,

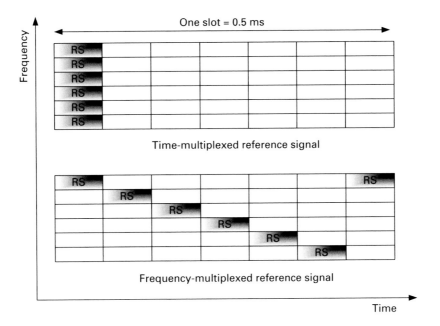

Figure 9.8. An illustration of time-multiplexed and frequency-multiplexed reference signals.

UE will know if it is scheduled or not in the current subframe. In the case where the UE determines that it is scheduled in the current frame, it will continue buffering the remaining transmission within the subframe. In the case where the UE determines that it is not scheduled in the current subframe, it can turn off its receiver thus enabling battery-power savings.

However, the drawback of the TDM reference signal approach is its inability to share transmission power between the reference signals and data transmission. The power sharing is particularly important for cases where the reference signal power needs to be boosted relative to the data power to enable better channel estimates. On the other hand, the FDM approach, which is also referred to as the fully scattered reference signals scheme, provides the flexibility to share and balance power between reference signal and data transmissions. The reference signal power spectral density (PSD) can be boosted by lowering the data power or data PSD. However, the obvious downside of the fully scattered approach is that the UE needs to receive the whole subframe before channel estimation can be performed for the decoding of the control information. This way the UE needs to continuously receive downlink transmissions and hence the UE battery power saving promised by micro-sleep cannot be achieved.

After much debate a hybrid TDM/FDM approach for reference signal multiplexing was selected for downlink reference signals in the LTE system. With the hybrid approach, the reference signals are frequency multiplexed with data transmission in a few OFDM symbols within a subframe. In the hybrid approach, the channel estimates for TDM control information can be derived based on reference signals in the first two OFDM symbols thus enabling the possibility of micro-sleep operation.

Three types of downlink reference signals are defined. For non-MBSFN transmissions, two types of reference signals are specified: cell-specific reference signals and UE-specific reference signals. For MBSFN transmissions, MBSFN area specific reference signals are specified. An antenna port is identified by a reference signal. For cell-specific reference signals, up to a maximum of four antenna ports are supported. For UE-specific reference signals and MBSFN reference signals, a single antenna port is supported.

9.6.1 Cell-specific reference signals

The cell-specific reference signals are used for various downlink measurements as well as for demodulation of non-MBSFN transmissions. The measurements performed using cell-specific reference signals include channel quality estimation, MIMO rank calculation, MIMO precoding vector/matrix selection and measurements for handoff. The non-MBSFN transmissions that depend upon cell-specific reference signals for channel estimation include downlink control channels and PDSCH transmissions not using UE-specific reference signals. A maximum of four antenna ports defined by reference signals $[R_0, R_1, R_2, R_3]$ and shown in Figure 9.9 can be used for non-MBSFN transmissions. The notation R_p is used here to denote a resource element used for reference signal transmission on antenna port p.

Two types of sequences were considered for cell-specific reference signals. The first approach to reference signals sequence design is based on using orthogonal sequences for three cells (sectors) within an eNB. The orthogonal sequences are further scrambled by a PN sequence. In the second approach, a simple cell-specific PN-sequence is used as a reference signal sequence without any spreading using orthogonal sequences.

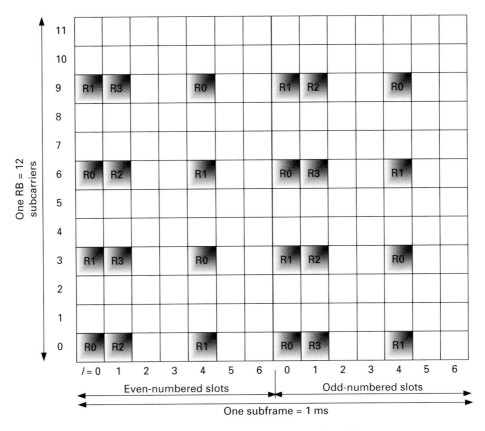

Figure 9.9. Mapping of downlink reference signals for normal cyclic prefix.

In the orthogonal sequence design approach, cell specific reference signals use a two-dimensional sequence whose elements are given by:

$$r_{mn}(i) = r_{mn}^{\mathrm{PRS}}(i) \cdot r_{mn}^{\mathrm{OS}} = r_{mn}^{\mathrm{PRS}} \cdot \left.\begin{bmatrix} S_j \\ S_j \\ \vdots \\ S_j \end{bmatrix}\right\} \; 72 \text{ entries}, \qquad (9.32)$$

where $r_{mn}^{\mathrm{PRS}}(i)$ is a two-dimensional pseudo-random sequence in slot i and r_{mn}^{OS} represents an orthogonal sequence which is obtained by 72 repetitions of matrix $S_j, j = 0, 1, 2$ given as below:

$$S_0 = \begin{bmatrix} 1 & 1 \\ 1 & 1 \\ 1 & 1 \end{bmatrix} \quad S_1 = \begin{bmatrix} 1 & e^{j4\pi/3} \\ e^{j2\pi/3} & 1 \\ e^{j4\pi/3} & e^{j2\pi/3} \end{bmatrix} \quad S_2 = \begin{bmatrix} 1 & e^{j2\pi/3} \\ e^{j4\pi/3} & 1 \\ e^{j2\pi/3} & e^{j2\pi/3} \end{bmatrix}. \qquad (9.33)$$

It can be noted that the first columns of these matrices are columns of a 3×3 DFT matrix while the second columns are a cyclic shift of the first column by one element. A 3×3 DFT matrix is given below:

$$DFT(3) = \begin{bmatrix} 1 & 1 & 1 \\ 1 & e^{j2\pi/3} & e^{j4\pi/3} \\ 1 & e^{j4\pi/3} & e^{j2\pi/3} \end{bmatrix}. \tag{9.34}$$

In the orthogonal sequence design approach, the sequence for a cell can be configured by the higher layers. Assuming matrix S_1 is configured in a cell, the two-dimensional reference signals sequence can be written as below.

$$R = r_{mn}(i) = r_{mn}^{PRS}(i) \cdot r_{mn}^{OS} = r_{mn}^{PRS} \cdot \begin{bmatrix} 1 & e^{j4\pi/3} \\ e^{j2\pi/3} & 1 \\ e^{j4\pi/3} & e^{j2\pi/3} \\ 1 & e^{j4\pi/3} \\ e^{j2\pi/3} & 1 \\ e^{j4\pi/3} & e^{j2\pi/3} \\ \vdots & \vdots & \vdots \\ 1 & e^{j4\pi/3} \\ e^{j2\pi/3} & 1 \\ e^{j4\pi/3} & e^{j2\pi/3} \\ 1 & e^{j4\pi/3} \end{bmatrix} \quad \begin{array}{l} m = 0,1,2,\ldots,219 \\ n = 0,1. \end{array}$$

$$\tag{9.35}$$

The size of the matrix R is 220×2, which means that the last two rows after 74 repetitions of S_i are dropped. A reference signal sequence length of 220 can accommodate a maximum of 110 resource blocks as each antenna port has two reference symbols within a resource block. A total of 510 cell IDs are obtained by combining 170 two-dimensional pseudo-random sequences with three two-dimensional orthogonal sequences.

An example of two-dimensional reference signal sequence mapping to resource elements allocated for cell-specific reference signals for four antenna ports within a slot is shown in Figure 9.10. We assumed the reference signal sequence given by Equation (9.35). The reference signals for antenna ports 0 and 1 ($p = 0, 1$) in OFDM symbol 0 ($l = 0$) use the first column of the matrix R while the reference signals for antenna ports 0 and 1 ($p = 0, 1$) in OFDM symbol 4 ($l = 4$) use the second column of matrix R. The reference signals for antenna ports 2 and 3 ($p = 2, 3$) in OFDM symbol 1 ($l = 1$) use the first column of the matrix R.

The number of rows of the matrix R used for the reference signal (that is the reference signal sequence length) depends upon the system bandwidth. The system bandwidth is defined in terms of the number of resource blocks. A resource block consists of 12 subcarriers with subcarrier spacing of 15 kHz. An example of reference signal sequence elements used for some bandwidths is shown in Figure 9.11. The starting row number is determined by using the following relationship:

$$m' = m + 110 - N_{RB}^{DL} \quad m = 0,1,2,\ldots,(2 \times N_{RB}^{DL} - 1), \tag{9.36}$$

where N_{RB}^{DL} represents the number of resource blocks within the downlink system bandwidth. The example shown in Figure 9.11 shows the rows of matrix R used for the reference signal sequence for N_{RB}^{DL} values of 110, 100, 50 and 25.

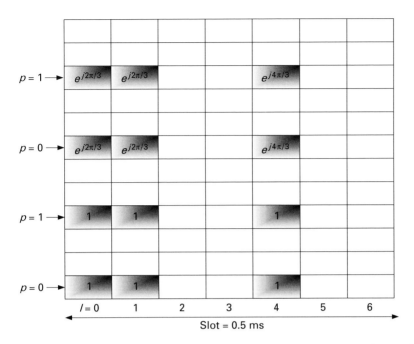

Figure 9.10. An example of reference signal sequence mapping to resource elements for the case of four antenna ports.

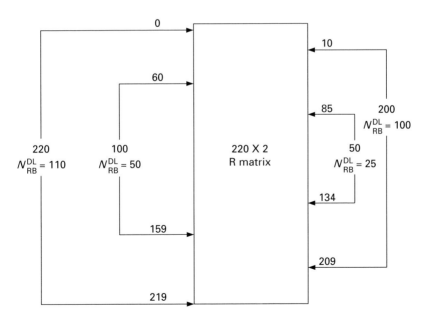

Figure 9.11. Rows of matrix R used for the reference signal sequence for different bandwidths.

A problem with orthogonal reference signal sequences is the loss of orthogonality in a frequency selective channel. This is because each element of the length-three DFT sequence being separated by six subcarriers can experience different fading in a high-delay-spread environment (resulting in a smaller coherence bandwidth). It was therefore decided to use only PN sequence without orthogonal sequences as the cell-specific reference signal sequence in the downlink.

The cell-specific reference-signal sequence $r_{l,n_s}(m)$ is then defined as:

$$r_{l,n_s}(m) = \frac{1}{\sqrt{2}}(1 - 2 \cdot c(2m)) + j\frac{1}{\sqrt{2}}(1 - 2 \cdot c(2m+1)),\ m = 0, 1, \ldots, 2N_{\mathrm{RB}}^{\mathrm{max,DL}} - 1,$$

(9.37)

where n_s and l are the slot number within a radio frame and the OFDM symbol number within the slot respectively. The pseudo-random sequence generator is initialized with:

$$c_{\mathrm{init}} = 2^{10} \cdot (7 \cdot (n_s + 1) + l + 1) \cdot \left(2 \cdot N_{\mathrm{ID}}^{\mathrm{cell}} + 1\right) + 2 \cdot N_{\mathrm{ID}}^{\mathrm{cell}} + N_{\mathrm{CP}}$$

(9.38)

at the start of each OFDM symbol where $N_{\mathrm{ID}}^{\mathrm{cell}}$ is the cell identity and $N_{\mathrm{CP}} = 1, 0$ for the normal and extended cyclic prefix respectively. Since each cell initializes the PN-sequence generator with a different value, the inter-cell interference on reference signals is randomized.

The LTE system allows reference signal boosting where power on reference signal symbols can be different from the data symbols power. However, in a synchronized system operation, if all cells in the system boost reference signal power, the reference signals are going to experience higher interference which can undermine the benefit of reference signal power boosting. In order to avoid reference signals collisions among the neighboring cells, a cell-specific frequency shift is applied to reference signal mapping to resource elements as below:

$$v_{\mathrm{shift}} = N_{\mathrm{ID}}^{\mathrm{cell}} \bmod 6.$$

(9.39)

The reference signal mapping given in Figure 9.9 assumes $v_{\mathrm{shift}} = 0$. In Figure 9.12, we show the mapping of downlink cell-specific reference signals for normal cyclic prefix with $v_{\mathrm{shift}} = 4$. We note that when one cell uses $v_{\mathrm{shift}} = 0$ as in Figure 9.9 and the neighboring cell uses $v_{\mathrm{shift}} = 4$ as in Figure 9.12, the reference signals of the two cells are mapped to different resource elements and hence do not collide. This means that when the reference signals power is boosted relative to data power, the reference symbols SINR is improved at the expense of more interference to data symbols.

We also note that the positions of antenna ports 0 and 1 ($p = 0, 1$) reference signals is shifted by three resource elements between the two OFDM symbols containing reference signals for ($p = 0, 1$) in a slot. On the other hand, reference signals for antenna ports 2 and 3 ($p = 2, 3$) exist on a single OFDM symbol within a slot. Therefore, the positions of antenna ports 2 and 3 ($p = 2, 3$) are shifted between the two slots within a subframe. This shift within a cell allows more accurate channel estimates for a highly frequency selective channel. This is because the effective separation of reference signals in frequency within a subframe becomes three resource elements. However, this is only true when the channel is static in time within a subframe or a slot.

Another observation that we can make is that reference signal density in the time-domain for antenna ports 0 and 1 ($p = 0, 1$) is two times higher than the reference signal density for antenna ports 2 and 3 ($p = 2, 3$). This imbalance in reference signal density leads to more accurate channel estimates for antenna ports 0 and 1 ($p = 0, 1$) relative to channel estimates for antenna ports 2 and 3 ($p = 2, 3$). This imbalance has certain implications and was one of the deciding factors in SFBC symbol pairs mapping to antenna ports 0 and 2 ($p = 0, 2$) and antenna ports 1 and 3 ($p = 1, 3$) in the four antenna ports SFBC-FSTD transmit diversity scheme.

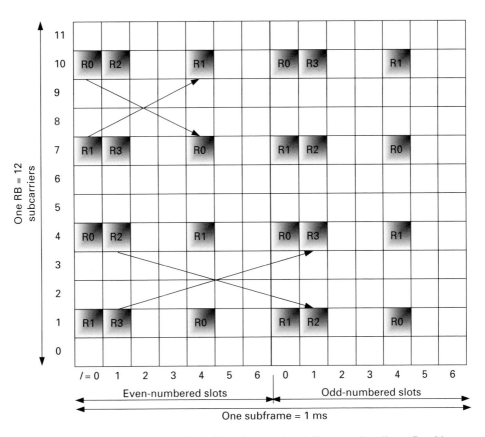

Figure 9.12. Mapping of downlink cell-specific reference signals for normal cyclic prefix with $v_{\text{shift}} = 4$.

In the case of the extended cyclic prefix, each slot contains one fewer OFDM symbol relative to the case of the normal cyclic prefix. The reference signal mapping for the extended cyclic prefix is shown in Figure 9.13 for $v_{\text{shift}} = 0$. It should be noted that the resource elements where the reference signal is transmitted from one antenna port is left blank in all the other antenna ports.

9.6.2 Downlink MBSFN reference signals

A single port number 4 is defined for MBSFN reference signals that are transmitted in subframes allocated for MBSFN transmissions. Since MBSFN transmissions generally experience larger delay spread due to signals received from multiple cells the MBSFN reference signals are defined for the extended cyclic prefix only. The MBSFN reference-signal sequence $r_{l,n_s}(m)$ is given by:

$$r_{l,n_s}(m) = \frac{1}{\sqrt{2}} \left(1 - 2 \cdot c(2m)\right) + j\frac{1}{\sqrt{2}} \left(1 - 2 \cdot c(2m + 1)\right), \quad m = 0, 1, \ldots, 6N_{\text{RB}}^{\text{max,DL}} - 1,$$

$$(9.40)$$

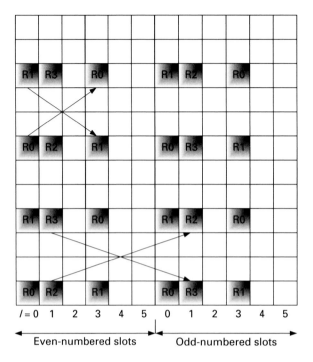

Figure 9.13. Mapping of downlink reference signals for extended cyclic prefix for $v_{\text{shift}} = 0$.

where n_s is the slot number within a radio frame and l is the OFDM symbol number within the slot. The PN-sequence $c(i)$ generator is initialized at the start of each OFDM symbol as:

$$c_{\text{init}} = 2^9 \cdot (7 \cdot (n_s + 1) + l + 1) \cdot \left(2 \cdot N_{\text{ID}}^{\text{MBSFN}} + 1\right) + N_{\text{ID}}^{\text{MBSFN}}, \tag{9.41}$$

where $N_{\text{ID}}^{\text{MBSFN}}$ is the MBSFN area identity. It should be noted that the PN sequence is initialized with the same value in all the cells in the MBSFN area. This is to enable composite multi-cell MBSFN channel estimates.

Since control information for non-MBSFN transmissions can be carried in the MBSFN subframes the first two OFDM symbols within the subframe can carry cell-specific reference signals. Therefore, the MBSFN reference signals are mapped starting with the third OFDM symbol ($l = 2$). The goal is to keep the MBSFN and non-MBSFN reference signals orthogonal in time to allow reference signal boosting. If the MBSFN and non-MBSFN reference signals are carried in the same OFDM symbol and the non-MBSFN reference signal power is boosted, this power needs to come from MBSFN reference signals, which is not a desirable situation. The mapping of MBSFN reference signals for subcarrier spacing of $\Delta f = 15$ and 7.5 kHz is shown in Figures 9.14 and 9.15 respectively. We note that the reference signal separation in both cases is 30 kHz, which is smaller than the case of non-MBSFN transmissions. This is because a multi-cell MBSFN channel is generally more frequency-selective than a single-cell non-MBSFN channel.

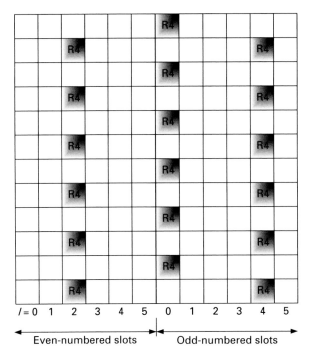

Figure 9.14. Mapping of downlink reference signals for MBSFN in the case of the extended cyclic prefix for subcarrier spacing of $\Delta f = 15\,\mathrm{KHz}$.

9.6.3   UE-specific reference signals

The UE-specific reference signals are provisioned to support rank-1 beam-forming on the downlink. We note that beam-forming can be supported using two or four antenna ports codebook along with cell-specific reference signals. However, by using UE-specific reference signals, beam-forming can be supported for more than four transmit antennas. In this case, however, the beam-forming weights need to be estimated based on uplink transmission as there is no codebook or feedback mechanism defined for more than four antenna ports. A major advantage of using UE-specific reference signals is that the beam-forming weights can also be applied to the reference signals and hence reference signals can also experience beam-forming gains.

The UE-specific reference signals are transmitted on antenna port 5. The UE is informed by higher layers whether the UE-specific reference signal is present and is a valid phase reference for PDSCH demodulation or not. Moreover, the UE-specific reference signals are transmitted only on the resource blocks allocated for PDSCH transmission. The UE-specific reference-signal sequence $r(m)$ is given by:

$$r(m) = \frac{1}{\sqrt{2}}\left(1 - 2 \cdot c(2m)\right) + j\frac{1}{\sqrt{2}}\left(1 - 2 \cdot c(2m+1)\right), \; m = 0, 1, \ldots, 12N_{\mathrm{RB}}^{\mathrm{PDSCH}} - 1,$$

$$(9.42)$$

where $N_{\mathrm{RB}}^{\mathrm{PDSCH}}$ denotes the bandwidth of the corresponding PDSCH transmission in resource blocks. The PN-sequence generator is initialized at the start of each subframe by a cell-specific

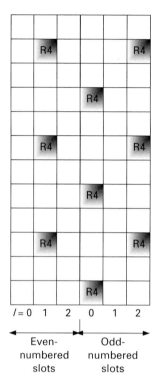

Figure 9.15. Mapping of downlink reference signals for MBSFN in the case of the extended cyclic prefix and subcarrier spacing of $\Delta f = 7.5\,\mathrm{KHz}$.

and a UE-specific value as below:

$$c_{\mathrm{init}} = \left(\lfloor n_{\mathrm{s}}/2 \rfloor + 1\right) \cdot \left(2 N_{\mathrm{ID}}^{\mathrm{cell}} + 1\right) \cdot 2^{16} + n_{\mathrm{RNTI}}, \tag{9.43}$$

where $N_{\mathrm{ID}}^{\mathrm{cell}}$ and n_{RNTI} represent the cell identity and RNTI respectively. The mapping for UE-specific reference signals for the normal cyclic prefix and the extended cyclic prefix is given in Figures 9.16 and 9.17 respectively. We note that for the case of the normal cyclic prefix, the OFDM symbols used for UE-specific reference signals and cell-specific reference signals are orthogonal in time. This is to avoid conflict between the two reference signal types as cell-specific reference signals use a cell-specific shift. In the case of extended cyclic prefix, however, the UE specific reference signals ($p = 5$) and cell-specific reference signals for antenna ports 2 and 3 ($p = 2, 3$) use the same second OFDM symbol ($l = 1$) in the second slot within the subframe. This can create a conflict between the cell-specific reference signals using cell-specific shift as in Equation (9.39) and UE-specific reference signals using no cell-specific shift. In order to avoid this conflict for the case of four antenna ports configuration, a cell-specific frequency shift is also applied to the UE-specific reference signals as below:

$$v_{\mathrm{shift}} = N_{\mathrm{ID}}^{\mathrm{cell}} \bmod 3. \tag{9.44}$$

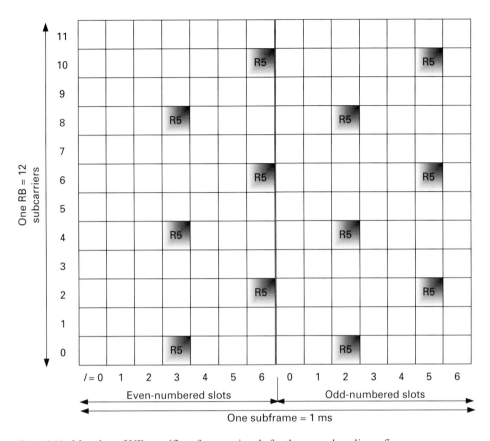

Figure 9.16. Mapping of UE-specific reference signals for the normal cyclic prefix.

However, unlike the six possible shifts for cell-specific reference signals only three shifts are defined for UE-specific reference signals. This is because UE-specific reference signals use every third subcarrier for the case of the extended cyclic prefix as shown in Figure 9.17.

9.7 Uplink reference signals

We noted that the downlink reference signals employ a hybrid TDM/FDM approach. In principle, a similar approach can be envisaged for uplink reference signals. However, the uplink access is based on single-carrier FDMA and frequency-multiplexing reference signals with data transmissions would violate the single-carrier property resulting in increased PAPR/CM. Therefore, the uplink reference signals are strictly time-multiplexed to assure the single-carrier property of SC-FDMA.

Another difference relative to downlink reference signals is that uplink reference signals are always UE-specific. The two types of reference signals supported on the uplink include demodulation reference signals (DMRS) used for channel estimation for PUSCH or PUCCH demodulation and sounding reference signals (SRS) used to measure the uplink channel quality for channel sensitive scheduling. The uplink demodulation reference signals for normal cyclic prefix and extended cyclic prefix cases are shown in Figures 9.18 and 9.19 respectively. The

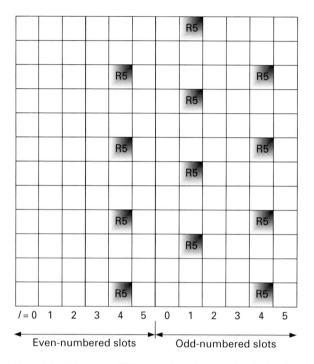

Figure 9.17. Mapping of UE-specific reference signals for the extended cyclic prefix.

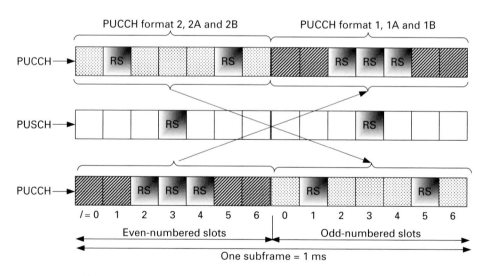

Figure 9.18. Uplink demodulation reference signals for the normal cyclic prefix.

demodulation reference signals for PUSCH are transmitted in the middle of the slot in symbol with $l = 2, 3$ for the extended and normal cyclic prefix respectively. For PUCCH format 1, 1A and 1B, the demodulation reference signals are transmitted in three symbols ($l = 2, 3, 4$) for the normal cyclic prefix and two symbols ($l = 2, 3$) for the extended cyclic prefix. For PUCCH format 2, 2A and 2B, the demodulation reference signals are transmitted in two

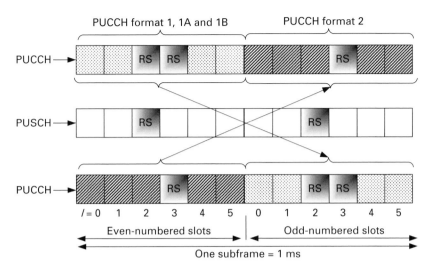

Figure 9.19. Uplink demodulation reference signals for the extended cyclic prefix.

symbols ($l = 1, 5$) for the normal cyclic prefix. For PUCCH format 2, the demodulation reference signal is transmitted in a single symbol ($l = 3$) for the extended cyclic prefix case.

9.7.1 Uplink reference signal sequences

In order to enable low PAPR/CM, constant amplitude reference signal sequences are selected for the uplink. The same set of base sequences is used for demodulation and sounding reference signals. The reference signal sequence $r_{u,v}^{(\alpha)}(n)$ is defined by a cyclic shift α of a base sequence $\bar{r}_{u,v}(n)$ according to:

$$r_{u,v}^{(\alpha)}(n) = e^{j\alpha n}\bar{r}_{u,v}(n), \quad 0 \le n < M_{\text{sc}}^{\text{RS}} = mN_{\text{sc}}^{\text{RB}}, \tag{9.45}$$

where $M_{\text{sc}}^{\text{RS}}$ is the length of the reference signal sequence with $1 \le m \le N_{\text{RB}}^{\text{max,UL}}$. Multiple reference signal sequences are defined from a single base sequence by using different values of the cyclic shift α.

The base sequences $\bar{r}_{u,v}(n)$ are divided into 30 groups, where $u \in \{0, 1, \ldots, 29\}$ is the group number and v is the base sequence number within the group. For sequence lengths up to five resource blocks, that is $M_{\text{sc}}^{\text{RS}} = mN_{\text{sc}}^{\text{RB}}, 1 \le m \le 5$, each sequence group contains a single base sequence ($v = 0$). For sequence lengths greater than five resource blocks, that is $M_{\text{sc}}^{\text{RS}} = mN_{\text{sc}}^{\text{RB}}, 6 \le m \le N_{\text{RB}}^{\text{max,UL}}$, each sequence group contains two base sequences ($v = 0, 1$). The sequence group number u and the sequence number v within the group can be hopped in time to randomize the interference.

The definition of the base sequence $\bar{r}_{u,v}(0), \ldots, \bar{r}_{u,v}(M_{\text{sc}}^{\text{RS}} - 1)$ depends on the sequence length $M_{\text{sc}}^{\text{RS}}$. For sequence lengths of up to two resource blocks, computer-generated CAZAC sequences are used as reference signal sequences. For sequence lengths of more than two resource blocks, ZC sequences are used as the uplink reference signal sequences.

Therefore, for $M_{sc}^{RS} \geq 3N_{sc}^{RB}$, the base sequence $\bar{r}_{u,v}(0), \ldots, \bar{r}_{u,v}(M_{sc}^{RS} - 1)$ is given by an extended ZC sequence as below:

$$\bar{r}_{u,v}(n) = x_q(n \bmod N_{ZC}^{RS}), \quad 0 \leq n < M_{sc}^{RS} , \tag{9.46}$$

where $\bmod\ N_{ZC}^{RS}$ operation extends the Zadoff–Chu sequence length N_{ZC}^{RS} to match the reference signal sequence length M_{sc}^{RS}. The qth root ZC sequence is defined by:

$$x_q(m) = e^{-j\frac{\pi qm(m+1)}{N_{ZC}^{RS}}}, \quad 0 \leq m \leq N_{ZC}^{RS} - 1 , \tag{9.47}$$

with q given by:

$$q = \left\lfloor \frac{N_{ZC}^{RS} \cdot (u+1)}{31} + \frac{1}{2} \right\rfloor + v \cdot (-1)^{\left\lfloor \frac{2N_{ZC}^{RS} \cdot (u+1)}{31} \right\rfloor}. \tag{9.48}$$

The length N_{ZC}^{RS} of the Zadoff–Chu sequence is given by the largest prime number such that $N_{ZC}^{RS} < M_{sc}^{RS}$. We note that for three resource blocks with $M_{sc}^{RS} = 36$ ZC, sequence length is $N_{ZC}^{RS} = 31$ as 31 is the largest prime number such that $N_{ZC}^{RS} < M_{sc}^{RS}$. We also know that with a ZC sequence length a prime number N_{ZC}^{RS}, a total of $(N_{ZC}^{RS} - 1)$ root sequences are available. With the smallest ZC sequence length of $N_{ZC}^{RS} = 31$ for the three resource blocks, a total of 30 root sequences are available. It is for this reason that the number of sequence groups is limited to 30.

The mapping of sequence group number u and sequence index v to ZC sequence root index q for $M_{sc}^{RS} = mN_{sc}^{RB}$, $3 \leq m \leq 6$, is given in Table 9.6. We know that for sequence lengths up to five resource blocks, that is $M_{sc}^{RS} = 60$, each sequence group contains a single base sequence ($v = 0$). However, when $M_{sc}^{RS} = 72$, the number of available ZC root sequences is 71, as 71 is the largest prime number such that $N_{ZC}^{RS} < M_{sc}^{RS}$. Given that we have 30 sequence groups, it now becomes possible to have more than one base sequence in a group. It should be noted that for $M_{sc}^{RS} = 60$, $N_{ZC}^{RS} = 59$ and, therefore, two sequences cannot be provided within a group as this requires a total of at least 60 base or root sequences.

For $M_{sc}^{RS} = N_{sc}^{RB}$ and $M_{sc}^{RS} = 2N_{sc}^{RB}$, the computer-generated CAZAC base sequence $\bar{r}_{u,v}(n)$ is given by:

$$\bar{r}_{u,v}(n) = e^{j\phi(n)\pi/4}, \quad 0 \leq n \leq M_{sc}^{RS} - 1 , \tag{9.49}$$

where the values of $\phi(n)$ are given in Table 9.7 for $M_{sc}^{RS} = N_{sc}^{RB}$. The values $\phi(0), \phi(1), \ldots, \phi(23)$ for $M_{sc}^{RS} = 2N_{sc}^{RB} = 24$ can be found in [3]. We note that the sequences are based on a constant amplitude QPSK alphabet with the following four alphabets:

$$\bar{r}_{u,v}(n) = e^{j\pi/4} = \frac{1}{\sqrt{2}} + j\frac{1}{\sqrt{2}}, \phi(n) = 1$$

$$\bar{r}_{u,v}(n) = e^{-j\pi/4} = \frac{1}{\sqrt{2}} - j\frac{1}{\sqrt{2}}, \phi(n) = -1 \tag{9.50}$$

Table 9.6. Mapping of sequence group number u and sequence index v to ZC sequence root index q for $M_{sc}^{RS} = mN_{sc}^{RB}, 3 \leq m \leq 6$.

Sequence group number u	ZC sequence root index q				
	$M_{sc}^{RS} = 36$ $N_{ZC}^{RS} = 31$	$M_{sc}^{RS} = 48$ $N_{ZC}^{RS} = 47$	$M_{sc}^{RS} = 60$ $N_{ZC}^{RS} = 59$	$M_{sc}^{RS} = 72$ $N_{ZC}^{RS} = 71$	
	$v=0$	$v=0$	$v=0$	$v=0$	$v=1$
0	1	2	2	2	3
1	2	3	4	5	4
2	3	5	6	7	6
3	4	6	8	9	10
4	5	8	10	11	12
5	6	9	11	14	13
6	7	11	13	16	17
7	8	12	15	18	19
8	9	14	17	21	20
9	10	15	19	23	22
10	11	17	21	25	26
11	12	18	23	27	28
12	13	20	25	30	29
13	14	21	27	32	33
14	15	23	29	34	35
15	16	24	30	37	36
16	17	26	32	39	38
17	18	27	34	41	42
18	19	29	36	44	43
19	20	30	38	46	45
20	21	32	40	48	49
21	22	33	42	50	51
22	23	35	44	53	52
23	24	36	46	55	54
24	25	38	48	57	58
25	26	39	49	60	59
26	27	41	51	62	61
27	28	42	53	64	65
28	29	44	55	66	67
29	30	45	57	69	68

$$\bar{r}_{u,v}(n) = e^{j3\pi/4} = -\frac{1}{\sqrt{2}} + j\frac{1}{\sqrt{2}}, \phi(n) = 3$$

$$\bar{r}_{u,v}(n) = e^{-j3\pi/4} = -\frac{1}{\sqrt{2}} - j\frac{1}{\sqrt{2}}, \phi(n) = -3.$$

In order to randomize the inter-cell interference, sequence group hopping can be enabled by higher layers. In this case, the sequence-group number u in slot n_s is defined by a group-hopping pattern $f_{gh}(n_s)$ and a sequence-shift pattern f_{ss} according to:

$$u = (f_{gh}(n_s) + f_{ss}) \bmod 30 \tag{9.51}$$

Table 9.7. Definition of $\phi(n)$ for $M_{sc}^{RS} = N_{sc}^{RB}$.

u	$\phi(0),\ldots,\phi(11)$											
0	−1	1	3	−3	3	3	1	1	3	1	−3	3
1	1	1	3	3	3	−1	1	−3	−3	1	−3	3
2	1	1	−3	−3	−3	−1	−3	−3	1	−3	1	−1
3	−1	1	1	1	1	−1	−3	−3	1	−3	3	−1
4	−1	3	1	−1	1	−1	−3	−1	1	−1	1	3
5	1	−3	3	−1	−1	1	1	−1	−1	3	−3	1
6	−1	3	−3	−3	−3	3	1	−1	3	3	−3	1
7	−3	−1	−1	−1	1	−3	3	−1	1	−3	3	1
8	1	−3	3	1	−1	−1	−1	1	1	3	−1	1
9	1	−3	−1	3	3	−1	−3	1	1	1	1	1
10	−1	3	−1	1	1	−3	−3	−1	−3	−3	3	−1
11	3	1	−1	−1	3	3	−3	1	3	1	3	3
12	1	−3	1	1	−3	1	1	1	−3	−3	−3	1
13	3	3	−3	3	−3	1	1	3	−1	−3	3	3
14	−3	1	−1	−3	−1	3	1	3	3	3	−1	1
15	3	−1	1	−3	−1	−1	1	1	3	1	−1	−3
16	1	3	1	−1	1	3	3	3	−1	−1	3	−1
17	−3	1	1	3	−3	3	−3	−3	3	1	3	−1
18	−3	3	1	1	−3	1	−3	−3	−1	−1	1	−3
19	−1	3	1	3	1	−1	−1	3	−3	−1	−3	−1
20	−1	−3	1	1	1	1	3	1	−1	1	−3	−1
21	−1	3	−1	1	−3	−3	−3	−3	−3	1	−1	−3
22	1	1	−3	−3	−3	−3	−1	3	−3	1	−3	3
23	1	1	−1	−3	−1	−3	1	−1	1	3	−1	1
24	1	1	3	1	3	3	−1	1	−1	−3	−3	1
25	1	−3	3	3	1	3	3	1	−3	−1	−1	3
26	1	3	−3	−3	3	−3	1	−1	−1	3	−1	−3
27	−3	−1	−3	−1	−3	3	1	−1	1	3	−3	−3
28	−1	3	−3	3	−1	3	3	−3	3	3	−1	−1
29	3	−3	−3	−1	−1	−3	−1	3	−3	3	1	−1

where the group-hopping pattern $f_{gh}(n_s)$ is given by:

$$f_{gh}(n_s) = \left(\sum_{i=0}^{7} c(8n_s + i) \cdot 2^i\right) \bmod 30, \qquad (9.52)$$

where the PN-sequence $c(i)$ is given by (9.4) and the PN-sequence generator is initialized at the beginning of each radio frame with a cell-specific value as:

$$c_{init} = \left\lfloor \frac{N_{ID}^{cell}}{30} \right\rfloor. \qquad (9.53)$$

When the group hopping is not enabled $f_{gh}(n_s)$ is set to zero, that is $f_{gh}(n_s) = 0$. A total of 510 hopping patterns via 17 group-hopping patterns and 30 sequence-shift patterns can be obtained. However, only a total of 504 hopping/shift patterns are used as depicted in Figure 9.20. This is to align the hopping/shift patterns to the physical layer cell identities

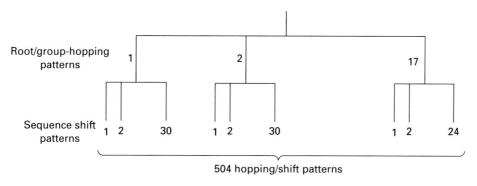

Figure 9.20. Uplink reference signal hopping/shift patterns.

(see Section 9.4). This allows for planning of hopping/shift patterns in a way similar to the planning for physical layer cell identities.

The group-hopping pattern $f_{gh}(n_s)$ is the same for PUSCH and PUCCH. However, the sequence-shift pattern f_{ss} definition differs between PUCCH and PUSCH:

$$f_{ss}^{PUCCH} = N_{ID}^{cell} \bmod 30$$

$$f_{ss}^{PUSCH} = \left(f_{ss}^{PUCCH} + \Delta_{ss}\right) \bmod 30, \quad \Delta_{ss} \in \{0, 1, \dots, 29\}, \tag{9.54}$$

where Δ_{ss} is configured by higher layers.

Sequence hopping can be applied to reference signals of length $M_{sc}^{RS} \geq 6N_{sc}^{RB}$. For reference signals of length $M_{sc}^{RS} < 6N_{sc}^{RB}$, there is a single base sequence with $v = 0$ and hence sequence hopping is not possible. When the sequence hopping is enabled (and group hopping disabled), the base sequence number v within the base sequence group in slot n_s is given by:

$$v = c(n_s), \tag{9.55}$$

where the PN-sequence $c(i)$ is given by Equation (9.4) and takes values of either zero or one, which in turn determines the base sequence with $v = 0$ or the base sequence with $v = 1$. The PN-sequence generator is initialized at the beginning of each radio frame with a cell-specific value as:

$$c_{init} = \left\lfloor \frac{N_{ID}^{cell}}{30} \right\rfloor \cdot 2^5 + f_{ss}^{PUSCH}. \tag{9.56}$$

In other cases when sequence hopping is not enabled or group hopping is enabled, the base sequence number v within the base sequence group in slot n_s is always set to zero, that is $v = 0$.

9.7.2 Demodulation reference signal

The demodulation reference signals are specified for PUSCH and PUCCH. The demodulation reference signal sequence $r^{PUSCH}(\cdot)$ for PUSCH is defined by:

$$r^{PUSCH}\left(m \cdot M_{sc}^{RS} + n\right) = r_{u,v}^{(\alpha)}(n), \quad m = 0, 1 \quad n = 0, \dots, \left(M_{sc}^{RS} - 1\right), \tag{9.57}$$

Table 9.8. Mapping of cyclic shift field to $n_{\text{DMRS}}^{(2)}$ values.

Cyclic shift field	$n_{\text{DMRS}}^{(2)}$
000	0
001	2
010	3
011	4
100	6
101	8
110	9
111	10

where $r_{u,v}^{(\alpha)}(n)$ is the reference signal sequence given by Equation (9.45) and $m = 0, 1$ represents the first and second slot within a subframe. The length of the reference signal sequence is equal to the number of resource elements or subcarriers used for PUSCH transmission, that is $M_{\text{sc}}^{\text{RS}} = M_{\text{sc}}^{\text{PUSCH}}$. The cyclic shift α in a slot is given as:

$$\alpha = \frac{2\pi\left[\left(n_{\text{DMRS}}^{(1)} + n_{\text{DMRS}}^{(2)} + n_{\text{PRS}}\right) \bmod 12\right]}{12}, \tag{9.58}$$

where $n_{\text{DMRS}}^{(1)}$ is broadcast as part of the system information. The values of $n_{\text{DMRS}}^{(2)}$ are signaled in the uplink scheduling assignment and are given in Table 9.8. n_{PRS} is given as:

$$n_{\text{PRS}} = \sum_{i=0}^{7} c(i) \times 2^{i}, \tag{9.59}$$

where the PN-sequence $c(i)$ is given in Equation (9.4). The PN-sequence generator is initialized at the beginning of each radio frame with a cell-specific value according to (9.55).

The demodulation reference signal sequence for PUSCH $r^{\text{PUSCH}}(\cdot)$ is multiplied with the amplitude scaling factor β_{PUSCH} and mapped in sequence starting with $r^{\text{PUSCH}}(0)$ to the same set of physical resource blocks as used for the corresponding PUSCH transmission.

The demodulation reference signal sequence for PUCCH $r^{\text{PUCCH}}(\cdot)$ is defined by:

$$r^{\text{PUCCH}}\left(m' N_{\text{RS}}^{\text{PUCCH}} M_{\text{sc}}^{\text{RS}} + m M_{\text{sc}}^{\text{RS}} + n\right) = \bar{w}(m) z(m) r_{u,v}^{(\alpha)}(n)$$

$$m = 0, \ldots, \left(N_{\text{RS}}^{\text{PUCCH}} - 1\right), \quad n = 0, \ldots, \left(M_{\text{sc}}^{\text{RS}} - 1\right), \quad m' = 0, 1. \tag{9.60}$$

The sequence $r_{u,v}^{(\alpha)}(n)$ is given by (9.49) with $M_{\text{sc}}^{\text{RS}} = 12$ where the expression for the cyclic shift α is determined by the PUCCH format as described in [3]. $m' = 0, 1$ represent the first and second slot within a subframe. $N_{\text{RS}}^{\text{PUCCH}}$ is the number of reference symbols per slot for PUCCH and $z(m)$ is given as:

$$\begin{aligned} z(m) &= d(10) \quad m = 1, \text{ PUCCH format 2A and 2B} \\ z(m) &= 1 \qquad \text{otherwise.} \end{aligned} \tag{9.61}$$

For PUCCH formats 2A and 2B, the bit(s) $b(20), \ldots, b(M_{\text{bit}} - 1)$ representing one or two bits ACK/NACK are modulated according to Table 9.9. This results in a single modulation symbol $d(10)$, which is used for the generation of the reference signal for PUCCH format

Table 9.9. Modulation symbol $d(10)$ for PUCCH formats 2A and 2B.

PUCCH format	$b(20), \ldots, b(M_{\text{bit}} - 1)$	$d(10)$
2A	0	-1
	1	1
2B	00	-1
	01	j
	10	$-j$
	11	1

Table 9.10. Orthogonal sequences $\left[\bar{w}(0), \bar{w}(1), \ldots, \bar{w} \left(N_{\text{RS}}^{\text{PUCCH}} - 1 \right) \right]$ for different PUCCH formats.

PUCCH format	Sequence index $\bar{n}_{\text{oc}}(n_s)$	Normal cyclic prefix	Extended cyclic prefix
1, 1A and 1B	0	$\begin{bmatrix} 1 & 1 & 1 \end{bmatrix}$	$\begin{bmatrix} 1 & 1 \end{bmatrix}$
	1	$\begin{bmatrix} 1 & e^{j2\pi/3} & e^{j4\pi/3} \end{bmatrix}$	$\begin{bmatrix} 1 & -1 \end{bmatrix}$
	2	$\begin{bmatrix} 1 & e^{j4\pi/3} & e^{j2\pi/3} \end{bmatrix}$	N/A
2, 2A and 2B	NA	$\begin{bmatrix} 1 & 1 \end{bmatrix}$	$[1]$

2A and 2B. Therefore, ACK/NACK in the case of PUCCH format 2A and 2B is in effect transmitted using the reference signals.

The orthogonal sequences $\left[\bar{w}(0), \bar{w}(1), \ldots, \bar{w} \left(N_{\text{RS}}^{\text{PUCCH}} - 1 \right) \right]$ for different PUCCH formats are given in Table 9.10. We know that PUCCH formats 1, 1A and 1B carrying a scheduling request or ACK/NACK are block-wise spread with orthogonal sequences to create more ACK/NACK channels. Similarly, block-wise spreading of the corresponding reference signals is required to match the number of reference signals to the number of possible ACK/NACKs. The PUCCH formats 2, 2A and 2B do not employ any block-wise spreading and hence corresponding reference signals are not spread either. The number and location of reference signals for formats 1, 1A and 1B as well as formats 2, 2A and 2B are given in Figures 9.18 and 9.19 for the normal and extended cyclic prefix respectively. We note that the length of the orthogonal sequences $\left[\bar{w}(0), \bar{w}(1), \ldots, \bar{w} \left(N_{\text{RS}}^{\text{PUCCH}} - 1 \right) \right]$ in Table 9.10 matches the number of reference signal symbols within a slot.

The reference signal sequence for PUCCH $r^{\text{PUCCH}}(\cdot)$ is multiplied with the amplitude-scaling factor β_{PUCCH} and mapped in sequence starting with $r^{\text{PUCCH}}(0)$ to resource elements (k, l). The mapping is in increasing order of first k and then l and finally the slot number. The same set of values for k as for the corresponding PUCCH transmission is used.

9.7.3 Sounding reference signal

The sounding reference symbol (SRS) is mainly used for uplink channel quality measurements for channel-sensitive scheduling and link adaptation. The SRS can also be used for uplink timing estimation and uplink power control. For frequency selective scheduling, SRS should provide a reliable channel estimate for each of the scheduling subbands. It is generally preferred to sound the whole bandwidth by SRS transmission in a single shot. However, the UEs near the cell edge may not have enough power for transmitting wideband SRS. On the other hand, a single shot transmission of wideband SRS is possible for UEs near the cell center. This requires

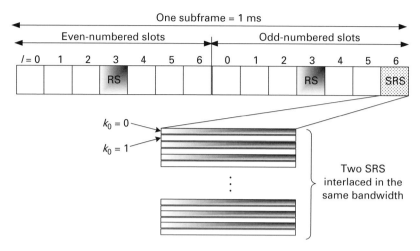

Figure 9.21. SRS mapping and transmission comb.

that multiple SRS bandwidths are provisioned such that SRS bandwidth can be configured in a UE-specific fashion.

The SRS is transmitted using distributed FDMA (see Chapter 4) with a repetition factor (RPF) of two in the last OFDM symbol within a subframe as depicted in Figure 9.21. The distributed FDMA allows one to enhance the multiplexing capability of SRS as SRS from two UEs can be multiplexed within the same bandwidth. However, in distributed-FDMA-based multiplexing, the efficient usage of SRS resources is not achieved if SRSs with different bandwidths are assigned randomly. Therefore, a systematic approach is required for multiplexing SRSs of different bandwidths.

The SRS parameters include starting subcarrier assignment, starting physical resource block assignment, duration of SRS transmission: single or indefinite (until disabled), periodicity of SRS transmissions: {2, 5, 10, 20, 40, 80, 160, 320} ms, frequency hopping (enabled or disabled) and cyclic shift. These SRS parameters are configured by higher layers for each UE in a semi-static fashion.

The SRS transmission bandwidth does not include the PUCCH region. In the case where the UE supports transmit antenna selection in the uplink, the SRS transmission antenna alternates sequentially between successive SRS transmission subframes. This allows eNB to estimate the channel quality on each of the transmit antennas. This information is then used for scheduling a UE on the best antenna in the uplink. A UE does not transmit SRS whenever SRS and CQI transmissions happen to coincide in the same subframe. Similarly, a UE does not transmit SRS whenever SRS and scheduling request (SR) transmissions happen to coincide in the same subframe.

The sounding reference signal sequence is given as:

$$r^{\text{SRS}}(n) = r_{u,v}^{(\alpha)}(n), \quad n = 0, 1, \ldots, \left(M_{\text{sc},b}^{\text{RS}} - 1\right), \tag{9.62}$$

where $r_{u,v}^{(\alpha)}(n)$ is given by Equation (9.45). The sequence index is derived from the PUCCH base sequence index and the cyclic shift of the sounding reference signal α^{SRS} is given as:

$$\alpha = 2\pi \frac{n_{\text{SRS}}}{8}, \tag{9.63}$$

where $n_{\text{SRS}} = 0, 1, 2, 3, 4, 5, 6, 7$ is configured by higher layers for each UE. The sequence $r^{\text{SRS}}(0), \ldots, r^{\text{SRS}}(M^{\text{RS}}_{\text{sc},b} - 1)$ is multiplied with the amplitude scaling factor β_{SRS} and mapped in sequence starting with $r^{\text{SRS}}(0)$ to resource elements (k, l) according to:

$$a_{(2k+k_0),l} = \begin{cases} \beta_{\text{SRS}} r^{\text{SRS}}(k) & k = 0, 1, \ldots \left(M^{\text{RS}}_{\text{sc},b} - 1 \right) \\ 0 & \text{otherwise,} \end{cases} \tag{9.64}$$

where k_0 is the frequency-domain starting position of the sounding reference signal and $M^{\text{RS}}_{\text{sc},b}$ is the length of the sounding reference signal sequence defined as:

$$M^{\text{RS}}_{\text{sc},b} = \frac{m_{\text{SRS},b} N^{\text{RB}}_{\text{sc}}}{2}. \tag{9.65}$$

The maximum separation between eight cyclic shifts requires that the SRS sequence length is divisible by eight. This is the case when the SRS bandwidth $\left(m_{\text{SRS},b} \times N^{\text{RB}}_{\text{sc}} \right)$ is a multiple of four resource blocks. The $m_{\text{SRS},b}$ is given in Table 9.11 for uplink bandwidth of $6 \leq N^{\text{UL}}_{\text{RB}} \leq 40$. For other uplink bandwidths $N^{\text{UL}}_{\text{RB}}$, $m_{\text{SRS},b}$ values can be found in [3]. In Table 9.11, the parameter SRS bandwidth configuration is cell-specific and represents that maximum SRS bandwidth. On the other hand, the parameter SRS bandwidth, b, is a UE-specific value, $b = B_{\text{SRS}}$.

The frequency-domain starting position k_0 is defined by:

$$k_0 = k_0' + \sum_{b=0}^{B_{\text{SRS}}} 2M^{\text{RS}}_{\text{sc},b} n_b, \tag{9.66}$$

where

$$k_0' = \left(\lfloor N^{\text{UL}}_{\text{RB}}/2 \rfloor - \frac{m_{\text{SRS},0}}{2} \right) N^{\text{RB}}_{\text{SC}} + k_{\text{TC}}, \tag{9.67}$$

where $k_{\text{TC}} \in \{0, 1\}$ is a UE-specific offset value depending on SRS transmission comb and n_b is the frequency position index for SRS bandwidth value b. When frequency hopping of the sounding reference signal is enabled, the SRS hopping bandwidth value $b_{\text{hop}} \in \{0, 1, 2, 3\}$ is given by higher layers.

Table 9.11. $m_{\text{SRS},b}$ and N_b values for the uplink bandwidth of $6 \leq N^{\text{UL}}_{\text{RB}} \leq 40$.

SRS bandwidth configuration	SRS bandwidth $b = 0$		SRS bandwidth $b = 1$		SRS bandwidth $b = 2$		SRS bandwidth $b = 3$	
	$m_{\text{SRS},b}$	N_b	$m_{\text{SRS},b}$	N_b	$m_{\text{SRS},b}$	N_b	$m_{\text{SRS},b}$	N_b
0	36	1	12	3	N/A	1	4	3
1	32	1	16	2	8	2	4	2
2	24	1	N/A	1	N/A	1	4	6
3	20	1	N/A	1	N/A	1	4	5
4	16	1	N/A	1	N/A	1	4	4
5	12	1	N/A	1	N/A	1	4	3
6	8	1	N/A	1	N/A	1	4	2
7	4	1	N/A	N/A	N/A	N/A	N/A	N/A

If frequency hopping of the sounding reference signal is not enabled (i.e., $b_{hop} \geq B_{SRS}$), the frequency position index n_b remains constant (unless re-configured) and is defined by:

$$n_b = \lfloor 4n_{RRC}/m_{SRS,b} \rfloor \bmod N_b, \tag{9.68}$$

where frequency-domain position n_{RRC} is a UE-specific value given by the higher layers. If frequency hopping of the sounding reference signal is enabled (i.e., $b_{hop} < B_{SRS}$), the frequency position indexes n_b are defined by:

$$n_b = \begin{cases} \lfloor 4n_{RRC}/m_{SRS,b} \rfloor \bmod N_b & b \leq b_{hop} \\ \{F_b(n_{SRS}) + \lfloor 4n_{RRC}/m_{SRS,b} \rfloor\} \bmod N_b & \text{otherwise.} \end{cases} \tag{9.69}$$

We note that for the case of full bandwidth SRS transmission with $b = 0$, SRS hopping is not applicable. N_b is given in Table 9.11 for uplink bandwidth of $6 \leq N_{RB}^{UL} \leq 40$. $F_b(n_{SRS})$ in Equation (9.69) is given as:

$$F_b(n_{SRS}) = \begin{cases} (N_b/2) \left\lfloor \dfrac{n_{SRS} \bmod \Pi_{b'=b_{hop}}^{b} N_{b'}}{\Pi_{b'=b_{hop}}^{b-1} N_{b'}} \right\rfloor + \left\lfloor \dfrac{n_{SRS} \bmod \Pi_{b'=b_{hop}}^{b} N_{b'}}{2\Pi_{b'=b_{hop}}^{b-1} N_{b'}} \right\rfloor & \text{if } N_b \text{ even} \\[4mm] \lfloor N_b/2 \rfloor \left\lfloor \dfrac{n_{SRS}}{\Pi_{b'=b_{hop}}^{b-1} N_{b'}} \right\rfloor & \text{if } N_b \text{ odd,} \end{cases} \tag{9.70}$$

where n_{SRS} counts the number of UE-specific SRS transmissions and is given as:

$$n_{SRS} = \lfloor (n_f \times 10 + \lfloor n_s/2 \rfloor)/T_{SRS} \rfloor, \tag{9.71}$$

where n_f is system frame number and $T_{SRS} \in \{2, 5, 10, 20, 40, 80, 160, 320\}$ is UE-specific periodicity of SRS transmission.

An example of SRS bandwidth configurations 0 and 2 for uplink bandwidth of $6 \leq N_{RB}^{UL} \leq 40$ is depicted in Figure 9.22. We note that a tree-structure is used within each bandwidth configuration while keeping the minimum SRS bandwidth of four resource blocks. In SRS bandwidth configurations 0, for example, two UEs can transmit SRS with maximum bandwidth of 36 resource blocks within a subframe by using transmission comb offset values of $k_0' = 0, 1$. Another possibility is that one UE transmit SRS over maximum bandwidth of 36 resource blocks with $k_0' = 0$ while three other UEs transmit SRS over bandwidth of 12 resource blocks with $k_0' = 1$. Similarly, other combinations of various SRS bandwidths of 4, 12 and 36 resource blocks can be considered.

Let us consider the case of SRS bandwidth configuration 0 at level $b = 2$ with $m_{SRS,b} = 4$ and $b_{hop} = 0$ to demonstrate the frequency hopping of SRS as shown in Figure 9.23. In this case, the values of $F_b(n_{SRS})$ for different SRS transmission indices n_{SRS} are given as:

$$\begin{aligned} F_{b=0}(n_{SRS}) &= 0, 0, 0, 0, 0, 0, 0, 0, 0 \\ F_{b=1}(n_{SRS}) &= 0, 1, 2, 3, 4, 5, 6, 7, 8, 9 \quad n_{SRS} = 0, 1, 2, \cdots . \\ F_{b=2}(n_{SRS}) &= 0, 0, 0, 1, 1, 1, 2, 3, 3 \end{aligned} \tag{9.72}$$

Figure 9.22. SRS bandwidth configurations 0 and 2 for uplink bandwidth of $6 \leq N_{RB}^{UL} \leq 40$.

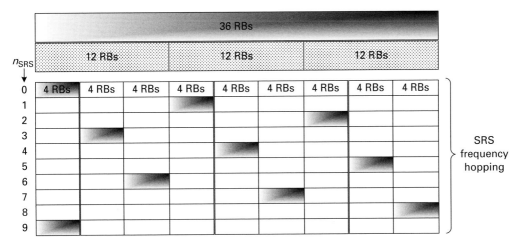

Figure 9.23. SRS frequency hopping for frequency bandwidth scanning.

By using Equation (9.69), we obtain values for n_b for different SRS transmission indices n_{SRS} as:

$$n_0 = 0, 0, 0, 0, 0, 0, 0, 0, 0$$
$$n_1 = 0, 1, 2, 0, 1, 2, 0, 1, 2 \quad n_{SRS} = 0, 1, 2, \cdots. \tag{9.73}$$
$$n_2 = 0, 0, 0, 1, 1, 1, 2, 3, 3$$

By assuming $n_{SRS} = 2$ in (9.69) and $k_0' = 0$ in (9.66), the frequency-domain starting position k_0 for $b = 0$ and for different SRS transmission indices n_{SRS} is:

$$k_0 = 0, 12N_{sc}^{RB}, 24N_{sc}^{RB}, 4N_{sc}^{RB}, \cdots n_{SRS} = 0, 1, 2, \cdots. \tag{9.74}$$

This means that the whole sounding bandwidth of 36 resource blocks can be scanned in nine SRS transmissions with each SRS scanning four resource blocks as shown in Figure 9.23. We also note that for $b = 1$, the whole sounding bandwidth of 36 resource blocks can be scanned in three SRS transmissions with each SRS scanning 12 resource blocks in a sequential fashion.

Table 9.12. SRS subframe configurations.

Configuration	Binary representation	Configuration Period, T_{SFC} (subframes)	Transmission offset Δ_{SFC} (subframes)
0	681	1	{0}
1	0001	2	{0}
2	0010	2	{1}
3	0011	5	{0}
4	0100	5	{1}
5	0101	5	{2}
6	0110	5	{3}
7	0111	5	{0,1}
8	1000	5	{2,3}
9	1001	10	{0}
10	1010	10	{1}
11	1011	10	{2}
12	1100	10	{3}
13	1101	10	{0,1,2,3,4,6,8}
14	1110	10	{0,1,2,3,4,5,6,8}
15	1111	Inf	N/A

The cell-specific subframe configuration period T_{SFC} and the cell specific subframe offsets Δ_{SFC} relevant for the transmission of sounding reference signals are given in Table 9.12. The SRS subframes satisfy the following condition:

$$(C_{SFC} \bmod T_{SFC}) \in \Delta_{SFC}, \tag{9.75}$$

where C_{SFC} represents the subframe counter. An example showing SRS transmission in SRS subframe configuration 7 and subframe configuration 8 is given in Figure 9.24. Since $T_{SFC} = 5$ the SRS is transmitted every fifth subframe with $\Delta_{SFC} = 0, 1$ and $\Delta_{SFC} = 2, 3$ for SRS subframe configuration 7 and subframe configuration 8 respectively. In SRS subframe configuration 7, subframes numbered $0, 1, 5, 6, \ldots$ contain SRS while in subframe configuration 8 subframes numbered $2, 3, 7, 8, \ldots$ contain SRS as these subframes satisfy the condition in (9.75). We also note that if subframe configuration 7 is used in one cell and subframe configuration 8 is used in a neighboring cell, the SRS in the two cells will be transmitted in different subframes. This would avoid interference from SRS transmission to SRS transmission. The SRS subframe configurations allow to control the overhead for SRS in a given cell. The SRS subframe configurations with smaller configuration periods T_{SFC} mean more SRS overhead. For example, in SRS configuration 0, SRS is transmitted in every subframe resulting in 7.14% (1/14) overhead for the case of the normal cyclic prefix. Note that when SRS is transmitted in every subframe in subframe configuration 0, no offset can be applied. The offset is only valid when the duty cycle of SRS is less than or equal to 50% ($T_{SFC} \geq 2$). The SRS subframe configuration 15 with $T_{SFC} = \infty$ means no SRS transmission in the cell.

9.8   Reference signals overhead

The downlink reference signal overhead estimates are given in Table 9.13. We note that the relative overheads for non-MBSFN transmissions are higher for the case of an extended cyclic

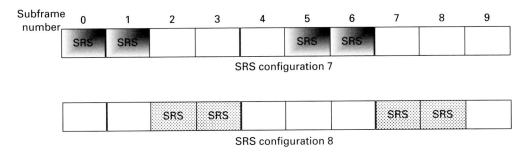

Figure 9.24. SRS subframe configuration 7 and subframe configuration 8.

Table 9.13. Downlink reference signal overhead estimates.

	Number of antenna ports	Reference signal overhead [%]			
		Unicast only subframes	MBSFN only subframes	MBSFN subframes containing unicast control	UE-specific reference signals
Normal cyclic prefix	1	4.76			7.14
	2	9.52			
	4	14.29			
Extended cyclic prefix	1	5.56	12.50	13.89	8.33
	2	11.11		15.28	
	4	16.67		18.06	
7.5 kHz	1	NA	12.50	NA	

prefix. This is because the number of reference symbols in a subframe are the same between the normal and the extended cyclic prefix, while the total resources within a subframe for the extended cyclic prefix is 85.7% (12/14) of the normal cyclic prefix resources. The MBSFN reference signals are defined for the case of an extended cyclic prefix only as the multi-cell MBSFN channel experiences a larger delay spread relative to a single-cell non-MBSFN channel. The UE-specific reference signals are defined for a single antenna port with 7.14% (1/14) and 8.33% (1/12) overhead for normal and extended cyclic prefix respectively.

We note that the UE-specific reference signal overhead is more than a single antenna port cell-specific reference signal overhead. The reason for this is that cell specific reference signals are transmitted on all the resource blocks and all the subframes (first two OFDM symbols in the case of MBSFN subframes), which allow performing interpolation in both the frequency and time domains. However, UE-specific reference signals are only transmitted in the resources and subframes allocated for PDSCH transmission to a UE, which limits the interpolation that can be performed in frequency and time. In order to compensate for this loss, the density of UE specific reference signals is made higher than cell-specific reference signals. Another point to note, however, is that UE-specific reference signals also benefit from beam-forming gains, which can actually improve the channel estimation performance.

Table 9.14. Uplink demodulation reference signal overhead estimates.

Uplink channel	Format	Reference signal overhead [%]	
		Normal cyclic prefix	Extended cyclic prefix
PUSCH		14.3%	16.7%
PUCCH	1, 1A and 1B	42.8%	33.3%
	2, 2A and 2B	28.6%	16.7%

The uplink demodulation reference signals for the normal cyclic prefix and extended cyclic prefix cases are shown in Figures 9.18 and 9.19 respectively. The demodulation reference signals for PUSCH are transmitted in the middle of the slot in symbol with $l = 2, 3$ for the extended and normal cyclic prefix respectively. The DMRS overheads for PUSCH and PUCCH are summarized in Table 9.14. The DMRS overhead for PUSCH is 14.3% (1/7) and 16.7% (1/6) for the normal and extended cyclic prefix respectively. We note that the uplink reference signal overhead is much larger than the downlink reference signals case. A reason for this is that the uplink reference signals are UE-specific and transmitted in the resource blocks and subframes allocated for uplink PUSCH transmission only, which limit the interpolation in frequency and time.

For PUCCH format 1, 1A and 1B, the demodulation reference signals are transmitted in three symbols ($l = 2, 3, 4$) for the normal cyclic prefix and two symbols ($l = 2, 3$) for the extended cyclic prefix. This represents a DMRS overhead of 42.8% (3/7) and 33.3% (2/6) for the normal and extended cyclic prefix respectively. It should be noted that PUCCH format 1 is used for 1-bit scheduling request transmission and PUCCH formats 1A and 1B are used for BPSK and QPSK ACK/NACK symbol respectively. For coherent detection of a single modulation symbol, the energy is generally split equally between the reference signal and the data symbol. Moreover, block-wise spreading of the ACK/NACK modulation symbol is used in PUCCH formats 1A and 1B. In order to enable the same number of reference signals as the number of ACK/NACK channels, block-wise spreading also needs to be applied to the reference signals. Since the number of orthogonal sequences used in block-wise spreading is three, the number of symbols within a slot required for reference signals is also three.

For PUCCH format 2, 2A and 2B, the demodulation reference signals are transmitted in two symbols ($l = 2, 3$) for a normal cyclic prefix and a single symbol ($l = 3$) for an extended cyclic prefix. This leads to a DMRS overhead of 28.6% and 16.7% for a normal and an extended cyclic prefix respectively. It should be noted that PUCCH formats 2, 2A and 2B are used for channel quality feedback using a block code. Therefore the reference signals overhead does not need to be as large as in the case of a single modulation symbol detection. We note that for the extended cyclic prefix, the DMRS overhead for PUCCH format 2, 2A and 2B is the same as the DMRS overhead for PUSCH.

9.9   Summary

The cell search design in LTE supports scalable bandwidths with a minimum bandwidth of 1.08 MHz. The synchronization signals and broadcast control information are transmitted in the center 1.08 MHz bandwidth irrespective of the system bandwidth. This allows UEs to acquire time and frequency synchronization with the network without knowledge of the system bandwidth, which makes the cell search more efficient and faster. After achieving the timing and frequency synchronization, the UE can obtain the remaining cell/system-specific information from the broadcast control channel. The information on the eNB antenna configuration is also encoded in the cyclic redundancy check on the broadcast control channel. Therefore, after successful reception of the broadcast control and CRC check, the UE also acquires information on the number of antenna ports supported by the eNB.

Three types of reference signals namely cell-specific reference signal, MBSFN reference signals and UE-specific reference signals are defined for the downlink. All the reference signals are multiplexed with data within a subframe using a hybrid TDM / FDM approach. In this hybrid TDM / FDM approach, a fraction of the subcarriers in a few OFDM symbols within a subframe is used for reference signals transmission. The number of cell-specific reference signals equals the number of antenna ports supported by the eNB. On the other hand, both UE-specific and MBSFN reference signals are defined with a single port which means that only rank-1 transmissions can be performed for MBSFN and for UEs scheduled using UE-specific reference signals.

The cell-specific reference signals are used for various downlink measurements as well as for demodulation of non-MBSFN transmissions. The UE-specific and MBSFN reference signals are used for demodulation purposes only. The cell-specific reference signals are scrambled using a cell-specific PN sequence. The MBSFN reference signals are scrambled using an MBSFN area specific PN sequence, which allows composite MBSFN channel estimates for MBSFN demodulation. The UE-specific reference signals are scrambled using a cell-specific and UE-specific PN sequence. All the PN sequences are based on a Gold sequence.

In the uplink, reference signals are defined for demodulation and channel sounding. As the LTE system does not support transmit diversity or spatial multiplexing in the uplink, reference signals are defined for a single antenna port only. In addition, the uplink reference signals are always UE-specific due to the many-to-one nature of the uplink. In many-to-one transmissions, the concept of cell-specific reference signals does not exist. The uplink reference signals use low PAPR/CM constant amplitude sequences. The uplink reference signals are strictly time-multiplexed with other transmissions to ensure the single-carrier property of the uplink transmissions. If reference signals were frequency-multiplexed with data, the uplink transmission would behave like a multi-carrier transmission with increased PAPR/CM. The uplink sounding reference signals are used for uplink channel quality measurement for channel-sensitive scheduling as well as for uplink power and timing control. The bandwidth of sounding reference signals can be configured in a UE-specific manner. On the other hand, the time-domain frequency of sounding reference signals is configured in a cell-specific manner. The cell-specific configuration of the time-domain frequency of sounding reference signals allows one to control the total SRS overhead in the cell.

References

[1] Gold, R., "Optimal binary sequences for spread spectrum multiplexing (Corresp.)," *IEEE Transactions on Information Theory*, vol. IT-13, no. 4, pp. 619–621, 1967.

[2] Chu, D. C., "Polyphase codes with good periodic correlation properties," *IEEE Transactions on Information Theory*, vol. IT-18, pp. 531–532, July 1992.

[3] 3GPP TSG RAN v8.3.0, Evolved Universal Terrestrial Radio Access: Physical Channels and Modulation (3GPP TS 36.211).

[4] 3GPP TSG RAN v8.3.0, Evolved Universal Terrestrial Radio Access: Multiplexing and Channel Coding (3GPP TS 36.212).

[5] 3GPP TSG RAN v8.3.0, Evolved Universal Terrestrial Radio Access: Physical Layer Procedures (3GPP TS 36.213).

10 Random access

Random access is generally performed when the UE turns on from sleep mode, performs handoff from one cell to another or when it loses uplink timing synchronization. At the time of random access, it is assumed that the UE is time-synchronized with the eNB on the downlink. Therefore, when a UE turns on from sleep mode, it first acquires downlink timing synchronization. The downlink timing synchronization is achieved by receiving primary and secondary synchronization sequences and the broadcast channel as discussed in Chapter 9. After acquiring downlink timing synchronization and receiving system information including information on parameters specific to random access, the UE can perform the random access preamble transmission. Random access allows the eNB to estimate and, if needed, adjust the UE uplink transmission timing to within a fraction of the cyclic prefix. When an eNB successfully receives a random access preamble, it sends a random access response indicating the successfully received preamble(s) along with the timing advance (TA) and uplink resource allocation information to the UE. The UE can then determine if its random access attempt has been successful by matching the preamble number it used for random access with the preamble number information received from the eNB. If the preamble number matches, the UE assumes that its preamble transmission attempt has been successful and it then uses the TA information to adjust its uplink timing. After the UE has acquired uplink timing synchronization, it can send uplink scheduling or a resource request using the resources indicated in the random access response message as depicted in Figure 10.1.

10.1 Random access preamble formats

Since the random access (RA) mechanism is used by the UE when it is not synchronized on the uplink a guard time (GT) needs to be introduced to avoid collisions with other transmissions. The duration of the guard time needs to account for the round trip propagation time as shown in Figure 10.2 and is dependent upon the cell size supported. With propagation speed of 1 km/3.33 μs, approximately 6.7 μs of guard time per kilometer is required to accommodate the round-trip time. In order to support cell size up to 100 km as required for LTE, the guard time should be in the range of 670 μs. However, this large guard time would be an undesired overhead when the system is deployed with smaller cells, which are most common. Therefore, multiple random access preamble formats with both small and large guard times are defined as shown in Figure 10.3. In order to enable simple frequency-domain processing, the random access preamble also uses a cyclic prefix. The length of the cyclic prefix (CP) accounts for both the propagations delays as well as the channel delay spread. The random access preamble length is always 0.8 ms. In format 0, both the CP and GT are equal to approximately

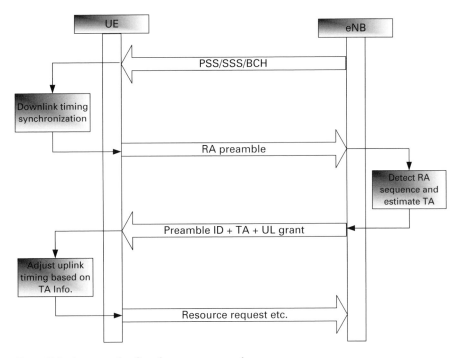

Figure 10.1. An example of random access procedure.

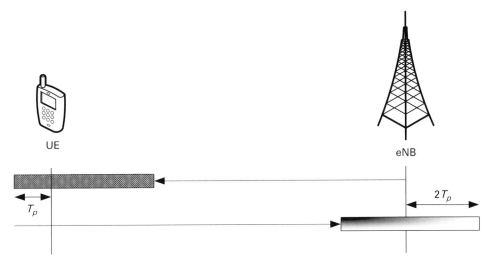

Figure 10.2. Round trip propagation time.

0.1 ms, which are sufficient to support cell size of up to approximately 15 km. In format 1, the CP and GT are respectively 0.68 s and 0.52 ms, which are sufficient to support cell size of up to approximately 78 kilometers. Another aspect to consider is if preamble length of 0.8 ms provides enough energy for it to be successfully detected at the eNB. Therefore, to provide energy gain, in formats 2 and 3, preamble is repeated once. In format 2, both the CP

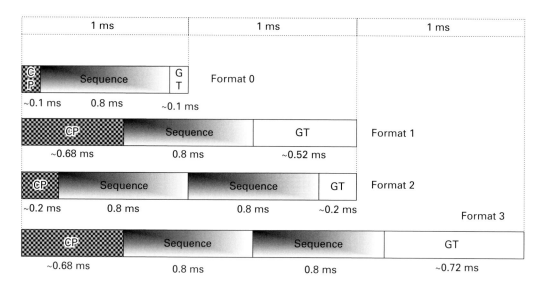

Figure 10.3. Random access preamble formats.

Table 10.1. Random access preamble parameters.

Preamble format	T_{CP}	T_{SEQ}	T_{GT}	$T_{CP} + T_{SEQ} + T_{GT}$
0	$3168 \cdot T_s \approx 0.1\,\text{ms}$	$24576 \cdot T_s = 0.8\,\text{ms}$	$2976 \cdot T_s \approx 0.1\,\text{ms}$	$3072 \cdot T_s = 1\,\text{ms}$
1	$21024 \cdot T_s \approx 0.68\,\text{ms}$	$24576 \cdot T_s = 0.8\,\text{ms}$	$15840 \cdot T_s \approx 0.52\,\text{ms}$	$2.3072 \cdot T_s = 2\,\text{ms}$
2	$6240 \cdot T_s \approx 0.2\,\text{ms}$	$2.24576 \cdot T_s = 1.6\,\text{ms}$	$6048 \cdot T_s \approx 0.2\,\text{ms}$	$2.3072 \cdot T_s = 2\,\text{ms}$
3	$21024 \cdot T_s \approx 0.68\,\text{ms}$	$2.24576 \cdot T_s = 1.6\,\text{ms}$	$21984 \cdot T_s \approx 0.72\,\text{ms}$	$3.3072 \cdot T_s = 3\,\text{ms}$

and GT are equal to approximately 0.2 ms, which is sufficient to support cell size of up to approximately 30 kilometers. In format 2, CP and GT are approximately equal to 0.68 and 0.72 ms respectively, which are sufficient to support cell size of over 100 km.

The detailed parameters for the random access preamble formats are summarized in Table 10.1. All the time durations are defined in terms of the sample period $T_s = 1/f_s$, where $f_s = 30.72\,\text{M samples/sec}$.

10.2   RA sequence length and resource mapping

In a frequency-domain, similar to PSS and SSS, random access transmission uses six resource blocks with total bandwidth equal to 1.08 MHz. We note that RA preamble length needs to be selected as a prime number in order to maximize the number of available sequences. However, with the six resource block allocation for random access, a total of 72 data subcarriers with 15 KHz subcarrier spacing are available. With random access subcarrier spacing of $\Delta f_{RA} = 1.25\,\text{kHz}$, the number of available subcarriers is 864, which is not a prime number. The prime numbers around 864 are 829, 839, 853, 857, 859, 863 and 877. A preamble length larger than 864 would lead to unavoidable interference to/from data subcarriers and therefore

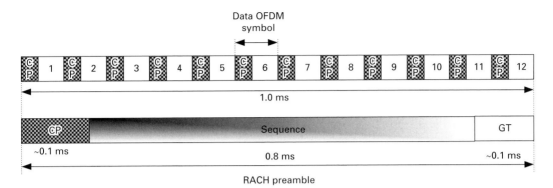

Figure 10.4. Waveform discontinuities between data and random access transmissions.

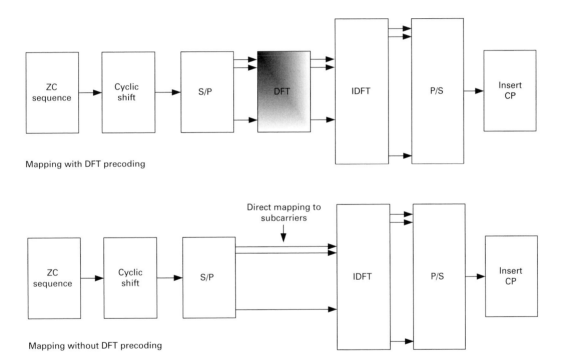

Mapping with DFT precoding

Mapping without DFT precoding

Figure 10.5. Random access preamble mapping schemes.

is not considered. At first instance, a sequence length of 863 appears a good choice. However, the data OFDM symbols and random access preamble transmissions are not aligned as shown in Figure 10.4. This would lead to interference between data transmission and random access transmissions. Therefore, a small guard band can be introduced between data and random access transmissions. A guard band consisting of a total of 25 guard subcarriers with subcarrier spacing of $\Delta f_{RA} = 1.25\,\text{kHz}$ was considered sufficient. This led to the selection of a random access preamble length of $N_{ZC} = 839$.

Two different schemes with and without DFT precoding as shown in Figure 10.5 can be envisioned for random access preamble mapping. However, DFT precoding of a ZC sequence converts the sequence to another ZC sequence with a different root index. Therefore, unlike data transmission using SC-FDMA, DFT precoding is an unnecessary step for uplink transmission of random access preamble. Therefore, a simpler structure without DFT precoding with direct mapping of ZC sequence samples to subcarriers is used in the LTE system.

With no DFT precoding, the time-continuous random access signal can be written as below:

$$s(t) = \beta \sum_{k=0}^{(N_{ZC}-1)} \sum_{n=0}^{(N_{ZC}-1)} x_{u,v}(n) \cdot e^{\frac{-j2\pi nk}{N_{ZC}}} \cdot e^{j2\pi\left(k+\varphi+\frac{\Delta f}{\Delta f_{RA}}\left(k_0+\frac{1}{2}\right)\right)\Delta f_{RA}(t-T_{CP})}, \quad (10.1)$$

where $x_{u,v}(n)$ is the uth root Zadoff–Chu sequence with cyclic shift v (see Section 10.4). Also, $0 \le t < (T_{SEQ} - T_{CP})$ and β_{PRACH} is an amplitude scaling factor. The random access preamble location in the frequency domain is controlled by the parameter k_0 given as:

$$k_0 = \left(k_{RA}N_{SC}^{RB}\right) - \left(\frac{N_{RB}^{UL}N_{SC}^{RB}}{2}\right), \quad (10.2)$$

where $0 \le k_{RA} \le \left(N_{RB}^{UL} - 6\right)$ is the physical resource block number configured by higher layers. Δf and Δf_{RA} are subcarrier spacings for data and random access respectively and the factor $\frac{\Delta f}{\Delta f_{RA}}$ accounts for the difference in subcarrier spacing. φ is a fixed offset determining the frequency-domain location of the random access preamble within the physical resource blocks. The frequency scaling factor k_0 as a function of k_{RA} for $N_{RB}^{UL} = 50$ and $N_{SC}^{RB} = 12$ is plotted in Figure 10.6. In this case there is a total of 600 data subcarriers and the starting position for random access can be between subcarrier number -300 and subcarrier number 228. We note that $k_0 = 0$ for $k_{RA} = \frac{N_{RB}^{UL}}{2}$.

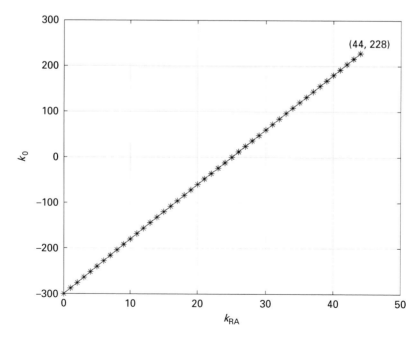

Figure 10.6. Frequency scaling factor k_0 as a function of k_{RA} for $N_{RB}^{UL} = 50$ and $N_{SC}^{RB} = 12$.

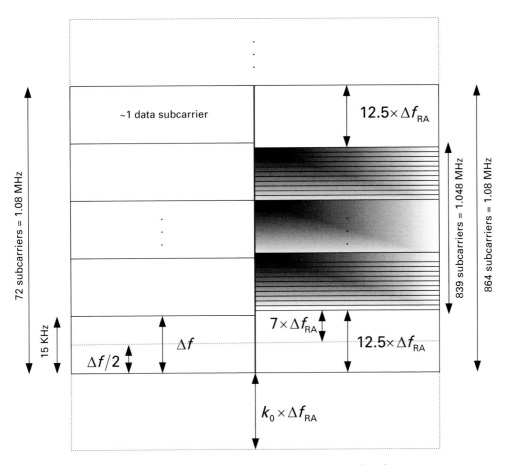

Figure 10.7. Random access preamble structure in the frequency domain.

The random access preamble structure in the frequency domain is shown in Figure 10.7. With $\varphi = 7$ assumed in the LTE system, a guard band of approximately one data subcarrier $\sim \Delta f$ or 12.5 random access subcarriers $\Delta f \approx 12.5 \Delta f_{RA}$ is provided on each side of the random access preamble transmission.

10.3 Random access configurations

In the time domain, a single random access resource is allowed at most. A total of 16 random access configurations are defined for each of the few preamble formats (see Table 10.2). The parameter PRACH configuration index in Table 10.2 is set by higher layers. The start of the random access preamble is aligned with the start of the corresponding uplink subframe at the UE assuming a timing advance of zero. This is because the round trip propagation delay is already accounted for in the guard time. For random access configuration using even system frame numbers, that is configurations 0, 1, 2, 15, 16, 17, 18, 31, 32, 33, 34, 47, 48, 49, 50 and 63, the UE may for handover purposes assume an absolute value of the relative time difference between radio frame i in the current cell and the target cell of less than $153\,600 \cdot T_s =$

Table 10.2. Random access configurations for preamble format 0−3.

PRACH config- uration index	Preamble format	System frame number	Subframe number	PRACH config- uration index	Preamble format	System frame number	Subframe number
0	0	Even	1	32	2	Even	1
1	0	Even	4	33	2	Even	4
2	0	Even	7	34	2	Even	7
3	0	Any	1	35	2	Any	1
4	0	Any	4	36	2	Any	4
5	0	Any	7	37	2	Any	7
6	0	Any	1, 6	38	2	Any	1, 6
7	0	Any	2, 7	39	2	Any	2, 7
8	0	Any	3, 8	40	2	Any	3, 8
9	0	Any	1, 4, 7	41	2	Any	1, 4, 7
10	0	Any	2, 5, 8	42	2	Any	2, 5, 8
11	0	Any	3, 6, 9	43	2	Any	3, 6, 9
12	0	Any	0, 2, 4, 6, 8	44	2	Any	0, 2, 4, 6, 8
13	0	Any	1, 3, 5, 7, 9	45	2	Any	1, 3, 5, 7, 9
14	0	Any	0, 1, 2, 3, 4, 5, 6, 7, 8, 9	46	N/A	N/A	N/A
15	0	Even	9	47	2	Even	9
16	1	Even	1	48	3	Even	1
17	1	Even	4	49	3	Even	4
18	1	Even	7	50	3	Even	7
19	1	Any	1	51	3	Any	1
20	1	Any	4	52	3	Any	4
21	1	Any	7	53	3	Any	7
22	1	Any	1, 6	54	3	Any	1, 6
23	1	Any	2, 7	55	3	Any	2, 7
24	1	Any	3, 8	56	3	Any	3, 8
25	1	Any	1, 4, 7	57	3	Any	1, 4, 7
26	1	Any	2, 5, 8	58	3	Any	2, 5, 8
27	1	Any	3, 6, 9	59	3	Any	3, 6, 9
28	1	Any	0, 2, 4, 6, 8	60	N/A	N/A	N/A
29	1	Any	1, 3, 5, 7, 9	61	N/A	N/A	N/A
30	N/A	N/A	N/A	62	N/A	N/A	N/A
31	1	Even	9	63	3	Even	9

0.5 ms. The overhead for random access can be calculated from the allowed time-domain frequency of random access transmissions. For example, if the system bandwidth is 50 resource blocks and configuration 9 with random access allowed in all system frame numbers, the overhead is:

$$OH = \frac{6}{50} \times \frac{3}{10} = 3.6\%. \tag{10.3}$$

Note that configuration 9 allocates three out of a total of 10 subframes for random access transmissions. It can be noted that for the same time-domain frequency, the random access overhead will be larger for smaller system bandwidths. On the other hand, it can be argued that if the system bandwidth is smaller, the number of users supported will also be smaller resulting in a smaller number of random access attempts. This would favor reducing time-domain frequency of random access transmissions using configurations with fewer subframes allocated for random access transmissions for smaller system bandwidths.

10.4   RA preamble cyclic shifts

In order to meet random access coverage requirements only 6-bit information is transmitted using a preamble. The preamble waveforms should have good detection probability while maintaining low false alarm rate, low collision probability, low peak-to-average power ratio (PAPR) or signal peakiness, and allow accurate timing estimation. Some examples of sequences that meet these requirements are Zadoff–Chu sequences [1] and generalized chirp-like (GCL) sequences [2]. These sequences have an advantage relative to pseudo random (PN) sequences due to their low PAPR property that is important for uplink transmissions due to limited transmit power of the UE. In the LTE system, ZC sequences were selected for uplink random access preamble transmission.

Since the random access preamble length selected in the LTE system is an odd number ($N_{ZC} = 839$), the uth root Zadoff–Chu sequence is given by:

$$x_u(n) = e^{-j\frac{\pi un(n+1)}{N_{ZC}}}, \quad 0 \leq n \leq N_{ZC} - 1. \tag{10.4}$$

From the uth root Zadoff–Chu sequence, random access preambles with zero correlation zones (ZCZ) of length ($N_{CS} - 1$) are defined by cyclic shifts according to

$$x_{u,v}(n) = x_u((n + C_v)_{N_{ZC}}), \tag{10.5}$$

where the modulo operation $(p)_N$ is defined as:

$$(p)_N = \begin{cases} \text{Rem}\left[\dfrac{p}{N}\right] & p \geq 0 \\[2em] \text{Rem}\left[\dfrac{N - \text{Rem}\left[\dfrac{-p}{N}\right]}{N}\right] & p < 0 \end{cases} \tag{10.6}$$

and Rem[ ] takes the remainder of the ratio of two integers.

In the LTE system, two sets of cyclic shifts namely unrestricted and restricted cyclic shifts are specified. In the restricted cyclic shifts, only a subset of the total cyclic shifts is allowed to support high mobility scenarios.

10.4.1   Unrestricted cyclic shifts

In the unrestricted case, the allowed cyclic shift value C_v is given by

$$C_v = v \; N_{CS} \quad v = 0,1,\ldots, \lfloor N_{ZC}/N_{CS} \rfloor - 1. \tag{10.7}$$

A total of 16 configurations with different values for N_{CS} as given in Table 10.3 are supported. A larger N_{CS} value allows larger delay spreads at the expense of reduced number of

Table 10.3. Cyclic shifts N_{CS} and number of available cyclic shifts.

N_{CS} configuration	Unrestricted		Restricted	
	N_{CS}	number of available cyclic shifts	N_{CS}	number of available cyclic shifts
0	0	Inf	15	
1	13	64	18	
2	15	55	22	
3	18	46	26	
4	22	38	32	
5	26	32	38	
6	32	26	46	
7	38	22	55	
8	46	18	68	
9	59	14	82	
10	76	11	100	Depends on
11	93	9	128	N_{CS} and ZC
12	119	7	158	root index
13	167	5	202	
14	279	3	237	
15	419	2	–	

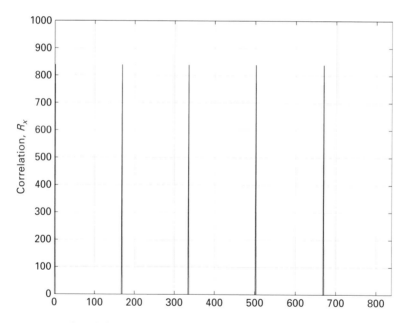

Figure 10.8. Correlation for N_{CS} configuration 13 with $N_{CS} = 167$. The number of available cyclic shifts is five.

available cyclic shifts. For example, N_{CS} configuration number 1 with $N_{CS} = 13$ provides 64 cyclic shifts and can support a delay spread of up to 12.4 microseconds ($13 \times 0.8\,\text{ms}/839$). In Figure 10.8, we show correlations for N_{CS} configuration number 13 with $N_{CS} = 167$. In this case, the number of available cyclic shifts is only five and, therefore, multiple root indices need to be used to provide a total of 64 required RA sequences in a cell.

10.4.2 Restricted cyclic shifts

A frequency error due to Doppler frequency shift affects the detection performance and false alarm rate due to the autocorrelation side lobes or the aliases of the main lobe. Let the received RA preamble without frequency offset be denoted as $r_u(n)$ and the frequency offset is $\Delta\omega$, then the received RA preamble with frequency offset is given by:

$$\hat{r}_u(n) = r_u(n) \cdot e^{j\Delta wn} = r_u(n) \cdot e^{j2\pi\left(\frac{\Delta f}{f_s}\right)n}, \tag{10.8}$$

where Δf is the frequency offset in H_z and f_s is the sampling rate of the RA preamble. Note that in the case of critical sampling, f_s is the bandwidth of the RACH preamble. With the above definitions and ignoring noise and interference, the auto-correlation of $\hat{r}_u(n)$ at the zeroth lag can be obtained by setting $r_u(n) = x_u(n)$ as :

$$R_x(0) = \sum_{n=0}^{(N_{ZC}-1)} \hat{r}_u(n) \cdot x_u^*(n)$$

$$= \sum_{n=0}^{(N_{ZC}-1)} e^{-j\frac{\pi un(n+1)}{N_{ZC}}} \cdot e^{j2\pi\left(\frac{\Delta f}{f_s}\right)n} \cdot e^{j\frac{\pi un(n+1)}{N_{ZC}}}$$

$$= \sum_{n=0}^{(N_{ZC}-1)} e^{j2\pi\left(\frac{\Delta f}{f_s}\right)n}. \tag{10.9}$$

We note that with no frequency error $\Delta f = 0$, the correlation at zeroth lag is equal to the ZC sequence length:

$$R_x(0) = \sum_{n=0}^{(N_{ZC}-1)} e^{j2\pi\left(\frac{0}{f_s}\right)n} = N_{ZC}. \tag{10.10}$$

The correlation is zero for all lags other than zero. This means that when a ZC sequence is received with no frequency errors, the sequence can be detected with high reliability.

The correlation at zeroth lag $R_x(0)$ as a function of frequency error Δf is shown in Figure 10.9. We note that the autocorrelation degrades in the presence of frequency errors. In particular, when the frequency error is an integer multiple of 1.25 kHz, the correlation at zeroth lag is zero and hence the detection is impossible. This is assuming RACH bandwidth of 1.048 MHz (839×1.25 kHz) for transmission of length 839 sequence with random access subcarrier separation Δf_{RA} of:

$$\Delta f_{RA} = \frac{1}{T_{SEQ}} = \frac{1}{0.5 \text{ ms}} = 1.25 \text{ kHz.} \tag{10.11}$$

Assuming frequency errors happen due to Doppler shift at higher UE speeds, then an absolute Doppler frequency shift of $f_D = 1/T_{SEQ} = \Delta f_{RA} = 1.25$ kHz gives rise to correlation peaks in the receiver's bank of correlators corresponding to the cyclic shifts $(vN_{CS} + d_u)_{N_{ZC}}$ or $(vN_{CS} - d_u)_{N_{ZC}}$ for $f_D = 1/T_{SEQ}$ and $f_D = -1/T_{SEQ}$ respectively, where d_u

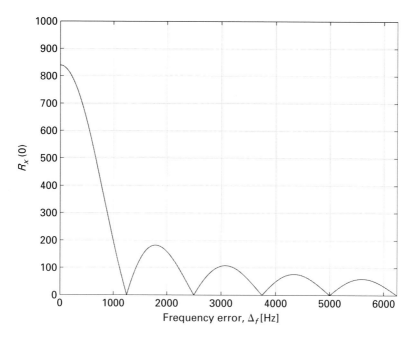

Figure 10.9. Correlation at zeroth lag as a function of frequency error Δf for $N_{ZC} = 839$.

is given as [3]:

$$d_u = \begin{cases} \dfrac{(mN_{ZC} - 1)}{u} & 0 \le \dfrac{(mN_{ZC} - 1)}{u} < N_{ZC}/2 \\[3mm] N_{ZC} - \dfrac{(mN_{ZC} - 1)}{u} & \text{otherwise,} \end{cases} \quad (10.12)$$

where m is the smallest positive integer for which d_u is an integer. The delay offset d_u values for ZC sequence indices with $N_{ZC} = 839$ are provided in Figure 10.10.

An alternative way of writing d_u is:

$$d_u = \begin{cases} u^{-1} \bmod N_{ZC} & 0 \le u^{-1} \bmod N_{ZC} < N_{ZC}/2 \\[3mm] N_{ZC} - u^{-1} \bmod N_{ZC} & \text{otherwise,} \end{cases} \quad (10.13)$$

where, u^{-1} is a positive integer for which the following condition holds:

$$u \times u^{-1} \bmod N_{ZC} = 1. \quad (10.14)$$

If we consider a ZC sequence root index $u = 300$, the delay offset d_u due to Doppler shift can be obtained:

$$d_u = \frac{(mN_{ZC} - 1)}{u} = \frac{(59 \times 839 - 1)}{300} = 165. \quad (10.15)$$

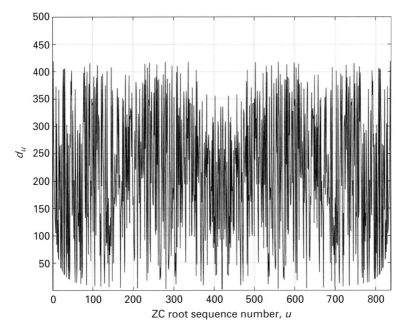

Figure 10.10. d_u for various root ZC indices for $N_{ZC} = 839$.

Here $m = 59$ is the smallest positive integer for which d_u is an integer. Using the alternative method, we first obtain the integer u^{-1}:

$$u \times u^{-1} \bmod N_{ZC} = 1$$
$$(300 \times 674) \bmod 839 = 1$$
$$\Rightarrow u^{-1} = 674. \tag{10.16}$$

Since $u^{-1} = 674 > N_{ZC} = 839$, d_u is given as:

$$d_u = N_{ZC} - u^{-1} \bmod N_{ZC} = 839 - 674 = 165. \tag{10.17}$$

The delay offset d_u due to Doppler shift can be obtained by using yet another condition as below:

$$u \times u^{-1} \bmod N_{ZC} = N_{ZC} - 1$$
$$(300 \times 165) \bmod 839 = 838$$
$$\Rightarrow d_u = 165. \tag{10.18}$$

Figure 10.11 shows correlation performance due to Doppler shift of $f_D = 1/T_{SEQ}$ and $f_D = -1/T_{SEQ}$ for $N_{ZC} = 839$, $v = 1N_{CS} = 93$, $u = 300$ and $d_u = 165$. In this case, the locations of the two correlation peaks are given as:

$$(vN_{CS} + d_u)_{N_{ZC}} = (93 + 165)_{839} = 258, \quad f_D = \frac{1}{T_{SEQ}}$$
$$(vN_{CS} - d_u)_{N_{ZC}} = (93 - 165)_{839} = 767, \quad f_D = \frac{-1}{T_{SEQ}}. \tag{10.19}$$

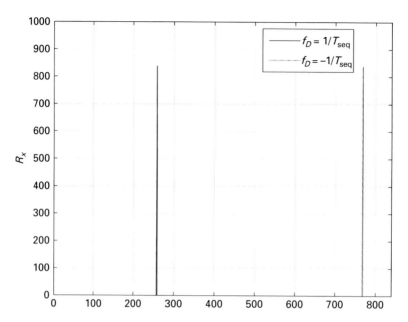

Figure 10.11. Correlation due to $f_D = 1.25$ and -1.25 kHz for $N_{ZC} = 839$, $N_{CS} = 93$, $u = 300$ and $d_u = 165$.

In the case of a Doppler shift of zero $f_D = 0$, the correlation peak would have been at $vN_{CS} = 93$. However, for Doppler shift of $f_D = \pm 1/T_{SEQ}$, correlation at $vN_{CS} = 93$ is zero. This means that the transmitted sequence will be detected wrongly and hence a false alarm is unavoidable.

In Figure 10.12, we show the correlation performance for $f_D = 1/(2T_{SEQ})$. We observe two equal size peaks at $vN_{CS} = 93$ and $(vN_{CS} + d_u)_{N_{ZC}} = (93 + 165)_{839} = 258$. This shows that with a smaller Doppler shift, a peak at the original location can be observed. However, another peak at a second location can lead to a false detection. For $f_D > 1/(2T_{SEQ})$, the peak at the original location $vN_{CS} = 93$ is smaller than the side peak at $(vN_{CS} + d_u)_{N_{ZC}} = (93 + 165)_{839} = 258$. A Doppler frequency shift of $f_D = 1/(2T_{SEQ}) = 0.625$ kHz occurs at quite high UE speeds, for example, at 337.5 km/h for 2.0 GHz carrier frequency. However, in addition to the Doppler frequency shift, the round trip time and delay spread may cause additional offset of the correlation peaks.

A possible solution to reduce the effect of Doppler shift would be to use shorter length ZC sequences and perform sequence repetitions. The correlation at zeroth lag as a function of frequency error Δf or Doppler shift f_D for $N_{ZC} = 419$ is given in Figure 10.13. We note that with smaller $T_{SEQ} = 0.4$ ms, the Doppler shift $f_D = \pm 1/T_{SEQ}$ for which correlation at the original location is zero increases to 2.5 kHz. A Doppler frequency shift of $f_D = 1/(2T_{SEQ}) = 1.25$ kHz now occurs at UE speed of 675 km/h for 2.0 GHz carrier frequency. However, a shorter ZC sequence length results in reduction of the available sequences as well as the processing gain. Note that with sequence length N_{ZC} selected as a prime number, there are $(N_{ZC} - 1)$ sequences available.

Another solution is based on restrictions on the root sequence and cyclic shift indices. The problem we are facing is that with Doppler shift, an RA preamble is falsely identified

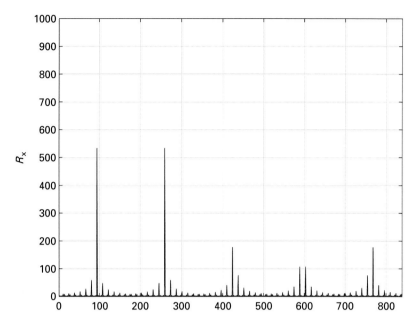

Figure 10.12. Correlation due to $f_D = 0.625\,\text{kHz}$ for $N_{ZC} = 839, N_{CS} = 93, u = 300$ and $d_u = 165$.

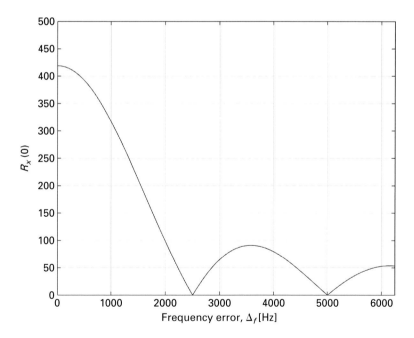

Figure 10.13. Correlation at zeroth lag as a function of frequency error Δf for $N_{ZC} = 419$.

in the receiver's bank of correlators as the RA preamble defined by one of the cyclic shifts $vN_{CS} \pm \lfloor d_u / N_{CS} \rfloor \times N_{CS}$ or $vN_{CS} \pm \left(1 + \lfloor d_u / N_{CS} \rfloor\right) \times N_{CS}$ of the same root sequence. In the case of $d_u < N_{CS}$ or $d_u > (N - N_{CS})$, the correlation peaks may lead to identification of the correct RA preamble but with erroneous delay and therefore sequences with such values of d_u

should not be used. If d_u is in the range $\left[\frac{(N_{ZC}-N_{CS})}{2}, \frac{(N_{ZC}+N_{CS})}{2}\right]$, then it is not possible for the detector to distinguish whether a correlation peak is due to a positive or a negative Doppler shift and hence it is not possible to obtain the correct time of arrival estimate and therefore these sequences should not be used. This approach of root sequence and cyclic shift indices restriction also results in reduction in the number of available sequences and an increase in complexity for RA preamble selection. Nevertheless, this approach was selected in the LTE system for supporting high mobility scenarios.

In order to avoid the cases of misdetection of the RA preamble at higher UE speeds, a restricted set of cyclic shifts for a given root sequence $x_u(k)$ is defined for generating RA preambles. These allowed cyclic shifts are related to the cyclic offset d_u of $x_u(k)$ that would be produced by the Doppler frequency shift $f_D = 1/T_{SEQ} = 1.25\,\text{kHz}$.

In the restricted case, the allowed cyclic shift value C_v is given by:

$$C_v = d_{\text{start}}\lfloor v/n_{\text{shift}}^{\text{RA}}\rfloor + (v \bmod n_{\text{shift}}^{\text{RA}})N_{CS} \quad v = 0, 1, \ldots, n_{\text{shift}}^{\text{RA}}n_{\text{group}}^{\text{RA}} + \bar{n}_{\text{shift}}^{\text{RA}} - 1.$$

$$(10.20)$$

The parameters for restricted sets of cyclic shifts depend on d_u and are given by:

$$
= \begin{cases}
\begin{aligned}
n_{\text{shift}}^{\text{RA}} &= \lfloor d_u/N_{CS}\rfloor \\
d_{\text{start}} &= 2d_u + n_{\text{shift}}^{\text{RA}}N_{CS} \\
n_{\text{group}}^{\text{RA}} &= \lfloor N_{ZC}/d_{\text{start}}\rfloor \\
\bar{n}_{\text{shift}}^{\text{RA}} &= \max\left(\left\lfloor(N_{ZC} - 2d_u \right.\right. \\
&\quad \left.\left. -n_{\text{group}}^{\text{RA}}d_{\text{start}})/N_{CS}\right\rfloor, 0\right)
\end{aligned} & N_{CS} \le d_u < N_{ZC}/3 \\
\\
\begin{aligned}
n_{\text{shift}}^{\text{RA}} &= \lfloor(N_{ZC} - 2d_u)/N_{CS}\rfloor \\
d_{\text{start}} &= N_{ZC} - 2d_u + n_{\text{shift}}^{\text{RA}}N_{CS} \\
n_{\text{group}}^{\text{RA}} &= \lfloor d_u/d_{\text{start}}\rfloor \\
\bar{n}_{\text{shift}}^{\text{RA}} &= \min\left(\max\left(\left\lfloor(d_u \right.\right.\right. \\
&\quad \left.\left.\left. -n_{\text{group}}^{\text{RA}}d_{\text{start}})/N_{CS}\right\rfloor, 0\right), n_{\text{shift}}^{\text{RA}}\right)
\end{aligned} & N_{ZC}/3 \le d_u \le (N_{ZC} - N_{CS})/2.
\end{cases}
$$

$$(10.21)$$

For all other values of d_u, there are no cyclic shifts in the restricted set.

In the case of unrestricted cyclic shifts, the allowed cyclic shift values are determined by the N_{CS} value only and are independent of the root sequence index. For the restricted cyclic shifts, however, the allowed cyclic shift values are also functions of the root sequence index. Let us consider an example for N_{CS} configuration 1 with $N_{CS} = 18$ and root sequence index of 300 with $d_u = 165$. The parameters for restricted cyclic shifts are then given as:

$$n_{\text{shift}}^{\text{RA}} = \lfloor d_u/N_{CS}\rfloor = \lfloor 165/18\rfloor = 9$$

$$d_{\text{start}} = 2d_u + n_{\text{shift}}^{\text{RA}}N_{CS} = 2 \times 165 + 9 \times 18 = 492$$

$$n_{\text{group}}^{\text{RA}} = \lfloor N_{ZC}/d_{\text{start}}\rfloor = \lfloor 839/492\rfloor = 1$$

$$\bar{n}_{shift}^{RA} = \max\left(\left\lfloor (N_{ZC} - 2d_u - n_{group}^{RA}d_{start})\Big/N_{CS}\right\rfloor, 0\right)$$
$$= \max\left(\left\lfloor (839 - 2 \times 165 - 1 \times 492)/18\right\rfloor, 0\right) = 0. \tag{10.22}$$

The cyclic shift values C_v are given as:

$$C_v = 492\lfloor v/9\rfloor + (v \bmod 9)18 \quad v = 0, 1, \ldots, 8. \tag{10.23}$$

We note that in this case a total of nine cyclic shifts $C_v = v \times 18, v = 0, 1, \ldots, 8$ are available.

Let us now consider another example for the same N_{CS} configuration with $N_{CS} = 18$ and root sequence index of 631 with $d_u = 359$. The parameters for restricted cyclic shifts in this case are given as:

$$n_{shift}^{RA} = \left\lfloor (N_{ZC} - 2d_u)/N_{CS}\right\rfloor = \left\lfloor (839 - 2 \times 359)/18\right\rfloor = 6$$
$$d_{start} = N_{ZC} - 2d_u + n_{shift}^{RA}N_{CS} = 839 - 2 \times 359 + 6 \times 18 = 229$$
$$n_{group}^{RA} = \left\lfloor d_u/d_{start}\right\rfloor = 1$$
$$\bar{n}_{shift}^{RA} = \min\left(\max\left(\left\lfloor (d_u - n_{group}^{RA}d_{start})\Big/N_{CS}\right\rfloor, 0\right), n_{shift}^{RA}\right)$$
$$= \min\left(\max\left(\left\lfloor (359 - 229)/18\right\rfloor, 0\right), 9\right) = 7. \tag{10.24}$$

The cyclic shift values C_v are given as:

$$C_v = d_{start}\left\lfloor v/n_{shift}^{RA}\right\rfloor + (v \bmod n_{shift}^{RA})N_{CS} \quad v = 0, 1, \ldots, n_{shift}^{RA}n_{group}^{RA} + \bar{n}_{shift}^{RA} - 1. \tag{10.25}$$

We note that in this case a total of 13 cyclic shifts with

$$C_v = [0 \quad 18 \quad 36 \quad 54 \quad 72 \quad 90 \quad 229 \quad 247 \quad 265 \quad 283 \quad 301 \quad 319 \quad 458] \tag{10.26}$$

are available. We remark that N_{CS} configuration 1 provides a total of 55 cyclic shifts for all root sequences in the unrestricted case. As expected, the cyclic shifts available for the restricted case are smaller.

10.4.3 Sequence and cyclic shift selection

Since the RACH preamble carries 6 bits a total of 64 preambles need to be available in each cell. The set of 64 preamble sequences in a cell is determined by including first, in the order of increasing cyclic shift, all the available cyclic shifts of a root Zadoff–Chu sequence with the logical index broadcast as part of the system information on BCH. In case 64 preambles cannot be generated from a single root Zadoff–Chu sequence, additional preamble sequences are obtained from the root sequences with the consecutive logical indexes until all the 64 sequences are found. The logical root sequence order is cyclic, which means that the logical index 0 is consecutive to 837. The relationship between a logical root sequence index and physical root sequence index u for the LTE system is given in Table 10.4.

Table 10.4. Mapping of logical root sequence numbers to physical root sequence numbers.

Logical root sequence number	Physical root sequence number u (in increasing order of the corresponding logical sequence number)
0–23	129, 710, 140, 699, 120, 719, 210, 629, 168, 671, 84, 755, 105, 734, 93, 746, 70, 769, 60, 779, 2, 837, 1, 838
24–29	56, 783, 112, 727, 148, 691
30–35	80, 759, 42, 797, 40, 799
36–41	35, 804, 73, 766, 146, 693
42–51	31, 808, 28, 811, 30, 809, 27, 812, 29, 810
52–63	24, 815, 48, 791, 68, 771, 74, 765, 178, 661, 136, 703
64–75	86, 753, 78, 761, 43, 796, 39, 800, 20, 819, 21, 818
76–89	95, 744, 202, 637, 190, 649, 181, 658, 137, 702, 125, 714, 151, 688
90–115	217, 622, 128, 711, 142, 697, 122, 717, 203, 636, 118, 721, 110, 729, 89, 750, 103, 736, 61, 778, 55, 784, 15, 824, 14, 825
116–135	12, 827, 23, 816, 34, 805, 37, 802, 46, 793, 207, 632, 179, 660, 145, 694, 130, 709, 223, 616
136–167	228, 611, 227, 612, 132, 707, 133, 706, 143, 696, 135, 704, 161, 678, 201, 638, 173, 666, 106, 733, 83, 756, 91, 748, 66, 773, 53, 786, 10, 829, 9, 830
168–203	7, 832, 8, 831, 16, 823, 47, 792, 64, 775, 57, 782, 104, 735, 101, 738, 108, 731, 208, 631, 184, 655, 197, 642, 191, 648, 121, 718, 141, 698, 149, 690, 216, 623, 218, 621
204–263	152, 687, 144, 695, 134, 705, 138, 701, 199, 640, 162, 677, 176, 663, 119, 720, 158, 681, 164, 675, 174, 665, 171, 668, 170, 669, 87, 752, 169, 670, 88, 751, 107, 732, 81, 758, 82, 757, 100, 739, 98, 741, 71, 768, 59, 780, 65, 774, 50, 789, 49, 790, 26, 813, 17, 822, 13, 826, 6, 833
264–327	5, 834, 33, 806, 51, 788, 75, 764, 99, 740, 96, 743, 97, 742, 166, 673, 172, 667, 175, 664, 187, 652, 163, 676, 185, 654, 200, 639, 114, 725, 189, 650, 115, 724, 194, 645, 195, 644, 192, 647, 182, 657, 157, 682, 156, 683, 211, 628, 154, 685, 123, 716, 139, 700, 212, 627, 153, 686, 213, 626, 215, 624, 150, 689
328–383	225, 614, 224, 615, 221, 618, 220, 619, 127, 712, 147, 692, 124, 715, 193, 646, 205, 634, 206, 633, 116, 723, 160, 679, 186, 653, 167, 672, 79, 760, 85, 754, 77, 762, 92, 747, 58, 781, 62, 777, 69, 770, 54, 785, 36, 803, 32, 807, 25, 814, 18, 821, 11, 828, 4, 835
384–455	3, 836, 19, 820, 22, 817, 41, 798, 38, 801, 44, 795, 52, 787, 45, 794, 63, 776, 67, 772, 72, 767, 76, 763, 94, 745, 102, 737, 90, 749, 109, 730, 165, 674, 111, 728, 209, 630, 204, 635, 117, 722, 188, 651, 159, 680, 198, 641, 113, 726, 183, 656, 180, 659, 177, 662, 196, 643, 155, 684, 214, 625, 126, 713, 131, 708, 219, 620, 222, 617, 226, 613
456–513	230, 609, 232, 607, 262, 577, 252, 587, 418, 421, 416, 423, 413, 426, 411, 428, 376, 463, 395, 444, 283, 556, 285, 554, 379, 460, 390, 449, 363, 476, 384, 455, 388, 451, 386, 453, 361, 478, 387, 452, 360, 479, 310, 529, 354, 485, 328, 511, 315, 524, 337, 502, 349, 490, 335, 504, 324, 515
514–561	323, 516, 320, 519, 334, 505, 359, 480, 295, 544, 385, 454, 292, 547, 291, 548, 381, 458, 399, 440, 380, 459, 397, 442, 369, 470, 377, 462, 410, 429, 407, 432, 281, 558, 414, 425, 247, 592, 277, 562, 271, 568, 272, 567, 264, 575, 259, 580

Table 10.4. (Continued)

Logical root sequence number	Physical root sequence number u (in increasing order of the corresponding logical sequence number)
562–629	237, 602, 239, 600, 244, 595, 243, 596, 275, 564, 278, 561, 250, 589, 246, 593, 417, 422, 248, 591, 394, 445, 393, 446, 370, 469, 365, 474, 300, 539, 299, 540, 364, 475, 362, 477, 298, 541, 312, 527, 313, 526, 314, 525, 353, 486, 352, 487, 343, 496, 327, 512, 350, 489, 326, 513, 319, 520, 332, 507, 333, 506, 348, 491, 347, 492, 322, 517
630–659	330, 509, 338, 501, 341, 498, 340, 499, 342, 497, 301, 538, 366, 473, 401, 438, 371, 468, 408, 431, 375, 464, 249, 590, 269, 570, 238, 601, 234, 605
660–707	257, 582, 273, 566, 255, 584, 254, 585, 245, 594, 251, 588, 412, 427, 372, 467, 282, 557, 403, 436, 396, 443, 392, 447, 391, 448, 382, 457, 389, 450, 294, 545, 297, 542, 311, 528, 344, 495, 345, 494, 318, 521, 331, 508, 325, 514, 321, 518
708–729	346, 493, 339, 500, 351, 488, 306, 533, 289, 550, 400, 439, 378, 461, 374, 465, 415, 424, 270, 569, 241, 598
730–751	231, 608, 260, 579, 268, 571, 276, 563, 409, 430, 398, 441, 290, 549, 304, 535, 308, 531, 358, 481, 316, 523
752–765	293, 546, 288, 551, 284, 555, 368, 471, 253, 586, 256, 583, 263, 576
766–777	242, 597, 274, 565, 402, 437, 383, 456, 357, 482, 329, 510
778–789	317, 522, 307, 532, 286, 553, 287, 552, 266, 573, 261, 578
790–795	236, 603, 303, 536, 356, 483
796–803	355, 484, 405, 434, 404, 435, 406, 433
804–809	235, 604, 267, 572, 302, 537
810–815	309, 530, 265, 574, 233, 606
816–819	367, 472, 296, 543
820–837	336, 503, 305, 534, 373, 466, 280, 559, 279, 560, 419, 420, 240, 599, 258, 581, 229, 610

10.5 Signal peakiness of RA sequences

In Chapter 5, we discussed the need for low signal peakiness for uplink transmissions due to limited UE transmit power. Low signal peakiness results in higher power efficiency due to smaller back off required for the UE power amplifier. Higher power amplifier efficiency in turn allows achieving larger coverage for the same power amplifier power rating. One of the desired characteristics of ZC sequences is that they exhibit low signal peakiness. We provide cubic metric (see Section 5.1.2) performance for different ZC root sequence indices in Figures 10.14 and 10.15. In Figure 10.14, the root index is defined in the time domain, which is the case for the scheme with DFT precoding in Figure 10.5. We note that cubic metric roughly increases with increasing root sequence index. We also observe the symmetry property with the rth and ($N_{ZC} - r$)th sequences exhibiting similar cubic metric performance. Figure 10.15 depicts cubic metric performance when the ZC sequence index is defined in the frequency-domain, which is the case for the scheme without DFT precoding in Figure 10.5. Again, we observe a symmetry which, however, is not as obvious as in the first case. Again, it should be understood that DFT precoding of a ZC sequence results in another ZC sequence with a different root index.

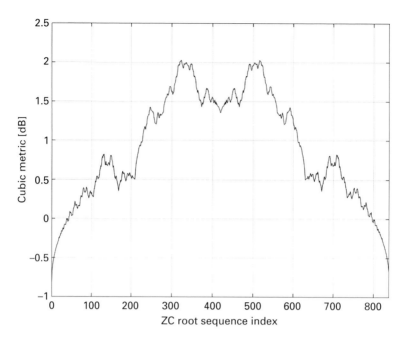

Figure 10.14. Cubic metric for the case of ZC root indices defined in the time domain.

10.6   Random access MAC procedures and formats

10.6.1   Random access MAC procedures

A random access procedure is initiated by a PDCCH (physical downlink control channel) order or by the MAC sublayer. The PDCCH order is used when the data for the UE arrives at eNB and eNB assumes the UE is out of synchronization. In all other cases, MAC initiates the procedure due to data coming from RRC (Radio Resource Control) or in the user plane. The PDCCH order or RRC can optionally indicate a random access preamble and PRACH resource to the UE. Before the random access procedure can be initiated, the UE has access to the information that includes the available set of PRACH resources and their corresponding RA-RNTIs (random access radio network temporary identifiers), the groups of random access preambles and the set of available access preambles in each group, the thresholds required for selecting one of the two groups of preambles, the parameters required to derive the TTI window, the power-ramping factor POWER_RAMP_STEP, the parameter PREAMBLE_TRANS_MAX, the initial preamble power PREAMBLE_ INITIAL_RECEIVED_TARGET_POWER, etc. The RA-RNTI is used on the PDCCH when random access response (RAR) messages are transmitted. It unambiguously identifies which time-frequency resource was utilized by the UE to transmit the Random Access preamble. A TTI window is a time period over which UE monitors random access response messages from the eNB. Once the UE is ready to perform a random access, it sets the PREAMBLE_TRANSMISSION_COUNTER to one and the backoff parameter value to 0 ms and proceeds to the selection of the random access resource.

  If the random access preamble and PRACH resource have been explicitly signaled by the eNB and the random access preamble expiration time (if configured) has not expired, the UE does not need to perform resource selection and can directly proceed to the transmission of

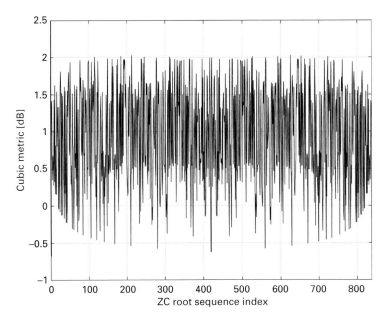

Figure 10.15. Cubic metric for the case of ZC root indices defined in the time domain.

the preamble. If the random access preamble and PRACH resource is not explicitly signaled, the UE first selects one of the two preamble groups. The groups' selection decision can be based on the size of the message to be transmitted on the uplink or the requested resource blocks as well as the radio conditions. Once the preamble group has been selected, the UE randomly selects a random access preamble within the selected group.

The UE then sets the preamble transmit power parameter as:

$$\text{PREAMBLE\_RECEIVED\_TARGET\_POWER} =$$
$$\text{PREAMBLE\_INITIAL\_RECEIVED\_TARGET\_POWER} +$$
$$(\text{PREAMBLE\_TRANSMISSION\_COUNTER}-1) \times$$
$$\text{POWER\_RAMP\_STEP}. \tag{10.27}$$

This makes sure that the preamble transmit power is increased by POWER_RAMP_STEP for every random access retry.

The UE then determines the next available random access occasion and transmits a preamble using the selected PRACH resource, preamble index and PREAMBLE_RECEIVED_TARGET_POWER. The MAC layer in the UE also informs the physical layer of the RA-RNTI corresponding to the PRACH resource.

In case the maximum number of preamble transmissions attempts is reached, that is:

$$\text{PREAMBLE\_TRANSMISSION\_COUNTER} =$$
$$\text{PREAMBLE\_TRANS\_MAX} + 1 \tag{10.28}$$

the MAC layer in the UE indicates a random access problem to upper layers.

After the UE has transmitted random access preamble, it monitors the PDCCH associated with the RA-RNTI within a TTI window for random access response(s) identified by the RA-RNTI. The RA-RNTI associated with the PRACH resource in which the preamble is transmitted is computed as:

$$RA\_RNTI = t_{ID} + 10 \times f_{ID}, \tag{10.29}$$

where $0 \leq t_{ID} < 10$ is the index of the first subframe of the specified PRACH resource, and $0 \leq f_{ID} < 6$ is the index of the specified PRACH resource within that subframe, in ascending order of frequency domain. The UE may stop monitoring for random access response(s) after successful reception of a random access response (RAR) corresponding to the preamble transmission. When the physical layer informs the MAC sublayer of reception of a random access response, the action taken by the MAC depends upon the response type. If the RAR contains a backoff indicator (BI) subheader (see Section 10.6.2), the MAC sets the backoff parameter value in the UE to the value indicated by the BI field as given in Table 10.6 below. If there is no BI indication subheader, the backoff parameter value in the UE is set to a default value of zero ms.

If the random access response contains a random access preamble identifier corresponding to the transmitted random access preamble, the UE assumes successful reception of the random access response and processes the received timing alignment value and the received UL grant value. Moreover, if the random access preamble was explicitly signaled (i.e., not selected by the MAC), the UE considers that the random access procedure has been successfully completed. This is because with explicit signaling of the preamble, there is no ambiguity that another UE might have used the same preamble for its random access attempt. On the other hand, if the random access preamble was selected by the UE MAC, the UE sets the temporary C-RNTI to the value received in the random access response message and proceeds to contention resolution. This is because at this stage it is not clear if more than one UE used the same preamble sequence for random access. When an uplink transmission is required, for example, for contention resolution, the eNB makes sure that the uplink grant is for at least 80 bits transmission.

If no random access response is received within the TTI window, or if all received random access responses contain random access preamble identifiers that do not match with the transmitted preamble, the random access response reception is considered unsuccessful. In this case, if the random access procedure was initiated by the MAC sublayer itself or by a PDCCH order and

$$PREAMBLE\_TRANSMISSION\_COUNTER <$$
$$PREAMBLE\_TRANS\_MAX \tag{10.30}$$

the UE increments the preamble transmission counter:

$$PREAMBLE\_TRANSMISSION\_COUNTER =$$
$$PREAMBLE\_TRANSMISSION\_COUNTER + 1. \tag{10.31}$$

Moreover, if the random access preamble was selected by the MAC or the random access preamble and PRACH resource were explicitly signaled and will expire before the next available random access occasion, the UE (based on the backoff parameter) computes and applies a backoff value and proceeds to the selection of a random access resource. This backoff value determines when a new random access transmission is attempted.

For simplicity, we only describe contention resolution for the case when the uplink message contains the C-RNTI MAC control element. This message is transmitted in response to a random access response indicating a matched preamble. After transmission of the C-RNTI MAC control element in the uplink, the UE starts the contention resolution timer and monitors the PDCCH until the contention resolution timer expires. When a PDCCH transmission addressed to the C-RNTI is received, the UE considers the contention resolution successful, stops the contention resolution timer and discards the temporary C-RNTI. On the other hand, if the contention resolution timer expires and no PDCCH transmission addressed to the C-RNTI is received, the UE considers the contention resolution is not successful. In this case, if the random access procedure was initiated by the MAC sublayer itself or by a PDCCH order and

$$\text{PREAMBLE\_TRANSMISSION\_COUNTER} < \\ \text{PREAMBLE\_TRANSMISSION\_MAX} \tag{10.32}$$

the UE increments the preamble transmission counter:

$$\text{PREAMBLE\_TRANSMISSION\_COUNTER} = \\ \text{PREAMBLE\_TRANSMISSION\_COUNTER} + 1 \tag{10.33}$$

and discards the temporary C-RNTI. Moreover, the UE computes and applies a backoff value and proceeds to the selection of a random access resource for a new random access request.

10.6.2 Random access MAC PDU format

A random access response MAC PDU is sent in response to random access by the UEs and consists of a MAC header and one or more MAC random access responses (MAC RAR) as shown in Figure 10.16. The MAC header is of variable size and consists of the extension (E) and type (T) fields. The extension field is a 1-bit flag indicating if more fields are present in the MAC header or not. The type field is another 1-bit flag indicating whether the MAC subheader contains a random access ID (RAID) or a Backoff Indicator (BI). The size of the RAID field is 6-bits indicating one of the 64 random access preambles. On receiving the backoff indication, the UE updates the random access backoff parameter value as given in Table 10.6. When the type flag indicates backoff indicator type, 2 bits are not used and are reserved (R).

A MAC RAR consists of three fields that include timing advance (TA), uplink (UL) grant and T-CRNTI. The timing advance field indicates the required adjustment to the uplink transmission timing for uplink timing synchronization. The uplink grant field indicates the resources to be used on the uplink. The temporary CRNTI (cell radio network temporary identifier) field indicates the temporary identity that is used by the UE until contention resolution succeeds or another random access procedure is initiated. The size of the MAC header and MAC RAR fields is given in Table 10.5.

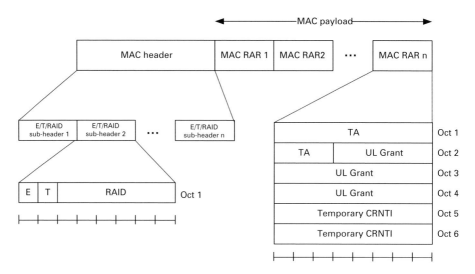

Figure 10.16. MAC PDU consisting of a MAC header and MAC RARs.

Table 10.5. Random access MAC header and MAC RAR fields.

Subheader/payload	Field	Bits
E/T/RAID (8-bits)	Extension	1
	Type	1
	Random access ID (RAID)	6
E/T/R/R/BI (8-bits)	Extension	1
	Type	1
	Reserved	2
	Backoff indicator (BI)	4
MAC RAR (48-bits)	Timing advance (TA)	11
	Uplink (UL) grant	21
	T-CRNTI	16

10.7   Summary

A random access mechanism is used by a UE to acquire uplink synchronization and also to inform the network when it needs to send or receive data from the network. Since random access is used by the UE when it is not synchronized on the uplink a guard time is introduced to avoid collisions with uplink data transmissions. Moreover, to enable simple frequency-domain processing, the random access preamble also uses a cyclic prefix. The guard period and cyclic prefix need to account for round trip propagation delay and hence a larger guard period is required for larger cells. However, a large guard time would be an unnecessary overhead when the system is deployed with smaller cell sizes. Therefore, four random access preamble formats with both small and large guard times and with and without preamble repetitions are defined to allow operation in vastly different deployments.

Table 10.6. Random access backoff parameter values.

Index	Backoff parameter value [ms]
0	0
1	10
2	20
3	30
4	40
5	60
6	80
7	120
8	160
9	240
10	320
11	480
12	960

The resources allocated for random access depend upon the random access load in a cell. A too small amount of resources would lead to an increased number of collisions and hence increase the system access delays. A too large amount of resources, on the other hand, would be an unnecessary overhead. In general, larger system bandwidths can support a larger number of users and hence the random access load for larger bandwidths is larger than the load for smaller bandwidths. Moreover, in high mobility scenarios where the handoff frequency is higher, more random access resources need to be provisioned for handoff users who need to perform synchronization to their new target cells. In the time domain, a single random access resource is allowed at most. Therefore, the amount of random access resource in a cell is controlled by changing the number of subframes that contain random access resources. In the LTE system, a total of 16 random access configurations with different degrees of time-domain frequency of random access occurrences are defined. This allows one to adjust random access resources to diverse load conditions.

The Zadoff–Chu sequences are used as preambles for random access due to their low signal peakiness as well as good correlation properties. Moreover, the length of the sequence is selected as a prime number that maximizes the number of available sequences. A large number of preamble sequences is desired, as neighboring cells should preferably use different sets of preambles to avoid false alarms in the neighboring cells. A large number of sequences, therefore, simplifies the network-planning task.

A frequency error due to Doppler frequency shift affects the detection and false alarm performance of the ZC sequence due to the autocorrelation side lobes. This effect is more pronounced at higher UE speeds when the Doppler frequency shift is comparable to the random access subcarrier spacing. This problem is overcome by using a restricted set of allowed cyclic shifts at the expense of reduction in the number of available random access sequences.

References

[1] Chu, D. C., "Polyphase codes with good periodic correlation properties," *IEEE Transactions on Information Theory*, vol. IT-18, pp. 531–532, July 1992.

[2] Popovic, B. M., "Generalized chirp-like polyphase sequences with optimum correlation properties," *IEEE Transactions on Information Theory*, vol. IT-38, no. 4, pp. 1406–1409, July 1992.

[3] 3GPP Tdoc R1-071560, Using Restricted Preamble Set for RACH in High Mobility Environments.

11 Channel coding

Like other 3G systems, the current HSPA system uses turbo coding as the channel-coding scheme. The LTE system supports peak data rates that are an order of magnitude higher than the current 3G systems. It is therefore fair to ask the question, can the turbo coding scheme scale to data rates in excess of 100 Mb/s supported by LTE, while maintaining reasonable decoding complexity? This question is particularly important as other coding schemes, which offer inherent parallelism and therefore provide very high decoding speeds such as Low Density Parity Check (LDPC) codes, have recently become available. A major argument against turbo coding schemes is that they are not amenable to parallel implementations thus limiting the achievable decoding speeds. The problem, in fact, lies in the turbo code internal interleaver used in the current HSPA system, which creates memory contention among processors in parallel implementation. Therefore, if the turbo code internal interleaver can somehow be made contention free, it becomes possible for turbo code to benefit from parallel processing and hence achieve high decoding speeds.

11.1 LDPC codes

Similar to turbo codes, LDPC codes are near-Shannon limit error correcting codes. More recently, LDPC codes have been adopted in standards including IEEE 802.16e wireless MAN [1], IEEE 802.11n wireless LAN and digital video broadcast DVB-S2. The LDPC codes allow an extremely flexible code design that can be tailored to achieve efficient encoding and decoding. The interest in LDPC codes comes from their potential to achieve very high throughput (due to the inherent parallelism of the decoding algorithm) while maintaining good error-correcting performance and low decoding complexity.

The LDPC codes were first introduced by Gallager in his doctoral dissertation [2]. However, due to the computational effort in implementing coder and decoder for such codes and the introduction of Reed–Solomon codes, they were mostly ignored until about ten years ago. One notable exception is Tanner, who wrote an important paper in 1981 [3], which generalized LDPC codes and introduced a graphical representation of LDPC codes, now called Tanner graphs. Apparently independent of Gallager's work, LDPC codes were re-invented in the mid 1990s by MacKay, Luby and others [4–7].

Basically there are two different possibilities to represent LDPC codes. First, like all linear block codes they can be described via matrices. Let us consider a low-density parity-check matrix for a $(8, 4)$ code as below:

$$H = \begin{bmatrix} 0 & 1 & 0 & 1 & 1 & 0 & 0 & 1 \\ 1 & 1 & 1 & 0 & 0 & 1 & 0 & 0 \\ 0 & 0 & 1 & 0 & 0 & 1 & 1 & 1 \\ 1 & 0 & 0 & 1 & 1 & 0 & 1 & 0 \end{bmatrix}. \tag{11.1}$$

We can now define two numbers describing this matrix, w_r for the number of 1's in each row and w_c for the number of 1's in each column. For a matrix to be called low-density the two conditions $w_r \ll n$ and $w_c \ll m$ must be satisfied, where n and m are respectively the number of columns and number of rows of the parity-check matrix H. In order to achieve low-density, the parity check matrix should generally be very large and, therefore, the parity-check matrix H in (11.1) would not qualify as a low-density matrix.

A second representation for LDPC uses Tanner graphs [3]. A Tanner graph not only provides a complete representation of the code but also helps in describing the decoding algorithms. Tanner graphs are bipartite graphs, which means that the nodes of the graph are separated into two distinctive sets and edges are only connecting nodes of two different types. The two types of nodes in a Tanner graph are called variable nodes (v-nodes) and check nodes (c-nodes). A Tanner graph has m check nodes, one for each row of H (the number of parity bits) and n variable nodes, one for each column of H (the number of bits in a codeword) as shown in Figure 11.1. A check node f_i is connected to variable node c_j if the element h_{ij} of H is a 1.

An LDPC code is called regular if w_c is constant for every column and

$$w_r = w_c \left(\frac{n}{m} \right) \tag{11.2}$$

is also constant for every row. The example matrix from Equation (11.1) is regular with $w_c = 2$ and $w_r = 4$. We can also notice the regularity of this code by looking at the graphical representation in Figure 11.1. There are the same number of incoming edges for every v-node and every c-node. If H has low density but the numbers of 1's in each row or in each column are not constant the code is referred to as irregular LDPC code.

We do not plan to go into details of decoding algorithms used for LDPC codes and the interested reader is referred to [8] for an overview of the decoding algorithms. We pointed out that LDPC codes are attractive from a computational efficiency point of view because they offer a high level of inherent parallelism. This is because they are composed of parity-check equations that can be updated independently. While interconnecting complexity is high, all check nodes can be updated simultaneously and all extrinsic information can be passed to the variable nodes simultaneously. Similarly, all variable nodes can be updated simultaneously. This is in contrast to turbo codes, where within a constituent code, all the information bits are related via one trellis, and inherently each trellis stage needs to obtain state information from the adjacent trellis stages.

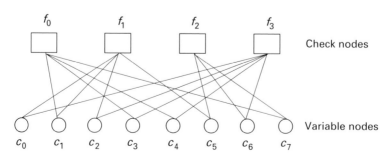

Figure 11.1. Tanner graph for an LDPC code with matrix H from Equation (11.1).

To facilitate implementation, structured LDPC codes can be defined without sacrificing performance. The H matrix of a structured LDPC code is constructed starting with a small binary base matrix of size $m_b \times n_b$. The 1's in the base matrix are then replaced by all-zero or shifted identity matrices of size $z \times z$. Using structured LDPC codes, both encoding/decoding can be performed based on a much smaller base matrix and vectors of bits/LLRs, with each vector having size z. Structured LDPC codes can be defined to allow a low complexity encoding/decoding algorithm with high throughput. For example, layered belief propagation can be used to decode one vector row at a time. Further, the vector rows may be pipelined to increase the throughput.

We note that LDPC codes offer inherent parallelism that facilitates high throughput decoding implementations. However, as we will discuss later in the chapter, turbo code can also make use of parallel implementations if turbo internal interleavers can be made contention-free. Therefore, it can be argued that LDPC codes do not provide any compelling advantage over turbo code as bit error performance of both coding schemes is comparable. An argument in favor of turbo code for the LTE system was that UMTS release 6 HSPA also uses turbo code. Also, for backward compatibility reasons, dual-mode LTE terminals will need to implement turbo code and therefore some decoding hardware can be reused. On the other hand, if a completely different new coding scheme such as LDPC code were introduced, this would result in increased implementation complexity as the terminals have to support two different coding schemes. Based on these arguments and also the fact that there was some sympathy towards the existing coding scheme used in 3GPP systems, turbo code was selected over LDPC for LTE.

11.2   Channel coding schemes in LTE

The major channel coding schemes used for different transport channels in the LTE system are summarized in Figure 11.2. The turbo coding is used for large data packets, which is the case for downlink and uplink data transmission, paging and broadcast multicast (MBMS) transmissions. A rate 1/3 tail biting convolutional coding is used for downlink control and uplink control as well as broadcast control channel (BCH).

In Figure 11.3, we show transport channel processing for DL-SCH, PCH and MCH, which use turbo coding in downlink transmissions. The processing steps that we will describe in detail later include adding CRC to the transport block, codeblock segmentation, codeblock CRC attachment, turbo coding, rate matching and codeblock concatenation. It should be noted that no separate channel interleaver is employed for downlink transmissions as subblock interleaving and interlacing that is performed in the rate-matching step partly achieves the goal of channel interleaving. A channel interleaver, however, is specified for uplink data transmissions to exploit the frequency diversity when hopping is enabled in the uplink.

The transport channel processing for BCH and DCI is illustrated in Figure 11.4. Since convolutional coding is used for small information block sizes used on BCH and downlink control there is no need for codeblock segmentation and codeblock concatenation steps in the transport channel processing. In addition to the difference in channel coding type, there are also some differences in rate-matching schemes between turbo and convolutional coding as will be discussed later.

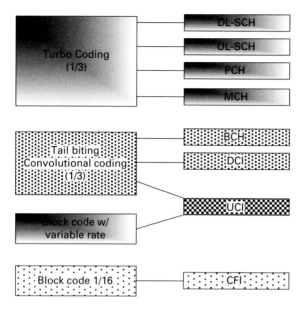

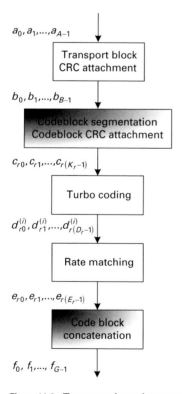

Figure 11.2. Channel coding schemes in the LTE system.

Figure 11.3. Transport channel processing for DL-SCH, PCH and MCH.

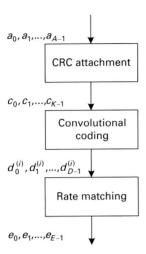

Figure 11.4. Transport channel processing for BCH and DCI.

11.3   Cyclic redundancy check

The error detection in the LTE system is performed using a cyclic redundancy check (CRC). An n-bit CRC, applied to a data block of arbitrary length, will typically detect any single error burst of length n bits or less and will detect a fraction $(1 - 2^{-n})$ of all longer error bursts.

Assume we use an L-bit CRC polynomial to generate the CRC. Denote the CRC generation polynomial by:

$$g(D) = g_0 D^L + g_1 D^{L-1} + \cdots + g_{L-1}D + g_L. \tag{11.3}$$

In general, for a message

$$m(D) = m_0 D^{M-1} + m_1 D^{M-2} + \cdots + m_{M-2}D + m_{M-1} \tag{11.4}$$

the CRC encoding is performed in a systematic form. Denote the CRC parity bits of the message as $p_0, p_1, \ldots, p_{L-1}$, which can also be represented as a polynomial of

$$p(D) = p_0 D^{L-1} + p_1 D^{L-2} + \ldots + p_{L-2}D + p_{L-1}. \tag{11.5}$$

The polynomial

$$
\begin{aligned}
m(D) \cdot D^L - p(D) = {} & m_0 D^{M+L-1} + m_1 D^{M+L-2} + \cdots \\
& + m_{M-2}D^{L+1} + m_{M-1}D^L + p_0 D^{L-1} + p_1 D^{L-2} \\
& + \cdots + p_{L-2}D + p_{L-1}
\end{aligned}
\tag{11.6}
$$

yields a remainder equal to 0 when divided by $g(D)$.

Note that if each bit in the message is binary, the message can be represented as a polynomial defined on GF(2). In that case, the operation of '+' and '−' is the same. In other words, if the message bits are binary, the message with CRC attached can be represented by either $m(D) \cdot D^L + p(D)$ or $m(D) \cdot D^L - p(D)$.

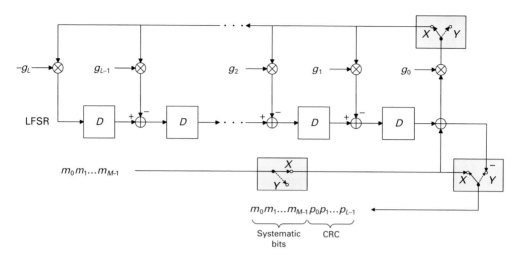

Figure 11.5. Using LFSR for CRC computation.

One reason for the popularity of CRC is its simplicity in implementation. The CRC calculation can be easily implemented by a linear feedback shift register (LFSR). The LFSR can be used as a circuit for polynomial division [9]. Assume an L-bit CRC as shown in Figure 11.5, where the LFSR has L shift registers. The switches are initially placed at position X. The message bit $m_0, m_1, \ldots,$ and m_{M-1} are fed into the shift register one at a time in order of increasing index. After the last bit (m_{M-1}) has been fed into the LFSR, the switches are moved to position Y. The LFSR is shifted by another L times to produce the CRC at the output of the rightmost register.

In general, the cyclic redundancy codes use a generator polynomial of the form $g(D) = (D + 1)p(D)$, where $p(D)$ is a primitive polynomial.

11.3.1   Early stopping and codeblock CRC

A codeblock consists of a set of data bits that are encoded together. The maximum codeblock size in LTE is limited to $Z = 6144$ bits. A transport block is a data block delivered by the MAC layer to the physical layer for transmission in a single subframe of one millisecond. In multi-codeword MIMO transmission in LTE, up to a maximum of two transport blocks also referred to as codewords in MIMO context can be transmitted in a single subframe. The transport blocks larger than the maximum codeblock size need to be segmented into multiple codeblocks as depicted in Figure 11.6. The maximum codeblock size is limited for reasons of turbo code internal interleaver size as well as decoding complexity.

The codeblocks are mapped to time-frequency resources in a frequency-first fashion as is also depicted in Figure 11.6. This allows for pipelining of codeblock decoding reducing decoding complexity. A drawback of frequency-first mapping is that time-diversity is not fully captured as codeblock transmissions can be localized within a subframe particularly for a larger number of codeblocks. However, with a short subframe duration of one millisecond, the channel is almost static within a subframe at low to medium speeds of interest. At higher UE speeds, we should expect some degradation of performance due to a lack of time-diversity with time-first mapping. However, we remark that the goal of the LTE system is to optimize

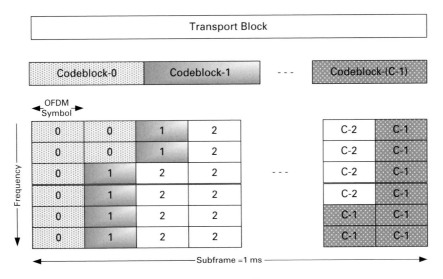

Figure 11.6. Transport block segmentation and mapping.

performance at low to medium speeds. Therefore, frequency-first mapping offers a good tradeoff of reduced complexity with some performance loss at high UE speeds.

The early stopping strategies can be used to further reduce the operational complexity and also the power consumption of turbo decoders. While many detection methods for early stopping have been studied [10], attachment and checking of CRC bits turns out to be a simpler and more reliable approach. A single acknowledgment (ACK) or negative acknowledgment (NACK) per transport block is provided for hybrid ARQ retransmissions. Therefore, in transmissions with multiple codeblocks, the receiver will NACK the transmission as long as one of the codeblocks is in error after the maximum number of iterations. If we can introduce a CRC per codeblock, the decoder can stop decoding after one codeblock is in error, thus saving power that could have been wasted in decoding the rest of the codeblocks. From a power-saving perspective, a small 8-bit codeblock CRC, which gives a miss detection rate of 0.4%, will be sufficient. Note that even if a miss detection occurs, the only negative impact is the receiver will proceed to decode the rest of the codeblocks and waste the decoding power with 0.4% of probability. Here, we assume that transport block CRC of length 24 bits is available to catch miss detections from the codeblock CRC to ensure transport block integrity.

However, when codeblock CRC is used for the early stopping of the iterative decoding, a high misdetection is undesirable. This is because if codeblock CRC declares a codeblock correct due to misdetection, the transport block CRC will fail, resulting in a transport block error event. If the decoder had continued until the maximum number of iterations, the codeblock might have been successfully decoded. Therefore, when used for early stopping, the CRC should be of sufficient length to limit the erroneous early stopping.

In multi-codeword MIMO transmission in LTE, two MIMO codewords are transmitted, each of which can carry multiple codeblocks. For each MIMO codeword or transport block, a CRC is computed based on all information bits in the transport block, that is based on all the codeblocks in the MIMO codeword. In the case when there is no codeblock CRC, the receiver has to wait until all the codeblocks in the first codeword are decoded before it can cancel this codeword and proceed to decoding of codeblocks in the second codeword as illustrated in Figure 11.7. This is because transport block CRC check cannot be performed until all the codeblocks within the transport block are received. A CRC check is necessary before canceling the first codeword to make sure that only the successfully received codeword is cancelled. A cancellation of an unsuccessful codeword may degrade successive interference cancellation (SIC) receiver performance due to residual interference. With a codeblock CRC, on the other hand, SIC operation can be performed on a codeblock basis as is illustrated in Figure 11.7. This helps to reduce the codewords decoding time and buffering complexity required for SIC.

11.3.2   CRC attachment schemes

We discuss three CRC attachment methods for multiple codeblocks shown in Figure 11.8. A transport block (TB) is first segmented into C codeblock segments. In CRC attachment scheme A, a CRC is computed for and attached to each segment independently. In scheme B, CRC computation for the first $(C - 1)$ codeblock segments is different from that for the last one. For the first $(C - 1)$ segments, CRC is computed for and attached to each segment independently. The CRC bits attached to the last segment are computed based on all information bits of that transport block. In scheme C, a TB-level CRC computed based on all information bits of that transport block is attached to the TB. The entire transport block along with the TB-level CRC is then segmented into multiple codeblock segments. A CRC is then computed for and attached to each segment independently.

In attachment scheme B and scheme C, a TB-level CRC attachment of $L = 24$ bits is retained. The probability of misdetection of an erroneous TB is roughly

$$P_m = 2^{-24} = 6 \times 10^{-8}. \tag{11.7}$$

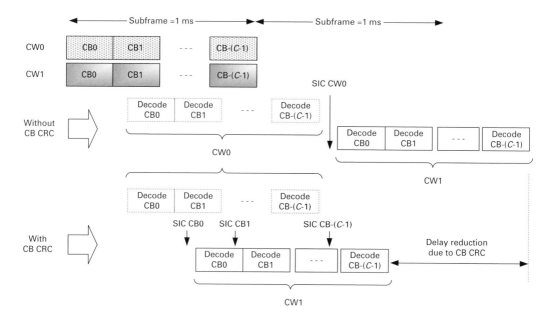

Figure 11.7. Codeblock CRC reduces delays and MIMO SIC complexity.

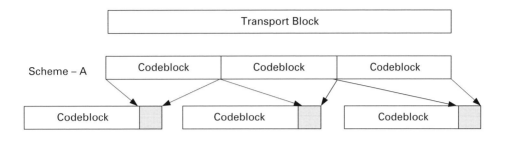

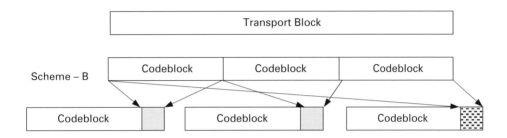

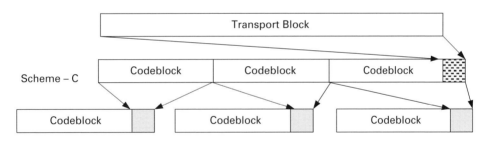

Figure 11.8. CRC attachment schemes.

For scheme A, the correctness of the TB is derived from the XOR of all segment-level CRC. If the length of per-segment CRC attachment in scheme A remains $L = 24$, the miss probability given by

$$P_m = C \times 6 \times 10^{-8} \tag{11.8}$$

would then increase with size of the TB, that is with increasing number of codeblocks C. For instance, the misdetection probability with $C = 25$ codeblocks would be roughly 1.5×10^{-6}. This may be too high a residual packet loss rate as seen by the higher layers.

In scheme A and scheme C, the scope of each segment-level CRC attachment is limited to the individual segment. This allows for the simple implementation of CRC checking within a turbo decoder to apply early stopping check after every decoding iteration. In scheme B, the last CRC attachment is computed from the entire transport block. On the receiver side, the implication is that early stopping rules cannot be used on the last segment unless all the first $(C - 1)$ segments have been decoded. As a consequence, a more complicated CRC checker implementation is needed to support early stopping for the last codeblock segment. Another drawback of this approach is that it is less friendly to receiver implementations based on multiple independent turbo decoders. Assume, for instance, a receiver with four independent decoders. If there are four codeblock segments left, early stopping rules cannot be applied to the last segment.

We noted that both scheme A and scheme C allow simple implementation of CRC checking for early stopping of decoding. We also noticed that scheme A can result in increased misdetection with increasing number of codeblocks which can affect performance of transport and application layer protocols. The drawback of scheme C relative to scheme A is increased CRC overhead because scheme C requires an additional 24 bits CRC. However, this overhead is generally very small because this additional overhead only appears for multiple codeblocks that happen for very large transport block sizes. Since the maximum codeblock size is limited to 6144 bits more than one codeblock is used for fairly large transport block sizes. The worst-case overhead due to 24 additional bits is therefore less than 1%. This small increase in overhead was considered acceptable given the low misdetection probability advantage of scheme C, which was eventually selected as the CRC attachment scheme in the LTE system.

11.3.3 CRC generator polynomials

In regard to the size for codeblock CRC, a similar overhead argument as for the additional CRC above can be made. This is to say that codeblock CRC is used when multiple codeblocks are present and then the codeblock size is fairly large. This is because the transport blocks smaller than the maximum codeblock size permitted would not be segmented into multiple codeblocks. The overhead difference between, for example, a 16-bit codeblock CRC and a 24-bit codeblock CRC is negligible given the large codeblock size. The advantage of 24-bits CRC is low misdetection probability for early stopping as well as MIMO SIC. Therefore, similar to transport block CRC, the codeblock CRC size is selected as 24-bits in LTE.

An issue of similar size for codeblock and transport block CRC, however, is that if identical generator polynomials are used for both CB and TB level CRCs, an error sequence passing the CB-level CRC checking will also pass the TB-level CRC checking. This is because with identical CRC generator polynomial on both levels, an error sequence that is divisible by the CB-level CRC generator will be divisible by the TB-level CRC generator as well. That is, the additional CRC checking on the TB level is redundant. The ACK/NACK feedback of the TB is equivalent to being determined by the CB-level CRC attachments only. The probability of undetected TB error events, in this case, is dependent upon codeblock CRC misdetection only and is simply:

$$P_{m-\text{TB}} = 1 - (1 - P_{m-\text{CB}})^C. \tag{11.9}$$

If the primitive polynomials of the CB and TB CRCs are different, an error sequence can pass both CRC checking only if it is divisible by both. Hence, the TB CRC can further reduce the misdetection probability. Noting that $(D + 1)$ is common in both CRC generators, the TB error miss rate is given by

$$P_{m-\text{TB}} = \left[1 - (1 - P_{m-\text{CB}})^C\right] \times 2^{-24}. \tag{11.10}$$

We plot the TB error miss rates for the cases of identical and different polynomials for CB- and TB-level CRC in Figure 11.9. We note that the TB error miss rates are reduced by almost

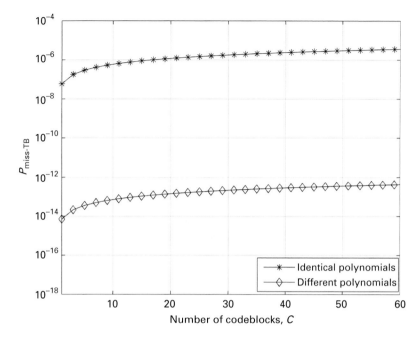

Figure 11.9. Transport error miss rates for identical and different polynomials for CB and TB-level CRC.

seven orders of magnitude by using two different polynomials for CB-level and TB-level CRC.

The two cyclic generator polynomials for $L = 24$ referred to as $g_{CRC24A}(D)$ and $g_{CRC24B}(D)$ for transport block CRC and codeblock CRC respectively are given as:

$$
\begin{aligned}
g_{CRC24A}(D) &= D^{24} + D^{23} + D^{18} + D^{17} + D^{14} \\
&\quad + D^{11} + D^{10} + D^7 + D^6 + D^5 + D^4 + D^3 + D + 1 \\
g_{CRC24B}(D) &= D^{24} + D^{23} + D^6 + D^5 + D + 1.
\end{aligned}
\tag{11.11}
$$

The first polynomial g_{CRC24A} is used for transport block CRC calculation on UL-SCH, DL-SCH, PCH and MCH. The second polynomial g_{CRC24B} is used for codeblock CRC calculation. Additionally, cyclic generator polynomials for $L = 16$ are defined as:

$$
\begin{aligned}
g_{CRC16}(D) &= (D + 1)\left(D^{15} + D^{14} + D^{13} + D^{12} + D^4 + D^3 + D^2 + D^1 + 1\right) \\
g_{CRC16}(D) &= D^{16} + D^{12} + D^5 + 1.
\end{aligned}
\tag{11.12}
$$

This polynomial is used for CRC calculation for BCH and DCI transport channels.

11.4   Codeblock segmentation

We discussed earlier that the maximum codeblock size denoted as Z is limited to 6144 bits. When the transport block size denoted as B (or total size after transport block concatenation) is larger than 6144 bits, segmentation of the input bit sequence is performed and an additional codeblock CRC (CB-CRC) sequence of $L = 24$ bits is attached to each codeblock. The input bit sequence to the codeblock segmentation is denoted by:

$$b_0, b_1, b_2, b_3, \ldots, b_{B-1} \quad B > 0. \tag{11.13}$$

The total number of codeblocks C is determined as below:

$$L = 0, \ C = 1, \ B' = B, \quad B \leq Z$$
$$L = 24, C\lfloor B/(Z-L)\rfloor, \quad B' = B + C \cdot L, \quad B > Z. \tag{11.14}$$

It should be noted that when there is a single codeblock, that is $C = 1$, a single transport block CRC (TB-CRC) of 24 bits is used. However, when there is more than one codeblock, that is $B > Z$, an additional CB-CRC of length $(L = 24)$ is attached to each codeblock, in addition to the transport block TB-CRC. The bits output from codeblock segmentation are denoted by:

$$c_{r0}, c_{r1}, c_{r2}, c_{r3}, \ldots, c_{r(K_r-1)}, \tag{11.15}$$

where r is the codeblock number, and K_r is the number of bits for the codeblock number r.

For the purposes of reducing complexity, a certain fixed number of turbo interleaver sizes is supported as given in Table 11.2 below. In particular, the granularity between two adjacent interleaver sizes is 8-bits for small codeblocks and goes up to 64 bits for the largest codeblock size. When the transport block size is not matched to the turbo interleaver size, filler bits are added. The reason for a coarser granularity of interleaver sizes for larger code blocks is that a larger number of filler bits is still a small fraction of the codeblock size when the codeblock size is large.

Let us assume a transport block size of 19 000 bits segmented into four codeblocks as shown in Figure 11.10. The last three segments are of maximum size 6144 bits and the first segment is of size 576 bits. The first segment size is actually 568 bits and is matched to the nearest interleaver size of 576 bits in Table 11.2 below. This results in a filler bits overhead of 8-bits ($576 - 568 = 8$ bits). However, there is a problem with this segmentation approach because vastly different codeblock sizes would result in different turbo code performance. This is because turbo code performance improves with increasing codeblock size. Therefore,

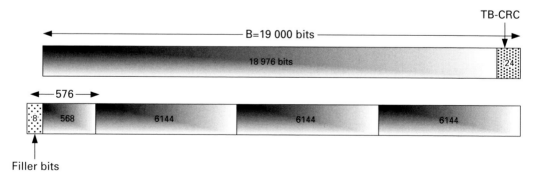

Figure 11.10. Transport block segmentation into unequal size codeblocks.

for a given SNR (signal to noise ratio), it would always be more likely that codeblock with the smallest size is in error. As we discussed early a single ACK/NACK is provided per transport block (set of codeblocks). Therefore, the error performance of transport block would be limited by the performance of the smallest codeblock size because the receiver will NACK the transport block as long as there is a single codeblock in error.

In order to reduce the number of filler bits while keeping the codeblock sizes approximately the same, the LTE system uses two adjacent interleaver sizes, a larger size K_+ and the next smaller size K_-. The first segmentation size denoted as K_+ is minimum K in Table 11.2 (see p. 272) such that

$$C \times K \geq B'. \tag{11.16}$$

The second segmentation size denoted as K_- is maximum K in Table 11.2 (p. 272) such that $K < K_+$. The difference in the two adjacent segmentation sizes that are also turbo interleaver sizes denoted as Δ_K is given as:

$$\Delta_K = K_+ - K_-. \tag{11.17}$$

Let us denote the number of codeblocks of size K_+ and size K_- as C_+ and C_- respectively. When the total number of codeblocks $C = 1$, the codeblock size is K_+ and $C_+ = C = 1$.

The number of filler bits K is obtained as:

$$F = C_+ \cdot K_+ + C_- \cdot K - B'. \tag{11.18}$$

The filler bits are added to the beginning of the first codeblock with $r = 0$.

An example of codeblock segmentation is shown in Figure 11.11. The total number of bits input to the codeblock segmentation is assumed as $B = 19\,000$ bits that include 1876 data bits and 24 bits transport block CRC (TB-CRC). Since $B > Z = 6144$ the total number of code blocks C is determined as below:

$$C = \lceil B/(Z - L) \rceil = \lceil 19\,000/(6144 - 24) \rceil = 4. \tag{11.19}$$

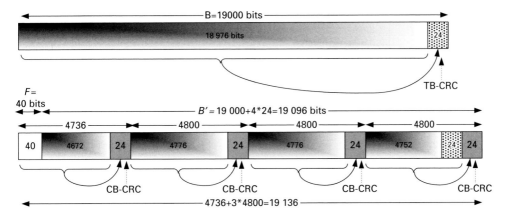

Figure 11.11. An example of codeblock segmentation.

Also, the new total size B' is given as:

$$B' = B + C \times L = 19\,000 + 4 \times 24 = 1996. \tag{11.20}$$

Since $C > 1$, the first and second segment sizes are determined by using the procedure described above. The first segmentation size $K_+ = 4800$ is minimum K in Table 11.2 (see p. 272) that results in $C \cdot K \ge B$. The second segment size is one size smaller than $K_+ = 4800$ which is $K_- = 4736$ from Table 11.2, on p. 272.

The difference between the two segment sizes is

$$\Delta K = K_+ - K_- = 4800 - 4736 = 64. \tag{11.21}$$

The number of codeblocks of size K_- and K_+ are given as:

$$\begin{aligned} C_- &= \left\lfloor \frac{C \cdot K_+ - B'}{\Delta K} \right\rfloor = \left\lfloor \frac{4 \times 4800 - 19\,096}{64} \right\rfloor = 1 \\ C_+ &= C - C_- = 4 - 1 = 3. \end{aligned} \tag{11.22}$$

The number of filler bits F is:

$$\begin{aligned} F &= C_+ \cdot K_+ + C_- \cdot K_- - B' \\ &= 3 \times 4800 + 1 \times 4736 - 19\,096 = 40. \end{aligned} \tag{11.23}$$

We note that the number of filler bits is more than the segmentation scheme in Figure 11.10. However, as pointed out earlier the scheme of Figure 11.10 suffers from the problem of vastly different codeblock sizes. We also note that if two adjacent interleaver sizes were not allowed and a single interleaver size with $K = 4800$ was used, the number of filler bits would have been greater as calculated below:

$$F = C \cdot K = 4 \times 4800 - 19\,096 = 104. \tag{11.24}$$

Therefore, using two adjacent interleaver sizes allows keeping the codeblocks size within a transport block approximately the same while reducing the number of filler bits. However, the two adjacent interleaver sizes decision was made under the assumption that these bits are transmitted over the air and hence represent an unnecessary overhead. In the later stages of LTE standard development, it was agreed that filler bits are mostly removed after channel coding as discussed in Section 11.5. Therefore, the overhead represented by filler bits is effectively smaller than if all the coded filler bits are transmitted over the air.

11.5 Turbo coding

The fundamental turbo code encoder is built using two identical recursive systematic convolutional (RSC) codes with parallel concatenation [12] as shown in Figure 11.12. An RSC encoder is typically a rate 1/2 encoder and is termed a constituent encoder. The input to the second constituent encoder is interleaved using an internal turbo code interleaver. Only one of the systematic outputs from the two component encoders is used. This is because the systematic output from the other component encoder is just a permuted version of the chosen systematic output.

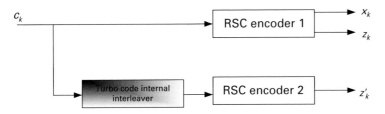

Figure 11.12. Turbo code encoder.

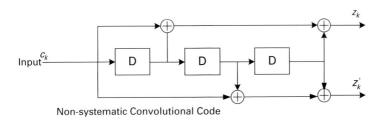

Non-systematic Convolutional Code

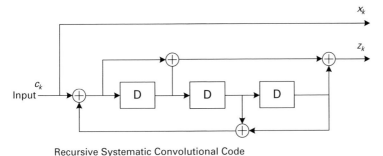

Recursive Systematic Convolutional Code

Figure 11.13. Conventional convolutional encoder and equivalent RSC encoder.

An example of a conventional convolutional encoder shown in Figure 11.13 is represented by the generator sequences $g_0(D) = 1 + D^2 + D^3$ and $g_1(D) = 1 + D + D^3$. It can equivalently be represented in a more compact form as $G = [g_0(D), g_1(D)]$. The RSC encoder of this conventional convolutional encoder is represented as $G(D) = \left[1, \; \frac{g_1(D)}{g_0(D)} \right]$, where the first output represented by $g_0(D)$ is fed back to the input. In the above representation, 1 denotes the systematic output, $g_1(D)$ denotes the feed-forward output and $g_0(D)$ is the feedback to the input of the RSC encoder. It was suggested in [12] that good codes can be obtained by setting the feedback of the RSC encoder to a primitive polynomial, because the primitive polynomial generates maximum-length sequences, which adds randomness to the turbo code.

The turbo coding used in the LTE system employs a parallel concatenated convolutional code (PCCC) with two 8-state rate 1/2 constituent encoders. The input to the second constituent encoder is interleaved using an internal turbo code interleaver as shown in Figure 11.14. The transfer function for the constituent coders is given as:

$$G(D) = \left[1, \; \frac{g_1(D)}{g_0(D)} \right], \quad g_0(D) = 1 + D^2 + D^3, \quad g_1(D) = 1 + D + D^3. \quad (11.25)$$

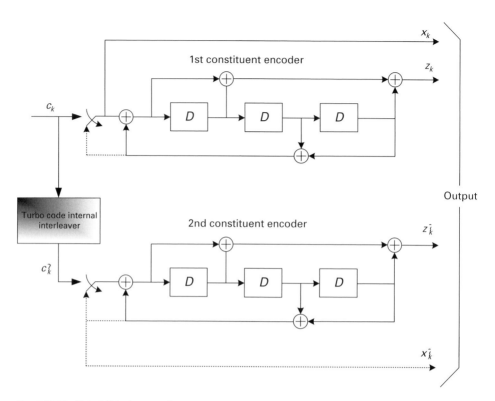

Figure 11.14. Rate 1/3 turbo encoder.

The initial value of the shift registers of the 8-state constituent encoders is set to all zeros state before starting to encode the input bits. The output from the turbo encoder is given as:

$$d_k^{(0)} = x_k, \quad d_k^{(1)} = z_k, \quad d_k^{(2)} = z_k' \quad k = 0, 1, 2, \ldots, K - 1. \tag{11.26}$$

The bits input to the turbo encoder are denoted by $c_0, c_1, c_2, c_3, \ldots, c_{K-1}$, and the bits output from the first and second constituent encoders are denoted respectively by $z_0, z_1, z_2, z_3, \ldots, z_{K-1}$ and $z_0', z_1', z_2', z_3', \ldots, z_{K-1}'$. The bits output from the turbo code internal interleaver are denoted by $c_0', c_1', \ldots, c_{K-1}'$, and these bits are input to the second constituent encoder.

11.5.1   Filler bits removal

The filler bits that go into systematic bits $d_k^{(0)} = x_k$ and second stream from first constituent encoder $d_k^{(1)} = z_k$ are removed before transmission. This is achieved by setting:

$$\begin{aligned} c_k &= 0 \\ d_k^{(0)} &= \langle \text{NULL} \rangle \quad k = 0, \ldots, (F - 1), \\ d_k^{(1)} &= \langle \text{NULL} \rangle \end{aligned} \tag{11.27}$$

where F is the total number of filler bits.

We note that the filler bits that go into the second constituent encoder are not removed. It is difficult to track these bits due to the fact that the input to the second constituent encoder is the interleaved version of the input stream to turbo code.

11.5.2 Trellis termination

For the conventional convolutional encoder, the trellis is terminated by inserting additional zero bits after the input sequence. These additional bits drive the conventional convolutional encoder to the all-zero state (trellis termination). However, this strategy is not possible for the RSC encoder due to the presence of feedback. The additional termination bits for the RSC encoder depend on the state of the encoder and are very difficult to predict [14]. Furthermore, even if the termination bits for one of the component encoders are found, the other component encoder may not be driven to the all-zero state with the same termination bits. The reason for this is the presence of the interleaver between the constituent encoders. A simple strategy developed in [14] can overcome this problem. The idea is that for encoding the input sequence, the switch in Figure 11.4 is turned on to a higher position and for terminating the trellis the switch is turned on to a lower position for each constituent encoder separately. The trellis termination is therefore performed by taking the tail bits from the shift register feedback after all information bits are encoded. The first three tail bits are used to terminate the first constituent encoder while the second constituent encoder is disabled. The last three tail bits are used to terminate the second constituent encoder while the first constituent encoder is disabled. The transmitted bits for trellis termination are then:

$$
\begin{aligned}
d_K^{(0)} &= x_K & d_{K+1}^{(0)} &= z_{K+1} & d_{K+2}^{(0)} &= x_K' & d_{K+3}^{(0)} &= z_{K+1}' \\
d_K^{(1)} &= z_K & d_{K+1}^{(1)} &= x_{K+2} & d_{K+2}^{(1)} &= z_K' & d_{K+3}^{(1)} &= x_{K+2}' \\
d_K^{(2)} &= x_{K+1} & d_{K+1}^{(2)} &= z_{K+2} & d_{K+2}^{(2)} &= x_{K+1}' & d_{K+3}^{(2)} &= z_{K+2}'.
\end{aligned}
\tag{11.28}
$$

11.5.3 Turbo code internal interleaver

LTE supports peak data throughputs in excess of 100 Mb/s. A good approach for achieving higher decoder speed is by parallelizing the log-MAP algorithm inside each constituent log-MAP decoder of the turbo decoder [15–19]. This method is referred to as parallel windowed decoding or parallelized log-MAP processing. In the parallel-windowed method, a codeblock of size $K = MW$ is divided into M windows of size W trellis steps. Then M log-MAP processors (one per window) operate in parallel to produce the extrinsic data over the windows. For a given codeblock size, increasing M (and decreasing W) increases the throughput in direct proportion to M. Each log-MAP processor can have its own dedicated input LLR and a priori memory banks, so that fetching data for processing can occur simultaneously for all processors.

Improving log-MAP decoder throughput translates into turbo decoding speed improvements only when turbo code internal interleaving/deinterleaving is not a bottleneck. With a contention-free interleaver, both interleaver and de-interleaver can be implemented by having M extrinsic values written to or read from M memory banks concurrently. This memory access procedure requires a contention-free interleaver designed for parallelization factor M.

When the interleavers are not designed properly, memory access contentions, wherein two or more processors attempt to read from (or write to) the same memory bank concurrently, can lead to extra delays as the processors have to access the same memory bank one after the other. These contentions may cause significant speed loss (due to delay buffers). Moreover complex hardware design is required for handling such contentions [20].

Contention-free interleaver

For a codeblock of size $K = MW$, the exchange and processing of the extrinsic information symbols between subblocks of the iterative decoder can be parallelized by M processors working on window sizes of length W in each subblock without contending for memory access provided that the following condition holds for both the interleaver, $f(x), 0 \leq x < K$, and the de-interleaver $g(x) = f^{-1}(x)$.

$$\lfloor \pi(j + tW)/W \rfloor \neq \lfloor \pi(j + vW)/W \rfloor, \tag{11.29}$$

where $0 \leq j < W, 0 \leq t < v < K/W$ and $\pi(\cdot)$ is either $f(\cdot)$ or $g(\cdot)$.

This means that information in M different memory banks (each of size W) can be accessed by M different processors simultaneously without contention. In fact, it can be further shown that an identical address is used to access the information within all memory banks:

$$\Pi(x + tW) \bmod W = \left[f_1(x + tW) + f_2(x + tW)^2 \right] \bmod W$$

$$= \left[\left(f_1 x + f_2 x^2 \right) + \left(f_1 t + 2 f_2 tx + f_2 t^2 W \right) W \right] \bmod W \tag{11.30}$$

$$= [\Pi(x)] \bmod W.$$

If an interleaver is contention free for all window size W dividing the interleaver length K, it is called a maximum contention-free interleaver.

Almost regular permutation interleaver

An "almost regular" permutation (ARP) interleaver [15] is given by the following expression:

$$\pi(i) = (iP + A + d(i)) \bmod K, \tag{11.31}$$

where $0 \leq i \leq (K - 1)$ is the sequential index of the bit positions after interleaving, $\pi(i)$ is the bit index before interleaving corresponding to position i, K is the codeblock size, P is a number that is relatively prime to K, A is a constant known as the offset and $d(i)$ is a "dither" vector of cycle length $C = 4, 8, 12, \ldots$, which is generally chosen as a multiple of 4. For an ARP interleaver of length K and cycle length C, any window size W, where W is a multiple of C and a factor of K, can be used for high-speed decoding without memory access contentions.

Quadratic permutation polynomial (QPP) interleaver

The relationship between the input and output bits of the QPP interleaver is given as follows:

$$x_i' = x_{\pi(i)} \quad i = 0, 1, \ldots, (K - 1), \tag{11.32}$$

where K is the number of input bits to the turbo code internal (QPP) interleaver. The relationship between the output index i and the input index $\Pi(i)$ satisfies the following quadratic
form:

$$\Pi(i) = \left(f_1 \cdot i + f_2 \cdot i^2\right) \bmod K, \tag{11.33}$$

where $0 \le i, f_1, f_2 < K$.

The general design guidelines for f_1 and f_2 are that the greatest common divisor of f_1 and
K should be 1 and any prime factor of K should also divide f_2. The QPP interleavers can be
made maximum contention-free by proper selection of f_1 and f_2.

The QPP interleaver allows calculating the next interleaved position from the current position recursively without multiplication or modulo operations [19] as illustrated by a simple
example below:

$$\begin{aligned}
\Pi(x+1) &= \left[f_1(x+1) + f_2(x+1)^2\right] \bmod K \\
&= \left[(f_1 x + f_2 x^2) + (f_1 + f_2 + 2f_2 x)\right] \bmod K \\
&= \Pi(x) + g(x),
\end{aligned} \tag{11.34}$$

where

$$g(x) = [f_1 + f_2 + 2f_2 x] \bmod K. \tag{11.35}$$

Furthermore, $g(x)$ can also be computed recursively as below:

$$\begin{aligned}
g(x+1) &= [f_1 + f_2 + 2f_2(x+1)] \bmod K \\
&= [g(x) + 2f_2] \bmod K.
\end{aligned} \tag{11.36}$$

As both $\Pi(x)$ and $g(x)$ are less than K, the modulo operations in both the equations can be
replaced by simple comparison operations.

Let us assume a codeblock size of 40 bits ($K = 40$), which is decoded by four parallel
processors using four memory banks. This results in a window size or memory bank size of
$W = 10$. Furthermore, we assume $f_1 = 3$ and $f_2 = 10$ for the QPP interleaver as below:

$$\Pi(i) = \left(3 \cdot i + 10 \cdot i^2\right) \bmod 40. \tag{11.37}$$

The logical addresses for the four processors along with intra-block and inter-block permutations are given in Table 11.1. The inter-block permutation can be seen as the memory bank
accessed by the corresponding processor. For example, during the seventh tick, processors P1,
P2, P3 and P4 access memory bank number 3, 0, 1 and 2 respectively. As expected, during a
total of 40 ticks, each processor is able to access all the addresses without any contention. This
is further illustrated in Figure 11.15 where the full sequence of logical addresses is shown for
each of the four processors. We also note that each processor performs a memory access in a
different memory bank but using the same logical address. This is shown by relative cyclic
shift of addresses by the window size ($W = 10$) between processors (P_0, P_1), (P_1, P_2) and
(P_2, P_3).

The reference [21] shows that for many QPP interleavers, the de-interleaver is also a QPP
interleaver. One advantage of an interleaver being its own inverse is that same algorithm and
hardware can be used for both interleaving and de-interleaving.

Table 11.1. Logical addresses for the four processors along with intra-block and inter-block permutations.

Tick	Logical addresses				Equivalent intra-block permutation	Equivalent intra-block permutation			
	PO	P1	P2	P3					
0	0	10	20	30	0	0	1	2	3
1	13	23	33	3	3	1	2	3	0
2	6	16	26	36	6	0	1	2	3
3	19	29	39	9	9	1	2	3	0
4	12	22	32	2	2	1	2	3	0
5	25	35	5	15	5	2	3	0	1
6	18	28	38	8	8	1	2	3	0
7	31	1	11	21	1	3	0	1	2
8	24	34	4	14	4	2	3	0	1
9	37	7	17	27	7	3	0	1	2
⋮	⋮	⋮	⋮	⋮	⋮	⋮	⋮	⋮	⋮
34	22	32	2	12	2	2	3	0	1
35	35	5	15	25	5	3	0	1	2
36	28	38	8	18	8	2	3	0	1
37	1	11	21	31	1	0	1	2	3
38	34	4	14	24	4	3	0	1	2
39	7	17	27	37	7	0	1	2	3

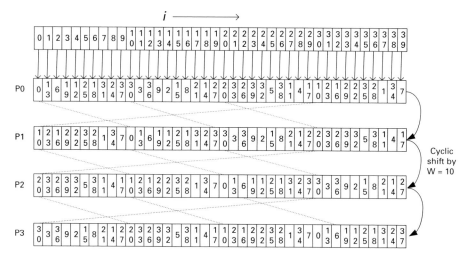

Figure 11.15. Addresses for four processors assuming QPP interleaver using $\Pi(i) = 3 \cdot i + 10 \cdot i^2$ mod 40.

It can be shown that the de-interleaver for the interleaver $\Pi(i) = (3 \cdot i + 10 \cdot i^2)$ mod 40 is the following QPP interleaver as is also shown in Figure 11.16:

$$i = \Pi^{-1}(j) = (7 \cdot j + 30 \cdot j^2) \bmod 40 , \tag{11.38}$$

where $j = \Pi(i)$.

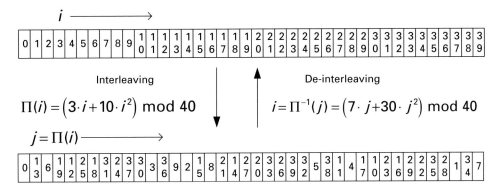

Figure 11.16. An example of interleaving and de-interleaving where both the interleaver and the de-interleaver are QPP interleavers.

The contention free interleaver used in the LTE system is based on the QPP principle. The bits input to the turbo code internal interleaver are denoted by $c_0, c_1, \ldots, c_{K-1}$, where K is the number of input bits. The bits output from the turbo code internal interleaver are denoted by $c'_0, c'_1, \ldots, c'_{K-1}$.

The relationship between the input and output bits is as follows:

$$c'_i = c_{\Pi(i)} \quad i = 0, 1, \ldots, (K-1), \tag{11.39}$$

where the relationship between the output index i and the input index $\Pi(i)$ satisfies the quadratic form in (11.33). The parameters f_1 and f_2 depend on the block size K and are summarized in Table 11.2. The QPP interleaver size granularity becomes coarser as the interleaver size increases. For $40 \leq K \leq 512$, K contains all multiples of 8, for $512 \leq K \leq 1024$, K contains all multiples of 16, for $1024 \leq K \leq 2048$, K contains all multiples of 32 and for $2048 \leq K \leq 6144$, K contains all multiples of 64.

11.6   Tail-biting convolutional code

As discussed earlier, the conventional convolutional codes add tail bits to the encoded sequence to ensure the encoder ends in an all-zero state. The tail ensures that the encoder begins and ends in a known state, a fact that helps in improving the decoder performance. However, adding tail bits has several disadvantages that include increased overheads and a waste of transmit power. The tail bits overhead can be quite significant for small block size transmissions, which is generally the case for control channels.

One way of eliminating the tail bits of a convolutional code is to simply discard the tail bits after encoding the information block. However, this can lead to significant performance degradation. Another possibility is to use tail biting [22], which ensures that the final state (after encoding an information block) of a convolutional encoder is the same as the initial state. This is done by initializing the shift register state with the last bits of the information block. The tail-biting convolutional decoder can be implemented using the same low complexity decoding algorithms that exist to traverse the trellis, such as the Viterbi algorithm.

Table 11.2. Turbo code internal interleaver parameter.

i	K_i	f_1	f_2	i	K_i	f_1	f_2	i	K_i	f_1	f_2	i	K_i	f_1	f_2
1	40	3	10	48	416	25	52	95	1120	67	140	142	3200	111	240
2	48	7	12	49	424	51	106	96	1152	35	72	143	3264	443	204
3	56	19	42	50	432	47	72	97	1184	19	74	144	3328	51	104
4	64	7	16	51	440	91	110	98	1216	39	76	145	3392	51	212
5	72	7	18	52	448	29	168	99	1248	19	78	146	3456	451	192
6	80	11	20	53	456	29	114	100	1280	199	240	147	3520	257	220
7	88	5	22	54	464	247	58	101	1312	21	82	148	3584	57	336
8	96	11	24	55	472	29	118	102	1344	211	252	149	3648	313	228
9	104	7	26	56	480	89	180	103	1376	21	86	150	3712	271	232
10	112	41	84	57	488	91	122	104	1408	43	88	151	3776	179	236
11	120	103	90	58	496	157	62	105	1440	149	60	152	3840	331	120
12	128	15	32	59	504	55	84	106	1472	45	92	153	3904	363	244
13	136	9	34	60	512	31	64	107	1504	49	846	154	3968	375	248
14	144	17	108	61	528	17	66	108	1536	71	48	155	4032	127	168
15	152	9	38	62	44	5	68	109	1568	13	28	156	4096	31	64
16	160	21	120	63	560	227	420	110	1600	17	80	157	4160	33	130
17	168	101	84	64	576	65	96	111	1632	25	102	158	4224	43	264
18	176	21	44	65	592	19	74	112	1664	183	104	159	4288	33	134
19	184	57	46	66	608	37	76	113	1696	55	954	160	4352	477	408
20	192	23	48	67	624	41	234	114	1728	127	96	161	4416	35	138
21	200	13	50	68	640	39	80	115	1760	27	110	162	4480	233	280
22	208	27	52	69	656	185	82	116	1792	29	112	163	4544	357	142
23	216	11	36	70	672	43	252	117	1824	29	114	164	4608	337	480
24	224	27	56	71	688	21	86	118	1856	57	116	165	4672	37	146
25	232	85	58	72	704	155	44	119	1888	45	354	166	4736	71	444
26	240	29	60	73	720	79	120	120	1920	31	120	167	4800	71	120
27	248	33	62	74	736	139	92	121	1952	59	610	168	4864	37	152
28	256	15	32	75	752	23	94	122	1984	185	124	169	4928	39	462
29	264	17	198	76	768	217	48	123	2016	113	420	170	4992	127	234
30	272	33	68	77	784	25	98	124	2048	31	64	171	5056	39	158
31	280	103	210	78	800	17	80	125	2112	17	66	172	5120	39	80
32	288	19	36	79	816	127	102	126	2176	171	136	173	5184	31	96
33	296	19	74	80	832	25	52	127	2240	209	420	174	5248	113	902
34	304	37	76	81	848	239	106	128	2304	253	216	175	5312	41	166
35	312	19	78	82	864	17	48	129	2368	367	444	176	5376	251	336
36	320	21	120	83	880	137	110	130	2432	265	456	177	5440	43	170
37	328	21	82	84	896	215	112	131	2496	181	468	178	5504	21	86
38	336	115	84	85	912	29	114	132	2560	39	80	179	5568	43	174
39	344	193	86	86	928	15	58	133	2624	27	164	180	5632	45	176
40	352	21	44	87	944	147	118	134	2688	127	504	181	5696	45	178
41	360	133	90	88	960	29	60	135	2752	143	172	182	5760	161	120
42	368	81	46	89	976	59	122	136	2816	43	88	183	5824	89	182
43	376	45	94	90	992	65	124	137	2880	29	300	184	5888	323	184
44	384	23	48	91	1008	55	84	138	2944	45	92	185	5952	47	186
45	392	243	98	92	1024	31	64	139	3008	157	188	186	6016	23	94
46	400	151	40	93	1056	17	66	140	3072	47	96	187	6080	47	190
47	408	155	102	94	1088	171	204	141	3136	13	28	188	6144	263	480

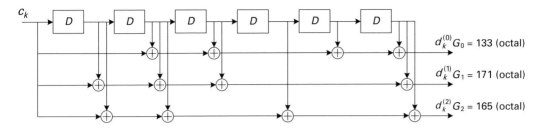

Figure 11.17. Rate 1/3 tail biting convolutional encoder.

Tail biting can reduce overheads at the expense of an increase in decoding complexity and slight performance degradation. In principle, the tail-biting scheme can be applied to turbo coding as well. However, the relative tail bits overhead for turbo code is very small given the large codeblocks sizes used for data transmission. Therefore, the overhead reduction versus complexity and small performance loss trade off appears worthwhile for control channels (with small block sizes) only. For these reasons, the tail-biting scheme was adopted in the LTE system for convolutional coding only and not for turbo coding.

A tail biting convolutional code with constraint length 7 and coding rate 1/3 as shown in Figure 11.17 is defined for LTE. The initial value of the shift register of the encoder is set to the values corresponding to the last 6 information bits in the input stream so that the initial and final states of the shift register are the same. Therefore, denoting the shift register of the encoder by $s_0, s_1, s_2, \ldots, s_5$, then the initial value of the shift register is set as:

$$s_i = c_{(K-1-i)} \quad i = 0, 1, \ldots, 5. \tag{11.40}$$

The encoder output streams $d_k^{(0)}$, $d_k^{(1)}$ and $d_k^{(2)}$ correspond to the first, second and third parity streams, respectively, as shown in Figure 11.17.

11.7 Circular-buffer rate matching for turbo code

An asynchronous and adaptive hybrid ARQ scheme is used for downlink data transmissions in LTE (see Chapter 12). Also, a synchronous adaptive hybrid ARQ is employed for uplink transmissions. With adaptive schemes, modulation, coding and resource allocation can change on retransmissions and therefore require different numbers of coded bits to be transmitted from the original transmission. Another constraint with turbo coding is that systematic bits should preferably be carried in the first transmission attempt, as the turbo code performance is sensitive to the systematic bits. This requires a highly flexible and adaptable rate matching scheme. In order to meet this goal, a circular-buffer-based rate-matching scheme is used in LTE as illustrated in Figure 11.18. The rate 1/3 turbo code generates a stream of systematic bits, a stream of parity bits from the first constituent convolutional code (parity 1 bits), and a stream of parity bits from the second constituent convolutional code (parity 2 bits). Each of these three streams is interleaved separately by subblock interleavers. Furthermore, the interleaved parity 1 and 2 bits are interlaced. During the hybrid ARQ rate-matching procedure, each

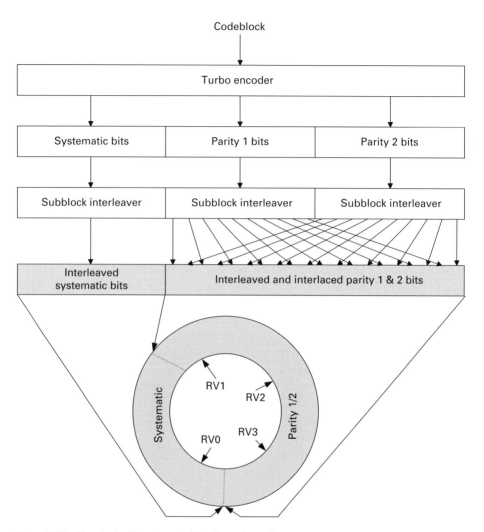

Figure 11.18. Circular-buffer rate matching for turbo code.

transmission reads bits from the buffer, starting from an offset position and increasing the bit index. If the bit index reaches a certain maximum number, the bit index is reset to the first bit in the buffer. In other words, the buffer is circular.

11.7.1 Subblock interleaving

The bits input to the block interleaver are denoted by:

$$d_0^{(i)}, d_1^{(i)}, d_2^{(i)}, \ldots, d_{D-1}^{(i)}, i = 0, 1, 2, \qquad (11.41)$$

where $D = K + 4$ is the number of bits for each of systematic, parity 1 and parity 2 streams. Note that K is the number of bits within a codeblock with bits $x_k, k = 0, 1, 2, \ldots, K - 1$, and trellis termination adds four bits to each of systematic, parity 1 and parity 2 streams.

The subblock interleaving is achieved by writing row-wise in a rectangular matrix, applying matrix columns permutations and finally reading from the matrix column-wise. The number of columns in the matrix is fixed to 32, that is

$$C_{\text{subblock}}^{\text{TC}} = 32. \tag{11.42}$$

The number of rows of the matrix $R_{\text{subblock}}^{\text{TC}}$ are determined by finding the minimum integer $R_{\text{subblock}}^{\text{TC}}$ such that:

$$R_{\text{subblock}}^{\text{TC}} \times C_{\text{subblock}}^{\text{TC}} \geq D. \tag{11.43}$$

When the number of bits D does not completely fill the $\left(R_{\text{subblock}}^{\text{TC}} \times C_{\text{subblock}}^{\text{TC}} \right)$ rectangular matrix, dummy bits are padded to fully fill the matrix as below:

$$
\begin{aligned}
y_k &= \langle NULL \rangle & k &= 0, 1, \ldots, N_D - 1 \\
y_{N_D + k} &= d_k^{(i)} & k &= 0, 1, \ldots, D - 1.
\end{aligned}
\tag{11.44}
$$

Note that the maximum number of dummy bits is limited to $\left(C_{\text{subblock}}^{\text{TC}} - 1 \right)$ and these bits are added to the beginning of the stream. Also, note that when $R_{\text{subblock}}^{\text{TC}} \times C_{\text{subblock}}^{\text{TC}} = D$, no dummy bits need to be added as the total D bits fully fill the matrix in this case. The input bit sequence is then written into the $\left(R_{\text{subblock}}^{\text{TC}} \times C_{\text{subblock}}^{\text{TC}} \right)$ rectangular matrix row by row starting with bit y_0 in column 0 of row 0 as below:

$$
\begin{bmatrix}
y_0 & y_1 & y_2 & \cdots & y_{C_{\text{subblock}}^{\text{TC}} - 1} \\
y_{C_{\text{subblock}}^{\text{TC}}} & y_{C_{\text{subblock}}^{\text{TC}} + 1} & y_{C_{\text{subblock}}^{\text{TC}} + 2} & \cdots & y_{2C_{\text{subblock}}^{\text{TC}} - 1} \\
\vdots & \vdots & \vdots & \ddots & \vdots \\
y_{(R_{\text{subblock}}^{\text{TC}} - 1) \times C_{\text{subblock}}^{\text{TC}}} & y_{(R_{\text{subblock}}^{\text{TC}} - 1) \times C_{\text{subblock}}^{\text{TC}} + 1} & y_{(R_{\text{subblock}}^{\text{TC}} - 1) \times C_{\text{subblock}}^{\text{TC}} + 2} & \cdots & y_{(R_{\text{subblock}}^{\text{TC}} \times C_{\text{subblock}}^{\text{TC}} - 1)}
\end{bmatrix}
$$

$$\tag{11.45}$$

For systematic and parity 1 bits $d_k^{(0)}$ and $d_k^{(1)}$, inter-column permutation for the matrix is performed based on the bit-reversal-order (BRO) pattern given in Table 11.3 and also shown in Figure 11.19, where $P(j)$ is the original column position of the jth permuted column. As an example, for $j = 3$ [00011], $P(j) = 24$ [11000], where the bits in binary notation are reversed. After permutation of the columns, the inter-column permuted $\left(R_{\text{subblock}}^{\text{TC}} \times C_{\text{subblock}}^{\text{TC}} \right)$ rectangular matrix is equal to:

Table 11.3. Inter-column permutation pattern for subblock interleaver for turbo code.

j	$P(j)$	j	$P(j)$	j	$P(j)$	j	$P(j)$
0	0	8	2	16	1	24	3
1	16	9	18	17	17	25	19
2	8	10	10	18	9	26	11
3	24	11	26	19	25	27	27
4	4	12	6	20	5	28	7
5	20	13	22	21	21	29	23
6	12	14	14	22	13	30	15
7	28	15	30	23	29	31	31

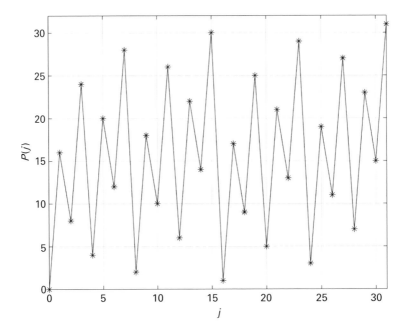

Figure 11.19. Inter-column permutation pattern for subblock interleaver for turbo code.

$$
\begin{bmatrix}
y_{P(0)} & y_{P(1)} & \cdots & y_{P(C_{\text{subblock}}^{\text{TC}}-1)} \\
y_{P(0)+C_{\text{subblock}}^{\text{TC}}} & y_{P(1)+C_{\text{subblock}}^{\text{TC}}} & \cdots & y_{P(C_{\text{subblock}}^{\text{TC}}-1)+C_{\text{subblock}}^{\text{TC}}} \\
\vdots & \vdots & \ddots & \vdots \\
y_{P(0)+(R_{\text{subblock}}^{\text{TC}}-1)\times C_{\text{subblock}}^{\text{TC}}} & y_{P(1)+(R_{\text{subblock}}^{\text{TC}}-1)\times C_{\text{subblock}}^{\text{TC}}} & \cdots & y_{P(C_{\text{subblock}}^{\text{TC}}-1)+(R_{\text{subblock}}^{\text{TC}}-1)\times C_{\text{subblock}}^{\text{TC}}}
\end{bmatrix}.
$$

$$(11.46)$$

The output of the block interleaver is the bit sequence read out column by column from the inter-column permuted matrix in (11.46). The bits after subblock interleaving $v_0^{(i)}, v_1^{(i)}, v_2^{(i)}, \ldots, v_{K_\Pi-1}^{(i)}$ are set as:

$$v_0^{(i)} = y_{P(0)}$$

$$v_1^{(i)} = y_{P(0)+C_{\text{subblock}}^{\text{TC}}}$$

$$\vdots$$

$$i = 0, 1, \qquad (11.47)$$

$$v_{K_\Pi-1}^{(i)} = y_{P(C_{\text{subblock}}^{\text{TC}}-1)+(R_{\text{subblock}}^{\text{TC}}-1)\times C_{\text{subblock}}^{\text{TC}}}$$

where $k_\Pi = R_{\text{subblock}}^{\text{TC}} \times C_{\text{subblock}}^{\text{TC}} = D$.

For parity 2 stream $d_k^{(2)}$, the output of the subblock interleaver denoted $v_0^{(2)}, v_1^{(2)}, v_2^{(2)}, \ldots, v_{K_\Pi-1}^{(2)}$ is given by:

$$v_k^{(2)} = y_{\pi(k)}, \qquad (11.48)$$

where

$$\pi(k) = \left(P\left(\left\lfloor \frac{k}{R_{\text{subblock}}^{\text{TC}}} \right\rfloor \right) + C_{\text{subblock}}^{\text{TC}} \times \left(k \bmod R_{\text{subblock}}^{\text{TC}} \right) + 1 \right) \bmod K_\Pi. \qquad (11.49)$$

The permutation function P in (11.49) is the same as before and is defined in Table 11.3.

11.7.2 Subblock interlacing

The circular buffer length is $K_w = 3K_\Pi$, where K_Π is the number of interleaved bits in each of systematic, parity 1 and parity 2 streams. The bit stream in the circular buffer is denoted as $w_0, w_1, \ldots, w_{(K_w-1)}$ and is given as:

$$
\begin{aligned}
w_k &= v_k^{(0)} & k &= 0, 1, \ldots, (K_\Pi - 1) \\
w_{K_\Pi+2k} &= v_k^{(1)} & k &= 0, 1, \ldots, (K_\Pi - 1) \\
w_{K_\Pi+2k+1} &= v_k^{(2)} & k &= 0, 1, \ldots, (K_\Pi - 1).
\end{aligned}
\qquad (11.50)
$$

It should be noted that the subblock interlacing is only performed between parity 1 and 2 bits as shown in Figure 11.20. The systematic bits are not interlaced. The reason is that systematic bits are generally part of the first hybrid ARQ transmission. In response to hybrid ARQ NACK, for example, subblock interlacing guarantees that an equal amount of parity 1 and 2 bits are transmitted.

11.7.3   Hybrid ARQ soft buffer limitation

The soft buffer size for the rth code block N_{cb} is given as:

$$N_{cb} = \begin{cases} \min\left(\left\lfloor \dfrac{N_{IR}}{C} \right\rfloor, K_w\right) & \text{downlink} \\ K_w & \text{uplink,} \end{cases} \qquad (11.51)$$

where C is the number of codeblocks within the transport block and K_w is the circular buffer size for the rth codeblock. N_{IR} is soft buffer size per codeword per hybrid ARQ process (see Chapter 12) available at the UE and is given as:

$$N_{IR} = \left\lfloor \frac{N_{\text{soft}}}{K_{\text{MIMO}} \cdot \min\left(M_{\text{DL\_HARQ}}, M_{\text{limit}}\right)} \right\rfloor, \qquad (11.52)$$

where N_{soft} is the total soft buffer size, which is set by higher layers. $K_{\text{MIMO}} = 1, 2$ for the case of single codeword and dual-codeword MIMO spatial multiplexing respectively. $M_{\text{DL\_HARQ}} = 8$ is the maximum number of hybrid ARQ processes and $M_{\text{limit}} = 9$.

We note that the soft buffer limitation only applies for the downlink due to soft buffering concerns for the UE receiver. In the uplink, there is no soft buffer limitation for the eNB and hence incremental redundancy can always be used. The soft buffer size is directly proportional to the supported data rate and is inversely proportional to the turbo coding rate. The idea with soft buffer limitation is that if UE has a certain buffer size dimensioned for a given data rate and a given coding rate then it can support either higher data rates with increasing coding rate (weaker code) or lower data rates with a stronger code. Therefore, UE can support incremental redundancy in most cases while falling back to Chase combining when it approaches peak data rates if its soft buffer is dimensioned to support Chase combining only at the peak data rate. In other cases, when the soft buffer is dimensioned to support the peak data rate with incremental redundancy in mind, full redundant bits can always be transmitted providing performance advantage. Therefore, the soft buffer approach provides a tool to balance the UE soft buffering complexity against incremental redundancy gains.

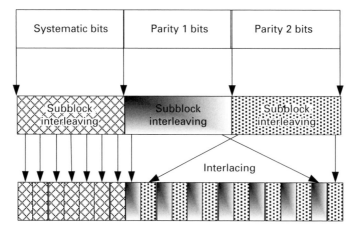

Figure 11.20. Subblock interlacing.

11.7.4 RV starting points

The bits from a single codeblock belonging to a transport block can be transmitted within a resource element (e.g. one subcarrier within one OFDM symbol for downlink, see Chapter 8). This means that the bits from two codeblocks from the same transport block cannot be mixed in the same modulation symbol including the spatial dimension. However, for two codewords MIMO transmission where two separate transport blocks are transmitted, bits from two codeblocks belonging to different transport blocks can possibly be mixed in the same physical resource element.

The transmission of bits from two codeblocks from the same transport block within a single resource element is avoided by first defining G' as:

$$G' = \frac{G}{(N_L \times Q_m)}, \tag{11.53}$$

where G is the total number of bits available for the transmission of one transport block and $Q_m = 2, 4, 6$ for QPSK, 16-QAM and 64-QAM respectively. $N_L = 1$ for transport blocks mapped onto one MIMO transmission layer and $N_L = 2$ for transport blocks mapped onto two or four MIMO transmission layers. It should be noted that SFBC transmit diversity is considered as a two-layer transmission while the SFBC-FSTD transmit diversity scheme is considered a four-layer transmission. Therefore, setting $N_L = 2$ for transmit diversity means that an even number of resource elements or modulation symbols is used for each codeblock in transmit diversity mode.

Let us now set:

$$\gamma = G' \bmod C. \tag{11.54}$$

The rate-matching output sequence of length E for the rth coded block is then given as:

$$E = \begin{cases} N_L \cdot Q_m \cdot \lfloor G'/C \rfloor & r \le C - \gamma - 1 \\ N_L \cdot Q_m \cdot \lceil G'/C \rceil & \text{otherwise.} \end{cases} \tag{11.55}$$

We note that some codeblocks may need to use one fewer resource element and some others one more resource element to avoid mixing of bits in the same resource element from two codeblocks from the same transport block. It should also be noted that the rate-matching output sequence length E is determined independently of the codeblock size. We noted earlier in Section 11.4 that two adjacent codeblock sizes can be used with the difference in size $\Delta_K = K_+ - K_-$ as large as 64 bits. Therefore, the codeblocks with the larger size will experience a slightly higher coding rate than the codeblocks with the smaller size. We also know that the codeblocks with the smaller size use lower index and from (11.55) we also note that the codeblocks with lower index $r \le C - \gamma - 1$ may use one fewer resource element than the codeblocks with higher index $r > C - \gamma - 1$. However, one resource element generally carries a much smaller number of bits than the maximum difference in codeblock sizes of 64 bits. For example, 16-QAM with $Q_m = 4$ using two MIMO layers ($N_L = 2$) carries 8 bits per resource element. Hence using one fewer resource element for smaller codeblocks does not compensate for the coding difference. Moreover, the number of codeblocks using one fewer resource element as given by (11.55) is not the same as the number of codeblocks with smaller codeblock size as these two quantities are determined independently. However, it can be argued that 64 bits difference in codeblock size only applies when multiple codeblocks are used, in which case the codeblock size is generally larger than 3072 bits. Therefore, the worst-case difference in the number of coded bits between different codeblocks can be around 2%, which is rather a small difference to make any significant difference in codeblock performance.

The rate-matching output bit sequence is:

$$e_k = w_{(k_0+j) \bmod N_{cb}} \quad k = 0, 1, \ldots, (E-1) \quad j = 0, 1, \ldots, (K_w - 1). \tag{11.56}$$

Note that the bit positions with $w_{(k_0+j) \bmod N_{cb}} = \langle NULL \rangle$, which denote dummy bits in the circular buffer, a total of $3N_D = (K_w - E)$, are ignored and not included in the transmission. The Redundancy Version (RV) starting point k_0 is given as:

$$k_0 = R_{\text{subblock}}^{\text{TC}} \cdot \left(2 \cdot \left\lceil \frac{N_{cb}}{8 R_{\text{subblock}}^{\text{TC}}} \right\rceil \cdot rv_{idx} + 2 \right) \quad rv_{idx} = 0, 1, 2, 3. \tag{11.57}$$

Where $rv_{idx} = 0, 1, 2, 3$. The operation $(k_0 + j) \bmod N_{cb}$ in (11.56) makes sure that the bit index is reset to the first bit in the buffer when the index reaches the maximum index of N_{cb}, which is the idea of a circular buffer.

Let us consider the last codeblock of size $K = 4800$ bits from Figure 11.11. This codeblock does not contain any filler bits and hence the number of coded bits at the output of the turbo decoder, after adding 4 bits for trellis termination, for each of systematic, parity 1 and parity 2 streams is $D = K + 4 = 4804$. Since 4804 is not divisible by $C_{\text{subblock}}^{\text{TC}} = 32$, 28 dummy bits are added to each of the streams separately before subblock interleaving, resulting in $K_\Pi = 4832$ and $K_w = 3K_\Pi = 14\,496$. Assuming no soft buffer limitation, we can set:

$$N_{cb} = K_w = 3K_\Pi = 14\,496 \text{ bits.} \tag{11.58}$$

We can then obtain redundancy version starting points $k_0 = 302, 3926, 7551, 11\,174$ for $rv_{idx} = 0, 1, 2, 3$ respectively as given in Figure 11.21. We note that spacing between RVs

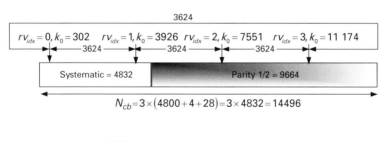

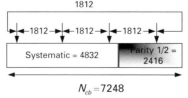

Figure 11.21. Redundancy version starting point k_0.

starting points is uniform at 3624 bits. It should be noted that the RV index only indicates RV starting points. There is no RV ending point and as many bits can be transmitted as determined by rate matching. Let us now assume that $N_{cb} < K_w$ due to soft buffer limitation:

$$N_{cb} = \frac{1}{2}K_w = \frac{3}{2}K_\Pi = 7248 \text{ bits}. \tag{11.59}$$

We again obtain redundancy version starting points $k_0 = 302, 2114, 3926, 5738$ for $rv_{idx} = 0, 1, 2, 3$ respectively as also shown in Figure 11.21. Once again spacing between RVs starting points is uniform at 1812 bits, which is half the spacing for the case of $N_{cb} = 14\,496$ bits.

We make a few observations. Firstly, when RV0 is used, some systematic bits are skipped from the beginning. Actually, two columns' worth of systematic bits from a total of 32 columns worth of bits are skipped as we can see by setting $rv_{idx} = 0$ in (11.57):

$$k_0 = 2 \times R_{\text{subblock}}^{\text{TC}}, \quad rv_{idx} = 0. \tag{11.60}$$

This is equivalent to skipping approximately 6.25% (2/32) systematic bits when RV0 is used. The rationale for skipping some systematic bits from possibly a first hybrid ARQ transmission using RV0 is that this provides a slight performance advantage at very high coding rates. On the other hand, when a lower coding rate is used, skipping some systematic bits is generally not an issue because performance is not sensitive to the presence of systematic bits when a stronger code is used. In the intermediate coding rates, however, systematic bits puncturing can lead to a small degradation in performance. Therefore, the overall benefit of systematic bits puncturing is not evident.

Secondly when $N_{cb} < K_w$ due to soft buffer limitation, all systematic bits are kept in the buffer. However, some parity 1 and 2 bits are dropped. This is because when the soft buffer is limited, the effective coding rate becomes larger than the turbo code base rate of 1/3, resulting in a weaker code, which benefits from transmission of systematic bits. We also note that when N_{cb} becomes smaller relative to K_w, less redundancy is transmitted reducing the coding gains and hence degrading hybrid ARQ performance.

11.8 Circular-buffer rate matching for convolutional code

The circular buffer rate matching scheme for convolutional code is illustrated in Figure 11.22. We note that in this case there are no systematic bits as the outputs of the convolutional code in Figure 11.17 are three parity streams $d_k^{(0)}$, $d_k^{(1)}$ and $d_k^{(2)}$. Similar to turbo code, subblock interleaving is performed separately on each of the output streams. Another difference relative to turbo code is that interleaved parity bits are not interlaced. Also, there are no redundancy version starting points since hybrid ARQ operation is not used for convolutionally coded transport channels.

11.8.1 Subblock interleaving

The encoder output streams $d_k^{(0)}$, $d_k^{(1)}$ and $d_k^{(2)}$ corresponding to the first, second and third parity streams, respectively, as shown in Figure 11.17 are interleaved separately using a subblock interleaver similar to the one used for the turbo coding case. The bits input to the

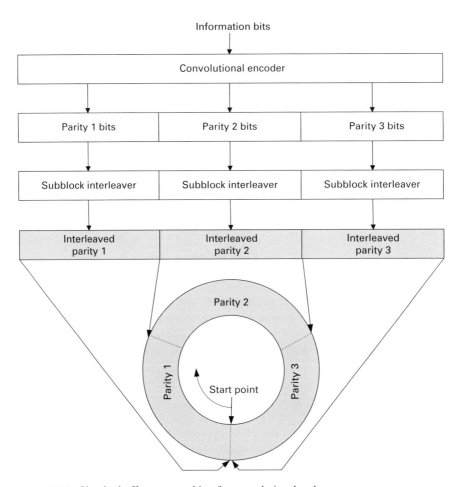

Figure 11.22. Circular buffer rate matching for convolutional code.

block interleaver are denoted by:

$$d_0^{(i)}, d_1^{(i)}, d_2^{(i)}, \ldots, d_{D-1}^{(i)}, \quad i = 0, 1, 2, \tag{11.61}$$

where D is the number of bits in each of the first, second and third parity streams. It should be noted that no tail bits are added for trellis termination in this case thanks to the tail-biting convolutional code.

The subblock interleaving is achieved by writing row-wise in a rectangular matrix, applying matrix columns permutations and finally reading from the matrix column-wise. Similar to turbo coding, the number of columns in the matrix is fixed to 32, that is

$$C_{\text{subblock}}^{\text{CC}} = 32. \tag{11.62}$$

The number of rows of the matrix $R_{\text{subblock}}^{\text{CC}}$ are determined by finding the minimum integer $R_{\text{subblock}}^{\text{CC}}$ such that:

$$R_{\text{subblock}}^{\text{CC}} \times C_{\text{subblock}}^{\text{CC}} \geq D. \tag{11.63}$$

When the number of bits D does not completely fill the $R^{CC}_{subblock} \times C^{CC}_{subblock}$ rectangular matrix, dummy bits are padded to fully fill the matrix as below:

$$
\begin{aligned}
y_k &= \langle \text{NULL} \rangle & k = 0, 1, \ldots, N_D - 1 \\
y_{N_D+k} &= d^{(i)}_k & k = 0, 1, \ldots, D - 1.
\end{aligned}
\tag{11.64}
$$

Note that the maximum number of dummy bits is limited to $\left(C^{CC}_{subblock} - 1\right)$ and these bits are added to the beginning of each of the first, second and third parity streams. Also, note that when $R^{CC}_{subblock} \times C^{CC}_{subblock} = D$, no dummy bits need to be added as the total D bits fully fill the matrix in this case. The input bit sequence is then written into the $\left(R^{CC}_{subblock} \times C^{CC}_{subblock}\right)$ rectangular matrix row by row starting with bit y_0 in column 0 of row 0 as below:

$$
\begin{bmatrix}
y_0 & y_1 & y_2 & \cdots & y_{C^{CC}_{subblock}-1} \\
y_{C^{CC}_{subblock}} & y_{C^{CC}_{subblock}+1} & y_{C^{CC}_{subblock}+2} & \cdots & y_{2C^{CC}_{subblock}-1} \\
\vdots & \vdots & \vdots & \ddots & \vdots \\
y_{(R^{CC}_{subblock}-1)\times C^{CC}_{subblock}} & y_{(R^{CC}_{subblock}-1)\times C^{CC}_{subblock}+1} & y_{(R^{CC}_{subblock}-1)\times C^{CC}_{subblock}+2} & \cdots & y_{(R^{CC}_{subblock}\times C^{CC}_{subblock}-1)}
\end{bmatrix}.
\tag{11.65}
$$

For each of the first, second and third parity streams, inter-column permutation for the matrix is performed based on the pattern given in Table 11.4 and also shown in Figure 11.23, where $P(j)$ is the original column position of the jth permuted column. The difference in permutation pattern relative to the one used for turbo code subblock interleaving in Table 11.3 is that the first set of 16 and the last set of 16 entries are switched. After permutation of the columns, the inter-column permuted $\left(R^{CC}_{subblock} \times C^{CC}_{subblock}\right)$ rectangular matrix is equal to:

$$
\begin{bmatrix}
y_{P(0)} & y_{P(1)} & \cdots & y_{P(C^{CC}_{subblock}-1)} \\
y_{P(0)+C^{CC}_{subblock}} & y_{P(1)+C^{CC}_{subblock}} & \cdots & y_{P(C^{CC}_{subblock}-1)+C^{CC}_{subblock}} \\
\vdots & \vdots & \ddots & \vdots \\
y_{P(0)+(R^{CC}_{subblock}-1)\times C^{CC}_{subblock}} & y_{P(1)+(R^{CC}_{subblock}-1)\times C^{CC}_{subblock}} & \cdots & y_{P(C^{CC}_{subblock}-1)+(R^{CC}_{subblock}-1)\times C^{CC}_{subblock}}
\end{bmatrix}.
\tag{11.66}
$$

The output of the block interleaver is the bit sequence read out column by column from the inter-column permuted matrix in (11.66). The bits after subblock interleaving $v^{(i)}_0, v^{(i)}_1, v^{(i)}_2, \ldots, v^{(i)}_{K_\Pi-1}$ are set as:

$$
\begin{aligned}
v^{(i)}_0 &= y_{P(0)} \\
v^{(i)}_1 &= y_{P(0)+C^{CC}_{subblock}} \\
&\vdots \\
v^{(i)}_{K_\Pi-1} &= y_{P(C^{CC}_{subblock}-1)+(R^{CC}_{subblock}-1)\times C^{CC}_{subblock}}
\end{aligned}
\qquad i = 0, 1,
\tag{11.67}
$$

where $k_\Pi = R^{CC}_{subblock} \times C^{CC}_{subblock} = D$.

Table 11.4. Inter-column permutation pattern for subblock interleaver for convolutional code.

j	$P(j)$	j	$P(j)$	j	$P(j)$	j	$P(j)$
0	1	8	3	16	0	24	2
1	17	9	19	17	16	25	18
2	9	10	11	18	8	26	10
3	25	11	27	19	24	27	26
4	5	12	7	20	4	28	6
5	21	13	23	21	20	29	22
6	13	14	15	22	12	30	14
7	29	15	31	23	28	31	30

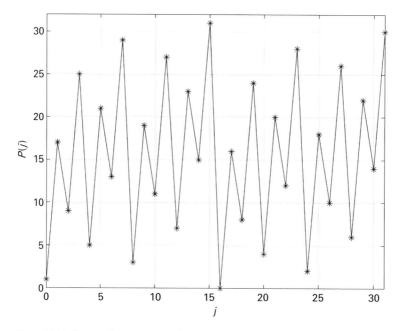

Figure 11.23. Inter-column permutation pattern for subblock interleaver for convolutional code.

11.8.2 Rate matching

The circular buffer length is $K_w = 3K_\Pi$, where K_Π is the number of interleaved bits in each of the first, second and third parity streams. The bit stream in the circular buffer is denoted as $w_0, w_1, \ldots, w_{(K_w-1)}$ and is given as:

$$
\begin{aligned}
w_k &= v_k^{(0)} & k &= 0, 1, \ldots, (K_\Pi - 1) \\
w_{K_\Pi+k} &= v_k^{(1)} & k &= 0, 1, \ldots, (K_\Pi - 1) \\
w_{2K_\Pi+k} &= v_k^{(2)} & k &= 0, 1, \ldots, (K_\Pi - 1).
\end{aligned}
\tag{11.68}
$$

If we compare (11.68) with (11.50), we note that there is no interlacing of interleaved parity bits for the convolutional coding case.

The rate matching output bit sequence is given as:

$$e_k = w_{(j) \bmod N_{cb}} \quad k = 0, 1, \ldots, (E-1) \quad j = 0, 1, \ldots, (K_w - 1). \tag{11.69}$$

Again, if we compare (11.69) with (11.56), we note that the transmission starting position is not staggered for convolutional code and is always at the first bit position in the first parity stream as is also shown in Figure 11.22. Also, no redundancy versions need to be defined as hybrid ARQ is not used in conjunction with convolutional code. Note that the bit positions with $w_{(k_0+j) \bmod N_{cb}} = \langle \text{NULL} \rangle$, which denotes dummy bits in the circular buffer are ignored and not included in the transmission.

11.9   Codeblock concatenation

The codeblock concatenation only needs to be done when the number of codeblocks is larger than one $(C > 1)$ for the turbo coding case. The input bit sequence for the codeblock concatenation and channel interleaving block are the sequences e_{rk}, for $r = 0, \ldots, C-1$ and $k = 0, \ldots, E_r - 1$. The output bit sequence from the codeblock concatenation and channel interleaving block is the sequence:

$$f_j = e_{rk}, \quad r = 0, \ldots, (C-1), \quad k = 0, \ldots, (E_r - 1) \quad j = 0, \ldots, (G-1), \tag{11.70}$$

where $G = (C \times E_r)$.

We note that the codeblock concatenation consists of sequentially concatenating the rate-matching outputs for the different codeblocks.

11.10   Channel interleaver

A frequency-first mapping without any channel interleaver is used on the downlink physical shared channel (PDSCH) to enable pipelining of codeblocks decoding reducing decoding complexity. However, for physical uplink shared channel (PUSCH), a simple interleaver where bits are written to a rectangular matrix row-by-row and read out column-by-column is defined. This channel interleaver in conjunction with the resource element mapping for PUSCH enables a time-first mapping of modulation symbols onto the transmit waveform. A time-first mapping for PUSCH allows capturing frequency diversity when slot-level hopping is applied as shown in Figure 11.24. This is particularly important when multiple codeblocks within a transport block are transmitted. In case a frequency-first mapping is applied, as is the case for downlink, then codeblocks transmitted within the first slot will experience different channel conditions from the codeblocks in the second time slot due to frequency hopping. As a hybrid ARQ NACK is sent if any of the codeblocks within the whole subframe (two slots) is in error, the performance is limited by the slot experiencing lower channel quality. On the other hand, with time-first mapping, all codeblocks within a transport block are expected to experience similar channel conditions. The decoding complexity is less of a concern for the uplink as eNB can generally afford higher complexity relative to a UE. Hence, the decoding pipelining argument that applies to downlink for a UE receiver is not necessarily critical for an eNB receiver in the uplink.

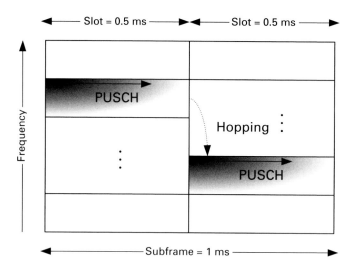

Figure 11.24. A time-first mapping on PUSCH captures frequency diversity when frequency hopping is enabled.

In the LTE uplink, when data and control need to be transmitted in the same subframe, multiplexing is performed at the input of transform (DFT) precoding in order to keep the single-carrier property of SC-FDMA. In addition, the control and data multiplexing is performed in such a way that hybrid ARQ ACK/NACK information is present on both slots in the subframe and is mapped to resources around the demodulation reference signals. As for the case of data transmission, it is important for hybrid ARQ ACK/NACK to gather frequency diversity when slot-level hopping is enabled. In addition, mapping of ACK/NACK to resources around reference signals enables better channel estimates for ACK/NACK reliable decoding. Furthermore, the multiplexing scheme ensures that control and data information are mapped to different modulation symbols.

The inputs to the data and control multiplexing are the coded bits of the control information denoted by $q_0, q_1, q_2, q_3, \ldots, q_{Q-1}$ and the coded bits of the UL-SCH denoted by $f_0, f_1, f_2, f_3, \ldots, f_{G-1}$. The output of the data and control multiplexing operation is denoted by:

$$\underline{g}_0, \underline{g}_1, \underline{g}_2, \underline{g}_3, \ldots, \underline{g}_{H'-1}, \quad H = (G + Q), \quad H' = \frac{H}{Q_m}, \tag{11.71}$$

where $\underline{g}_i, i = 0, \ldots, H' - 1$ are column vectors of length Q_m. Also, $Q_m = 2, 4, 6$ for QPSK, 16-QAM and 64-QAM respectively. $H = (G + Q)$ is the total number of coded bits for transmission including both the control and data bits.

The number of SC-FDMA symbols per subframe $N_{\text{symb}}^{\text{PUSCH}}$ for PUSCH transmission is given by:

$$N_{\text{symb}}^{\text{PUSCH}} = \left(2 \cdot \left(N_{\text{symb}}^{\text{UL}} - 1\right) - N_{\text{SRS}}\right), \tag{11.72}$$

where $N_{\text{symb}}^{\text{UL}}$ is the number of uplink SC-FDMA symbols per slot, $N_{\text{symb}}^{\text{UL}} = 6, 7$ for extended and normal cyclic prefix respectively. For instance, assuming normal cyclic prefix and one

SC-FDMA symbol for sounding reference signal (SRS), $N_{SRS} = 1$, we obtain $N_{symb}^{PUSCH} = 11$. It should be noted that one SC-FDMA symbol in each slot is reserved for demodulation reference signal and cannot be used for data transmission.

We first assign $C_{mux} = N_{symb}^{PUSCH}$ to be the number of columns of the matrix. The number of rows of the matrix is then obtained as:

$$R_{mux} = \frac{H}{C_{mux}}. \tag{11.73}$$

Also, we define $R'_{mux} = R_{mux}/Q_m$, where $Q_m = 2, 4, 6$ for QPSK, 16-QAM and 64-QAM respectively. The input vector sequence, i.e., $\underline{y}_k = \underline{g}_k$ for $k = 0, 1, \ldots, (H' - 1)$, is written into the $(R_{mux} \times C_{mux})$ matrix by sets of Q_m rows starting with the vector $\underline{y}_0$ in column 0 and row 0 to row $(Q_m - 1)$:

$$\begin{bmatrix} \underline{y}_0 & \underline{y}_1 & \underline{y}_2 & \cdots & \underline{y}_{C_{mux}-1} \\ \underline{y}_{C_{mux}} & \underline{y}_{C_{mux}+1} & \underline{y}_{C_{mux}+2} & \cdots & \underline{y}_{2C_{mux}-1} \\ \vdots & \vdots & \vdots & \ddots & \vdots \\ \underline{y}_{(R'_{mux}-1)\times C_{mux}} & \underline{y}_{(R'_{mux}-1)\times C_{mux}+1} & \underline{y}_{(R'_{mux}-1)\times C_{mux}+2} & \cdots & \underline{y}_{(R'_{mux}\times C_{mux}-1)} \end{bmatrix}. \tag{11.74}$$

When uplink ACK/NACK is transmitted, some modulation symbols are punctured. Writing bits into the matrix as sets of Q_m rows allows straightforward puncturing of modulation symbols when ACK/NACK needs to be transmitted in the same subframe as the data transmission. The output of the block interleaver is the bit sequence read out column by column from the $(R_{mux} \times C_{mux})$ matrix in (11.74).

11.11 Summary

Two major proposals were debated as candidate schemes for the LTE channel coding. The first scheme was based on one retaining the release 6 HSPA turbo code while replacing the turbo internal interleaver with a contention-free interleaver which offers parallel processing. The second scheme was to use a different coding scheme such as an LDPC code that has high parallelism built in. The decision went in favor of turbo code partly because it is the incumbent coding scheme in 3GPP. The argument was that if both turbo code and LDPC can offer similar performance and complexity then it is preferable to use the existing coding scheme as dual-mode terminals implementing HSPA and LTE anyway need to include turbo code scheme and some hardware can be reused between the two systems.

The contention-free interleaver is based on the quadratic permutation polynomial (QPP) principle. The QPP interleaver was selected over the almost regular permutation (ARP) interleaver. The QPP interleaver allows calculating the next interleaved position from the current position recursively without multiplication or modulo operations. Additionally, for many QPP interleavers, the de-interleaver is also a QPP interleaver. One advantage of an interleaver being its own inverse is that the same algorithm and hardware can be used for both interleaving and de-interleaving.

The maximum codeblock size (amount of information data input to the turbo code) is limited to 6144 bits for complexity reasons. Therefore, transport blocks larger than 6144 bits are segmented into codeblocks with each codeblock having its own CRC. The codeblock CRC allows early termination of turbo decoding iterations resulting in battery power savings. Moreover, for MIMO multi-codeword transmissions, codeblock CRC along with frequency-first mapping of codeblocks allows performing successive interference cancellation (SIC) at the codeblock level. This reduces both the SIC delay as well as buffering required. In addition to the codeblock CRC, a transport block level CRC is always used. Both codeblock and transport block CRCs are of the same length but use different generator polynomials to avoid the problem of missing the same error event.

A highly flexible yet simple rate-matching scheme based on the circular buffer concept is used to enable adaptive hybrid ARQ, which may require a different number of coded bits transmitted between different hybrid ARQ retransmissions. A subblock interleaver is used to separately interleave systematically two sets of parity bits. Furthermore, the parity bits from the first and second constituent encoder are interlaced, which ensures that equal numbers of parity bits from the first and second constituent encoders are transmitted. Only the redundancy version starting points need to be signaled and the total number of coded bits that can be transmitted on one attempt is not limited as the circular buffer can go around and around as many times as needed. The redundancy version starting points also allow for puncturing of a small fraction of systematic bits, which is claimed to provide a slight performance advantage at very high coding rates.

A soft buffer limitation for the downlink transmissions can be enabled due to buffering concerns for the UE. In the uplink, there is no soft buffer limitation for the eNB and hence incremental redundancy can always be used. The idea with a soft buffer limitation is that if UE has a certain buffer size dimensioned for a given data rate and a given coding rate then it can support either higher data rates with increasing coding rate (weaker code) or lower data rates with a stronger code. Therefore, the soft buffer approach provides a tool to balance the UE soft buffering complexity against incremental redundancy gains.

The rate-matching scheme used for convolutional coding is also based on the circular buffer concept with some differences relative to the scheme used for turbo code. Firstly, there are no systematic bits for the convolutional coding case. Secondly, the parity bits are not interlaced and no redundancy versions starting points need to be specified as hybrid ARQ is not used in conjunction with convolutional coding. A single starting point at the first bit of the first set of parity bits is always used.

A limited number of QPP interleaver sizes are defined for complexity reasons. In order to match the codeblock size to the supported turbo code internal interleaver (QPP) size, filler bits can be added to the codeblock. Similarly, dummy bits can be added to match the turbo code output subblock size to the supported subblock diagonal matrix size. All dummy bits are removed before the coded bits are transmitted. The filler bits from the systematic bits stream as well as the parity bit stream from the first constituent encoder are removed. However, the filler bits that go into parity bits from the second constituent encoder are not removed. It is difficult to track these bits due to the fact that the input to the second constituent encoder is an interleaved version of the input stream to turbo code.

No channel interleaving is defined for downlink transmissions as the subblock interleaving partly achieves the goal of channel interleaving. In addition, frequency-first mapping of coded bits is employed for decoding complexity reasons and hence multiple codeblocks cannot be interleaved in time anyway. The uplink transmissions can, however, employ slot-level frequency hopping and hence a channel interleaver along with resource mapping scheme ensures that the transmitted bits achieve time-first mapping. Using this scheme all the codeblocks within a transport block are transmitted over both the slots. Hence, frequency diversity can be captured when slot-level frequency hopping is enabled. Moreover, the receiver complexity is less of a concern for eNB to receive uplink transmissions and therefore frequency-first mapping is not crucial to enable codeblocks decoding pipelining in the uplink. This enables using a channel interleaver with time-first mapping in the uplink.

References

[1] IEEE Standard 802.16-2005, Part 16: Air Interface for Fixed and Mobile Broadband Wireless Access Systems.

[2] Gallager, R. G., *Low-Density Parity-Check Codes*, Cambridge, MA: M. I. T. Press, 1963. (Also, Gallager, R. G., "Low density parity-check codes," *IRE Transactions on Information Theory*, IT-8, pp. 21–28, Jan. 1962.)

[3] Tanner, R. M., "A recursive approach to low complexity codes," *IEEE Transactions on Information Theory*, vol. IT-27, pp. 533–547, Sept. 1981.

[4] MacKay, D. and Neal, R., "Good codes based on very sparse matrices", *Cryptography and Coding, 5th IMA Conference*, Ed. C. Boyd, *Lecture Notes in Computer Science*, Berlin: Springer-Verlag, 1995.

[5] Alon, N. and Luby, M., "A linear time erasure-resilient code with nearly optimal recovery," *IEEE Transactions on Information Theory*, vol. IT-42, pp. 1732–1736, Nov. 1996.

[6] Byers, J., Luby, M., Mitzenmacher, M. and Rege, A., "A digital fountain approach to reliable distribution of bulk data," *Proceedings of ACM SIGCOMM '98*, Vancouver, BC, Canada, Jan. 1998, pp. 56–67.

[7] MacKay, D. J. C., "Good error-correcting codes based on very sparse matrices," *IEEE Transactions on Information Theory*, vol. IT-45, no. 2, pp. 399–431, 1999.

[8] MacKay, D. J. C., *Information Theory, Inference and Learning Algorithms*, Cambridge: Cambridge University Press, 2003.

[9] Wicker, S. B., *Error Control Systems for Digital Communication and Storage*, New Jersey: Prentice Hall, 1995.

[10] Shao, R. Y., Fossorier, M. and Lin, S., "Two simple stopping criteria for iterative decoding," *Proceedings of the IEEE International Symposium on Information Theory '98*, p. 279, Aug. 1998.

[11] Shibutani, A., Suda, H. and Adachi, F., "Complexity reduction of turbo decoding," *Proceedings of IEEE Vehicle Technology Conference '99 Fall*, Oct. 1999.

[12] Berrou, C., Glavieux, A. and Thitimajshima, P., "Near Shannon limit error-correcting coding and decoding: turbo-codes(1)," *Proceedings of IEEE International Conference on Communications, ICC'93*, pp. 1064–1070, May 1993.

[13] Battail, G., Berrou, C. and Glavieux, A., "Psuedo-random recursive convolutional coding for near-capacity performance," *IEEE Global Communication Conference, Globecom 1993*, pp. 23–27, Dec. 1993.

[14] Divsalar, D. and Pollara, F., "Turbo codes for PCS applications," *Proceedings of ICC 1995*, Seattle, WA, pp. 54–59, June 1995.

[15] Blankenship, T. K., Classon, B. and Desai, V., "High-throughput turbo decoding techniques for 4G," *Proceedings of International Conference on 3G and Beyond*, 2002, pp. 137–142.

[16] Berrou, C., Saouter, Y., Douillard, C., Kerouedan, S. and Jezequel, M., "Designing good permutations for turbo codes: towards a single model," *Proceeings of ICC 2004*, vol. 1, pp. 341–345, June 2004.

[17] Thul, M. J., Gilbert, F. and Wehn, N., "Optimized concurrent interleaving architecture for high-throughput turbo-decoding," *Proceedings of International Conference on Electronics Circuits and Systems*, Dubrovnik, Croatia, Sept. 2002, pp. 1099–1102.

[18] Sun, J. and Takeshita, O. Y., "Interleavers for turbo codes using permutation polynomials over integer rings," *IEEE Transactions on Information Theory*, vol. IT-51, no. 1, pp. 101–119, Jan. 2005.

[19] Takeshita, O. Y., "On maximum contention-free interleavers and permutation polynomials over integer rings," *IEEE Transactions on Information Theory*, vol. IT-52, no. 3, pp. 1249–1253, Mar. 2006.

[20] Cheng, M. K., Moision, B. E., Hamkins, H. and Nakashima, M., "An interleaver implementation for the serially concatenated pulse-position modulation decoder," *Proceedings of IEEE ISCAS '06*, pp. 4244–4247, May 2006.

[21] Ryu, J. and Takeshita, O. Y., "On quadratic inverses for quadratic permutation polynomials over integer rings," *IEEE Transactions on Information Theory*, vol. IT-52, no. 3, pp. 1254–1260, Mar. 2006.

[22] Ma, H. H. and Wolf, J. K., "On tail biting convolutional codes," *IEEE Transactions on Communications,* vol. com-34, no. 2, pp. 104–111, 1986.

12 Scheduling, link adaptation and hybrid ARQ

A cellular radio system consists of a collection of fixed eNBs that define the radio coverage areas or cells. Typically, a non-line-of-sight (NLOS) radio propagation path exists between an eNB and a UE due to natural and man-made objects that are situated between the eNB and the UE. As a consequence, the radio waves propagate via reflections, diffractions and scattering. The arriving waves at the UE in the downlink direction (at the eNB in the uplink direction) experience constructive and destructive additions because of different phases of the individual waves. This is due to the fact that, at the high carrier frequencies typically used in the cellular wireless communication, small changes in the differential propagation delays introduce large changes in the phases of the individual waves. If the UE is moving or there are changes in the scattering environment, then the spatial variations in the amplitude and phase of the composite received signal will manifest themselves as the time variations known as Rayleigh fading or fast fading. Traditionally, the time-varying nature of the wireless channel was considered undesirable because it required very high signal-to-noise ratio (SNR) margins for providing the desired bit error or packet error reliability. Therefore, system design efforts focused on averaging out the signal variations due to fast fading by using various forms of diversity schemes such as space, angle, polarization, field, frequency, time or multi-path diversity.

At low to medium UE speeds targeted for the LTE system, the channel variations can be tracked at the eNB and fast channel variations due to Rayleigh fading can be used to our advantage. For downlink transmissions, UEs feed downlink channel quality information back to the eNB. Using a channel quality sensitive scheduler such as the proportional fair scheduler, eNB can serve a UE on time, frequency and spatial resources where it is experiencing the best conditions thus improving system capacity and throughput. It is well known that when channel-sensitive scheduling, also referred to as multi-user diversity, can be exploited, use of other forms of diversity such as transmit diversity degrades performance. This is because multi-user diversity relies on large variations in channel conditions, while the transmit diversity tries to average out the channel variations.

While the channel-sensitive scheduler selects a user experiencing good channel conditions on a given time, frequency and spatial resources, link adaptation is used to adapt the transmissions format including modulation and coding (MCS) as well as MIMO rank and precoding to the current channel conditions on the allocated time, frequency and spatial resources. Both channel sensitive scheduling and link adaptation rely on availability of accurate channel quality information at the eNB. This information can be obtained at the eNB either from uplink reference signal transmissions in a TDD system or UEs can feedback downlink channel quality information back to the eNB in an FDD system. In a practical situation, accurate channel quality knowledge is not feasible for various reasons that include channel quality measurement errors, errors due to channel quality feedback delays, interference variations and channel quality information transmission errors. These errors in channel quality

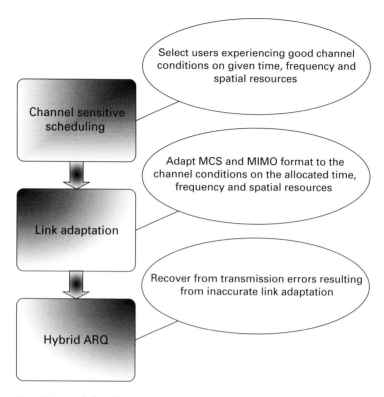

Figure 12.1. Relationship between scheduling, link adaptation and hybrid ARQ.

estimates would require large back offs in transmission data rates degrading system through-put. A hybrid automatic repeat-request (ARQ) mechanism allows one to correct errors in link adaptation by incremental redundancy transmission in the case of packet errors. This way, transmission data rates do not need to be reduced to account for channel quality estimation and prediction errors. The relationship between between scheduling, link adaptation and hybrid ARQ is schematically shown in Figure 12.1. Channel-sensitive scheduling, link adaptation and hybrid ARQ are integral parts of all modern cellular communication systems due to the potential benefit they provide in fast-fading environments. In this chapter, we describe each of these three features in detail.

12.1   Channel-sensitive scheduling

The concept of channel-sensitive scheduling, also referred to as multi-user diversity, [1] is to exploit the independent nature of fast fading across users. When there are many users that fade independently, it is highly likely to find a user with good channel conditions at a given time. A 100 ms snapshot of Rayleigh fading for four users each experiencing an average SNR of 0.0 dB is shown in Figure 12.2. We assumed a 10 km/h speed at 2 GHz frequency. It can be noted that the average SNR seen by the users can be increased if users are scheduled during the times when they experience upfades.

The sum capacity for channel-sensitive scheduling as a function of number of users for SNR of 0, 10 and −10 dB are shown in Figures 12.3–12.5 respectively. The capacities for

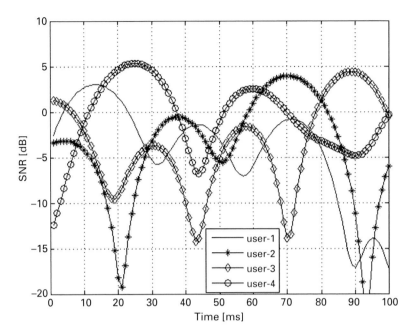

Figure 12.2. A snapshot of Rayleigh fading for four users.

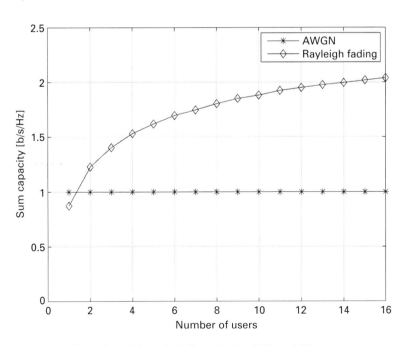

Figure 12.3. Channel-sensitive scheduling gains for SNR $= 0$ dB.

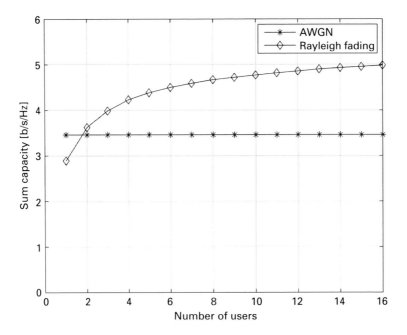

Figure 12.4. Channel-sensitive scheduling gains for SNR $= 10$ dB.

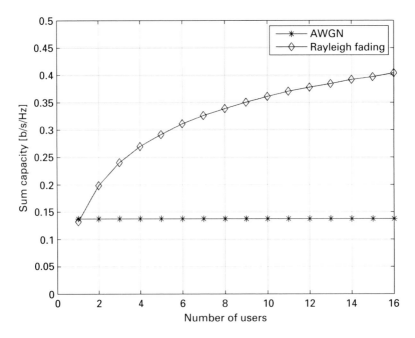

Figure 12.5. Channel-sensitive scheduling gains for SNR $= -10$ dB.

the AWGN channel are provided for reference. We note that the channel-sensitive scheduling gains increase with the number of users. This is because with increasing number of users, the selected user's channel becomes stronger and stronger. We remark that for an SNR of 0.0 dB and for 16 users, channel-sensitive scheduling can provide up to two times larger sum capacity relative to the AWGN channel. The gains from channel-sensitive scheduling also depend upon the operating SNR. At higher SNR, capacity scales logarithmically with SNR and incremental gains in SNR due to scheduling translate into relatively lower capacity gains. At SNR $= 10$ dB and for the 16 users case, the channel-sensitive scheduling provides 1.4 times the sum capacity for the AWGN channel as shown in Figure 12.4. As capacity scales almost linearly with SNR at low SNR, the scheduling gains are expected to be larger at these SNRs. This is because the SNR gain provided by scheduling translates into a larger capacity gain at lower SNR relative to the higher SNR case. At SNR $= -10$ dB and 16 users case, almost three times the gain is achieved as shown in Figure 12.5.

We can conclude that channel-sensitive scheduling is more beneficial for the weak users in the system experiencing low SNR. On the other hand, the channel quality feedback overhead also tends to be larger for weak users. This is because weak users need to transmit at higher power relative to a strong user for conveying the same amount of channel quality information to the eNB. Also, the SINRs for cell-edge users in the uplink are generally lower than the downlink SINR. This is because, in the downlink, the interference at the cell edge is experienced from a few base stations typically as far away from the user as its serving cell. However, in the uplink, many users in the neighboring cells can be transmitting to their own cells creating a large amount of interference to a cell-edge user in the cell of interest. Additionally, in coverage-limited situations the uplink SNR is generally lower than the downlink SNR due to lower UE transmission power relative to eNB transmission power. It is therefore expected that at system level, scheduling can provide larger gains on the uplink than on the downlink.

Until now, we have talked of scheduling gains in the context of a single-path frequency-flat Rayleigh fading channel. In WCDMA based HSPA (High Speed Packet Access) systems, multi-paths lead to channel averaging when a Rake receiver is used. This is because the total received SINR in a Rake receiver is simply the sum of SINRs on each finger (see Section 3.2.1). This channel averaging or channel hardening effect reduces the achievable scheduling gains. Let us now try to explain this effect.

The PDF of a Rayleigh distributed random variable is given as:

$$f_R(r) = \frac{r}{\sigma^2} e^{\left(-\frac{r^2}{2\sigma^2}\right)} \quad x \geq 0. \tag{12.1}$$

The Rayleigh distributed random variable R can be written as:

$$R = \sqrt{X_1^2 + X_2^2}, \tag{12.2}$$

where X_1 and X_2 are zero-mean statistically independent Gaussian random variables, each having a variance of σ^2. The Rayleigh distribution is closely related to the chi-square distribution. A random variable Y following chi-square distribution with n degrees of freedom is defined as:

$$Y = \sum_{i=1}^{n} X_i^2, \tag{12.3}$$

where $X_i, i = 1, 2, \ldots, n$ are zero-mean statistically independent Gaussian random variables each having a variance of σ^2. We note that the Rayleigh distributed random variable R is related to the 2-degrees-of-freedom chi-square random variable Y as below:

$$R = \sqrt{X_1^2 + X_2^2} = \sqrt{Y}. \qquad (12.4)$$

This means that signal power for a single-path fading channel follows a chi-square distribution with 2 degrees of freedom.

$$R^2 = X_1^2 + X_2^2 = Y. \qquad (12.5)$$

The PDF of a chi-square distributed random variable with n degrees of freedom is given as:

$$f_Y(y) = \frac{1}{\sigma^n 2^{n/2} \Gamma\left(\frac{n}{2}\right)} y^{(n/2-1)} e^{-y/2\sigma^2} \quad x \geq 0, \qquad (12.6)$$

where $\Gamma(p)$ is the gamma function, defined as:

$$\Gamma(p) = \int_0^\infty t^{(p-1)} e^{-t}\, dt \qquad p > 0$$

$$\Gamma(p) = (p-1)! \qquad p \text{ an integer}, p > 0 \qquad (12.7)$$

$$\Gamma\left(\frac{1}{2}\right) = \sqrt{\pi} \qquad\qquad \Gamma\left(\frac{3}{2}\right) = \frac{\sqrt{\pi}}{2}.$$

The PDF of a chi-square distributed random variable with various degrees of freedom is plotted in Figure 12.6. The mean and variance of the chi-square distribution are given as:

$$E(Y) = n\sigma^2$$

$$E\left(Y^2\right) - [E(Y)]^2 = 2n\sigma^4. \qquad (12.8)$$

In probability theory and statistics, the coefficient of variation (CV) is a normalized measure of dispersion of a probability distribution. The CV defined as the ratio of the standard deviation to the mean for the chi-square distribution is given as:

$$CV = \frac{\sqrt{2n\sigma^4}}{n\sigma^2} = \sqrt{\frac{2}{n}}. \qquad (12.9)$$

The coefficient of variation is important because the standard deviation of data must always be understood in the context of the mean of the data. The standard deviation of an exponential distribution is equal to its mean, so its coefficient of variation is equal to 1. The distributions with $CV < 1$ are considered low-variance, while those with $CV > 1$ are considered high-variance. We note that CV for the chi-square distribution with 2 degrees of freedom is 1. This is because the chi-square distribution with 2 degrees of freedom is actually an exponential distribution with PDF given as:

$$f_Y(y) = \frac{1}{2\sigma^2} e^{-y/2\sigma^2} \quad x \geq 0. \qquad (12.10)$$

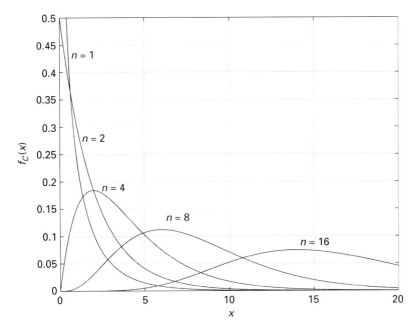

Figure 12.6. The PDF of a chi-square distributed random variable.

With increasing degrees of freedom n, the CV of the chi-square distribution decreases. This means that the distribution becomes less and less of a variant distribution.

Let us now try to answer the question of why scheduling gains in an HSPA system decrease in a multi-path fading channel. When multi-path signals are combined using a Rake receiver, we are actually adding signals on different paths. Assuming each multi-path component is statistically identical and fades independently, the overall received signal power actually follows a chi-square distribution with n degrees of freedom, where $n = 2 \times L$ with L being the number of paths or diversity order. The channel-sensitive scheduling gains depend upon the tail of the fading distribution. An increasing number of paths results in a distribution with lower CV and lighter tail and hence reduces achievable scheduling gains. It should be noted that multi-path diversity due to the Rake receiver could improve the outage performance when no channel state information is available at the transmitter. However, when the channel state information such as channel quality is available at the scheduler, the overall performance degrades with diversity. The sum capacity results for SNR $= 0$ dB and different orders of diversity, $L = 1, 2, 4, 8$ are shown in Figure 12.7. It should be noted that we are ignoring the SNR loss due to a Rake receiver in our comparison. This loss comes from the fact that a signal on a Rake finger experiences interference from the remaining Rake fingers as described in Section 3.2.1. The goal is to show the effect of diversity on channel-sensitive scheduling performance. We note that for the 16 users case, the scheduling gain reduces from two times the AWGN sum capacity for $L = 1$ to 1.8 times the AWGN capacity for $L = 2$. The scheduling gain is only 1.4 times the AWGN capacity for $L = 8$ at 16 users. It should be noted that for simplicity we assumed that each multi-path component is statistically identical with similar mean power. In practice, different multi-path components may have different average powers and thus can impact scheduling performance differently from that depicted in Figure 12.7. However, the general trend of decreasing scheduling gains with increasing diversity order should hold.

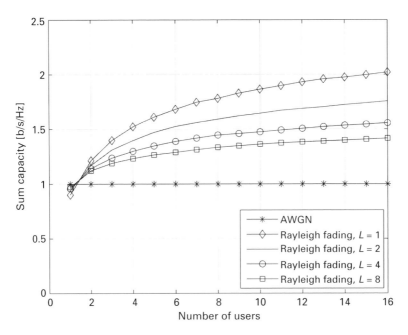

Figure 12.7. Channel-sensitive scheduling gains for SNR $= 0$ dB and different numbers of multi-paths, $L = 1, 2, 4, 8$.

12.2   Frequency-selective multi-user scheduling

We noted that in the HSPA system using a Rake receiver, multi-paths lead to channel harden-ing reducing scheduling gains. In OFDM systems, multi-paths lead to frequency selectivity introducing channel diversity. The larger the number of independent paths, the larger the diversity introduced. Therefore, if time-domain scheduling with selected users transmitting over the whole bandwidth is performed, the scheduling performance should be degraded with increasing number of multi-paths. A snapshot of time-frequency fading over 500 ms for aver-age SNR $= 0$ dB is depicted in Figure 12.8. We assumed a UE speed of 3 km/h, FFT size of 1024 (number of subcarriers) and four multi-paths with delays of 2, 4, 6 and 8 samples. It can be noted that if scheduling can be performed in both frequency and time, users can be scheduled on time-frequency resources where they experience upfades. We refer to channel-sensitive scheduling across time and frequency as frequency-selective scheduling (FSS). Also, the scheme where a user is scheduled over the whole bandwidth with time-domain scheduling only is referred to as non-frequency-selective scheduling (NFSS).

Figure 12.9 shows FSS and NFSS gains for SNR $= 0$ dB and the number of multi-paths equal to 2 and 4 ($L = 2, 4$). We note that the performance of the NFSS scheme degrades as the number of multi-paths is increased. This is because with increasing number of multi-paths, there is more frequency diversity, which reduces the average SNR (averaged over frequency) variations in time. We observe that for the same diversity order, scheduling gain from NFSS relative to AWGN in Figure 12.9 is smaller than the gain in Figure 12.7. This difference comes from the fact that the NFSS scheme suffers from additional channel capacity loss due to different SNR across the frequency band (see Section 3.3). On the other hand, the single-carrier WCDMA/HSPA system considered in Figure 12.7 assumes a single SNR at

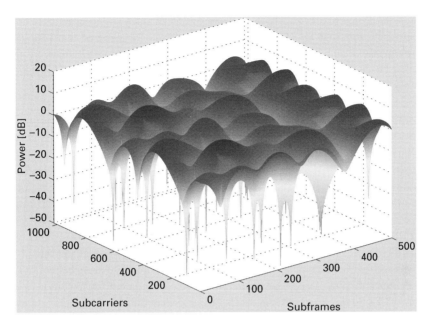

Figure 12.8. Frequency-selective fading.

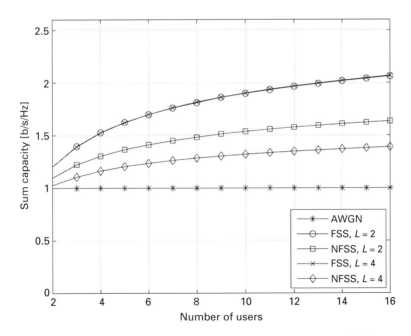

Figure 12.9. Frequency-selective scheduling gains for SNR $= 0$ dB and different numbers of multi-paths, $L = 2, 4$.

a given time. However, in a multi-path fading channel, the WCDMA/HSPA system experiences SNR loss that is not accounted for in Figure 12.7. We also note from Figure 12.9 that the performance of FSS is unaffected with increased frequency diversity. This is because FSS can select users hitting peaks in frequency and time and avoid extreme downfades in both frequency and time. The FSS gains over NFSS for the 16 users case are approximately 28% and 48% for $L = 2, 4$ respectively. It should be noted that FSS gains come at the expense of an additional control signaling overhead that includes both the frequency-selective channel quality information feedback as well as frequency resource allocation control information.

The larger number of multi-paths provides increased frequency selectivity and results in larger fading peaks. The reason for larger fading peaks is that frequency responses of different multi-path components add constructively at some frequency subcarriers. However, the downside is that the peaks also become narrower and narrower with increasing frequency diversity. We make the assumption that larger numbers of multi-paths also lead to a larger delay spread. The sum capacity for FSS for $L = 1, 2, 4$ is shown in Figure 12.10. The case for single-path $L = 1$ represents a frequency-flat fading channel. We note that the performance of FSS is only marginally better than the equivalent flat-fading channel case. We also remark that FSS gains increase slightly with increasing number of multi-paths.

Another benefit of FSS relative to NFSS is lower transmission delays when a user does not necessarily need transmission over the whole bandwidth. This may be the case for small packets transmission. A snapshot of frequency-selective fading for four users is shown in Figure 12.11. In the case of NFSS, user-1 experiencing the overall better channel over the whole bandwidth will be selected for transmission. This means that the remaining users will have to wait for an opportunity in the future time slots. In the case of slow-moving UEs, the channel changes slowly in time resulting in larger durations between channel peaks. This would lead to larger average transmission delays. However, in the case of FSS, a user can be scheduled over part of the frequency band where it experiences an upfade even though its average SNR over the whole bandwidth is not the largest. For example, user-4 experiences an

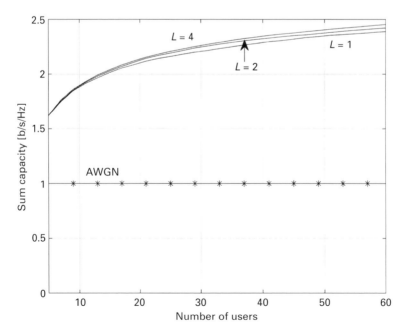

Figure 12.10. Frequency-selective scheduling gains for SNR $= 0$dB and different numbers of multi-paths, $L = 1, 2, 4$.

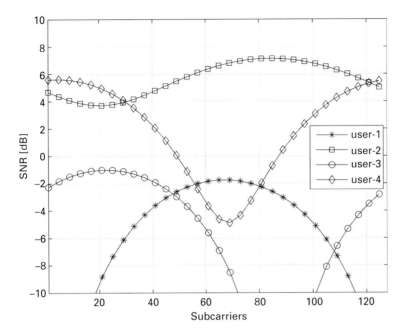

Figure 12.11. A snapshot of frequency-selective fading for four users.

average SNR (calculated over the whole frequency band) of 2.97 relative to 5.74 dB for user-1. However, when SNR is calculated over subcarriers 1–30, user-4 experiences overall better channel quality with an average SNR of 5.15 relative to 3.99 dB for user-1. This way user-4 can be scheduled over subcarriers 1–30 while user-1 is scheduled over subcarriers 31–128. The FSS scheme thus allows simultaneous transmission to multiple users in frequency multiplexing mode without degrading scheduling performance. If multiple frequency-multiplexed users were scheduled simultaneously in the NFSS scheme, scheduling performance would degrade as the users are not scheduled at their peak locations in the frequency domain.

12.3 Proportional fair scheduling

Let us first consider the time-domain scheduling such as is the case for the NFSS scheme. By using proportional fair scheduling (PFS), eNB transmits to the user m^* in the nth subframe [2]:

$$m^*(n) = \arg \max_{m=1,2,\ldots,M} \frac{R_m(n)}{T_m(n)}, \tag{12.11}$$

where $R_m(n), m = 1, 2, \ldots, M$ is the data rate for the mth user in the nth subframe. Also $T_m(n)$ is the average throughput for the mth user in a past window and is updated at each subframe according to:

$$T_m(n+1) = \begin{cases} \left(1 - \frac{1}{t_c}\right) T_m(n) + \left(\frac{1}{t_c}\right) R_m(n) & m = m^*(n) \\ \\ \left(1 - \frac{1}{t_c}\right) T_m(n) & m \neq m^*(n), \end{cases} \tag{12.12}$$

where t_c is window length that can be adjusted to maintain fairness over a predetermined time-horizon. We note that the PFS algorithm schedules a user when its channel quality is better than its average channel quality condition over the time scale t_c. A smaller value for t_c maintains fairness over short time periods, which may be the case for delay-sensitive services. For larger t_c, throughput is averaged over longer periods, which means that the scheduler can afford to wait longer before scheduling a user at its peak. We plot $1/T(n)$ as a function of time for $t_c = 50$, 100 and 200 subframes in Figure 12.12. We note that for smaller t_c, $1/T(n)$ increases faster as function of time improving the user's priority for scheduling. As the scheduler has little time to wait for peaks, a smaller value for t_c will make the user scheduled at relatively lower peaks reducing the scheduling gains. A larger t_c will allow the scheduler to wait for really high peaks and therefore results in improved system throughput at the expense of increased latency. The value of t_c can therefore be selected to strike a balance between latency and throughput.

For very large t_c (approaching ∞), the PFS algorithm maximizes [2]:

$$\sum_{m=1}^{M} \log{(T_m)}, \tag{12.13}$$

where T_m is the long-term average throughput for user m. Also, $\log{(T_m)}$ can be interpreted as the level of satisfaction or utility for user m. We can therefore define the PF algorithm in terms of the system utility function:

$$U(n) = \sum_{m=1}^{M} \log{[T_m(n)]}. \tag{12.14}$$

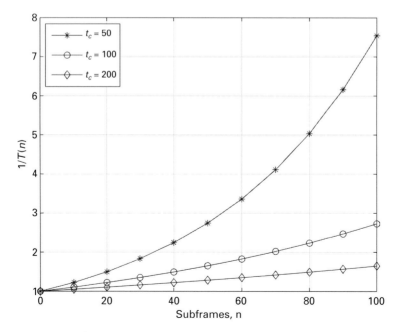

Figure 12.12. $\frac{1}{T(n)}$ as a function of time for $t_c = 50$, 100 and 200 subframes.

In this case, eNB transmits to the user m^* in the nth subframe:

$$m^*(n) = \arg \max_{m=1,2,\ldots,M} U(n+1\,|m),$$
(12.15)

where

$$U(n+1\,|m) = \sum_{m=1}^{M} \log[T_m(n+1\,|m)],$$
(12.16)

where $T_m(n+1|m)$ denotes $T_m(n+1)$ given that user m is scheduled is subframe n. Therefore, the PF algorithm schedules a user in subframe n that gives the largest instantaneous reward in the system utility function $U(n)$.

Let us write the scheduling expression in a more general form as:

$$m^*(n) = \arg \max_{m=1,2,\ldots,M} \frac{[R_m(n)]^\alpha}{[T_m(n)]^\beta}.$$
(12.17)

For $(\alpha = 0, \beta = 1)$ this algorithm reduces to a round-robin scheduling serving all users equally with no regard to users' channel quality or data rate R. On the other hand for $(\alpha = 1, \beta = 0)$ the algorithm schedules a user with the best channel conditions ignoring the user throughputs. This may be fine if all the users have the same average channel quality. But in a cellular system, the difference between average SINR of good and weak users can be as large as 20 dB. This would mean that peaks of the weak users may never overcome the channel quality experienced by good users and hence weak users may never be scheduled. While this scheme will maximize system throughput, the major drawback is the unfairness with weak users experiencing very low throughputs or no transmission opportunities at all. Obviously, a PF scheduler with $(\alpha = 1, \beta = 1)$ strikes a balance between fairness and throughput in the system. Other values of α and β can be used to achieve a desired fairness and system throughput tradeoff.

We have discussed the PFS algorithm in the context of time-domain scheduling. This algorithm can easily be extended to take into account frequency-selective scheduling for the OFDMA and SC-FDMA systems. Furthermore, spatial domain channel variations can be taken into account for time-frequency and spatial scheduling in MU-MIMO or SDMA schemes.

12.4 Link adaptation

We noted that the role of PF scheduling is to perform transmission to a user when its instantaneous channel conditions are better than its average channel condition. The scheduler therefore picks the time, frequency and spatial resources to be used for a user's transmission. The objective of link adaptation is to match the transmission parameters such as modulation and coding (MCS) scheme as well as MIMO transmission rank and precoding to the channel conditions on resources allocated by the scheduler.

12.4.1 Frequency-dependent modulation adaptation

A frequency-dependent modulation adaptation scheme can be employed in OFDMA systems. In localized transmission, channel quality can vary significantly over the scheduled bandwidth

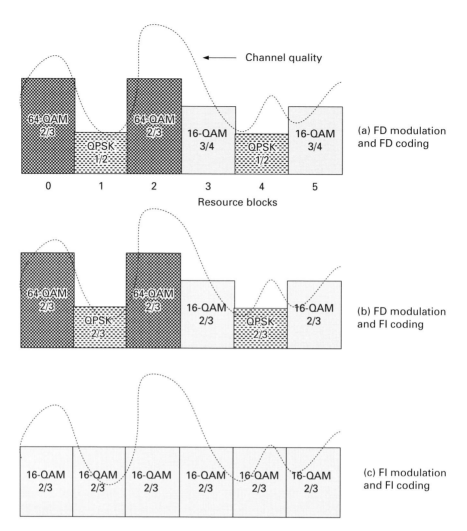

Figure 12.13. Link adaptation schemes with various degrees of frequency-domain adaptation.

when a user is allocated a bandwidth larger than the channel coherence bandwidth. In case of distributed transmissions, the transmission is generally spread over a larger bandwidth leading to channel quality variations over the allocated resources.

We can consider three types of link adaptation scheme with various degrees of frequency-domain adaptation as shown in Figure 12.13. In the first scheme referred to as frequency-dependent modulation and frequency-dependent (FD) coding, both modulation and coding can be adapted in the frequency domain. This scheme can be expected to provide the highest performance as both modulation and coding can be closely matched to the observed channel conditions. The drawback of this scheme, however, is increased overhead because information blocks transmitted using different modulation and coding schemes (MCS) need to be separately coded and MCS information needs to be signaled to the receiver.

In the second scheme modulation is frequency dependent while coding is common across the scheduled bandwidth. This scheme has relatively lower overhead because only frequency-dependent modulation information needs to be signaled to the receiver. However, the performance of this scheme is also expected to degrade relative to the first scheme as only modulation can be matched to the observed channel conditions. In the third scheme, a common modulation and coding is used across the scheduled bandwidth. This scheme has the lowest signaling overhead as a single MCS value for the whole bandwidth needs to be signaled to the receiver.

The schemes with modulation adaptation only or both modulation and coding adaptation are expected to provide some performance gain relative to a scheme where both modulation and coding are selected to match the average channel conditions over the scheduled bandwidth. However, these gains are limited for various reasons as described below.

- With frequency-selective scheduling, a user is scheduled on frequency resources where it experiences upfades, which reduces the channel quality variation across the scheduled resources. This is because extreme downfades are avoided in frequency-selective scheduling. The reduced variation across the allocated resources also reduces gains due to frequency-dependent modulation and coding.
- Similarly, when receive diversity is employed, the frequency-domain variations of the combined signal reduces which also reduces gains due to frequency-dependent modulation and coding. It should be noted that receive diversity also reduces frequency-selective scheduling gains as the tail of the SINR distribution becomes lighter.
- Even when a common modulation and coding is used across the allocated resource blocks, the channel coding and interleaving is performed across all of the resource blocks. This reduces the impact of non-accurate modulation and coding on certain resource blocks due to frequency-independent modulation and coding.
- The frequency-dependent modulation and coding schemes rely on accurate frequency-selective channel quality information. The performance of frequency-dependent modulation and coding schemes is more sensitive to channel quality errors due to measurement inaccuracies, feedback delays and transmission errors. Therefore, under imperfect channel quality information, the gains of frequency-dependent modulation and coding schemes would further reduce.

For reasons of increased signaling overhead and little potential for performance gains from frequency dependent modulation and coding, the LTE system uses common modulation and coding across the scheduled resource blocks. However, modulation and coding adaptation is permitted in the time domain across subframes and also in the spatial domain for multi-codeword (MCW) MIMO transmission. In addition, different users scheduled on different resource blocks in the same subframe can use different modulation and coding. This is because channel qualities of different users can be vastly different due not only to fast fading but also to user location and slow fading effects. We would also like to point out that for uplink SC-FDMA, all the modulation symbols experience similar SINR due to FFT-precoding. Therefore, frequency-dependent modulation and coding schemes cannot be envisaged for the SC-FDMA uplink. However, frequency-selective scheduling can still be used in the SC-FDMA uplink because different users still experience different channel qualities on different parts of the bandwidth in a frequency-selective channel.

12.4.2 LA based on channel frequency-selectivity

An important aspect to consider in the design of a good modulation and coding rate set for link adaptation is the modulation switching points i.e. at what coding rate the next higher level (or lower level) of modulation should be used. The answer may be simple if all the modulations symbols within a codeblock transmission experience the same channel gain. This is, for example, the case for uplink SC-FDMA based transmissions in both frequency-flat and frequency-selective channels. This is also the case for OFDM for single-path frequency-flat channels. In this case, the modulation switching points can be optimized based on an AWGN channel because of the static nature of the channel over the time-frequency resources used for the codeblock transmission. With a short subframe duration of 1 ms in the LTE system, the channel is expected to be *quasi-static* in time for low to medium speeds that are of most interest. However, multi-path channels can be highly frequency selective resulting in different channel gains for symbols transmitted on different OFDM subcarriers. This will be particularly true for distributed-mode transmissions where the codeblock symbols are preferably spread over a larger bandwidth to maximize the frequency diversity benefit. Also, as pointed out earlier, the codeblock symbols channel gains or SINR variance are expected to be smaller for frequency-selective multi-user scheduling in localized mode. Also, when a user is scheduled over a contiguous set of resource blocks, the channel gains tend to be correlated according to the channel coherence bandwidth.

We can therefore expect that the codeblock symbol SINR variance can differ between localized and distributed transmission. In general, when the codeblock symbol SINR variance is smaller, it is expected that modulation switching to the next higher modulation level would happen at a relatively higher coding rate. This is due to the fact that the majority of the symbols contribute to successful decoding of the codeblock because all symbols have SINR within a small range. On the other hand, for transmission over a highly frequency-selective channel, modulation switching to the next higher level of modulation should happen at a relatively lower coding rate. This is explained by the fact that with large symbol SINR variance, some of the modulation symbols experience very low SINR and therefore are unable to make a significant contribution to decoding. Therefore, the resulting coding rate experienced by the decoder becomes higher resulting in a penalty in the coding gain. In this case, it may be preferable to use a stronger code while using a higher order modulation. In short, switching to a higher order modulation can be expected to happen at a relatively higher coding rate for localized transmission or for cases with low SINR variations. Similarly, switching happens at a relatively lower coding rate for distributed transmission or for cases with high SINR variations.

Moreover, the symbol SINR variance can be even smaller for distributed transmission for smaller system bandwidths of 1.4 MHz and 3.0 MHz that experience relatively smaller frequency selectivity for a given channel scenario. We know that channel coherence bandwidth is determined by channel delay spread. The channel delay spread is dependent upon the deployment scenario and multi-path environment. For a given channel delay spread, the code block symbol SINR is going to be more correlated for smaller bandwidths than for larger bandwidths. This is because the smaller bandwidths may only cover a few coherence bandwidths while larger bandwidths can span many more coherence bandwidths. In other cases, a single path fading or Rician Channel (LOS) may also occur in some cases in macro-cell deployments resulting in a non-frequency-selective channel. Also, small femtocell indoor deployments exhibit typical path delay of the order of hundreds of nanoseconds (coherence bandwidth of the order of 10 MHz) exhibiting *flat-fading* for bandwidths up to 10 MHz. Therefore, optimal modulation switching points can also vary depending upon the system bandwidth and channel environment, etc.

The link performance of QPSK with 2/3 coding rate and 16-QAM with 1/3 coding rate is evaluated under both a frequency-flat fading channel and six paths frequency-selective

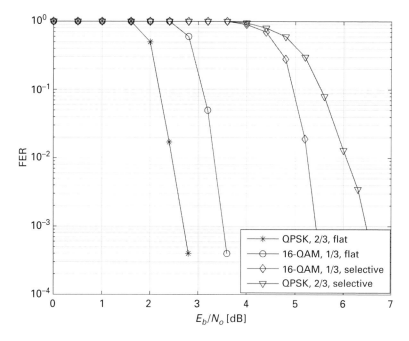

Figure 12.14. QPSK and 16-QAM link performance in frequency-flat and frequency-selective fading channels.

Typical Urban (TU) channel in Figure 12.14. It should be noted that QPSK with 2/3 coding rate and 16-QAM with 1/3 coding rate are equivalent as both schemes provide 3 b/s/Hz spectral efficiency. We note that QPSK with 2/3 coding rate outperforms 16-QAM by about 1 dB under single-path frequency-flat fading channel while 16-QAM with 1/3 coding rate has a 1 dB advantage in the TU channel. This result suggests that one can use QSPK modulation with a higher code rate in a flat-fading channel and can switch to 16-QAM with a lower code rate in a frequency-selective fading channel to maximize system performance.

In order to support channel frequency-selectivity-based link adaptation, the LTE system provides modulation overlap around switching points as given in the modulation and coding scheme (MCS) (Table 12.1) for downlink data transmissions. We note that MCS indices $I_{MCS} = 9, 10$ are mapped to the same transport block size (TBS) $I_{MCS} = 9$ resulting in the same data rate. However, $I_{MCS} = 9$ corresponds to QPSK modulation while $I_{MCS} = 10$ corresponds to a 16-QAM modulation. Similarly, MCS indices $I_{MCS} = 16, 17$ correspond to 16-QAM and 64-QAM modulation while being mapped to the same TBS index of $I_{TBS} = 15$ providing the same data rate or spectral efficiency.

The transport block size is determined implicity from the TBS index and the number of resource blocks allocated to the UE. For $29 \leq I_{MCS} \leq 31$, the TBS is assumed to be as determined from the previous scheduling grant for the same transport block using $0 \leq I_{MCS} \leq 28$. This allows for modulation adaptation on hybrid ARQ retransmission with $I_{MCS} = 29, 30, 31$ indicating QPSK, 16-QAM and 64-QAM modulation respectively.

The channel frequency-selectivity-based link adaptation does not apply in the uplink as the channel gain experienced by all the modulation symbols is the same due to SC-FDMA. However, a modulation overlap is defined at the switching points as indicated by $I_{MCS} = 10, 11$

Table 12.1. Modulation and TBS index for downlink.

MCS index I_{MCS}	Modulation order Q_m	TBS index I_{TBS}
0	2	0
1	2	1
2	2	2
3	2	3
4	2	4
5	2	5
6	2	6
7	2	7
8	2	8
9	2	9
10	4	9
11	4	10
12	4	11
13	4	12
14	4	13
15	4	14
16	4	15
17	6	15
18	6	16
19	6	17
20	6	18
21	6	19
22	6	20
23	6	21
24	6	22
25	6	23
26	6	24
27	6	25
28	6	26
29	2	
30	4	reserved
31	6	

and $I_{MCS} = 20, 21$ in Table 12.2. It should be noted that in the uplink the MCS index I_{MCS} not only indicates modulation but also the hybrid ARQ redundancy version. When the hybrid ARQ redundancy version of 1, 2 or 3 is indicated by using $I_{MCS} = 29, 30, 31$ respectively, the modulation order is assumed to be the one indicated in the initial grant. When modulation is explicitly indicated by using $0 \leq I_{MCS} < 29$, the hybrid ARQ redundancy version is always zero.

12.5 Hybrid ARQ

The term Hybrid ARQ is used to describe any combined FEC+ARQ scheme in which unsuccessful attempts are used in FEC decoding instead of being discarded.

Table 12.2. Modulation, TBS index and redundancy version for uplink.

MCS index I_{MCS}	Modulation order Q_m	TBS index I_{TBS}	Redundancy version rv_{idx}
0	2	0	0
1	2	1	0
2	2	2	0
3	2	3	0
4	2	4	0
5	2	5	0
6	2	6	0
7	2	7	0
8	2	8	0
9	2	9	0
10	2	10	0
11	4	10	0
12	4	11	0
13	4	12	0
14	4	13	0
15	4	14	0
16	4	15	0
17	4	16	0
18	4	17	0
19	4	18	0
20	4	19	0
21	6	19	0
22	6	20	0
23	6	21	0
24	6	22	0
25	6	23	0
26	6	24	0
27	6	25	0
28	6	26	0
29			1
30		reserved	2
31			3

12.5.1 ARQ protocols

Automatic Repeat-reQuest (ARQ) is an error-control method for data transmission, which makes use of acknowledgments and timeouts to achieve reliable data transmission. An acknowledgment is a message sent by the receiver to the transmitter to indicate whether it has correctly received a data frame or not. The ARQ protocols can be divided into three broad categories namely stop-and-wait (SAW), Go-Back-N and selective repeat protocols.

Stop-and-wait ARQ is the simplest type of ARQ scheme. A stop-and-wait ARQ transmitter sends one packet at a time. After sending a packet, the transmitter waits for acknowledgment (ACK) or negative acknowledgment (NACK) and does not send any new packets until it receives either an ACK or a NACK as shown in Figure 12.15. After successfully receiving a packet, the receiver sends an ACK signal. In case of a failed decoding of a packet, a NACK signal is sent. The transmitter sends a new packet after receiving an ACK signal. If a NACK signal

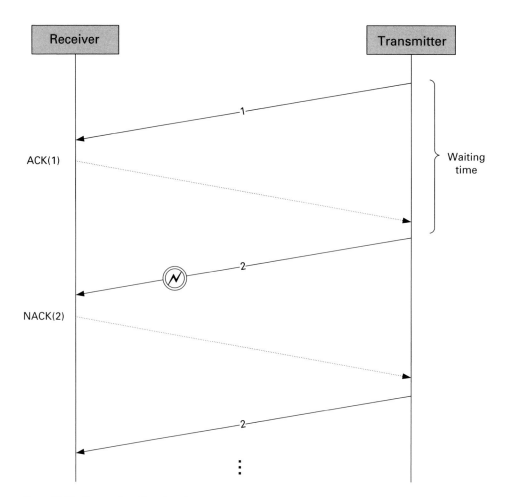

Figure 12.15. Stop-and-wait protocol.

is received, the transmitter retransmits the missing packet. It can be noted that the stop-and-wait protocol requires the receiver to buffer at most one packet that is currently being decoded.

A problem with the stop-and-wait protocol, however, is high transmission delays. This is because the transmitter has to wait for ACK/NACK feedback before proceeding with further transmissions. These waiting times can be quite long due to transmission delays as well as receiver processing times. In the LTE system, for example, the packet transmission time is only one subframe (1 ms) while it requires a seven subframe duration (7 ms) waiting time before the packet in error can be retransmitted using Hybrid ARQ. The major component of processing time at the receiver is generally turbo decoding processing times.

The Go-Back-N ARQ protocol solves the waiting time problem associated with the stop-and-wait protocol. In this case, the transmitter continues to send a number of packets specified by a *window size* even without receiving an ACK from the receiver as shown in Figure 12.16. When the transmitter receives a negative acknowledgment (NACK) for a missing packet, it starts retransmitting packets starting from the missing packet. Similar to the stop-and-wait protocol, Go-Back-N requires buffering of at most one packet at the receiver. Note that after sending a NACK, the receiver ignores all subsequent packets until it receives retransmission

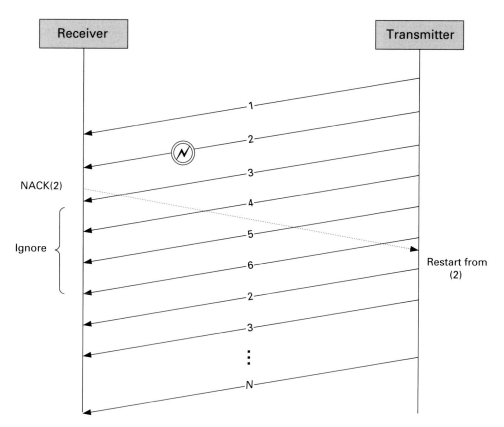

Figure 12.16. Go-Back-N protocol.

for the missing packet. The drawback of the Go-Back-N approach is duplicate transmissions as the transmitter retransmits some packets that are already successfully decoded at the receiver.

In selective repeat ARQ protocol, the sending process continues to send a number of packets specified by a *window size* even after a packet loss as shown in Figure 12.17. Unlike Go-Back-N, the receiving process will continue to accept and acknowledge packets sent after an initial error. We note that the selective repeat ARQ protocol retransmits only the missing packets and therefore avoids the duplicate transmissions problem of Go-Back-N. In addition, as the packets can be sent continuously, unlike stop-and-wait, there is no waiting time problem.

The operation of transmit and receive windows in the selective repeat protocol is depicted in Figure 12.18. The size of the sending and receiving windows is equal, and half the maximum sequence number N. The sender moves its window for every packet that is acknowledged. In the example of Figure 12.18, we assume that the receiver has successfully decoded packet 1 and therefore it moves its window forward by one packet and can accept a packet with sequence number $(N/2 + 1)$. However, the transmitter has not yet received acknowledgment for packet 1 or negative acknowledgment for packet 2 and therefore it cannot transmit a packet with sequence number $(N/2 + 1)$. When the transmitter receives the acknowledgment information from the receiver, it can also move its window one packet forward and therefore can transmit a packet with sequence number $(N/2 + 1)$. It can be noted that multiple packets in the receiver window can be in error at a given time. When the selective repeat protocol

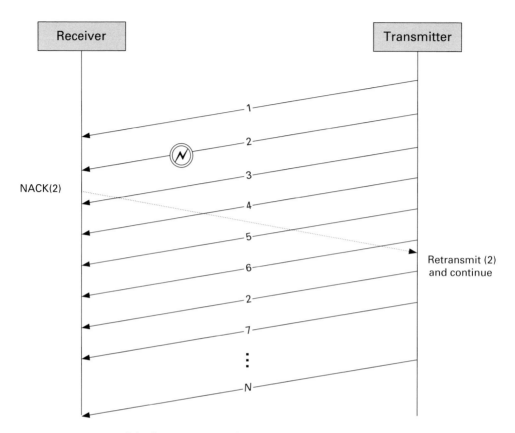

Figure 12.17. Selective repeat protocol.

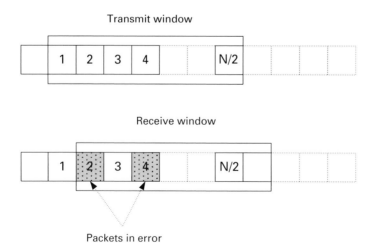

Figure 12.18. Transmit and receive windows in the selective repeat protocol.

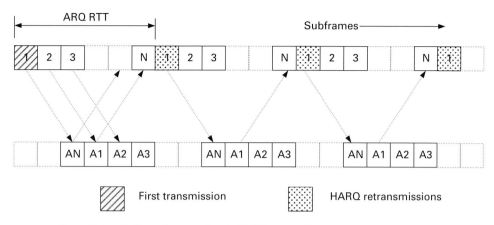

Figure 12.19. N-channel stop-and-wait (SAW) protocol.

is designed to provide in-sequence delivery of packets, the out-of-sequence packets, such as is the case for packet 3 in Figure 12.18, are not delivered to the higher layers. The receiver stores these packets in its buffer until in-sequence delivery can be made.

When sending acknowledgment information in the selective repeat protocol, the receiver should include the sequence numbers of the packets that are successfully received or are in error. A bitmap technique can also be used to provide ACK/NACK feedback with, for example, a starting sequence number and bitmap indicating successful and erroneous packets. In any case, the ACK/NACK overhead can be significant for large sequence numbers spaces to support larger window sizes. Another drawback of the selective repeat protocol is that the receiver needs to buffer a large number of packets.

12.5.2 N-channel stop-and-wait

The N-channel stop-and-wait protocol is a multi-channel variant of the stop-and-wait protocol and boasts some of the same benefits as the selective repeat protocol. Like the selective repeat protocol, the sending process can continuously send a number of packets while avoiding duplicate transmissions. An example of the N-channel stop-and-wait protocol is given in Figure 12.19. Each channel within the N channels operates like a simple stop-and-wait protocol. When the transmitter is waiting for acknowledgment on one SAW channel, it can start transmission on the remaining SAW channels. The number of channels required for continuous transmission depends upon the round-trip time (RTT), which is defined as the period from the time a packet is sent to the time when ACK or NACK is received at the transmitter. When the number of SAW channels N is equal to RTT, a continuous transmission can be guaranteed. A single-bit ACK/NACK indication is required on each SAW channel because the timing of ACK/NACK relative to the packet transmission is pre-determined and fixed. This avoids sending the packet sequence number in the ACK/NACK. The maximum buffering required for N-channel SAW is N packets.

The N-channel stop-and-wait protocol is used as the hybrid ARQ protocol in the LTE system due to its desirable features of simplicity, low buffering requirement and low ACK/NACK feedback overhead. It should be noted that buffering of the soft values of the coded bits is required for hybrid ARQ, which is generally much larger than the buffering required for information packets. Also, the sequence numbers and ACK/NACK need to be carried on separate physical channels, which is generally more costly than the data bits cost. This is because data transmission benefits from many of the advanced techniques such as turbo coding, scheduling, link adaptation and hybrid ARQ, which generally do not apply to physical channels. For RLC (radio link control, see Chapter 2) retransmissions at layer 2, LTE employs a selective repeat protocol. The sequence numbers and ACK/NACK overhead are less of a concern for RLC as this information is generally carried in-band as part of the data packet. In addition, RLC buffering is relatively small because only information bits need to be stored at the layer 2 level.

12.5.3 Hybrid ARQ combining

The simplest form of hybrid ARQ shceme was proposed by Chase [5] in 1985. The Chase combining involves the retransmission by the transmitter of the same coded data packet. The decoder at the receiver combines these multiple copies of the transmitted packet. Another form of hybrid ARQ combining is called incremental redundancy. In this scheme, instead of sending simple repeats of the coded data packet, progressive parity packets are sent in each subsequent transmission of the packet. The decoder then combines all the transmissions and decodes the packet at a lower code rate. These two combining schemes have been extensively studied and it is generally agreed that incremental redundancy can provide superior performance due to coding gain at retransmissions. However, this gain comes at the expense of additional UE complexity because the buffering required in the case of incremental redundancy is higher than in the case of Chase combining. This is because redundant transmissions in the case of incremental redundancy need to be buffered separately. In the case of Chase combining, the combined output after receiving each retransmission can be stored and hence a larger number of retransmissions does not necessarily mean larger buffering.

The LTE system supports both Chase combining and incremental redundancy by introducing the hybrid ARQ soft buffer limitation concept as described in Chapter 11, Section 11.7.3.

12.5.4 Scheduling control information

In the LTE system, the time-frequency resources can be allocated in either a persistent fashion or a non-persistent fashion. The persistent allocation is used for applications with regular packet arrivals such as is the case for the Voice-over-IP (VoIP) application. In persistent allocation, the time-frequency resources over multiple subframes periods are allocated to a UE in a semi-static fashion. In dynamic channel-sensitive scheduling, resources are allocated on a subframe basis with a scheduling message sent with every downlink or uplink grant. The scheduling grant carries various types of control fields as shown in Figure 12.20. The UE ID indicates the UE (or group of UEs) for which the data transmission is intended. The new data indicator (NDI) is used to indicate if a subblock belongs to a new packet transmission

UE ID	NDI	Resource assign.	Payload	Mod.	HARQ info.	MIMO info.

Figure 12.20. Scheduling grant control message contents.

or to retransmission for a previous packet. The resource assignment indicates which time-frequency resource units are allocated to the UE. The modulation indicates one of the supported modulations such as QPSK, 16-QAM or 64-QAM. The payload size or transport block size gives the data information block size. This information can be derived from the modulation and coding indication and number of allocated resource blocks. The hybrid ARQ information consists of hybrid ARQ process number, redundancy version and new data indicator. The MIMO control information includes information on transmission rank and precoding, etc.

12.5.5 Hybrid ARQ timing and adaptation

Based on retransmission timing and adaptation, hybrid ARQ schemes can be classified into four categories namely synchronous non-adaptive, synchronous adaptive, asynchronous non-adaptive and asynchronous adaptive as shown schematically in Figure 12.21. The synchronous timing means that the retransmissions happen at fixed time intervals relative to the initial transmission. In case of asynchronous timing, similar to initial transmission, retransmissions can be scheduled at any time after a NACK signal is received. In the non-adaptive case, frequency resource allocation, MCS and MIMO format, etc. are the same as the initial transmissions. With adaptive HARQ, one or more of the retransmissions parameters such as frequency resource allocation, MCS and MIMO format, etc. can be different from the original transmission.

An example of the N-channel stop-and-wait synchronous non-adaptive hybrid ARQ protocol is shown in Figure 12.22 for the case of $N = 8$. In the case of a synchronous HARQ protocol, the retransmissions happen at fixed time intervals relative to the initial transmission. With $N = 8$, if the first subblock (SB) is transmitted in time subframe #0, the first retransmission attempt can only take place in subframe #8 and similarly the second retransmission in subframe #16 as shown in Figure 12.22. One of the benefits of synchronous non-adaptive HARQ is that the control information needs to be transmitted along with the first subblock only. The control information is not transmitted with retransmitted subblocks as the timing of the retransmissions is predetermined. However, a major drawback of synchronous non-adaptive HARQ is that the retransmitted subblocks cannot be scheduled at time-frequency resources experiencing good channel conditions at the time of retransmissions. Moreover, the MCS and resource format cannot be adapted at the time of retransmission according to the prevailing channel conditions. Another downside is that synchronous retransmissions can potentially lead to conflicts with persistently allocated resources in which case either the synchronous retransmission or the persistent transmission has to be dropped.

A synchronous adaptive hybrid ARQ protocol shown in Figure 12.23 allows one to change the resource allocation and MCS information for retransmissions. Similar to the synchronous

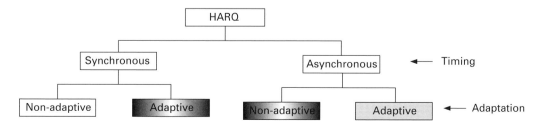

Figure 12.21. Hybrid ARQ classification based on timing and adaptation.

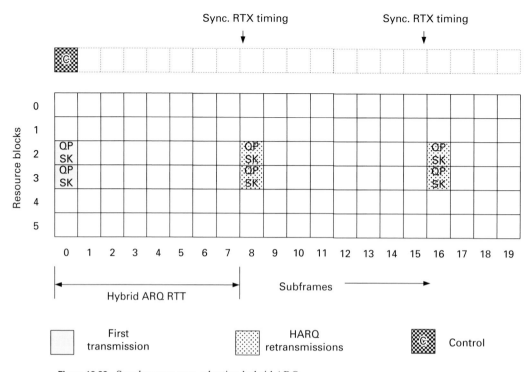

Figure 12.22. Synchronous non-adaptive hybrid ARQ.

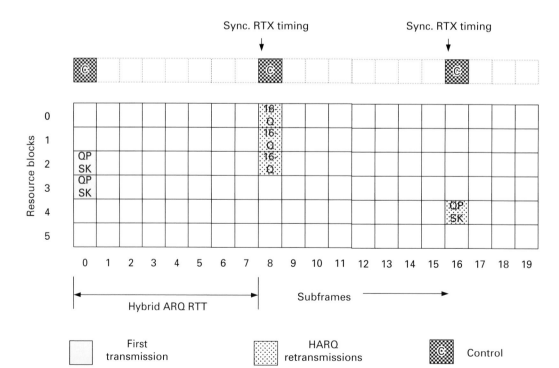

Figure 12.23. Synchronous adaptive hybrid ARQ.

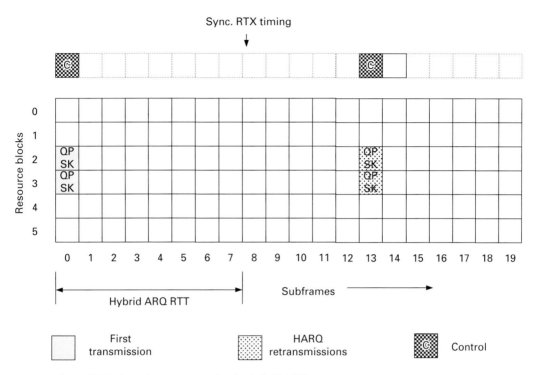

Figure 12.24. Asynchronous non-adaptive hybrid ARQ.

non-adaptive hybrid ARQ scheme, the retransmission timing is fixed. Since the resource allocation, MCS and MIMO precoding can change for retransmissions, the control information carrying these fields is sent with retransmission. It should be noted that the UE ID field does not need to be carried with retransmission as this information can be derived from the retransmission timing. The synchronous adaptive hybrid ARQ scheme thus allows scheduling retransmissions at frequency resources experiencing good channel conditions at the time of the retransmissions and hence recuperates some frequency-selective scheduling gains. In addition, resource conflicts with persistent allocation can be avoided by scheduling retransmissions around persistently allocated resources.

An asynchronous non-adaptive hybrid ARQ protocol allows scheduling retransmissions in time as shown in Figure 12.24. In this case, resource allocation, MCS and MIMO formats kept the same as the initial transmission. The control information carrying UE ID, hybrid ARQ process and redundancy version is carried with retransmissions. The time-domain channel-sensitive scheduling can be performed for retransmissions as these can be scheduled when the prevailing channel quality is good. The resource conflicts with persistent allocations can be avoided by scheduling retransmissions in time around persistent allocations. The drawback of this scheme is limited flexibility as the retransmissions resource allocation, MCS and MIMO formats cannot be adapted.

An asynchronous adaptive hybrid ARQ scheme [6] provides full flexibility for retransmissions adaptation as it treats hybrid ARQ retransmissions the same way as the original transmissions as shown in Figure 12.25. Therefore, retransmission timing, resource allocation, MCS and MIMO formats can all be adapted according to the prevailing channel and resource

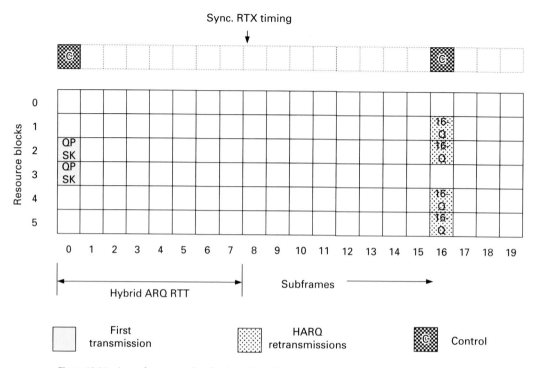

Figure 12.25. Asynchronous adaptive hybrid ARQ.

conditions at the time of retransmission. However, one major drawback of asynchronous and adaptive HARQ is that the full control information needs to be sent with retransmissions. It should be noted that control information in asynchronous adaptive HARQ needs to be transmitted even if the timing, resource allocation, MCS and MIMO formats for retransmissions are unchanged relative to the first transmission. This is because in asynchronous adaptive hybrid ARQ, the receiver only tries to decode a packet when it receives control information indicating the presence of a transmission.

Flexibility versus overhead trade off for various hybrid ARQ schemes is schematically shown in Figure 12.26. As discussed, the synchronous non-adaptive scheme requires the lowest overhead and provides the lowest flexibility. On the other hand, the asynchronous adaptive scheme provides the same flexibility for retransmissions as the first transmission at the expense of the largest overhead. The synchronous adaptive scheme and the asynchronous non-adaptive scheme provide some flexibility such as avoiding resource conflicts with persistent allocations with intermediate overhead. The overhead for the synchronous adaptive scheme is expected to be larger than the asynchronous adaptive scheme as resource allocation generally contributes the most to the total overhead. The synchronous non-adaptive scheme does not need to carry the UE ID with retransmissions. However, the UE ID is generally carried as part of the CRC by, for example, masking of the CRC with the UE ID. Therefore, the overhead required for the synchronous adaptive scheme is only marginally smaller than the full overhead in the AA hybrid ARQ scheme.

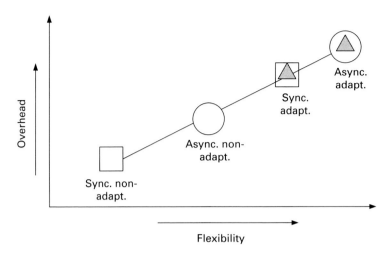

Figure 12.26. Flexibility versus control overhead trade off for various hybrid ARQ schemes.

12.5.6 Low overhead AA hybrid ARQ schemes

We noted that control information in asynchronous adaptive (AA) HARQ needs to be transmitted even if the timing, resource allocation, MCS and MIMO formats for retransmissions are unchanged relative to the original transmission. A straightforward way to reduce the overhead would be to send control information only if one or more of the timing, resource allocation, MCS and MIMO formats needs to be changed at the time of retransmission. This allows avoiding transmission of the control information when the retransmitted subblocks are sent in a synchronous and non-adaptive fashion. However, when retransmitted subblocks need to be sent in an asynchronous or adaptive fashion, control information accompanies the retransmitted subblocks. In this way the control information is only transmitted when necessary.

An example of such a low overhead asynchronous adaptive HARQ scheme is given in Figure 12.27. For illustration, we assumed hybrid ARQ RTT of four subframes and hence a maximum of four hybrid ARQ processes. In this example, the first subblock is transmitted in subframe #0. The control information is always transmitted along with the first subblock. However, the retransmission of the second subblock is delayed until subframe #11. The UE always tries to receive the retransmitted subblocks under the assumption of synchronous retransmissions. In this case, the receiver expects a retransmission of SB2 in subframe #4. Therefore, the receiver assumes that SB2 is transmitted in subframe #4. Since the transmitter has preempted the transmission of SB2 with transmission X to another UE, the UE expecting SB2 has no way of knowing if the transmission in subframe #4 is SB2 or some other transmission. Similarly, in subframe #8, the UE expects transmission of SB3 but the transmitter has performed another transmission Y to probably another UE. The receiver tries to decode the information packet by combining SB1 and X after receiving X in subframe #5 and by combining SB1 with X and Y after receiving a transmission in subframe #8. The decoding obviously fails because the receiver is combining the wrong subblocks. After three unsuccessful decoding attempts, first in subframe #0, second in subframe #4 and third in subframe #8, the receiver waits for transmission of SB4 in subframe #12. However, in subframe #11, the receiver decodes a control signal indicating transmission of SB2. Upon receiving

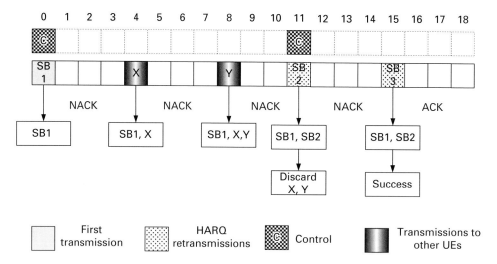

Figure 12.27. A low-overhead asynchronous adaptive hybrid ARQ scheme.

this control information, the receiver knows that transmissions in subframe #4 and subframe #8 were not for it and therefore it discards X and Y. In subframe #11, the receiver tries to decode the information packet by combining SB1 and SB2 only. The decoding fails and a retransmission of SB3 is now performed in a synchronous manner in subframe #15. In this case, no control information is transmitted along with SB3. After receiving SB3 in subframe #15, the receiver combines SB1, SB2 and SB3 and successfully decodes the information packet.

When the retransmission timing, resource allocation, MCS and MIMO formats are the same as the original transmission, the low-overhead AA hybrid ARQ scheme operates in a similar way to the synchronous non-adaptive scheme as shown in Figure 12.28.

In the low overhead AA hybrid ARQ approach, the UE always buffers the subblocks that are received along with the control information. This is because there is no confusion about the subblocks that are received along with the control information. However, on reception of an asynchronous retransmission, the UE discards the transmissions that are received at synchronous timing without control information. This is because reception of an asynchronous retransmission indicates preemption of previous subblocks to the UE on the same hybrid ARQ process at synchronous timing that are actually transmissions to other UEs. However, it is also possible to indicate via control with asynchronous retransmission after a preemption how many of the previous synchronous retransmissions are valid. Since the UE needs to buffer all the received subblocks separately, the buffering required in the low overhead AA hybrid ARQ scheme can be larger than in the conventional AA hybrid ARQ scheme. This is because in the conventional AA hybrid ARQ scheme when the retransmitted subblocks are repetitions of the previous subblocks, the UE can combine (chase combining) these subblocks and store a single combined subblock. In the case of incremental redundancy (IR) based hybrid ARQ, subblocks need to be stored separately even in the conventional AA hybrid ARQ scheme.

In addition, in the low overhead AA hybrid ARQ scheme, the number of NACKs may increase because when retransmissions to a UE are preempted, the UE tries to combine the wrong information and therefore sends a NACK signal. A possible solution to this problem

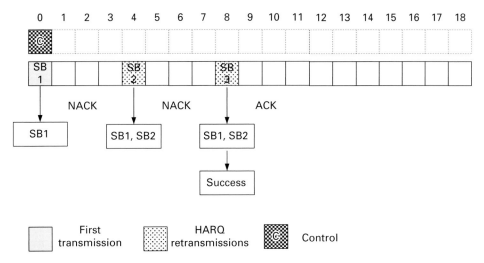

Figure 12.28. Synchronous non-adaptive hybrid ARQ mode of low-overhead asynchronous adaptive hybrid ARQ scheme.

is to code the NACK signal with no energy ('OFF') transmissions while an ACK is coded with an 'ON' signal. In the case when a packet transmission is not preempted, the transmitter interprets the absence of energy on the ACK/NACK signal as an indication of a NACK signal. This way additional NACKs from a preempted UE do not cause any overhead. A drawback of such an ON/OFF signaling is that it requires 3 dB higher energy for an ACK relative to a binary ACK/NACK signaling. However, for a NACK signal, ON/OFF signaling does not require any energy while binary ACK/NACK signaling requires the same energy for NACKs as for ACKs. In situations where the fraction of ACKs and NACKs is similar, both ON/OFF and binary ACK/NACK signaling leads to a similar overhead. In cases where the fraction of ACKs is higher than NACKs, binary signaling requires a lesser overhead than ON/OFF signaling. The higher energy requirement for ON/OFF signaling is a peak power limitation and coverage argument rather than average overhead concern.

A relative control overhead comparison for synchronous non-adaptive hybrid ARQ, AA HARQ and low overhead AA HARQ is given in Figure 12.29. The relative control overhead is shown as a function of preemption or retransmission adaptation probability. The overhead for synchronous non-adaptive HARQ is normalized and assumed as one indicating single control transmission per information block irrespective of the number of retransmission attempts. Also, we assume that synchronous non-adaptive hybrid ARQ requires no indication of the hybrid ARQ process number and also no NDI. The case with preemption or retransmission adaptation probability of zero corresponds to synchronous non-adaptive hybrid ARQ operation of low overhead AA HARQ. In this case, the overhead is only 1.1 times that of synchronous non-adaptive hybrid ARQ due to the hybrid ARQ process number and NDI that is transmitted in low overhead AA hybrid ARQ. The case with preemption or retransmission adaptation probability of one corresponds to conventional AA hybrid ARQ. This indicates the case that even if there is no preemption or retransmission adaptation in AA hybrid ARQ, the control information is sent with every retransmission. For example, if the average number of hybrid ARQ transmissions per packet is two, the overhead due to AA hybrid ARQ is 2.2 times that of synchronous non-adaptive hybrid ARQ. This accounts for approximately 10% additional

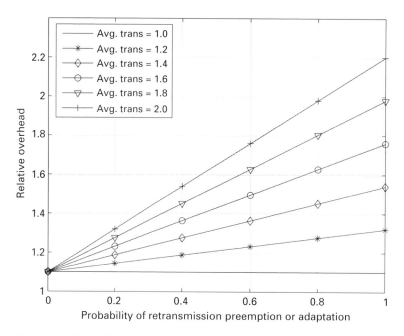

Figure 12.29. Control overhead comparison for various hybrid ARQ schemes.

overhead due to hybrid ARQ process number and NDI as well as an average of two control transmissions per information block.

The control overhead due to low overhead AA hybrid ARQ depends upon the operating scenario in terms of probability of preemption or adaptation of retransmissions as shown in Figure 12.29. In low overhead AA hybrid ARQ, the scheduling information is only sent when needed due to change in retransmission parameters such as timing, resource allocation, MCS or MIMO formats, etc. When the retransmission parameters do not change, there is no need to send the control information. As an example when the average number of hybrid ARQ transmissions per information packet is 1.8 and the probability of preemption or retransmission adaptation is 0.2, the overhead due to low overhead AA hybrid ARQ scheme is only 1.28 times the synchronous non-adaptive hybrid ARQ overhead. For the same preemption or retransmission adaptation probability, the overhead due to conventional AA hybrid ARQ is 1.98 times the synchronous non-adaptive hybrid ARQ overhead. This represents a saving of 45% in overhead with the low overhead AA hybrid ARQ scheme relative to the conventional AA hybrid ARQ scheme.

12.5.7 Hybrid ARQ for MIMO

For single-user MIMO operation in the downlink, the LTE system supports a maximum of four MIMO layers in a 4×4 antenna configuration (see Chapter 7). However, the maximum number of codewords is limited to two. Link adaptation and hybrid ARQ are performed independently on each of the codewords. This means that transport block size, modulation and coding information is signaled for each of the codewords in the downlink. Similarly, ACK/NACK feedback in the uplink is provided for the two codewords. The maximum number of codewords was limited to two to strike a balance between MIMO gains due to successive interference cancellation (SIC) and the signaling overhead. Codeword to layer mapping in the LTE system is shown in Figure 12.30. MIMO rank refers to the number of MIMO layers supported at given time-frequency resources. Since hybrid ARQ is performed independently on the two codewords each codeword is split into its own subblocks for hybrid ARQ incremental retransmissions as is also depicted in Figure 12.30.

Figure 12.31 depicts two codewords transmission for rank-2 MIMO using the synchronous non-adaptive hybrid ARQ scheme. We assume that CW1 is transmitted on layer-1 and CW2 on layer-2. When one of the codewords succeeds while the other one fails, the synchronous non-adaptive hybrid ARQ scheme leads to layer blanking. This is because no control information is sent with retransmissions and hence a new transport block transmission cannot be initiated on the layer with successful codeword. With an AA hybrid ARQ scheme, a new transport block transmission can be initiated on the layer with successful codeword thus making full utilization of spatial resources at all times.

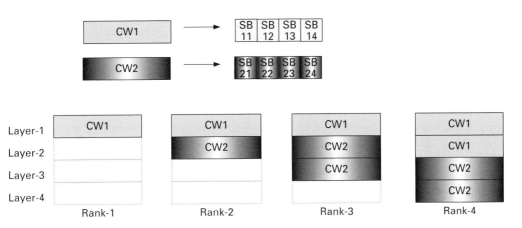

Figure 12.30. Codeword to layer mapping.

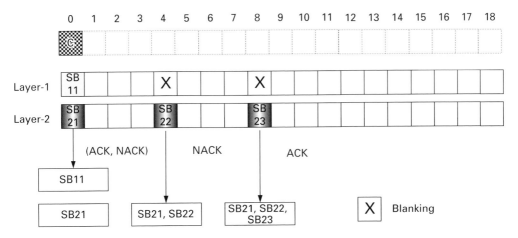

Figure 12.31. Blanking with the synchronous non-adaptive hybrid ARQ scheme.

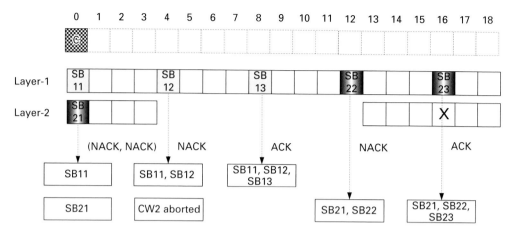

Figure 12.32. An illustration of limitations of synchronous non-adaptive hybrid ARQ during MIMO rank change.

Another issue with the synchronous non-adaptive hybrid ARQ scheme in the context of multi-codeword MIMO transmissions is its inability to efficiently handle situations where MIMO rank changes between original transmissions and retransmissions as depicted in Figure 12.32. For example, when MIMO rank changes from two codewords transmission to a single codeword transmission, transmission of the second codeword needs to be aborted as only a single codeword can be transmitted. The transmission of the second codeword can be continued after successful transmission of the first codeword. However, this will increase the codeword transmission delays. Another possibility would be to completely abort the transmission of the second codeword and start on a new hybrid ARQ process. This would degrade the system throughput because previously received hybrid ARQ subblocks for the second codeword are discarded. In this case, when rank changes from single codeword transmissions to one that can support two codewords transmission, MIMO layer blanking occurs resulting in efficiencies in spatial resource utilization. It can be noted that both of these problems can be avoided if the AA hybrid ARQ scheme is used.

Yet another issue with the synchronous non-adaptive hybrid ARQ scheme is that of persistent interference as illustrated in Figure 12.33. This situation happens when two beams in two neighboring cells targeted for UEs in the own cell collide on initial transmission. On beam collision, each of the two UEs experience high interference and transmissions to both UEs are likely to fail. The problem arises when synchronous non-adaptive hybrid ARQ retransmissions are performed because both beams collide again. This is because with synchronous non-adaptive hybrid ARQ, the retransmission time, resource and MIMO precoding is the same as the original transmission. As beams keep on colliding on retransmissions in a persistent fashion, the transport blocks at both UEs may not be recovered even after the maximum number of hybrid ARQ retransmission attempts are made triggering higher layer retransmission. In the case of AA hybrid ARQ, timing, resource allocation or precoding is likely to change during the course of multiple retransmissions attempts, which can avoid persistent interference.

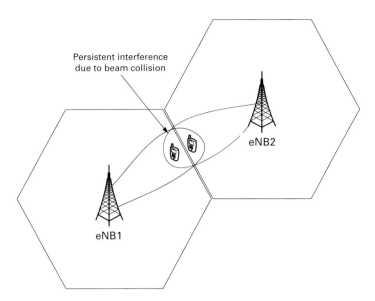

Figure 12.33. An illustration of the persistent interference issue with synchronous non-adaptive hybrid ARQ.

12.6 Hybrid ARQ in the LTE system

The LTE system employs incremental redundancy (IR) based hybrid ARQ with Chase combining as a special case of IR. In regard to timing and adaptivity, asynchronous adaptive (AA) hybrid ARQ is used in the downlink while synchronous adaptive hybrid ARQ is supported in the uplink. The new data indicator (NDI) field in the uplink scheduling grant is used to indicate if the grant is for an adaptive retransmission of a previous transmission or grant for a new transport block transmission. If the uplink scheduling assignment is received with the NDI bit toggled, this means that eNB is scheduling a new uplink transmission. On the other hand, if NDI is not toggled, this indicates adaptive retransmission of the previous transmission attempt. Moreover, if no uplink scheduling assignment is received while an ACK is received on the PHICH, this indicates successful transmission of the uplink transport block. It should be noted that the LTE system does not support MIMO in the uplink and therefore synchronous hybrid ARQ problems regarding multi-codeword MIMO do not exist.

12.6.1 Number of hybrid ARQ processes

The buffering and delays in the N-channel stop-and-wait protocol are directly proportional to the number of SAW channels also referred to as hybrid ARQ processes. The number of hybrid ARQ processes N_{HARQ} is given as:

$$N_{\mathrm{HARQ}} = \left\lceil \frac{\left(2T_p + T_{\mathrm{sb}} + T_{\mathrm{uep}} + T_{\mathrm{ack}} + T_{\mathrm{nbp}} \right)}{T_{\mathrm{sb}}} \right\rceil, \tag{12.18}$$

where T_p is the propagation time between eNB and the UE, T_{sb} subblock transmission time, T_{uep} UE processing time including the decoding times, T_{ack} ACK/NACK transmission time

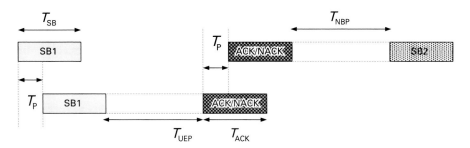

Figure 12.34. Stop-and-wait hybrid ARQ round-trip time (RTT).

and finally T_{nbp} eNB processing time. The relationship between these parameters is shown in Figure 12.34.

Since buffering and delay increase with increasing number of hybrid ARQ processes it is desirable to keep the number of hybrid ARQ processes to a minimum. The propagation time for cell sizes of most interest is generally much smaller than the subblock transmission and processing times and therefore can be neglected. A smaller subblock transmission time T_{sb} generally requires a larger overhead. This is because the smaller the subblock transmission time, the smaller the amount of information that can be carried in a transport block. This means a larger number of transport blocks each with corresponding control information needs to be transmitted to carry a given amount of application layer data. The subblock transmission time in the LTE system is one subframe (1 ms) which was selected as a compromise between latency and signaling overhead.

The UE processing time T_{uep} represents the total amount of time required to process and decode the subblock. A major component of UE processing time is turbo decoding delays for supporting peak data rates that exceeds 100 Mb/s in the LTE system. The UE complexity increases for smaller T_{uep} and therefore the value for UE processing time is selected as a compromise between complexity and latency.

The ACK/NACK transmission time T_{ack} is the time period over which the hybrid ARQ acknowledgment or negative acknowledgment feedback is transmitted. For reliability reasons, T_{ack} cannot be very short. This is because the shorter the ACK/NACK transmission time the smaller the amount of ACK/NACK signal energy for a given transmit power. Since the UE transmit power is generally small relative to eNB, the required ACK/NACK reliability can limit the uplink coverage for short T_{ack}. In the LTE system, the ACK/NACK transmission time is assumed equal to one subframe (1 ms).

The eNB processing time T_{nbp} accounts for the time required by the eNB to decode ACK/NACK and also the time required for the scheduling decision for new transport blocks. In the LTE system, both the UE and the eNB processing times T_{uep} and T_{nbp} are selected equal to three subframes (3 ms).

By using the values for hybrid ARQ RTT parameters in Table 12.3, the number of hybrid ARQ processes in the LTE system is calculated as:

$$N_{HARQ} = \left\lceil \frac{(0+1+3+1+3)}{1} \right\rceil = 8. \qquad (12.19)$$

With eight hybrid ARQ processes and RTT parameters in Table 12.3, an ACK/NACK response for a subblock received in subframe n is transmitted in subframe $n+4$ as shown

Table 12.3. Hybrid ARQ RTT parameters for the LTE system.

Parameter	Symbol	Value
Propagation time	T_p	Negligible
Subblock transmission time	T_{sb}	1 ms
UE processing time	T_{uep}	3 ms
ACK transmission time	T_{ack}	1 ms
eNB processing time	T_{nbp}	3 ms

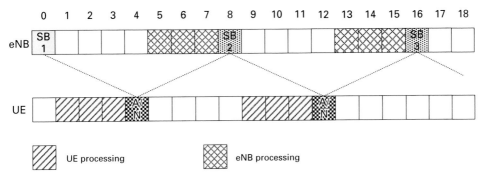

Figure 12.35. An illustration of a hybrid ARQ process in the LTE system.

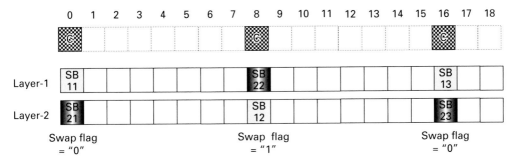

Figure 12.36. MIMO codeword swapping on hybrid ARQ retransmissions.

in Figure 12.35. The number of hybrid processes in both the uplink and the downlink is the same. For AA hybrid ARQ in the downlink, a 3-bit hybrid ARQ process identifier is carried in the downlink control information to indicate one of the eight processes.

12.6.2 MIMO codeword swapping

MIMO codeword swapping on hybrid ARQ retransmissions is enabled by a 1-bit hybrid ARQ flag in the downlink control information as shown in Figure 12.36. This approach allows both codewords to experience similar channel conditions after the hybrid ARQ retransmission.

12.7   Summary

The LTE system employs channel-sensitive scheduling, link adaptation and hybrid ARQ for enhanced performance in fast-fading environments. While the channel-sensitive scheduler selects a user experiencing good channel conditions on given time, frequency and spatial resources, link adaptation is used to adapt the transmissions format including modulation and coding (MCS) as well as MIMO rank and precoding to the current channel conditions on the allocated resources. Both channel sensitive scheduling and link adaptation rely on availability of accurate channel quality information at the eNB. In a practical situation, accurate channel quality knowledge is not feasible due to various reasons that include channel quality measurement errors, errors due to channel quality feedback delays, interference variations and channel quality information transmission errors. A hybrid automatic repeat-request (ARQ) mechanism allows one to correct errors in link adaptation by incremental redundancy transmission in the case of packet errors. This way, transmission data rates do not need to be reduced to account for channel quality estimation and prediction errors.

For reasons of increased signaling overhead and little potential for performance gains from frequency-dependent modulation and coding adaptation, the LTE system uses common modulation and coding across the scheduled resource blocks. However, for multi-codeword MIMO transmission, modulation and coding adaptation per codeword is supported.

The LTE system employs incremental redundancy (IR) based hybrid ARQ with chase combining as a special case of IR. In regard to timing and adaptivity, asynchronous adaptive (AA) hybrid ARQ is used in the downlink and synchronous adaptive hybrid ARQ in the uplink. Since the LTE system does not support MIMO in the uplink the problem regarding multi-codeword MIMO in synchronous hybrid ARQ does not exist. The hybrid ARQ round-trip time equals eight subframes (8 ms) that accounts for 3 ms processing for each of eNB and UE.

References

[1] Knopp, R. and Humblet, P., "Information capacity and power control in single-cell multiuser communication," *Proceedings of IEEE International Conference on Communication, ICC'95*, vol. 1, pp. 331–335, June 1995.

[2] Tse, D. and Viswanath, P., *Fundamentals of Wireless Communication*, New York: Cambridge University Press, 2005.

[3] Proakis, J. G., *Digital communications*, 4th ed., New York: McGraw-Hill, 2000.

[4] Borst, S. and Whiting, P., "Dynamic rate control algorithms for HDR throughput optimization," *Proceedings of Twentieth Annual Joint Conference of the IEEE Computer and Communications Societies, IEEE INFOCOM 2001*, vol. 2, pp. 976–985, 2001.

[5] Chase, D., "Code combining - A maximum-likelihood decoding approach for combining an arbitrary number of noisy packets," *IEEE Transactions on Communications*, vol. 33, pp. 385–393, May 1985.

[6] Das, A., Khan F. and Nanda, S., "A^2IR: an asynchronous and adaptive hybrid ARQ scheme for 3G evolution," *53rd IEEE VTS Vehicular Technology Conference, VTC 2001 Spring*, Vol. 1, pp. 628–632, 2001.

13 Power control

The goal of power control is to transmit at the right amount of power needed to support a certain data rate. Too much power generates unnecessary interference, while too little power results in an increased error rate requiring retransmissions and hence resulting in larger transmission delays and lower throughputs. In a WCDMA system, power control is important particularly in the uplink to avoid the near–far problem. This is because the uplink transmissions are non-orthogonal and very high signal levels from cell-center UEs can overwhelm the weak signals received from cell-edge UEs. Therefore, a very elaborate power control mechanism based on the fast closed-loop principle is used in the WCDMA system. Similarly, power control is used for the downlink of WCDMA systems to support the fixed rate delay-sensitive voice service. However, for high-speed data transmission in WCDMA/HSPA systems, transmissions are generally performed at full power and link adaptation is preferably used to match the data rate to the channel conditions.

The LTE uplink uses orthogonal SC-FDMA access and hence the near–far problem of WCDMA does not exist. However, high levels of interference from neighboring cells can still limit the uplink coverage if UEs in the neighboring cells are not power controlled. The cellular systems are generally coverage limited in the uplink due to limited UE transmit power. The increased levels of interference from neighboring cells increase Interference over Thermal (IoT) limiting coverage at the desired cell. Therefore, uplink power control is beneficial in an orthogonal uplink access as well. However, the power control mechanism does not need to be as elaborate as in the case of WCDMA because intra-cell interference is not an issue in the SC-FDMA uplink. In general, at least compensating for path-loss and shadowing on a slow basis is seen as necessary. By performing the slow power control scheme on each UE uplink transmission power, the average inter-cell interference level received at the eNB is effectively reduced.

In the downlink, the LTE system uses dedicated control signaling with some degree of link adaptation for uplink and downlink scheduling assignments. The power levels on the dedicated control channels can also be adjusted based on the channel conditions of the UE addressed in the scheduling grant. A simple power allocation based on downlink channel quality feedback from the UE is generally sufficient. Similarly, for the delay-sensitive VoIP (Voice over Internet Protocol) service, a combination of link adaptation and power control is used. The transmissions for high-speed data traffic are performed at the peak power. However, in order to provide inter-cell interference coordination (ICIC), different power levels can be allocated on different resource blocks in a semi-static fashion.

13.1 Uplink power control

For the LTE uplink, an event-based combined open-loop and closed-loop power control algorithm is employed. The scheme is referred to as event based because, unlike the WCDMA system, transmit power control commands (TPC) do not need to be transmitted at regular intervals. The uplink power control determines the average power over an SC-FDMA symbol in which the physical channel is transmitted. The uplink power control adjusts the transmit power of the different uplink physical channels.

13.1.1 PUSCH power control

The setting of the UE transmit power P_{PUSCH} for the physical uplink shared channel (PUSCH) transmission in subframe i is defined by:

$$P_{\text{PUSCH}}(i) = \min\{P_{\text{MAX}}, P_{\text{PUSCH-CALC}}(i)\}, \tag{13.1}$$

where P_{MAX} is the maximum allowed power that depends on the UE power class. $P_{\text{PUSCH-CALC}}(i)$ is the calculated power in subframe i and is given as:

$$P_{\text{CALC}}(i) = 10\log_{10}[M_{\text{PUSCH}}(i)] + P_{\text{O\_PUSCH}}(j) + \alpha \cdot PL + \Delta_{\text{TF}}(i) + f(i) \text{ dBm}, \tag{13.2}$$

where $M_{\text{PUSCH}}(i)$ is the size of the PUSCH resource assignment expressed as the number of resource blocks valid for subframe i. The transmit power is increased proportional to $M_{\text{PUSCH}}(i)$ in order to assure the same power spectral density irrespective of the number of resource blocks allocated for uplink transmission. PL is the downlink path-loss estimated at the UE. A scaling α of the path-loss allows one to account for the uplink/downlink path-loss mismatch due to feeder losses and other deployment aspects. The values $\alpha \in \{0, 0.4, 0.5, 0.6, 0.7, 0.8, 0.9, 1\}$ are indicated using a 3-bit cell-specific parameter given by higher layers.

The PUSCH power offset $P_{\text{O\_PUSCH}}(j)$ is given as:

$$P_{\text{O\_PUSCH}}(j) = P_{\text{O\_NOMINAL\_PUSCH}}(j) + P_{\text{O\_UE\_PUSCH}}(j), \tag{13.3}$$

where $P_{\text{O\_NOMINAL\_PUSCH}}(j)$ is the cell-specific nominal component signaled from higher layers for $j = 0, 1$ in the range of $[-126, 24]$ dBm with 1 dB increments. $P_{\text{O\_UE\_PUSCH}}(j)$ is a UE-specific component configured by RRC for $j = 0, 1$ in the range of $[-8, 7]$ dB with 1 dB resolution. $j = 0$ for PUSCH (re)transmissions corresponding to a configured scheduling grant. $j = 1$ for PUSCH (re)transmissions corresponding to a received PDCCH with DCI format 0 associated with a new packet transmission. $P_{\text{O\_NOMINAL\_PUSCH}}(j)$ and $P_{\text{O\_UE\_PUSCH}}(j)$ are signaled by using eight and four bits signaling respectively.

The modulation and coding scheme (MCS) or transport format (TF) compensation $\Delta_{\text{TF}}(i)$ is defined as:

$$\Delta_{TF}(i) = \begin{cases} 10\log_{10}\left(2^{MPR(i)\cdot K_S} - 1\right), & K_S = 1.25 \\ 0 & K_S = 0, \end{cases} \tag{13.4}$$

where K_S is a cell specific parameter given by RRC. MPR(i) is given as:

$$MPR(i) = \frac{TBS(i)}{M_{\text{PUSCH}}(i) \cdot N_{\text{sc}}^{\text{RB}} \cdot \left(2N_{\text{symb}}^{\text{UL}}\right)}, \tag{13.5}$$

where $TBS(i)$ is the transport block size for subframe i. The denominator in (13.5) represents the number of resource elements used for PUSCH transmission in subframe i. $N_{\text{sc}}^{\text{RB}} = 12$ is the number of subcarriers per resource block and $2N_{\text{symb}}^{\text{UL}}$ is the number of SC-FDMA symbols in a subframe. Therefore MPR(i) is effectively the spectral efficiency representing the number of information bits transmitted per resource element. A larger value of MPR(i) means a larger Δ_{TF} and hence a larger transmit power P_{PUSCH} for PUSCH transmission. We note that larger spectral efficiency means use of higher coding rate and/or higher order modulation which requires larger transmit power than the case of a stronger code and a lower order modulation.

We note that when $K_S = 0$, no MCS compensation is performed. When MCS compensation is performed $K_S = 1.25 \rightarrow \frac{1}{K_S} = 0.8$. This means only 80% of the uplink resource elements are assumed for MCS compensation. This is to account for the uplink resource elements used for the uplink reference signal and other uplink control information punctured on to the PUSCH. The effective spectral efficiency assumed for power control purposes is then MPR(i) $\times K_S$ as indicated in (13.4). The term $\left(2^{\text{MPR}(i) \cdot K_S} - 1\right)$ in (13.4) is simply the spectral efficiency translated into SNR using Shannon's channel capacity formula:

$$C = \log_2\left(1 + \text{SNR}\right) \rightarrow \text{SNR} = 2^C - 1. \tag{13.6}$$

The current PUSCH power control adjustment state is given by $f(i)$ which is defined by:

$$f(i) = f(i-1) + \delta_{\text{PUSCH}}(i-4), \quad f(0) = 0, \tag{13.7}$$

where δ_{PUSCH} is a UE specific correction value, also referred to as a TPC command, and is included in PDCCH in the downlink. $\delta_{\text{PUSCH}}(i-4)$ indicates correction signaled on PDCCH in subframe $(i-4)$ which represents four subframes delay between receiving a TPC command and when it is applied for uplink power adjustment. $\delta_{\text{PUSCH}} = 0$ dB for a subframe where no TPC command is decoded or where DRX occurs. The δ_{PUSCH} dB accumulated values signaled on PDCCH with DCI format 0 are $[-1, 0, 1, 3]$. The δ_{PUSCH} dB accumulated values signaled on PDCCH with DCI format 3/3A are one of $[-1, 1]$ or $[-1, 0, 1, 3]$ as semi-statically configured by higher layers.

We note from (13.7) that power-up and power-down commands are accumulated even when the UE has reached maximum power or minimum power respectively. The power-up does not affect the transmission power when the UE has reached maximum transmit power P_{MAX} as indicated by (13.1). A drawback of accumulating power-up commands when a UE is already transmitting at maximum power is that it might take a long time to reduce the output power once this is needed. This happens when a UE is transmitting at peak power for some time and receives and accumulates several power-up commands, and then either the path-gain rapidly improves (e.g. due to corner effects) or the scheduled bandwidth is reduced so that the power is sufficient to reach (and exceed) the target. In order to mitigate this problem the positive or power-up TPC commands are not accumulated when the maximum power has been reached, that is:

$$f(i) = f(i-1) + \min\{0, \delta_{\text{PUSCH}}(i-4)\}, \tag{13.8}$$

when

$$P_{MAX} \leq 10 \log_{10} [M_{PUSCH}(i)] + P_{O\_PUSCH}(j) + \alpha \cdot PL + \Delta_{TF}(i) + f(i-1). \qquad (13.9)$$

Note that negative TPC commands are still accumulated when the maximum power has been reached. Similarly, If UE has reached minimum power, negative TPC commands are not accumulated. Moreover, the accumulation is reset, $f(i) = 0$, when a UE changes cell, a UE enters or leaves RRC active state, a UE resynchronizes with the system after synchronization loss, a UE receives $P_{O\_UE\_PUSCH}(j)$ and when an absolute TPC command is received. The TPC commands accumulation is reset on cell-change because different cells may have different uplink/downlink path-loss mismatch due to feeder losses and other deployment aspects.

In addition to TPC command accumulation, absolute power adjustments are also supported. The type of correction (accumulation or current absolute) is a UE specific parameter that is given by RRC. In the case of absolute correction, $f(i)$ is updated as:

$$f(i) = \delta_{PUSCH}(i-4), \qquad (13.10)$$

where $\delta_{PUSCH}(i-4)$ represents adjustment as signaled on PDCCH in subframe $(i-4)$. The δ_{PUSCH} dB absolute values signaled on PDCCH with DCI format 0 are $[-4, -1, 1, 4]$. Also, $f(i) = f(i-1)$ for a subframe where no PDCCH with DCI format 0 is decoded or where DRX occurs.

13.1.2 Uplink power headroom

The power headroom report is transmitted using L2/L3 signaling in the same subframe that it refers to. The UE power headroom PH valid for subframe i is defined by:

$$PH(i) = (P_{MAX} - P_{PUSCH}(i)) \quad dB, \qquad (13.11)$$

where P_{MAX} is the maximum allowed power that depends on the UE power class and $P_{PUSCH}(i)$ is given by (13.1). The power headroom is rounded to the closest value in the -40 to -23 dB range with steps of 1 dB and is delivered by the physical layer to higher layers. A total of 64 power headroom values require 6-bits signaling.

13.1.3 PUCCH power control

The setting of the UE transmit power P_{PUCCH} for the physical uplink control channel (PUCCH) transmission in subframe i is defined by:

$$P_{PUCCH}(i) = \min \{P_{MAX}, P_{PUCCH-CALC}(i)\}, \qquad (13.12)$$

where P_{MAX} is the maximum allowed power that depends on the UE power class and $P_{PUCCH-CALC}(i)$ is the calculated PUCCH power in subframe i and is given as:

$$P_{PUCCH-CALC}(i) = P_{O\_PUCCH} + PL + \Delta_{F\_PUCCH}(F) + g(i) \quad dBm, \qquad (13.13)$$

where values of $\Delta_{F\_PUCCH}(F)$ for each PUCCH transport format are given by RRC. Each 2-bit signaled value of $\Delta_{F\_PUCCH}(F)$ corresponds to a PUCCH transport format relative to PUCCH format 0. P_{O_PUCCH} is given as:

$$P_{O\_PUCCH} = P_{O\_NOMINAL\_PUCCH} + P_{O\_UE\_PUCCH}, \qquad (13.14)$$

where $P_{O\_NOMINAL\_PUCCH}$ is a 5-bit cell specific parameter provided by higher layers with 1 dB resolution in the range of $[-127, -96]$ dBm and $P_{O\_UE\_PUCCH}$ is a 4-bit UE specific component configured by RRC in the range of $[-8, 7]$ dB with 1 dB resolution.

$g(i)$ is updated as:

$$g(i) = g(i-1) + \delta_{PUCCH}(i-4), \quad g(0) = 0. \qquad (13.15)$$

δ_{PUCCH} is a UE specific correction value, also referred to as a TPC command, included in a PDCCH with DCI format 1A/1/2 or sent jointly coded with other UE specific PUCCH correction values on a PDCCH with DCI format 3/3A. When the UE is in DRX or no PDDCH with a TPC command is received δ_{PUCCH} is set to zero, that is $\delta_{PUCCH} = 0$. The δ_{PUCCH} dB values signaled on PDCCH with DCI format 1A/1/2 are $[-1, 0, 1, 3]$. The δ_{PUCCH} dB values signaled on PDCCH with DCI format 3/3A are $[-1, 1]$ or $[-1, 0, 1, 3]$ as semi-statically configured by higher layers.

Similar to PUSCH power control, positive TPC commands are not accumulated when a UE has reached the maximum power and negative TPC commands are not accumulated when a UE has reached the minimum power. The TPC command accumulation is reset under similar conditions as in the case of PUSCH power control.

13.1.4 SRS power control

The sounding reference signal (SRS) is mainly used for uplink channel quality measurements for channel sensitive scheduling. The UE transmit power for sounding reference signal P_{SRS} transmitted in subframe i is defined by:

$$P_{SRS}(i) = \min \{P_{MAX}, P_{SRS\text{-}CALC}(i)\}, \qquad (13.16)$$

where P_{MAX} is the maximum allowed power that depends on the UE power class and $P_{SRS\text{-}CALC}(i)$ is the calculated SRS power in subframe i and is given as:

$$P_{SRS\text{-}CALC}(i) = P_{SRS\_OFFSET} + 10\log_{10}(M_{SRS}) + P_{O\_PUSCH}(j) + \alpha \cdot PL + f(i) \text{ dBm}, \qquad (13.17)$$

where M_{SRS} is the bandwidth of the SRS transmission in subframe i expressed in number of resource blocks. The transmit power is increased proportional to SRS bandwidth M_{SRS} in order to assure the same power spectral density irrespective of the SRS bandwidth used. $P_{O\_PUSCH}(j)$ is defined in Section 13.1.1 and $f(i)$ is the current power control adjustment state for the PUSCH as given in Section 13.1.1. P_{SRS_OFFSET} represents SRS power offset and its values are given in Table 13.1. The P_{SRS_OFFSET} is semi-statically configured by higher layers in a UE-specific manner using a 4-bit indicator.

Table 13.1. SRS power offset
P_{SRS_OFFSET} values [dB].

	SRS power offset [dB] P_{SRS_OFFSET}	
	$K_S = 1.25$	$K_S = 0$
0	−3	−10.5
1	−2	−9
2	−1	−7.5
3	0	−6
4	1	−4.5
5	2	−3
6	3	−1.5
7	4	0
8	5	1.5
9	6	3
10	7	4.5
11	8	6
12	9	7.5
13	10	9
14	11	10.5
15	12	12

13.2   Downlink power control

Unlike the uplink power control, there is no explicit feedback from the UEs to control the eNB transmit power for the downlink power control. The power levels for dedicated control channels such as PDCCH and PHICH can be determined based on downlink channel quality feedback from the UEs. The downlink channel quality feedback is provided by the UEs to support channel sensitive scheduling in the downlink. Therefore, the downlink transmit power control is fundamentally a power allocation scheme rather than a power control scheme. Since the control channels transmissions are spread over the whole system bandwidth, wideband channel quality information is used to determine the power levels for these control channels. It should be noted that various channel quality feedback formats are supported and wideband channel quality is always present in all the feedback formats to enable power control for the downlink control channels.

The downlink power control determines the energy per resource element (EPRE) prior to cyclic prefix insertion. The EPRE also denotes the average energy taken over all constellation points for the modulation scheme applied. The eNB determines the downlink transmit energy per resource element. A UE may assume the downlink reference symbol EPRE is constant across the downlink system bandwidth and constant across all subframes until different reference signal power information is received.

13.2.1   Downlink power imbalance

Let us first discuss the downlink power imbalance issue that arises due to TDM/FDM multiplexing of downlink reference signals within a subframe. An example of reference signals

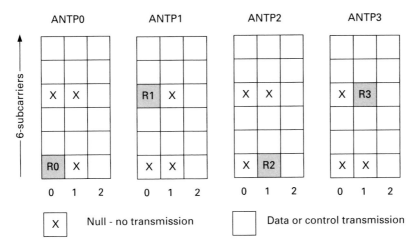

Figure 13.1. Mapping of downlink reference signals in OFDM symbol 0 and 1, $l = 0, 1$.

transmission over six subcarriers within the first three OFDM symbols from each of the four antenna ports is shown in Figure 13.1. It can be noted that the power available from each antenna port for subcarriers other than the reference signals varies from OFDM symbol to OFDM symbol as the reference signals are not transmitted in all the OFDM symbols. If the energy per resource element (EPRE) has to be kept constant on all the antenna ports in all the OFDM symbols, total power is used inefficiently because power level is limited to the minimum power level available from a given antenna port even though other ports may have extra power available.

Let us first consider the case of a single antenna port where the spatial-domain power imbalance issue does not occur. In this case, there is no a power imbalance issue as long as the reference signal power relative to data power is not boosted. However, if the reference signal power is boosted, less power is available on data resource elements in the OFDM symbols containing the reference signals. This is because when the reference signals power is boosted in OFDM symbols containing reference signals, this power comes from the data resource elements in the same OFDM symbol. Therefore, the data resource elements transmitted in OFDM symbols containing reference signals will have less power available than the resource elements not containing any reference signals. This creates a power imbalance issue across OFDM symbols. Similarly, for the case of two antenna ports, the power imbalance issue only happens in the time domain when reference signals are not boosted. However, for the case of four antenna ports, the time-domain power imbalance issue happens even when reference signals are not boosted. This is because reference signals for antenna ports $p = 0, 1$ and antenna ports $p = 2, 3$ are carried in different OFDM symbols.

Since higher order modulation such as 16-QAM and 64-QAM assumes fixed traffic to reference signal power ratio in all the resource elements varying power levels across OFDM symbols affects demodulation performance. A solution to fully utilize the transmit power and to keep the power constant across all OFDM symbols could be to puncture some data subcarriers in OFDM symbols containing reference signals. This means that when the reference signal is boosted, the data resource element from which the power is borrowed is left blank without any transmission. However, this approach results in the waste of subcarrier resources affecting system performance.

Let us assume the four antenna ports SFBC-FSTD transmission scheme to demonstrate the spatial and time-domain power imbalance and the benefit of non-constant EPRE among different transmit antennas. The transmission matrix of the SFBC-FSTD scheme described in Chapter 6 is given by:

$$T = \begin{bmatrix} S_1 & -S_2^* & 0 & 0 \\ 0 & 0 & S_3 & -S_4^* \\ S_2 & S_1^* & 0 & 0 \\ 0 & 0 & S_4 & S_3^* \end{bmatrix}. \tag{13.18}$$

We further denote the nominal PSD on each resource element (RE) by a. Then in the absence of reference signal power boosting, each reference symbol will have a power of $2a$. This is because an OFDM symbol contains reference signals for two antenna ports and each antenna leaves the REs used by the other antenna for reference signals as blank. Therefore, each reference signal symbol can use two times the power. Furthermore, without loss of generality we assume a case where each data resource element also has a power of $2a$. This is assuming the SFBC-FSTD scheme where each antenna transmits on two resource elements out of a total of four resource elements as indicated by the transmit matrix in (13.18) and also illustrated in Figure 13.2, with a focus on the first OFDM symbol. According to the constant

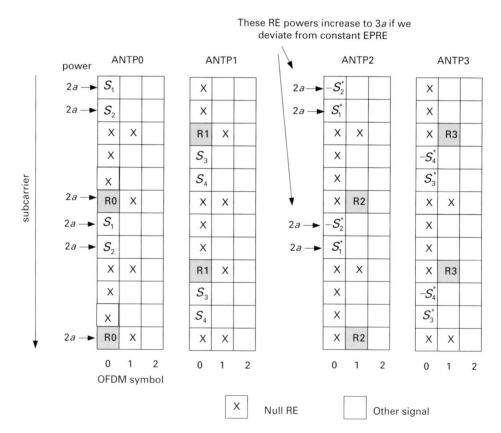

Figure 13.2. Transmission power assignment for an SFBC-FSTD scheme.

Table 13.2. Non-constant data to reference signal power ratio without reference signal boosting.

	$l \in \{0, 4, 7, 11\}$	$l \in \{1, 8\}$	$l \in \{2, 3, 5, 6, 9, 10, 12, 13\}$
$p \in \{0, 1\}$	$\dfrac{2a}{2a}$	$\dfrac{3a}{2a}$	$\dfrac{2a}{2a}$
$p \in \{2, 3\}$	$\dfrac{3a}{2a}$	$\dfrac{2a}{2a}$	$\dfrac{2a}{2a}$

EPRE assumption, all data resource elements on all antenna ports and all OFDM symbols have to use a power of $2a$. However, we note that this constant-EPRE requirement results in a power waste of $4a$ on the third and fourth transmit antenna ports in the first OFDM symbol. This power loss can be avoided if we simply allow the transmission power of each resource element to increase from $2a$ to $3a$ on the third and fourth transmit antenna ports in the first OFDM symbol.

To clearly see how the power of each resource element changes when we deviate from the constant EPRE assumption, we summarize the data to reference signal power ratios for resource elements on different transmit antenna ports and in different OFDM symbols within a subframe, see Table 13.2. The OFDM symbols $l \in \{0, 4, 7, 11\}$ contain reference signals for antenna port 0 and antenna port 1 ($p = 0, 1$). The OFDM symbols $l \in \{1, 8\}$ contain reference signals for antenna port 2 and antenna port 3 ($p = 2, 3$). In contrast, all entries would have been $\frac{2a}{2a}$ in this four-antenna-ports SFBC-FSTD scheme if we had stayed with the constant EPRE assumption. Therefore, allowing different powers in the spatial domain across antenna ports can improve the total eNB power utilization.

We proceed to consider the case where the Node-B decides to boost the reference signal power by 3 dB to $4a$, up from the $2a$ in the previous example. With the constant EPRE assumption, reference signal power boosting is achieved by puncturing several data or control REs as illustrated in Figure 13.3. Again, the focus is on OFDM symbol number 0, $l = 0$. It is clear that data puncturing results in loss of both resource elements and transmit power in some antenna port and OFDM symbol combinations. Instead of data/control RE puncturing, if we are allowed to vary the power scaling of each antenna and OFDM symbol combination, then we arrive at the alternative solution shown in Figure 13.4.

The resource element power assignments for each antenna port and OFDM symbol combination for the reference signal boosting with non-constant EPRE are summarized in Table 13.3. We note that full eNB power from all the transmit antennas across all the OFDM symbols can be used if non-constant EPRE is allowed.

13.2.2 Downlink power allocation

In the current release of the LTE system, power imbalance is only allowed in the time domain across OFDM symbols. The EPRE is kept constant in the spatial-domain across antenna ports. Two PDSCH to reference signal EPRE ratios are defined, one for the OFDM symbols containing reference signals and the other for the OFDM symbols not containing any reference signals.

For each UE, the PDSCH to reference signal EPRE ratio among PDSCH REs in all the OFDM symbols not containing the reference signal is equal and is denoted by ρ_A. The UE may assume that for 16-QAM or 64-QAM and also for rank two or greater spatial multiplexing as

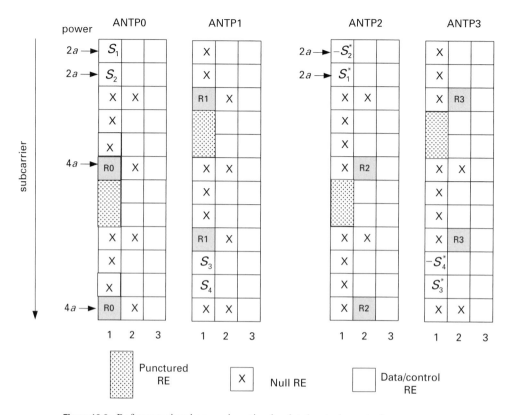

Figure 13.3. Reference signal power boosting by data/control puncturing.

well as for PDSCH transmissions:

$$\rho_A = P_A + \delta_{\text{power-offset}}, \qquad (13.19)$$

where $\delta_{\text{power-offset}}$ is 0 dB for all transmission modes except multi-user MIMO where it takes two values of 0 or -3 dB (see Chapter 7, Section 7.6.2). P_A is a UE specific semi-static parameter signaled by higher layers in the range of [3, 2, 1, 0, -1, -2, -3, -6] dB using 3-bits. For each UE, the PDSCH to reference signal EPRE ratio among PDSCH REs in all the OFDM symbols containing RS is equal and is denoted by ρ_B. The cell-specific ratio ρ_B/ρ_A is given by Table 13.4 according to the cell-specific parameter P_B signaled by higher layers and the number of cell-specific antenna ports configured at eNB.

For the MBSFN transmissions using PMCH with 16-QAM or 64-QAM, the UE may assume that the PMCH reference signal EPRE ratio is equal to 0 dB. This means that reference signal boosting is not supported for the MBSFN transmissions.

13.2.3 eNB power restrictions

The eNB power restrictions enable inter-cell interference coordination (ICIC) and load balancing (see Chapter 16). The interference experienced by a cell from a neighboring cell depends upon the neighbor cell's transmit power. Without ICIC, the transmit power spectral density is generally constant over the whole system bandwidth. In case of ICIC, however, the transmit

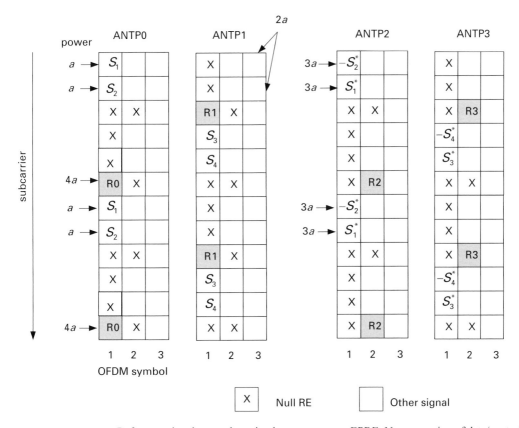

Figure 13.4. Reference signal power boosting by non-constant EPRE. No puncturing of data/control necessary.

Table 13.3. Non-constant data to reference signal power ratio with reference signal boosting.

	$l \in \{0, 4, 7, 11\}$	$l \in \{1, 8\}$	$l \in \{2, 3, 5, 6, 9, 10, 12, 13\}$
$p \in \{0, 1\}$	$\dfrac{a}{4a}$	$\dfrac{3a}{4a}$	$\dfrac{2a}{4a}$
$p \in \{2, 3\}$	$\dfrac{3a}{4a}$	$\dfrac{a}{4a}$	$\dfrac{2a}{4a}$

power is increased on certain parts of the frequency while it is decreased on certain other parts of the frequency. Note that the total eNB transmit power is fixed. Therefore, if power spectral density is increased on part of the bandwidth, less power is available for the remaining part of the bandwidth. The intention of power restrictions is to coordinate interference among cells in the frequency domain to increase throughput at the cell edges. In particular, neighboring cells may preferably schedule users in distinct sets of physical resource blocks to avoid mutual interference.

Transmissions to users located at the cell edge require larger transmit power than transmissions to users at the cell center. Therefore, the transmissions to users at the cell edge also

Table 13.4. Ratio of PDSCH to reference
signal EPRE in symbols with and without
reference symbols for $P = 1, 2, 4$ cell
specific antenna ports.

	ρ_B/ρ_A	
P_B	$P = 1$	$P = 2, 4$
0	1	5/4
1	4/5	1
2	3/5	3/4
3	2/5	1/2

cause more interference to the neighboring cells and are also prone to increased interference
from the neighboring cells. The transmissions to users at the cell center with reduced power
cause less interference to the neighboring cells in the physical resource blocks used for these
users.

A relative narrowband TX power (RNTP) indication is defined for exchange over the inter-
eNB X2 interface. The determination of reported RNTP X2 indicator bitmap RNTP(n_{PRB}) is
defined as follows:

$$RNTP(n_{PRB}) = \begin{cases} 0 & \text{if} \quad \dfrac{E_A(n_{PRB})}{E_{max\_nom}^{(p)}} \leq RNTP_{threshold} \\ 1 & \text{if} \quad \text{no promise about the upper limit of } \dfrac{E_A(n_{PRB})}{E_{max\_nom}^{(p)}} \text{ is made,} \end{cases}$$

(13.20)

where $E_A(n_{PRB})$ is the maximum intended EPRE of UE-specific PDSCH REs in OFDM symbols not containing RS in this physical resource block on antenna port p in the considered
future time interval. $n_{PRB} = 0, \ldots, (N_{RB}^{DL} - 1)$ is the physical resource block number. The
values for RNTP threshold $RNTP_{threshold}$ are:

$$RNTP_{threshold} \in \begin{Bmatrix} -\infty, -11, -10, -9, -8, -7, \ldots \\ -6, -5, -4, -3, -2, -1, 0, +1, +2, +3 \end{Bmatrix} \text{ dB.}$$

(13.21)

The threshold value $-\infty$ is used to indicate that the eNB intends not to use certain physical
resource blocks for UE-specific PDSCH transmission at all.

The nominal power spectral density $E_{max\_nom}^{(p)}$ in (13.20) is given as:

$$E_{max\_nom}^{(p)} = \frac{P_{max}^{(p)}}{\left(N_{RB}^{DL} \times \Delta f \times N_{SC}^{RB}\right)},$$

(13.22)

where $P_{max}^{(p)}$ is the eNB maximum output power and the denominator is total system bandwidth. An example of X2 indicator bitmap RNTP(n_{PRB}) for the case of 10 physical resource
blocks is depicted in Figure 13.5. We assumed that $\frac{E_A(n_{PRB})}{E_{max\_nom}^{(p)}} > RNTP_{threshold}$ for resource
blocks numbered $n_{PRB} = 0, 1, 2, 7$ while $\frac{E_A(n_{PRB})}{E_{max\_nom}^{(p)}} < RNTP_{threshold}$ for the remaining physical
resource blocks. The indicator bitmap RNTP(n_{PRB}) of size N_{RB}^{DL} and the 4-bit RNTP threshold

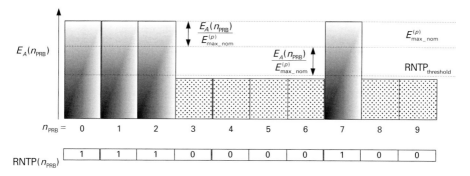

Figure 13.5. An example of X2 indicator bitmap RNTP(n_{PRB}) for $N_{\text{RB}}^{\text{DL}} = 10$.

RNTP$_{\text{threshold}}$ is exchanged in a single message on the X2 interface to provide the complete information.

13.3   Summary

The uplink power control in the LTE system is event driven in the sense that regular transmit power control (TPC) commands are not transmitted. The TPC commands for uplink power control are transmitted on the downlink PDCCH. The UE attempts to decode PDCCH formats containing TPC commands in every subframe. The power control mechanism itself is a combination of open-loop and closed-loop power control. Two types of uplink power adjustments namely accumulated and absolute power corrections are supported. The type of power adjustment is UE-specific and can be configured by RRC in a semi-static fashion.

In the downlink, dynamic power control is applied to dedicated control channels addressed to a single UE or a group of UEs. No feedback of TPC commands is provided on the uplink and power allocation is based on the downlink channel quality feedback from the UEs. Different power levels can be allocated to different resource blocks used for data transmission in a semi-static way to support inter-cell interference coordination (ICIC). Moreover, two different power levels can be used on OFDM symbols used for data transmission within a subframe to improve eNB power utilization. This is because the available power for data resource elements is different between OFDM symbols containing reference signals and OFDM symbols containing no reference signals. However, different power levels across antenna ports to resolve the spatial-domain power imbalance issue are not permitted.

References

[1] 3GPP TSG RAN v8.3.0, Evolved Universal Terrestrial Radio Access: Physical Channels and Modulation (3GPP TS 36.211).

[2] 3GPP TSG RAN v8.3.0 , Evolved Universal Terrestrial Radio Access: Multiplexing and Channel Coding (3GPP TS 36.212).

[3] 3GPP TSG RAN v8.3.0, Evolved Universal Terrestrial Radio Access: Physical Layer Procedures (3GPP TS 36.213).

14 Uplink control signaling

The LTE system supports fast dynamic scheduling on a per subframe basis to exploit gains from channel-sensitive scheduling. Moreover, advanced techniques such as link adaptation, hybrid ARQ and MIMO are employed to meet the performance goals. A set of physical control channels are defined in both the uplink and the downlink to enable the operation of these techniques. In order to support channel sensitive scheduling and link adaptation in the downlink, the UEs measure and report their channel quality information back to the eNB. Similarly, for downlink hybrid ARQ operation, the hybrid ARQ ACK/NACK feedback from the UE is provided in the uplink.

Two types of feedback information are required for MIMO operation, the first is MIMO rank information and the second is preferred precoding information. It is well known that even when a system supports $N \times N$ MIMO, rank-N or N MIMO layers transmission is not always beneficial. The MIMO channel experienced by a UE generally limits the maximum rank that can be used for transmission. In general, for weak users in the system, a lower rank transmission is preferred over a higher rank transmission. This is because at low SINR, the capacity is power limited and not degree-of-freedom limited and therefore multiple layers transmission is not helpful. Moreover, when the antennas are correlated, the channel matrix is rank deficient leading to a single layer or rank-1 transmission. Therefore, the system should support a variable number of MIMO layers transmission to maximize gains from MIMO. In order to support rank adaptation, the channel rank information needs to be provided to the eNB.

In a closed-loop MIMO system, the knowledge of channel state information at the transmitter can help improve overall system throughput. This is because with channel state information available, eNB can select transmit antenna weights that maximize the received signal at the UE. In a TDD system, the channel state information can be derived based on uplink signal transmissions thanks to channel reciprocity. However, in an FDD system, the UE needs to feedback the channel state information to the eNB. In order to reduce the channel state information feedback overhead, the channel state is generally quantized by using a precoding codebook. The UE then selects a preferred precoder from the codebook that matches to the observed channel condition. This information on this preferred precoder is reported back to the eNB.

Another important aspect for uplink signaling design is the amount of feedback overhead required. In an OFDM system, multiple feedback reports each covering a frequency subband are required to exploit frequency-selective multi-user scheduling gains. The introduction of multiple MIMO layers further increases the feedback overhead as the channel quality information for each subband now needs to be reported for each MIMO codeword separately. In order to keep the feedback overhead low, differential encoding of channel quality is supported for both subband channel quality and MIMO codeword channel quality feedback. In addition, multiple feedback modes are supported, each tailored to a particular transmission mode to match the feedback format to the transmission scheme. For example, when transmit diversity is used, the information on precoding is not required. Similarly, spatial differential CQI only needs to be provided when two codewords MIMO is configured and the rank is greater than one. The specification of multiple feedback formats help to keep the feedback overhead low at the expense of increased system complexity.

14.1   Data control multiplexing

With a frequency-multiplexing-based uplink access scheme as SC-FDMA, it is natural to think of frequency-multiplexing for control and data transmitted from a single UE. The frequency-multiplexing approach offers numerous advantages relative to a TDM approach such as power balancing between control and data transmissions. A higher power spectral density can be allowed on the control frequencies than the data frequencies thus improving the system coverage. It should be noted that the system coverage is generally limited by uplink and particularly by uplink control transmission. The power-balancing scheme in the FDM approach is similar to code-multiplexing in a WCDMA system that allows using different power levels on different codes transmitted. A drawback of the frequency-multiplexing approach, however, is that it violates the single-carrier property of SC-FDMA.

The time multiplexing of data and control retains the single carrier property, which allows one to maintain low PAPR/CM. However, a major drawback of time-multiplexing approach is that power cannot be shared between control and data. In the TDM approach, the data and control is time-multiplexed at the input of DFT as shown in Figure 14.1. In order to make the control reliable in power-limited situations when a UE reaches its maximum transmit power, the information needs to be repeated in time resulting in a waste of system resources. This is because if power cannot be increased, the transmission time needs to be increased to be able to transmit certain energy.

In the FDM data/control multiplexing approach, different power gains can be used on subcarriers belonging to data and control as shown in Figure 14.2. In general, multiple types of control information with different error requirements need to be transmitted on the uplink. For example, hybrid ARQ ACK/NACK feedback generally has a lower error rate requirement

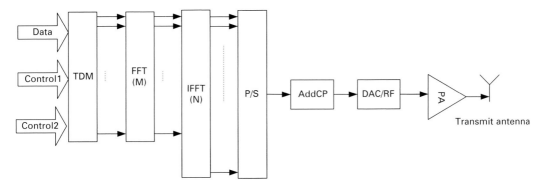

Figure 14.1. Time-division multiplexing of data control.

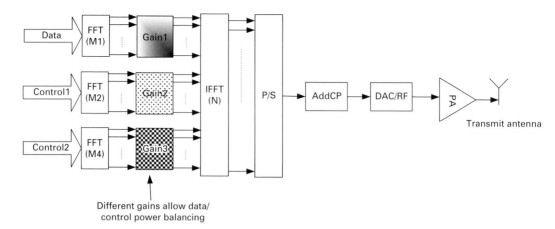

Figure 14.2. Frequency division multiplexing of data control.

than the channel quality information. The FDM approach further allows one to control power on different types of control information to meet the error requirements. It should be noted that if different power gains are used on control and data information that is time-multiplexed at the input of DFT as in Figure 14.1, the PAPR/CM of the time-domain signal at the output of IDFT increases significantly. This is because DFT and IDFT operations partially cancel each other and therefore the signal at the output of IDFT has characteristics of signal at the input of DFT as discussed in Chapter 4.

We analyze the PAPR performance of the FDM approach in Figure 14.3. We assumed an IFFT size of 512 and FFT precoding size of 64 ($M_1 = 64$) for data. Furthermore, we assumed FFT precoding size of 4 ($M_2 = 4$) for both controls 1 and 2. We also assumed that the power spectral density of the control subcarriers is 10 dB higher than the data subcarriers. The overall scenario can be seen as a worst case, where two types of control are transmitted in parallel with data with a large difference in transmit power. The PAPR performance for this scenario is plotted in Figure 14.3. We also show the e^{-x} reference curve, which represents the PAPR for an OFDM scheme (see Chapter 5). We note that the FDM of data and control increases PAPR with a higher increase when both controls 1 and 2 are frequency-multiplexed with data. However, we note that even after the increase, the PAPR at the 0.1% point is more than 1 dB better than OFDMA PAPR. Similarly, the cubic metric (CM) for SC-FDMA with frequency-multiplexed control 1 and control 2 is approximately 0.9 dB better than OFDMA. We should point out that the relative increase in PAPR due to FDM would be smaller when more bandwidth is allocated to data or when data uses higher-order modulations such as 16-QAM. Moreover, when the difference in control and data transmit power is relatively smaller, the increase in PAPR/CM is also smaller.

Both the TDM and FDM data/control multiplexing approaches were studied for the LTE system. The scheme finally adopted in the standard includes both FDM and TDM components. However, the FDM approach is only used when there is no uplink data transmission from a UE. This means that control from one UE can be frequency-multiplexed with data or control from other UEs. However, in the presence of uplink data transmission from a UE, the control is time-multiplexed at the input of DFT. The FDM control is transmitted using a standalone physical layer control channel referred to as the physical uplink control channel (PUCCH).

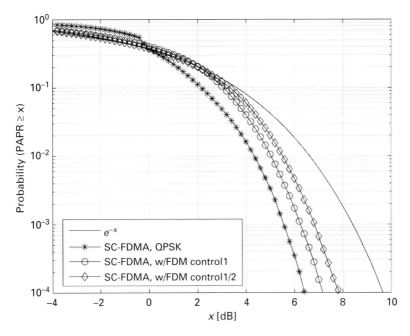

Figure 14.3. PAPR performance for frequency division multiplexing of data control.

In the presence of both control and data, the control is multiplexed on the same bandwidth as allocated for data on the uplink data channel referred to as the physical uplink shared channel (PUSCH).

Another question that arises for uplink control transmission is where to place the control in the frequency domain. A simple answer would be to distribute control over a larger bandwidth to exploit frequency diversity. However, another constraint arising from keeping the single-carrier property of SC-FDMA is that data allocations need to be contiguous in frequency. The distribution of control in the whole bandwidth would mean that data allocations would be fragmented with some subcarriers scattered in the band for control information transmission. This would mean an increase in data PAPR/CM when data transmissions happen on non-contiguous frequency resources as pointed out in Chapter 5. In order to address this issue, the PUCCH carrying control information is transmitted at the two edges of the allocated bandwidth as shown in Figure 14.4. Moreover, the PUCCH is hopped from one edge in the first slot within a subframe to the other frequency edge in the second slot within a subframe. The hopping approach allows one to capture some frequency-diversity while making a contiguous bandwidth available for data in the middle of the band in both the time slots within a subframe.

14.2 Control signaling contents

14.2.1 Channel quality indication

A CQI index is defined in terms of a channel coding rate value and modulation scheme (QPSK, 16-QAM, 64-QAM) as given in Table 14.1. In addition to 4-bit absolute CQI indices, three

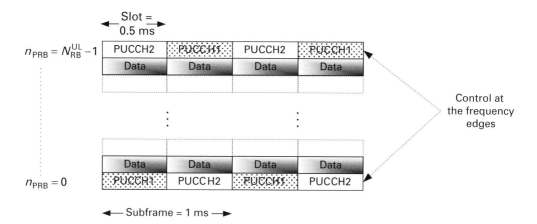

Figure 14.4. PUCCH multiplexing.

Table 14.1. 4-bit CQI table.

CQI index	Modulation	Coding rate	Efficiency [b/s/Hz]
0		Out of range	
1	QPSK	78/1024	0.1523
2	QPSK	120/1024	0.2344
3	QPSK	193/1024	0.3770
4	QPSK	308/1024	0.6016
5	QPSK	449/1024	0.8770
6	QPSK	602/1024	1.1758
7	16 QAM	378/1024	1.4766
8	16 QAM	490/1024	1.9141
9	16 QAM	616/1024	2.4063
10	64 QAM	466/1024	2.7305
11	64 QAM	567/1024	3.3223
12	64 QAM	666/1024	3.9023
13	64 QAM	772/1024	4.5234
14	64 QAM	873/1024	5.1152
15	64 QAM	948/1024	5.5547

differential CQI values are defined to reduce the CQI signaling overhead. In the case of two codewords MIMO, 3-bit spatial differential CQI represents the difference between CQI value for codeword 1 and CQI value for codeword 2 as:

$$CQI_{Diff\text{-}spatial} = CQI_{CW1} - CQI_{CW2}. \tag{14.1}$$

The set of $CQI_{Diff\text{-}spatial}$ values is $\{-4, -3, -2, -1, 0, +1, +2, +3\}$. The subband CQIs for each codeword are also encoded differentially with respect to their respective wideband CQI using 2 bits as defined by:

$$CQI_{Diff\text{-}sb}(i) = CQI_{sb}(i) - CQI_{WB}. \tag{14.2}$$

The set of possible subband differential CQI values is $\{-2, 0, +1, +2\}$. We note that two values are used to represent subband CQIs larger than the wideband CQI ($+1$ and $+2$) while only a single value (-2) is used to represent a subband CQI lower than the wideband CQI. This means that subband upfades are quantized with higher granularity than the subband downfades. When a proportional-fair scheduler is used for frequency-selective channel sensitive scheduling, a UE is more likely to be scheduled on subbands experiencing upfades than the subbands experiencing downfades. Quantizing subband upfades with higher granularity results in more accurate CQI estimates for the scheduled subbands.

The CQI value for the M selected subbands for each codeword is encoded differentially using 2 bits relative to the respective wideband CQI as:

$$\text{CQI}_{\text{Diff-}M} = \text{CQ}_{\text{avg-}M} - \text{CQI}_{\text{WB}}. \tag{14.3}$$

The possible differential CQI $\text{CQI}_{\text{Diff-}M}$ values are $\{+1, +2, +3, +4\}$. It should be noted that when best-M subbands are selected, the average CQI on these subbands is always larger than the wideband CQI, which is averaged over a larger bandwidth.

We will note that a UE always reports the wideband CQI even when it selects a subset of subbands. This is because wideband CQI is required for setting the power levels for downlink control channels that are transmitted in a frequency diverse transmission format over the wideband to exploit frequency diversity.

14.2.2 Rank and precoding indication

The precoding codebooks for two and four antenna ports are given in Tables 14.2 and 14.3 respectively. The codebook for the two-antenna ports is based on a DFT matrix, while the codebook for the four-antenna ports is based on the Householder principle (see Chapter 7, Section 7.4.2). The MIMO transmission rank can be either one or two for the case of two-antenna ports requiring single-bit rank indication (RI). The numbers of precoders for the two-antenna ports are four and two for rank-1 and rank-2 respectively. Therefore, the precoding matrix indication (PMI) requires two bits for rank-1 and a single bit for rank-2.

Table 14.2. Codebook for transmission on antenna ports $\{0, 1\}$.

Codebook index	Number of layers υ	
	1	2
0	$\frac{1}{\sqrt{2}}\begin{bmatrix} 1 \\ 1 \end{bmatrix}$	$\frac{1}{2}\begin{bmatrix} 1 & 1 \\ 1 & -1 \end{bmatrix}$
1	$\frac{1}{\sqrt{2}}\begin{bmatrix} 1 \\ -1 \end{bmatrix}$	$\frac{1}{2}\begin{bmatrix} 1 & 1 \\ j & -j \end{bmatrix}$
2	$\frac{1}{\sqrt{2}}\begin{bmatrix} 1 \\ j \end{bmatrix}$	—
3	$\frac{1}{\sqrt{2}}\begin{bmatrix} 1 \\ -j \end{bmatrix}$	—

Table 14.3. Codebook for transmission on antenna ports $\{0, 1, 2, 3\}$.

Codebook index	u_n	Number of layers v			
		1	2	3	4
0	$u_0 = [\,1 \ \ -1 \ \ -1 \ \ -1\,]^T$	$W_0^{\{1\}}$	$W_0^{\{14\}}\big/\sqrt{2}$	$W_0^{\{124\}}\big/\sqrt{3}$	$W_0^{\{1234\}}\big/2$
1	$u_1 = [\,1 \ \ -j \ \ 1 \ \ j\,]^T$	$W_1^{\{1\}}$	$W_1^{\{12\}}\big/\sqrt{2}$	$W_1^{\{123\}}\big/\sqrt{3}$	$W_1^{\{1234\}}\big/2$
2	$u_2 = [\,1 \ \ 1 \ \ -1 \ \ 1\,]^T$	$W_2^{\{1\}}$	$W_2^{\{12\}}\big/\sqrt{2}$	$W_2^{\{123\}}\big/\sqrt{3}$	$W_2^{\{3214\}}\big/2$
3	$u_3 = [\,1 \ \ j \ \ 1 \ \ -j\,]^T$	$W_3^{\{1\}}$	$W_3^{\{12\}}\big/\sqrt{2}$	$W_3^{\{123\}}\big/\sqrt{3}$	$W_3^{\{3214\}}\big/2$
4	$u_4 = [\,1 \ \ (-1-j)\big/\sqrt{2} \ \ -j \ \ (1-j)\big/\sqrt{2}\,]^T$	$W_4^{\{1\}}$	$W_4^{\{14\}}\big/\sqrt{2}$	$W_4^{\{124\}}\big/\sqrt{3}$	$W_4^{\{1234\}}\big/2$
5	$u_5 = [\,1 \ \ (1-j)\big/\sqrt{2} \ \ j \ \ (-1-j)\big/\sqrt{2}\,]^T$	$W_5^{\{1\}}$	$W_5^{\{14\}}\big/\sqrt{2}$	$W_5^{\{124\}}\big/\sqrt{3}$	$W_5^{\{1234\}}\big/2$
6	$u_6 = [\,1 \ \ (1+j)\big/\sqrt{2} \ \ -j \ \ (-1+j)\big/\sqrt{2}\,]^T$	$W_6^{\{1\}}$	$W_6^{\{13\}}\big/\sqrt{2}$	$W_6^{\{134\}}\big/\sqrt{3}$	$W_6^{\{1324\}}\big/2$
7	$u_7 = [\,1 \ \ (-1+j)\big/\sqrt{2} \ \ j \ \ (1+j)\big/\sqrt{2}\,]^T$	$W_7^{\{1\}}$	$W_7^{\{13\}}\big/\sqrt{2}$	$W_7^{\{134\}}\big/\sqrt{3}$	$W_7^{\{1324\}}\big/2$
8	$u_8 = [\,1 \ \ -1 \ \ 1 \ \ 1\,]^T$	$W_8^{\{1\}}$	$W_8^{\{12\}}\big/\sqrt{2}$	$W_8^{\{124\}}\big/\sqrt{3}$	$W_8^{\{1234\}}\big/2$
9	$u_9 = [\,1 \ \ -j \ \ -1 \ \ -j\,]^T$	$W_9^{\{1\}}$	$W_9^{\{14\}}\big/\sqrt{2}$	$W_9^{\{134\}}\big/\sqrt{3}$	$W_9^{\{1234\}}\big/2$
10	$u_{10} = [\,1 \ \ 1 \ \ 1 \ \ -1\,]^T$	$W_{10}^{\{1\}}$	$W_{10}^{\{13\}}\big/\sqrt{2}$	$W_{10}^{\{123\}}\big/\sqrt{3}$	$W_{10}^{\{1324\}}\big/2$
11	$u_{11} = [\,1 \ \ j \ \ -1 \ \ j\,]^T$	$W_{11}^{\{1\}}$	$W_{11}^{\{13\}}\big/\sqrt{2}$	$W_{11}^{\{134\}}\big/\sqrt{3}$	$W_{11}^{\{1324\}}\big/2$
12	$u_{12} = [\,1 \ \ -1 \ \ -1 \ \ 1\,]^T$	$W_{12}^{\{1\}}$	$W_{12}^{\{12\}}\big/\sqrt{2}$	$W_{12}^{\{123\}}\big/\sqrt{3}$	$W_{12}^{\{1234\}}\big/2$
13	$u_{13} = [\,1 \ \ -1 \ \ 1 \ \ -1\,]^T$	$W_{13}^{\{1\}}$	$W_{13}^{\{13\}}\big/\sqrt{2}$	$W_{13}^{\{123\}}\big/\sqrt{3}$	$W_{13}^{\{1324\}}\big/2$
14	$u_{14} = [\,1 \ \ 1 \ \ -1 \ \ -1\,]^T$	$W_{14}^{\{1\}}$	$W_{14}^{\{13\}}\big/\sqrt{2}$	$W_{14}^{\{123\}}\big/\sqrt{3}$	$W_{14}^{\{3214\}}\big/2$
15	$u_{15} = [\,1 \ \ 1 \ \ 1 \ \ 1\,]^T$	$W_{15}^{\{1\}}$	$W_{15}^{\{12\}}\big/\sqrt{2}$	$W_{15}^{\{123\}}\big/\sqrt{3}$	$W_{15}^{\{1234\}}\big/2$

For the case of four antenna ports, the MIMO transmission rank can be one, two, three or four requiring two bits rank indication. The number of precoders for each rank is 16 and therefore requires four bits PMI indication.

14.2.3 Hybrid ARQ ACK/NACK

The uplink ACK/NACK provides feedback for downlink hybrid ARQ transmissions. A maximum of two codewords are transmitted in the downlink when the rank is greater than one for MIMO spatial multiplexing. A single codeword is transmitted in all other cases. The hybrid ARQ ACK/NACK is provided independently for each codeword. The number of ACK/NACK bits are one and two for single codeword and two codewords hybrid ARQ transmission as given in Table 14.4.

14.3   Periodic reporting

For uplink control information, both periodic and aperiodic reporting are supported. The periodic reporting of CQI, PMI and RI is carried out using PUCCH while aperiodic reporting is done on the PUSCH data channel as depicted in Figure 14.5. The reporting types supported on PUCCH include wideband CQI and UE selected subband CQI. With aperiodic reporting on PUSCH, in addition to wideband CQI and UE selected subband CQI, higher layer configured subband CQI reporting is also supported.

In periodic reporting, a UE is semi-statically configured by higher layers in one of the modes given in Table 14.5. The modes without any PMI namely mode 1–0 and mode 2–0 are used for single transmit antenna port, transmit diversity and open-loop spatial multiplexing. The two modes with a single PMI (mode 1–1 and mode 2–1) are used for closed-loop spatial

Table 14.4. Uplink ACK/NACK.

	ACK bits	Reporting mode
One-bit ACK/NACK	1	Codeword 0 positive acknowledgment (ACK)
	0	Codeword 0 negative acknowledgment (NACK)
Two-bits ACK/NACK	11	Codeword 0 ACK codeword 1 ACK
	10	Codeword 0 ACK codeword 1 NACK
	01	Codeword 0 NACK codeword 1 ACK
	00	Codeword 0 NACK codeword 1 NACK

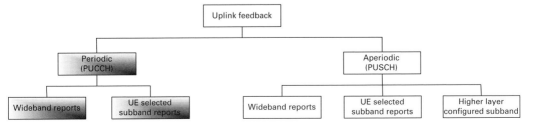

Figure 14.5. Uplink reporting schemes.

Table 14.5. Periodic CQI/PMI reporting modes.

	PMI feedback	Reporting mode
Wideband CQI	No PMI	Mode 1–0
	Single PMI	Mode 1–1
UE selected subband CQI	No PMI	Mode 2–0
	Single PMI	Mode 2–1

multiplexing. It should be noted that subband PMI is not supported with periodic reports on PUCCH.

For the UE-selected subband CQI, a CQI report in a certain subframe describes the channel quality in a particular part or in particular parts of the bandwidth referred to as bandwidth part (BP) or parts. The number of subbands N is given as:

$$N = \left\lceil N_{RB}^{DL}/k \right\rceil, \tag{14.4}$$

where $\left\lfloor N_{RB}^{DL}/k \right\rfloor$ subbands are of size k. Also, if $\left\lceil N_{RB}^{DL}/k \right\rceil - \left\lfloor N_{RB}^{DL}/k \right\rfloor > 0$ then one of the subbands is of size $\left(N_{RB}^{DL} - k \cdot \left\lfloor N_{RB}^{DL}/k \right\rfloor \right)$. A bandwidth part is frequency-consecutive and consists of N_J subbands given as:

$$N_J = \begin{cases} \left\lceil \dfrac{N_{RB}^{DL}}{k} \right\rceil & J = 1 \\[3ex] \left\lceil \dfrac{N_{RB}^{DL}}{J \times k} \right\rceil \quad or \quad \left\lfloor \dfrac{N_{RB}^{DL}}{J \times k} \right\rfloor & J > 1, \end{cases} \tag{14.5}$$

where J bandwidth parts span system bandwidth N_{RB}^{DL} as given in Table 14.6. Each bandwidth part j is scanned in sequential order according to increasing frequency as defined by:

$$j = N_{SF} \bmod J, \tag{14.6}$$

where N_{SF} is a counter that a UE increments after each subband report transmission for the bandwidth part. In case of UE selected subband feedback, a single subband out of N_J subbands of a bandwidth part is selected as shown in Figure 14.6. The CQI for this subband is then reported along with a corresponding L-bit label where $L = \left\lceil \log_2 N_J \right\rceil$.

In the case where RI and wideband CQI/PMI reporting are configured, RI and wideband CQI/PMI are time-multiplexed as shown in Figure 14.7. The RI reporting is done on a slower basis with RI reporting interval as an integer multiple of wideband CQI/PMI period with an offset applied. In Figure 14.7, the RI reporting is two times slower than CQI/PMI reporting and an offset of one subframe is applied to RI reporting. In the case when RI, wideband CQI/PMI and subband CQI reporting are configured, the same set of CQI reporting instances is shared by both wideband CQI/PMI and subband CQI reports in a time-multiplexed fashion.

The CQI and PMI payload sizes of each PUCCH reporting mode are given in Table 14.7. We note that four CQI/PMI and RI reporting *types* with distinct periods and offsets are supported.

In the four reporting *modes* in Table 14.5, namely modes 1–0, 1–1, 2–0 and 2–2, RI feedback is done in the same way. In the subframe where RI is reported, a UE determines a RI assuming

Table 14.6. Subband size (k) and bandwidth parts (J).

System bandwidth N_{RB}^{DL}	Subband size k (RBs)	Bandwidth parts (J)
6–7	Wideband CQI only	1
8–10	4	1
11–26	4	2
27–64	6	3
65–110	8	4

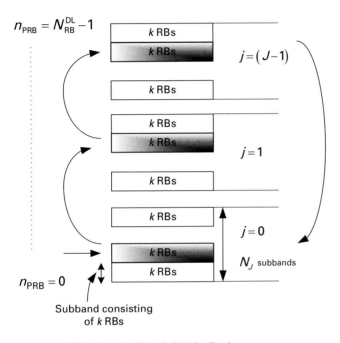

Figure 14.6. UE-selected subband CQI feedback.

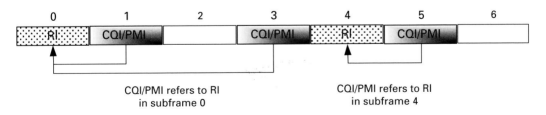

Figure 14.7. CQI/PMI and RI reporting.

transmission on set S subbands and reports back this RI using a type-3 report. In mode 1–0, a wideband CQI and RI is reported without any PMI feedback. In the subframe where CQI is reported, a UE reports a type-4 report consisting of one wideband CQI value, which is calculated assuming transmission on set S subbands. In mode 1–1, in the subframe where CQI/PMI is reported, a single precoding matrix is selected from the codebook subset assuming

Table 14.7. PUCCH report types and payload sizes.

PUCCH report type	Reported	Mode state	PUCCH reporting modes			
			Mode 1–1 (bits/BP)	Mode 2–1 (bits/BP)	Mode 1–0 (bits/BP)	Mode 2–0 (bits/BP)
1	Subband CQI	RI = 1	NA	4 + L	NA	4 + L
		RI > 1	NA	7 + L	NA	4 + L
2	Wideband CQI/PMI	2 TX antennas RI = 1	6	6	NA	NA
		4 TX antennas RI = 1	8	8	NA	NA
		2 TX antennas RI > 1	8	8	NA	NA
		4 TX antennas RI > 1	11	11	NA	NA
3	RI	2-layer spatial multiplexing	1	1	1	1
		4-layer spatial multiplexing	2	2	2	2
4	Wideband CQI	RI = 1	NA	NA	4	4

transmission on set S subbands and conditioned on the last reported RI. The set S subbands represent the system bandwidth N_{RB}^{DL}. The UE then feedback wideband CQI and PMI consisting of, for example for four antenna ports, four CQI bits and four PMI bits with a total of eight bits when the rank is one. The feedback additionally includes a 3-bit spatial differential CQI when the rank is greater than one with a total of 11 bits for the case of four antenna ports.

The reporting is done in a similar way for mode 2–0 and mode 2–1 with the difference that the UE selects the subbands over which CQI/PMI is calculated. This requires an additional L-bit label to indicate the selected subband. In mode 2–0, in the subframe where wideband CQI is reported, the UE reports a type 4 report on each respective successive reporting opportunity consisting of one wideband CQI value conditioned on the last reported RI. In the subframe where CQI for the selected subbands is reported, the UE selects the preferred subband within the set of N_j subbands in each of the J bandwidth parts where J is given in Table 14.6. The UE reports a type-1 report consisting of one CQI value reflecting transmission only over the selected subband of a bandwidth part determined along with the corresponding best subband L-bit label. A type-1 report consisting of $(L + 4)$ bits for each bandwidth part is in turn reported in respective successive reporting opportunities. The CQI represents channel quality across all layers irrespective of the rank information (RI).

In mode 2–1, in the subframe where wideband CQI/PMI is reported, the feedback is done in a way similar to that in mode 1–1. In the subframe where CQI for the selected subbands is reported, the UE selects the preferred subband within the set of N_j subbands in each of the J bandwidth parts given in Table 14.6. The UE reports a type-1 report per bandwidth part on each respective successive reporting opportunity consisting of a single CQI value reflecting transmission only over the selected subband of a bandwidth part along with the corresponding best subband L-bit label. When rank is greater than 1, an additional 3-bit spatial differential CQI is reported with a total of $(L + 7)$ bits per bandwidth part.

Table 14.8. CQI/PMI size for single antenna port, transmit diversity and open loop spatial multiplexing.

Feedback type	Field	Bitwidth
Wideband CQI	Wideband CQI	4
UE selected subband CQI reports	Subband CQI	4
	Subband label	1 or 2

Table 14.9. CQI/PMI size for closed loop spatial multiplexing.

		Bitwidth			
		2 antenna ports		4 antenna ports	
Feedback type	Field	Rank $= 1$	Rank $= 2$	Rank $= 1$	Rank > 1
Wideband CQI reports	Wideband CQI	4	4	4	4
	Spatial differential CQI	0	3	0	3
	Precoding matrix indication	2	1	4	4
	Total	6	8	8	11
UE selected subband CQI reports	Subband CQI	4	4	4	4
	Spatial differential CQI	0	3	0	3
	Subband label	1 or 2	1 or 2	1 or 2	1 or 2
	Total	5 or 6	8 or 9	5 or 6	8 or 9

The number of bits for the two CQI feedback types, namely wideband CQI and UE selected subband CQI for a single antenna port, transmit diversity and open loop spatial multiplexing are given in Table 14.8. Similarly, the number of CQI/PMI bits for the two feedback types for closed loop spatial multiplexing is summarized in Table 14.9.

14.3.1 Channel coding for periodic reporting

Three forms of channel coding are used for periodic reporting on PUCCH, one for CQI/PMI, another for hybrid ARQ ACK/NACK and scheduling request and finally one for a combination of CQI/PMI and hybrid ARQ ACK/NACK. The CQI/PMI bits input to the channel coding block are denoted by $a_0, a_1, a_2, a_3, \ldots, a_{A-1}$, where A is the number of bits and depends on the transmission format. These bits are coded using a variable Reed–Muller (RM) $(20, A)$ code whose codewords are a linear combination of the 13 basis sequences denoted $M_{i,n}$ as given in Table 14.10. The 13 basis sequences allow to code a maximum of 13 PUCCH payload bits, which is the case when two bits of ACK/NACK are jointly coded with a maximum of 11 bits of CQI/PMI from Table 14.9. The bit sequence after encoding denoted by $b_0, b_1, b_2, b_3, \ldots, b_{B-1}$ is given as:

$$b_i = \sum_{n=0}^{A-1} (a_n \cdot M_{i,n}) \bmod 2 \quad i = 0, 1, 2, \ldots, (B-1). \tag{14.7}$$

When CQI/PMI and hybrid ARQ ACK/NACK need to be transmitted simultaneously, ACK/NACK bits are either appended to the coded CQI/PMI sequence or coded jointly with CQI/PMI. The first approach is used for the normal cyclic prefix where two reference signals

Table 14.10. Basis sequences for $(20, A)$ code.

i	$M_{i,0}$	$M_{i,1}$	$M_{i,2}$	$M_{i,3}$	$M_{i,4}$	$M_{i,5}$	$M_{i,6}$	$M_{i,7}$	$M_{i,8}$	$M_{i,9}$	$M_{i,10}$	$M_{i,11}$	$M_{i,12}$
0	1	1	0	0	0	0	0	0	0	0	1	1	0
1	1	1	1	0	0	0	0	0	0	1	1	1	0
2	1	0	0	1	0	0	1	0	1	1	1	1	1
3	1	0	1	1	0	0	0	0	1	0	1	1	1
4	1	1	1	1	0	0	0	1	0	0	1	1	1
5	1	1	0	0	1	0	1	1	1	0	1	1	1
6	1	0	1	0	1	0	1	0	1	1	1	1	1
7	1	0	0	1	1	0	0	1	1	0	1	1	1
8	1	1	0	1	1	0	0	1	0	1	1	1	1
9	1	0	1	1	1	0	1	0	0	1	1	1	1
10	1	0	1	0	0	1	1	1	0	1	1	1	1
11	1	1	1	0	0	1	1	0	1	0	1	1	1
12	1	0	0	1	0	1	0	1	1	1	1	1	1
13	1	1	0	1	0	1	0	1	0	1	1	1	1
14	1	0	0	0	1	1	0	1	0	0	1	0	1
15	1	1	0	0	1	1	1	1	0	1	1	0	1
16	1	1	1	0	1	1	1	0	0	1	0	1	1
17	1	0	0	1	1	1	0	0	1	0	0	1	1
18	1	1	0	1	1	1	1	1	0	0	0	0	0
19	1	0	0	0	0	1	1	0	0	0	0	0	0

within a slot are available for PUCCH as shown in Figure 14.9 below. The ACK/NACK is carried by modulating the second reference signal within the slot. In the case of the extended cyclic prefix where one fewer SC-FDMA symbol is available within a slot, a single reference signal symbol is used within a slot. The ACK/NACK, in this case, is not modulated on the single reference signal due to concerns on channel estimation performance. Therefore, ACK/NACK is jointly coded with CQI/PMI. The joint coding, however, results in a weaker code which affects the performance of both CQI/PMI and ACK/NACK.

The number of hybrid ARQ ACK/NACK feedback bits is one or two for a single-codeword or two-codeword downlink transmission respectively. In the case of the normal cyclic prefix with one or two ACK/NACK bits appended to the coded CQI/PMI sequence, the resulting length of coded bit sequence is $B + 1 = 21$ or $B + 2 = 22$ bits. For the case of joint coding the number of bits at the input of the $(20, A)$ code is $(A + 1)$ or $(A + 2)$ for the case of a single bit ACK/NACK or two bits ACK/NACK respectively, where A is the number of CQI/PMI bits.

14.3.2 PUCCH formats

The physical uplink control channel supports multiple formats as shown in Table 14.11. The combinations of uplink control information formats supported on PUCCH include hybrid ARQ ACK/NACK using PUCCH format 1A or 1B, the Scheduling Request (SR) using PUCCH format 1, hybrid ARQ ACK/NACK and SR using PUCCH format 1A or 1B, CQI/PMI using PUCCH format 2, CQI/PMI and hybrid ARQ ACK/NACK using PUCCH format 2A or 2B for the normal cyclic prefix or format 2 for the extended cyclic prefix.

Formats 2A and 2B are supported for the normal cyclic prefix only. The formats 1A and 1B are used for standalone transmission of single-bit and two-bits ACK/NACK respectively.

Table 14.11. PUCCH formats.

PUCCH format	Modulation scheme	Number of bits per subframe M_{bit}
1	N/A	N/A
1A	BPSK	1
1B	QPSK	2
2	QPSK	20
2A	QPSK+BPSK	21
2B	QPSK+BPSK	22

The format 2 is used when CQI/PMI is transmitted without ACK/NACK multiplexing using a $(20, A)$ code or when CQI/PMI and ACK/NACK are jointly coded for the case of the extended cyclic prefix. It should be noted that the number of channel bits to be transmitted in both cases is 20. The formats 2A and 2B are used when CQI/PMI and ACK/NACK are transmitted simultaneously for the case of the normal cyclic prefix. We note that the number of channel bits in this case is 21 or 22 for single-bit and two-bits ACK/NACK feedback respectively. The modulation for single-bit ACK/NACK is BPSK while QPSK is used in all other cases.

All PUCCH formats use a cyclic shift of a sequence in each symbol, where $n_{cs}^{cell}(n_s, l)$ is used to derive the cyclic shift for different PUCCH formats. In order to randomize the inter-cell interference, the quantity $n_{cs}^{cell}(n_s, l)$ is varied with the symbol number l and the slot number n_s according to:

$$n_{cs}^{cell}(n_s, l) = \sum_{i=0}^{7} c\left(8N_{symb}^{UL} \cdot n_s + 8l + i\right) \cdot 2^i, \qquad (14.8)$$

where $c(i)$ is the pseudo-random sequence. The pseudo-random sequence generator is initialized with $c_{init} = N_{ID}^{cell}$ at the beginning of each radio frame.

The physical resources used for PUCCH depend on two parameters, $N_{RB}^{(2)}$ and $N_{cs}^{(1)}$, which are set by higher layers. The variable $N_{RB}^{(2)} \geq 0$ denotes the bandwidth in terms of resource blocks that are reserved exclusively for PUCCH formats 2/2A/2B transmission in each slot. The variable $N_{cs}^{(1)} \in \{0, 1, \ldots, 8\}$ denotes the number of cyclic shift used for PUCCH formats 1/1A/1B in a resource block used for a mix of formats 1/1A/1B and 2/2A/2B. The value of $N_{CS}^{(1)}$ is an integer multiple of Δ_{shift}^{PUCCH} within the range of 0, 1, ..., 8, where Δ_{shift}^{PUCCH} is set by higher layers and is given as:

$$\Delta_{shift}^{PUCCH} \in \begin{cases} \{1, 2, 3\} \text{ for normal cyclic prefix} \\ \{1, 2, 3\} \text{ for extended cyclic prefix.} \end{cases} \qquad (14.9)$$

No mixed resource block is present if $N_{cs}^{(1)} = 0$. At most one resource block in each slot supports a mix of formats 1/1A/1B and 2/2A/2B. The resources used for transmission of PUCCH format 1/1A/1B and 2/2A/2B are represented by the non-negative indices $n_{PUCCH}^{(1)}$ and $n_{PUCCH}^{(2)} < N_{RB}^{(2)} N_{sc}^{RB} + \left\lceil \frac{N_{cs}^{(1)}}{8} \right\rceil \cdot (N_{sc}^{RB} - N_{cs}^{(1)} - 2)$, respectively.

PUCCH format 1, 1A and 1B

The PUCCH format 1 is used to transmit the scheduling request (SR). The scheduling request is carried by the presence or absence of transmission (ON/OFF keying) of PUCCH from the UE. When both ACK/NACK and SR are transmitted in the same subframe a UE transmits the ACK/NACK on its assigned ACK/NACK PUCCH resource for a negative SR transmission and

Table 14.12. Modulation symbol $d(0)$ for PUCCH formats 1A and 1B.

PUCCH format	$b(0), \ldots, b(M_{\text{bit}} - 1)$	$d(0)$
1A	0	1
	1	-1
1B	00	1
	01	$-j$
	10	j
	11	-1

transmits the ACK/NACK on its assigned SR PUCCH resource for a positive SR transmission. When only an ACK/NACK or only an SR is transmitted, a UE uses PUCCH Format 1A or 1B for the ACK/NACK resource and PUCCH format 1 for the SR resource.

For PUCCH formats 1A and 1B, one or two explicit bits are transmitted using BPSK and QPSK modulation respectively resulting in a complex-valued symbol $d(0)$ as given in Table 14.12. The complex-valued symbol $d(0)$ is multiplied with a cyclically shifted length $N_{\text{seq}}^{\text{PUCCH}} = 12$ sequence $r_{u,v}^{(\alpha)}(n)$ according to:

$$y(n) = d(0) \cdot r_{u,v}^{(\alpha)}(n), \quad n = 0, 1, \ldots, \left(N_{\text{seq}}^{\text{PUCCH}} - 1 \right), \tag{14.10}$$

where $r_{u,v}^{(\alpha)}(n)$ is defined by a cyclic shift α of a base sequence $\bar{r}_{u,v}(n)$ according to:

$$r_{u,v}^{(\alpha)}(n) = e^{j\alpha n} \bar{r}_{u,v}(n), \quad n = 0, 1, \ldots, (N_{\text{seq}}^{\text{PUCCH}} - 1). \tag{14.11}$$

The cyclic shift α is varied between SC-FDMA symbols and slots within a subframe to randomize inter-cell interference. The sequence $r_{u,v}^{(\alpha)}(n)$ is a computer-generated constant-amplitude zero auto-correlation (CAZAC) sequence.

The block of complex-valued symbols $y(0), \ldots, y(N_{\text{seq}}^{\text{PUCCH}} - 1)$ is scrambled by $S(n_s)$ and block-wise spread with the orthogonal sequence $w_{n_{\text{oc}}}(i)$ according to:

$$z(m' \times N_{\text{SF}}^{\text{PUCCH}} \cdot N_{\text{seq}}^{\text{PUCCH}} + m \cdot N_{\text{seq}}^{\text{PUCCH}} + n) = w_{n_{\text{oc}}}(m) \cdot y(n)$$

$$m = 0, \ldots, (N_{\text{SF}}^{\text{PUCCH}} - 1), \quad n = 0, \ldots, (N_{\text{seq}}^{\text{PUCCH}} - 1), \quad m' = 0, 1, \tag{14.12}$$

where $N_{\text{SF}}^{\text{PUCCH}} = 4$ for both slots of PUCCH format 1 and normal PUCCH formats 1a/1b, and $N_{\text{SF}}^{\text{PUCCH}} = 4$ for the first slot and $N_{\text{SF}}^{\text{PUCCH}} = 3$ for the second slot of shortened PUCCH formats 1A/1B. The scrambling symbol $S(n_s)$ in a slot simply rotates the constellation of the ACK/NACK symbol to reduce inter-code interference.

In the case of simultaneous transmission of the sounding reference signal and PUCCH format 1A or 1B, one SC-FDMA symbol on PUCCH is punctured to accommodate the sounding reference signal. This creates shortened PUCCH formats 1A/1B with one fewer SC-FDMA symbol available. The shortened PUCCH formats 1A/1B are therefore used for the case when SRS is transmitted in the same subframe as the PUCCH. The sequences $w_{n_{\text{oc}}}(i)$ are given in Table 14.13. We note that only three out of a total of four Walsh sequences are used for the case of $N_{\text{SF}}^{\text{PUCCH}} = 4$. The rationale for this choice is to avoid or reduce performance degradation

Table 14.13. Orthogonal sequences $\left[w(0) \ldots w \left(N_{\mathrm{SF}}^{\mathrm{PUCCH}} - 1 \right) \right]$.

Sequence index $n_{\mathrm{oc}}(n_{\mathrm{s}})$	Spreading factor $N_{\mathrm{SF}}^{\mathrm{PUCCH}}$	Orthogonal sequences $\left[w(0) \quad \ldots \quad w \left(N_{\mathrm{SF}}^{\mathrm{PUCCH}} - 1 \right) \right]$		
0		$[+1$	$+1 \quad +1$	$+1 \ ]$
1	$N_{\mathrm{SF}}^{\mathrm{PUCCH}} = 4$	$[+1$	$-1 \quad +1$	$-1 \ ]$
2		$[+1$	$-1 \quad -1$	$+1 \ ]$
0		$[1$	$1 \quad 1 \ ]$	
1	$N_{\mathrm{SF}}^{\mathrm{PUCCH}} = 3$	$[1$	$e^{j2\pi/3} \quad e^{j4\pi/3} \ ]$	
2		$[1$	$e^{j4\pi/3} \quad e^{j2\pi/3} \ ]$	

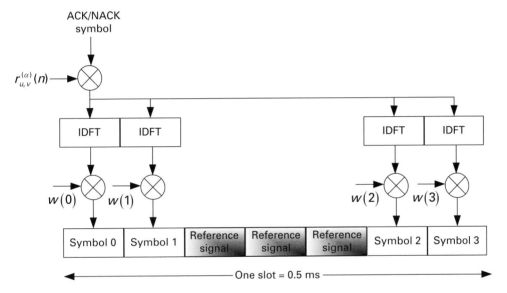

Figure 14.8. Block-wise spreading in PUCCH formats 1A/1B.

in cases where sequences may suffer from loss of orthogonality. The multiplication with a cyclically shifted length $N_{\mathrm{seq}}^{\mathrm{PUCCH}} = 12$ sequence and block-wise spreading operations are pictorially shown in Figure 14.8.

PUCCH formats 2, 2A and 2B
In PUCCH formats 2, 2A and 2B, the block of bits $b(0), \ldots, b(M_{\mathrm{bit}} - 1)$ is scrambled with a UE-specific scrambling sequence, resulting in a block of scrambled bits $\tilde{b}(0), \ldots, \tilde{b}(M_{\mathrm{bit}} - 1)$ according to:

$$\tilde{b}(i) = (b(i) + c(i)) \bmod 2, \tag{14.13}$$

where $c(i)$ is the pseudo-random scrambling sequence $c(i)$. The scrambling sequence generator is initialized at the start of each subframe as:

$$c_{\mathrm{init}} = (\lfloor n_{\mathrm{s}}/2 \rfloor + 1) \cdot \left(2N_{\mathrm{ID}}^{\mathrm{cell}} + 1 \right) \cdot 2^{16} + n_{\mathrm{RNTI}}. \tag{14.14}$$

The block of scrambled bits $\tilde{b}(0), \ldots, \tilde{b}(19)$ is modulated using QPSK modulation resulting in a block of complex-valued modulation symbols $d(0), \ldots, d(9)$. Each complex-valued symbol $d(0), \ldots, d(9)$ is multiplied with a cyclically shifted length $N_{\text{seq}}^{\text{PUCCH}} = 12$ sequence $r_{u,v}^{(\alpha)}(n)$ according to:

$$z(N_{\text{seq}}^{\text{PUCCH}} \cdot n + i) = d(n) \cdot r_{u,v}^{(\alpha)}(i) \quad n = 0, 1, \ldots, 9 \quad i = 0, 1, \ldots, \left(N_{\text{sc}}^{\text{RB}} - 1\right), \quad (14.15)$$

where $r_{u,v}^{(\alpha)}(i)$ is given by (14.10).

The PUCCH format 2 mapping in a slot is shown in Figure 14.9 for the case of the normal cyclic prefix. In the case of the extended cyclic prefix, the second reference signal within the slot is dropped leaving five SC-FDMA symbols available for PUCCH. Therefore, the number of SC-FDMA symbols available for CQI/PMI transmission within a subframe is 10 for the case of both normal and extended cyclic prefixes. This allows carrying a total of 20 bits using QPSK modulation.

For PUCCH formats 2A and 2B, the bit(s) $b(20), \ldots, b(M_{\text{bit}} - 1)$ representing one- or two-bits ACK/NACK bits are modulated similarly to ACK/NACK in formats 1A and 1B as described in Table 14.14. This results in a single modulation symbol $d(10)$, which is used for the generation of the reference-signal for PUCCH formats 2A and 2B. This modified reference signal carrying ACK/NACK is then mapped to the SC-FDMA symbol originally carrying the second reference signal within the slot as shown in Figure 14.9.

14.3.3 PUCCH mapping

The block of complex-valued symbols $z(i)$ is multiplied with the amplitude-scaling factor β_{PUCCH} according to the power control algorithm and mapped to resource elements. PUCCH uses one resource block in each of the two slots in a subframe. Within the physical resource block used for transmission, the mapping of $z(i)$ to resource elements (k, l) not used for

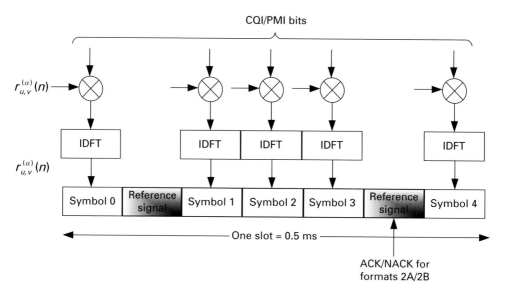

Figure 14.9. The PUCCH format 2, 2A and 2B structure.

Table 14.14. Modulation symbol $d(10)$ for PUCCH formats 2A and 2B.

PUCCH format	$b(20), \ldots, b(M_{\mathrm{bit}} - 1)$	$d(10)$
2A	0	1
	1	-1
2B	00	1
	01	$-j$
	10	j
	11	-1

transmission of reference signals is in increasing order of first k, then l and finally the slot number, starting with the first slot in the subframe.

The resources used for transmission of PUCCH formats 1, 1A and 1B are identified by a resource index $n_{\mathrm{PUCCH}}^{(1)}$ from which the orthogonal sequence index $n_{\mathrm{oc}}(n_s)$ and the cyclic shift $\alpha(n_s)$ are determined as described in [1]. Similarly, resources used for transmission of PUCCH formats 2/2A/2B are identified by a resource index $n_{\mathrm{PUCCH}}^{(2)}$ from which the cyclic shift α is determined. It should be noted that for PUCCH formats 2/2A/2B, the modulation symbols are not spread and hence no orthogonal sequence needs to be determined.

The physical resource block to be used for transmission of PUCCH in slot n_s is given by:

$$n_{\mathrm{PRB}} = \begin{cases} \left\lfloor \dfrac{m}{2} \right\rfloor & \text{if } (m + n_s \bmod 2) \bmod 2 = 0 \\ N_{\mathrm{RB}}^{\mathrm{UL}} - 1 - \left\lfloor \dfrac{m}{2} \right\rfloor & \text{if } (m + n_s \bmod 2) \bmod 2 = 1, \end{cases} \tag{14.16}$$

where the variable m depends on the PUCCH format. For formats 1, 1A and 1B, m is given as:

$$m = \begin{cases} N_{\mathrm{RB}}^{(2)} & \text{if } n_{\mathrm{PUCCH}}^{(1)} < c \cdot N_{\mathrm{cs}}^{(1)} \big/ \Delta_{\mathrm{shift}}^{\mathrm{PUCCH}} \\ \left\lfloor \dfrac{n_{\mathrm{PUCCH}}^{(1)} - c \cdot N_{\mathrm{cs}}^{(1)} \big/ \Delta_{\mathrm{shift}}^{\mathrm{PUCCH}}}{c \cdot N_{\mathrm{sc}}^{\mathrm{RB}} \big/ \Delta_{\mathrm{shift}}^{\mathrm{PUCCH}}} \right\rfloor + N_{\mathrm{RB}}^{(2)} + \left\lceil \dfrac{N_{\mathrm{cs}}^{(1)}}{8} \right\rceil & \text{otherwise,} \end{cases} \tag{14.17}$$

where $c = 3, 2$ for the normal and extended cyclic prefix respectively. $\Delta_{\mathrm{shift}}^{\mathrm{PUCCH}} \in \{1, 2, 3\}$ is set by higher layers.

For formats 2, 2A and 2B, m is given as:

$$m = \left\lfloor n_{\mathrm{PUCCH}}^{(2)} / N_{\mathrm{sc}}^{\mathrm{RB}} \right\rfloor. \tag{14.18}$$

The mapping of PUCCH to physical resource blocks is illustrated in Figure 14.10.

14.4 Aperiodic reporting

An aperiodic reporting of CQI/PMI/RI is performed on PUSCH and is triggered by a scheduling grant. The aperiodic report size and message format is given by RRC (radio resource

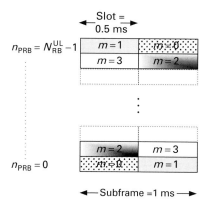

Figure 14.10. PUCCH mapping to physical resource blocks.

Table 14.15. CQI and PMI feedback types for PUSCH reporting modes.

CQI type	PMI granularity	Reporting mode
Wideband CQI	No PMI	NA
	Single PMI	NA
	Multiple PMI	Mode 1–2
UE selected subband CQI	No PMI	Mode 2–0
	Single PMI	NA
	Multiple PMI	Mode 2–2
Higher layer configured subband CQI	No PMI	Mode 3–0
	Single PMI	Mode 3–1
	Multiple PMI	NA

control). The minimum reporting interval for aperiodic reporting is one subframe. The different reporting modes are summarized in Table 14.15. The modes 2–0 and 3–0 without any PMI are used for single transmit antenna port transmission, transmit diversity and open-loop spatial multiplexing. The modes with single PMI (mode 3–1) or multiple PMI (modes 1–2 and 2–2) are used for closed-loop spatial multiplexing.

In the case of mode 1–2, a wideband CQI with multiple PMI is reported. A preferred precoding matrix is selected for each subband from the codebook subset assuming transmission only in this subband. Then a single wideband CQI value per codeword is calculated assuming the use of the corresponding selected precoding matrix in each subband and transmission on set S subbands according to Table 14.16. The wideband CQI and selected precoding matrix indicator for each set S subband is then reported to eNB.

For UE-selected subband feedback modes, the UE reports the positions of the M selected subbands using a combinatorial index r defined as:

$$r = \sum_{k=0}^{M-1} \binom{N - s_k}{M - k},$$

(14.19)

Table 14.16. Subband size (k) for different downlink system bandwidths.

System bandwidth (N_{RB}^{DL})	Subband size (k)
6–7	(wideband CQI only)
8–10	4
11–26	4
27–63	6
64–110	8

Table 14.17. Subband size (k) and M values for different downlink system bandwidths.

System bandwidth N_{RB}^{DL}	Subband size k (RBs)	M
6–7	(wideband CQI only)	(wideband CQI only)
8–10	2	1
11–26	2	3
27–63	3	5
64–110	4	6

where the set $\{s_k\}_{k=0}^{M-1}$, $(1 \leq s_k \leq N, \quad s_k < s_{k+1})$ contains the M sorted subband indices and

$$\left\langle \begin{array}{c} x \\ y \end{array} \right\rangle = \left\{ \begin{array}{ll} \left(\begin{array}{c} x \\ y \end{array} \right) & x \geq y \\ 0 & x < y \end{array} \right. \tag{14.20}$$

is the extended binomial coefficient, resulting in unique label $r \in \left\{ 0, \cdots, \binom{N}{M} - 1 \right\}$ requiring $L = \left\lceil \log_2 \binom{N}{M} \right\rceil$ signaling bits.

In UE-selected subband feedback mode 2-0, the UE selects a set of M preferred subbands of size k within the set of S subbands, where k and M are given in Table 14.17. We note that the subband sizes k in this case are half the sizes in Table 14.6 used for periodic reporting on PUCCH. The UE reports one CQI value reflecting transmission only over the M selected subbands. The CQI represents channel quality across all MIMO layers irrespective of computed or reported RI. In addition to M preferred subbands CQI, the UE reports one wideband CQI value. No PMI is reported in mode 2–0.

In UE-selected subband feedback mode 2–2, the UE performs joint selection of the set of M preferred subbands of size k within the set of S subbands and a preferred single precoding matrix selected from the codebook subset that is preferred to be used for transmission over the M selected subbands. The UE reports one CQI value per codeword reflecting transmission only over the selected M preferred best subbands and using the same selected single precoding matrix in each of the M subbands. In addition, the UE reports a wideband CQI value per codeword, which is calculated assuming the use of the single precoding matrix in set S

Table 14.18. CQI/PMI size for single antenna port, transmit diversity and open-loop spatial multiplexing.

Feedback type	Field	Bitwidth
UE selected subband CQI reports	Wideband CQI codeword 0	4
	Subband differential CQI	2
	Position of the M selected subbands	L
	Total	$(L + 6)$
Higher layer configured subband CQI reports	Wideband CQI codeword 0	4
	Subband differential CQI	$2N$
	Total	$(2N + 4)$

subbands. The UE also reports the selected single precoding matrix indicator for the set S subbands. The CQI value for the M selected subbands for each codeword is encoded differentially using 2 bits relative to its respective wideband CQI as given by (14.3). We note that a total of two CQIs and two PMIs are reported in mode 2–2.

In higher layer-configured subband feedback mode 3–0, a UE reports a wideband CQI value, which is calculated assuming transmission on set S subbands. The UE also reports one subband CQI value for each set S subband. The subband CQI value is calculated assuming transmission only in the subband. The CQI represents channel quality for the first codeword, even when RI > 1. No PMI is reported in mode 3–0.

In mode 3–1, a single precoding matrix is selected from the codebook subset assuming transmission on set S subbands. The UE reports one subband CQI value per codeword for each set S subband, which are calculated assuming the use of a single precoding matrix in all subbands. This single selected precoding matrix indicator is reported. In addition, the UE reports a wideband CQI value per codeword which is calculated assuming the use of the single precoding matrix and transmission on set S subbands. The subband CQI for each codeword is encoded differentially with respect to its respective wideband CQI using 2 bits as given by (14.2). The supported subband sizes k include those given in Table 14.16. In this case, the k is value which is a function of system bandwidth is semi-statically configured by higher layers.

The number of bits for the two CQI feedback types namely UE selected subband CQI and higher layer configured subband CQI reports for single antenna port, transmit diversity and open-loop spatial multiplexing are given in Table 14.18. We note from Table 14.15 that wideband CQI report without PMI is not supported for aperiodic feedback on PUSCH. The single antenna port, transmit diversity and open-loop spatial multiplexing transmission modes require no PMI. The number of CQI/PMI bits for the three CQI feedback types namely wideband CQI, UE selected subband CQI and higher layer configured subband CQI reports for closed loop spatial multiplexing are summarized in Table 14.19.

14.4.1   Channel coding for aperiodic reporting

The aperiodic reporting is performed by transmission of control information on PUSCH. The channel coding for hybrid ARQ ACK/NACK, rank indication (RI) and CQI/PMI is done

Table 14.19. CQI/PMI size for closed-loop spatial multiplexing.

Feedback type	Field	2 antenna ports Rank = 1	2 antenna ports Rank = 2	4 antenna ports Rank = 1	4 antenna ports Rank > 1
Wideband CQI reports	Wideband CQI codeword 0	4	4	4	4
	Wideband CQI codeword 1	0	4	0	4
	Precoding matrix indication	$2N$	N	$4N$	$4N$
	Total	$(2N+4)$	$(2N+8)$	$(2N+4)$	$(2N+8)$
UE selected subband CQIreports	Wideband CQI codeword 0	4	4	4	4
	Subband differential CQI codeword 0	2	2	2	2
	Wideband CQI codeword 1	0	4	0	4
	Subband differential CQI codeword 1	0	2	0	2
	Position of the M selected subbands	L	L	L	L
	Precoding matrix indication	4	2	8	8
	Total	$(L+10)$	$(L+14)$	$(L+14)$	$(L+20)$
Higher layer configured subband CQI reports	Wideband CQI codeword 0	4	4	4	4
	Subband differential CQI codeword 0	$2N$	$2N$	$2N$	$2N$
	Wide band CQI codeword 1	0	4	0	4
	Subband differential CQI codeword 1	0	$2N$	0	$2N$
	Precoding matrix indication	2	1	4	4
	Total	$(2N+6)$	$(4N+9)$	$(2N+8)$	$(4N+12)$

Table 14.20. Channel coding schemes for aperiodic reporting.

Channel coding scheme	Control information
Simplex (3,2) code	Two-bits ACK/NACK Two-bits rank information (RI)
Variable Reed–Muller (32, O) block code	CQI/PMI payload $\leq$ 11 bits
Tail-biting convolutional code	CQI/PMI payload > 11 bits

independently. The channel coding schemes used for aperiodic reporting are summarized in Table 14.20. The ACK/NACK information is one bit or two bits for one codeword downlink and two codewords downlink transmission respectively. Similarly, for two and four transmit antenna ports for maximum layers of two or four, the RI is one or two bits respectively. When ACK/NACK or RI is two bits, the coding is performed using a simplex (3,2) code where the parity bit o_2 is given as.

$$o_2 = (o_0 \oplus o_1), \tag{14.21}$$

where $\oplus$ represents XOR operation and $o_0 o_1$ are two ACK/NACK or RI bits.

The CQI/PMI bits input to the channel coding block are denoted by $o_0, o_1, o_2, o_3, \ldots, o_{O-1}$, where O is the number of bits. For CQI/PMI payload sizes greater than 11 bits, the information

Table 14.21. Basis sequences for variable Reed–Muller $(32, O)$ block code.

i	$M_{i,0}$	$M_{i,1}$	$M_{i,2}$	$M_{i,3}$	$M_{i,4}$	$M_{i,5}$	$M_{i,6}$	$M_{i,7}$	$M_{i,8}$	$M_{i,9}$	$M_{i,10}$
0	1	1	0	0	0	0	0	0	0	0	1
1	1	1	1	0	0	0	0	0	0	1	1
2	1	0	0	1	0	0	1	0	1	1	1
3	1	0	1	1	0	0	0	0	1	0	1
4	1	1	1	1	0	0	0	1	0	0	1
5	1	1	0	0	1	0	1	1	1	0	1
6	1	0	1	0	1	0	1	0	1	1	1
7	1	0	0	1	1	0	0	1	1	0	1
8	1	1	0	1	1	0	0	1	0	1	1
9	1	0	1	1	1	0	1	0	0	1	1
10	1	0	1	0	0	1	1	1	0	1	1
11	1	1	1	0	0	1	1	0	1	0	1
12	1	0	0	1	0	1	0	1	1	1	1
13	1	1	0	1	0	1	0	1	0	1	1
14	1	0	0	0	1	1	0	1	0	0	1
15	1	1	0	0	1	1	1	1	0	1	1
16	1	1	1	0	1	1	1	0	0	1	0
17	1	0	0	1	1	1	0	0	1	0	0
18	1	1	0	1	1	1	1	1	0	0	0
19	1	0	0	0	0	1	1	0	0	0	0
20	1	0	1	0	0	0	1	0	0	0	1
21	1	1	0	1	0	0	0	0	0	1	1
22	1	0	0	0	1	0	0	1	1	0	1
23	1	1	1	0	1	0	0	0	1	1	1
24	1	1	1	1	1	0	1	1	1	1	0
25	1	1	0	0	0	1	1	1	0	0	1
26	1	0	1	1	0	1	0	0	1	1	0
27	1	1	1	1	0	1	0	1	1	1	0
28	1	0	1	0	1	1	1	0	1	0	0
29	1	0	1	1	1	1	1	1	1	0	0
30	1	1	1	1	1	1	1	1	1	1	1
31	1	0	0	0	0	0	0	0	0	0	0

is coded using a tail-biting convolutional code (see Chapter 11). In this case, a length $L = 8$ cyclic redundancy check (CRC) is also computed and attached to the PMI/CQI report. For payload sizes of 11 bits or less, the CQI/PMI is coded using a variable Reed–Muller (RM) $(32, O)$ block code. The codewords of the $(32, O)$ block code are a linear combination of 11 basis sequences denoted $M_{i,n}$ and given in Table 14.21. The 11 basis sequences allow one to encode a maximum of 11 bits of CQI/PMI.

The encoded CQI/PMI block is denoted $b_0, b_1, b_2, b_3, \ldots, b_{B-1}$ and given as:

$$b_i = \sum_{n=0}^{O-1} \left(o_n \cdot M_{i,n} \right) \bmod 2 \quad i = 0,\, 1, 2, \ldots, (B-1) = 31. \tag{14.22}$$

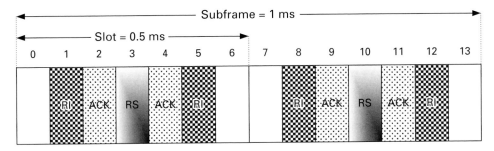

Figure 14.11. PUCCH mapping to physical resource blocks.

The output bit sequence $q_0, q_1, q_2, q_3, \ldots, q_{Q-1}$ is obtained by circular repetition of the encoded CQI/PMI block as follows:

$$q_i = b_{(i \bmod B)} \quad i = 0, 1, 2, \ldots, (Q-1). \tag{14.23}$$

The channel interleaver in conjunction with the resource element mapping for PUSCH results in a time-first mapping of modulation symbols onto the transmit waveform while ensuring that the hybrid ARQ ACK/NACK information is present on both slots in the subframe and is mapped to resources around the uplink demodulation reference signals as shown in Figure 14.11. The motivation for transmitting ACK/NACK next to reference signals is better channel estimates for ACK/NACK reliability. We also note that RI information is transmitted around the SC-FDMA symbols carrying ACK.

14.5 Summary

The uplink feedback control information consists of channel quality, MIMO channel rank, preferred precoding matrix and hybrid ARQ ACK/NACK. The channel quality information is used for channel sensitive scheduling as well as link adaptation in the downlink. The MIMO rank and precoding information is used to select a MIMO transmission format which includes the number of MIMO layers transmitted and antenna weights for beam-forming. The hybrid ARQ ACK/NACK is used to support downlink hybrid ARQ operation.

Multiple feedback formats are defined to match the downlink transmission modes such as single antenna port, transmit diversity, open-loop spatial multiplexing and closed-loop spatial multiplexing. Moreover, different levels of frequency-selective subband CQI feedback are defined that allow for the achievement of a given tradeoff between frequency-selective multiuser scheduling gains and the CQI feedback overhead. A wideband CQI covering the system bandwidth is reported in each feedback mode to allow power control of downlink control channels whose transmission is distributed over the system bandwidth. Similarly, various levels of granularity are provided for precoding information feedback that includes no PMI, a single PMI or multiple PMIs for the frequency subbands.

The feedback control and data multiplexing approach support both an FDM approach and a TDM approach. An FDM approach is used for periodic reporting on PUCCH where PUCCH transmission from one UE can be frequency multiplexed with control or data transmissions from other UEs. When a UE has uplink data transmission in a subframe, the control information is time-multiplexed with data at the input of DFT. The frequency-multiplexing of simultaneous transmission of control and data from a single UE is not permitted because it violates the single-carrier property of SC-FDMA and results in some increase in PAPR/CM. Moreover, the FDM control needs to be transmitted at the bandwidth edges to make contiguous bandwidth in the middle available for data transmission. This is because if data from a single UE is transmitted on non-contiguous frequency blocks, the PAPR/CM of SC-FDMA increases. In order to provide some frequency-diversity, the control is hopped from one edge of the band to the other edge between the two slots within a subframe.

The periodic reports and hybrid ARQ ACK/NACK on PUCCH are transmitted by using a cyclic shift of a computer-generated constant-amplitude zero auto-correlation (CAZAC) sequence. The PUCCH transmissions from multiple UEs are kept orthogonal by allo-cating different cyclic shifts of the CAZAC sequence. The cyclic shift is hopped from one SC-FDMA symbol to the next and from one slot to the next to randomize inter-ference. Moreover, for ACK/NACK transmissions, the CAZAC sequence over multiple SC-FDMA symbols is spread using orthogonal sequences to create orthogonal ACK/NACK channels. The orthogonal sequences are hopped at slot-level again to randomize inter-cell interference.

The number of CQI/PMI bits that can be carried on PUCCH for periodic reporting is limited. Therefore, frequency-selective PMI is not transmitted with periodic reporting. In order to provide subband CQI with periodic reporting on PUCCH, a time-cycling approach is used where a UE provides feedback for a single subband in a given subframe with cycling though multiple subbands in multiple subframes.

A differential encoding of CQI is used for both subband CQI as well as spatial CQI for the second codeword in order to keep the feedback overhead low.

The channel coding for the periodic CQI/PMI reports on PUCCH employ a variable Reed–Muller (RM) $(20, A)$ block code. In the case of the extended cyclic prefix, the ACK/NACK and CQI/PMI are jointly coded using the same $(20, A)$ block code when transmitted simultaneously. In the case of the normal cyclic prefix, when ACK/NACK and CQI/PMI are transmitted simul-taneously, the ACK/NACK is transmitted by modulating one of the two PUCCH reference signals.

In the case of aperiodic CQI/PMI reports on PUSCH, two-bits ACK/NACK and rank indication are coded using a simplex $(3,2)$ code. The CQI/PMI reports consisting of 11 bits or fewer employ a variable RM $(30, O)$ block code. The CQI/PMI reports larger than 11 bits are coded using a tail-biting convolutional code. In this case, an 8-bits CRC is also computed and attached to the CQI/PMI report.

In summary, the uplink control design in LTE employs multiple modes and formats to optimize performance for various transmission modes and deployment scenarios. However, a downside of this approach is increased complexity in terms of both implementation as well as system operation. A reason for a large number of formats in the uplink is that the single-carrier FDMA scheme does not allow simultaneous transmission of control and data in a frequency-multiplexed fashion. For example, if the uplink is based on OFDMA, supporting uplink control in both FDM and TDM fashion is not necessary.

References

[1] 3GPP TSG RAN v8.3.0, Evolved Universal Terrestrial Radio Access: Physical Channels and Modulation (3GPP TS 36.211).

[2] 3GPP TSG RAN v8.3.0, Evolved Universal Terrestrial Radio Access: Multiplexing and Channel Coding (3GPP TS 36.212).

[3] 3GPP TSG RAN v8.3.0, Evolved Universal Terrestrial Radio Access: Physical Layer Procedures (3GPP TS 36.213).

15 Downlink control signaling

With the exception of a scheduling request, all uplink control consists of feedback information to support downlink transmissions. The channel quality feedback is provided to support downlink channel-sensitive scheduling and link adaptation. The rank and precoding matrix indication is used for selecting a downlink MIMO transmission format. The ACK/NACK signaling provides feedback on downlink hybrid ARQ transmissions. In contrast to uplink control, the only feedback information on the downlink is ACK/NACK signaling to support uplink hybrid ARQ operation and transmission power control (TPC) commands to support uplink power control. The reason for this asymmetry is simply the fact that both the uplink and the downlink schedulers resides in the eNB. Therefore, the bulk of downlink signaling involves uplink and downlink scheduling grants that convey information on the transmission format and resource allocation for both the uplink and downlink transmissions. In order to support the uplink channel-sensitive scheduling, the uplink channel quality is estimated from the uplink sounding reference signal (SRS).

The three downlink control channels transmitted every subframe are physical control format indicator channel (PCFICH), physical downlink control channel (PDCCH) and physical hybrid ARQ indicator channel (PHICH). The PCFICH carries information on the number of OFDM symbols used for PDCCH. The PDCCH is used to inform the UEs about the resource allocation as well as modulation, coding and hybrid ARQ control information. Since multiple UEs can be scheduled simultaneously within a subframe in a frequency or space division multiplexed fashion multiple PDCCHs each carrying information for a single UE are transmitted. A maximum of three (or four for smaller bandwidths) OFDM symbols within a subframe can be used for PDCCH. With dynamic indication via PCFICH of the number of OFDM symbols used for PDCCH, the unused OFDM symbols among the three (or four) PDCCH OFDM symbols can be used for data transmission. The PHICH is used to carry hybrid ARQ ACK/NACK feedback for uplink transmissions.

A design goal for control information transmission is high reliability because data transmission fails when control information is in error. This is because control information carries UE identity as well as transmission format and resource allocation information. A simple approach for making control information reliable is to broadcast it with sufficiently high power so that all the UEs in the cell can receive it. This approach is generally used for control information such as broadcast control that is targeted for all the UEs in the cell. This is because when a UE wakes up from the sleep-mode, it needs to acquire system parameters by receiving synchronization and broadcast signals. Since the network may be unaware of this UE, the broadcast needs to be transmitted with sufficient power to reach the cell-edge user. However, when eNB is sending control information that is targeted for a single UE or a group of UEs and eNB has channel quality information for these UEs, the control information can be power

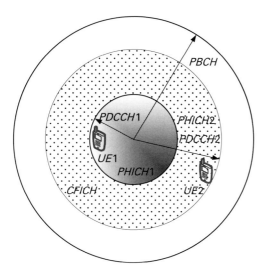

Figure 15.1. Coverage for different downlink control channels.

controlled with just the right amount of power allocated for successful reception of the control information.

The coverage diagram for downlink control channels is shown in Figure 15.1. We also show the physical broadcast channel (PBCH) as a reference, which is always transmitted with cell-edge coverage in mind. Since PCFICH carries information on the duration (in OFDM symbols) of PDCCH, it is power controlled to reach the UE with the worst channel conditions among the scheduled UEs in the subframe. The PHICH can also be power controlled according to the channel conditions for the UE for which it is carrying hybrid ARQ feedback. The PDDCH is also used to carry TPC (transmit power control) commands for uplink data and control transmission. In this case, the PDCCH is power controlled to reach the UE with the worst channel conditions having a TPC command for it in the PDCCH.

15.1   Data control multiplexing

Similar to data control multiplexing in the uplink, both time-multiplexing and frequency-multiplexing can be considered for the downlink. The rationale for the TDM approach for the downlink, however, is different from that in the uplink. In the uplink, the TDM approach allows keeping the single-carrier property of SC-FDMA and hence achieves low PAPR/CM. In the downlink, the multiple-access scheme is OFDMA and therefore the low PAPR/CM argument does not apply. Another argument in favor of the TDM approach in the downlink, however, is that a UE can decode control information in the beginning of the subframe and can go to a short-term sleep-mode if there is no data scheduled for it in the subframe. It is argued that such a micro-sleep-mode as shown in Figure 15.2 can save UE battery power consumption. However, the gains of micro-sleep mode are debatable given that decoding of control information cannot be complete until towards the middle of the subframe. As the UE needs to turn on again at the beginning of the next subframe, the total duration of micro-sleep can be only a small fraction of the subframe duration. Another benefit of the TDM approach is the small

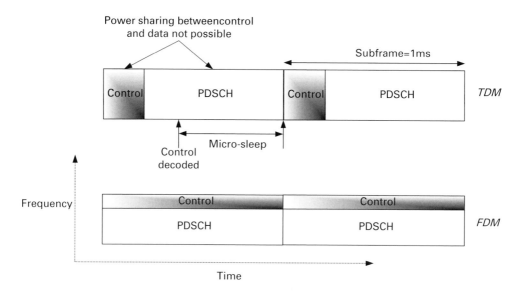

Figure 15.2. TDM and FDM data control multiplexing approaches for the downlink.

reduction in delay as the UE can decode control information before it starts decoding data transmission.

Similar to the uplink, a major drawback of the TDM approach is that it does not allow sharing power between data and control transmissions and hence can limit the control channel coverage. While the FDM approach allows power sharing, it does not enable micro-sleep as the control information is transmitted over the whole subframe in a frequency-multiplexed fashion. Moreover, the decoding delays in the FDM approach can be slightly larger than in the TDM approach because once a subframe is received, the UE first needs to decode the control information before the data decoding can start. Both the TDM and FDM approaches were considered and debated in detail for the LTE system and finally the TDM approach was selected as the data control multiplexing approach for the downlink.

15.2   Resource element groups

The downlink physical control channels are mapped to resource units smaller than a resource block. The motivation for doing this is to distribute the transmission over a larger bandwidth to capture frequency-diversity. We note that control information messages are generally much smaller than data messages and if resource blocks were used for transmission of control information, the transmissions would be localized in frequency, which is not desired from a control-channel-performance perspective. The unit of resource that is used for control information transmission is referred to as a resource element group (REG), which consists of four useful subcarriers (resource elements) within a resource block in an OFDM symbol. A resource block contains two or three REGs depending upon whether the resource block in the OFDM symbol carries reference signals or not as depicted in Figure 15.3. When reference signals are present in a resource block, 4 out of the 12 subcarriers are used for reference signals transmission. The remaining eight subcarriers then form two REGs. We note that the position

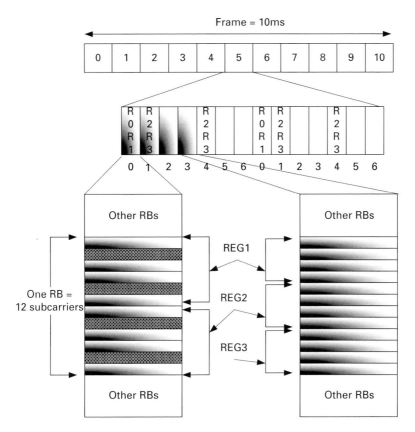

Figure 15.3. An illustration of resource element groups (REGs).

of reference signals subcarriers within a resource block is not fixed and is dependent upon the cell specific frequency shift applied.

A resource-element group is represented by the index pair (k', l') of the resource element with the lowest index k in the group with all resource elements in the group having the same value of l. The REGs can only exist on up to a maximum of four OFDM symbols (first four) within a subframe. This is because PDCCH can span a maximum of four OFDM symbols for smaller bandwidths. The first OFDM symbol always has one or two reference signals. Also, the second OFDM symbol contains two reference signals for antenna ports 2 and 3 in the case of four cell-specific reference signals. In the case of the first OFDM symbol and the second OFDM when the number of antenna ports is configured as four, two resource-element groups in physical resource block n_{PRB} consist of resource elements (k, l) with:

$$
\begin{aligned}
k &= k_0 + 0, k_0 + 1, \ldots, k_0 + 5, \quad \text{REG1} \\
k &= k_0 + 6, k_0 + 7, \ldots, k_0 + 11 \quad \text{REG2},
\end{aligned}
\tag{15.1}
$$

where k_0 indicates the first resource element in the resource block and is given as:

$$
k_0 = n_{\text{PRB}} \times N_{\text{sc}}^{\text{RB}}, \quad 0 \le n_{\text{PRB}} < N_{\text{RB}}^{\text{DL}}.
\tag{15.2}
$$

It should be noted that even when a single cell-specific reference signal is configured on antenna port 0, the resource elements reserved for reference signals of antenna port 1 are left unused. We note that each REG is defined over six resource elements as two resource elements are used for reference signals.

In the second OFDM symbol in the case of only one or two cell-specific reference signals configured, the third and fourth OFDM symbols, the three resource-element groups in physical resource block n_{PRB} consist of resource elements (k, l) with:

$$\begin{aligned}
k &= k_0 + 0, k_0 + 1, \ldots, k_0 + 3 & \text{REG1} \\
k &= k_0 + 4, k_0 + 5, \ldots, k_0 + 7 & \text{REG2} \\
k &= k_0 + 8, k_0 + 9, \ldots, k_0 + 11 & \text{REG3.}
\end{aligned} \tag{15.3}$$

The mapping of a symbol-quadruplet $\langle z(i), z(i+1), z(i+2), z(i+3) \rangle$ onto a resource-element group represented by resource-element (k', l') is defined such that elements $z(i)$ are mapped to resource elements (k, l) of the resource-element group not used for cell-specific reference signals in increasing order of i and k. A motivation for defining mapping in terms of symbol-quadruplets and hence using four resource elements for a REG is that control channels can use up to four layers SFBC-FSTD transmit diversity scheme (see Chapter 6).

15.3   Control format indicator channel

The control format indicator (CFI) channel is used to dynamically indicate resources used for downlink control information transmission on a subframe-by-subframe basis. The amount of resource needed for downlink control information within a subframe depends upon the number of users scheduled as well as the transmissions formats for these users. The CF indication is performed in units of OFDM symbols. The idea is that when the OFDM symbols needed for downlink control information are less than the maximum number of OFDM symbols that can be used for control, this information can be conveyed to the scheduled UEs and the unused OFDM symbols can instead be used for data transmission. Therefore, a UE first needs to decode CFI within a subframe and then decode downlink control information based on the number of OFDM symbols indicated in CFI. An alternative approach would have been that UEs could blindly decode the number of OFDM symbols used for control. However, this would have required additional complexity at the UE. In the LTE system, the downlink control information is dedicated to a UE and therefore can be power-controlled according to the UEs channel conditions. In contrast, the CFI needs to be reliably decoded by all the UEs having scheduling grants within the subframe. Therefore, if power control is applied to CFI, the power needs to be allocated in such a way that CFI transmission is reliably received by the scheduled UE experiencing the worst channel conditions in that subframe.

The number of OFDM symbols used for PDCCH in various subframe types is summarized in Table 15.1. The number of OFDM symbols for PDCCH in MBSFN subframes with four cell specific antenna ports configured, that is, $P = 4$, equals two. This is because the reference signals for the first and second antenna ports are carried in the first OFDM symbol while reference signals for the third and fourth antenna ports are carried in the second OFDM symbol (see Chapter 9, Section 9.6.1).

The CFI transmission chain processing is depicted in Figure 15.4. The CFI information arrives each subframe to the coding unit in the form of an indicator for the time span, in units

Table 15.1. Number of OFDM symbols used for PDCCH.

Subframe	Number of OFDM symbols for PDCCH when $N_{RB}^{DL} > 10$	Number fo OFDM symbols for PDCCH when $N_{RB}^{DL} \geq 10$
Subframe 1 and 6 for frame structure type 2	1,2	2
MBSFN subframes (one or two cell specific antenna ports, $P = 1, 2$)	1,2	2
MBSFN subframes (four cell specific antenna ports, $P = 4$)	2	2
MBSFN subframes on a carrier not supporting PDSCH	0	0
All other cases	1, 2, 3	2, 3, 4

Table 15.2. CFI codewords.

CFI	CFI codeword $\langle b_0, b_1, \ldots, b_{31} \rangle$
1	$\langle 0, 1, 1, 0, 1, 1, 0, 1, 1, 0, 1, 1, 0, 1, 1, 0, 1, 1, 0, 1, 1, 0, 1, 1, 0, 1, 1, 0, 1, 1, 0, 1 \rangle$
2	$\langle 1, 0, 1, 1, 0, 1, 1, 0, 1, 1, 0, 1, 1, 0, 1, 1, 0, 1, 1, 0, 1, 1, 0, 1, 1, 0, 1, 1, 0, 1, 1, 0 \rangle$
3	$\langle 1, 1, 0, 1, 1, 0, 1, 1, 0, 1, 1, 0, 1, 1, 0, 1, 1, 0, 1, 1, 0, 1, 1, 0, 1, 1, 0, 1, 1, 0, 1, 1 \rangle$
4 (reserved)	$\langle 0, 0, 0, 0, 0, 0, 0, 0, 0, 0, 0, 0, 0, 0, 0, 0, 0, 0, 0, 0, 0, 0, 0, 0, 0, 0, 0, 0, 0, 0, 0, 0 \rangle$

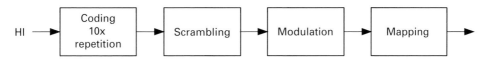

Figure 15.4. CFI transmission chain processing.

of OFDM symbols, of the downlink control information (DCI) in that subframe. The CFI takes three values CFI $= 1, 2, 3$ with CFI $= 4$ reserved. The four CFI codewords given in Table 15.2 are obtained by using a (3, 2) simplex code with 10 repetitions and appending 2 systematic bits at the end to match the 32 bits allowed on PCFICH (physical control format indicator channel). The minimum distance for the code in Table 15.2 is $d_{min} = 21$.

For bandwidths larger than 1.8 MHz $\left(N_{RB}^{DL} > 10 \right)$, CFI $= 1, 2, 3$ indicates that downlink control information (DCI) occupies one, two or three OFDM symbols respectively. For smaller bandwidths of 1.8 MHz $\left(N_{RB}^{DL} = 10 \right)$ or less, there is some concern that the DCI may need to be transmitted in more than three OFDM symbols because the amount of resources available in an OFDM symbol is less for smaller bandwidths relative to larger bandwidths. In general, one would expect that the number of users supported on a carrier is also less for smaller bandwidths and hence DCI resources needed would also be less. Nevertheless, for smaller bandwidths CFI $= 1, 2, 3$ indicates that DCI occupies two, three or four OFDM symbols (one OFDM symbol more than the case of larger bandwidths).

In order to randomize the inter-cell interference, the block of coded bits $b(0), \ldots,$ $b(31)$ from Table 15.2 is scrambled with a cell-specific sequence resulting in a block of scrambled bits $\tilde{b}(0), \ldots, \tilde{b}(31)$ according to:

$$\tilde{b}(i) = (b(i) + c(i)) \bmod 2. \tag{15.4}$$

The scrambling sequence generator is initialized at the start of each subframe as below:

$$c_{\text{init}} = \left(\lfloor n_s/2 \rfloor + 1 \right) \cdot \left(2N_{\text{ID}}^{\text{cell}} + 1 \right) \cdot 2^9 + N_{\text{ID}}^{\text{cell}}, \tag{15.5}$$

where $n_s = 0, 1, \ldots, 19$ is the slot number within a frame and $N_{\text{ID}}^{\text{cell}}$ is the physical-layer cell identity. The term $\left(\lfloor n_s/2 \rfloor + 1 \right)$ increments every other slot or every subframe as PCFICH is transmitted once per subframe. The scrambling sequence initialization based on the physical-layer cell identity $N_{\text{ID}}^{\text{cell}}$ guarantees that the neighboring cells use different scrambling sequences for PCFICH. The PCFICH always uses QPSK modulation. Therefore, 32 coded and scrambled bits are mapped to 16 QPSK symbols. These 16 symbols are further divided into 4 groups of 4 modulation symbols each for transmit diversity processing. The PCFICH uses the same number of antenna ports and the same transmit diversity scheme as the physical broadcast channel (PBCH). It should be noted that the number of antenna ports is blindly decoded by multiple decoding hypothesis testing on PBCH. Therefore, after successfully decoding the PBCH, the UEs have knowledge about the number of antenna ports used in the cell. Therefore, there is no need to explicitly signal the number of antenna ports for downlink control information transmission. The output of transmit diversity processing is four quadruplets with each quadruplet consisting of four complex-valued symbols:

$$z^{(p)}(i) = \langle y^{(p)}(4i), y^{(p)}(4i+1), y^{(p)}(4i+2), y^{(p)}(4i+3) \rangle, \tag{15.6}$$

where $z^{(p)}(i)$ denote symbol quadruplet $i = 0, 1, 2, 3$ for antenna port $p = 0, 1, \ldots, (P-1)$. For each of the antenna ports, symbol quadruplets are mapped in increasing order of i to the four resource-element groups in the first OFDM symbol in a downlink subframe with the representative resource-element given by:

$$k = \left(N_{\text{sc}}^{\text{RB}}/2 \right) \left[\left(N_{\text{ID}}^{\text{cell}} \bmod 2N_{\text{RB}}^{\text{DL}} \right) + \lfloor i \times N_{\text{RB}}^{\text{DL}}/2 \rfloor \right] i = 0, 1, 2, 3, \tag{15.7}$$

where the additions are modulo $N_{\text{RB}}^{\text{DL}} N_{\text{sc}}^{\text{RB}}$.

Let us consider an example assuming $N_{\text{RB}}^{\text{DL}} = 25$ and $N_{\text{sc}}^{\text{RB}} = 12$, which leads to $N_{\text{RB}}^{\text{DL}} N_{\text{sc}}^{\text{RB}} = 300$. For even-numbered physical-layer cell identity $k = 0, 72, 150, 222$ and for odd-numbered physical-layer cell identity $k = 150, 222, 0, 72$. We note that the four REGs used for CFI are almost uniformly distributed in frequency as illustrated in Figure 15.5, a desired effect to benefit from frequency-diversity.

15.4   Downlink resource allocation

In an SC-FDMA or OFDM system, the frequency-domain resource allocation information needs to be signaled to the UE. Because of the large number of resource blocks within the frequency band, the resource allocation is one of the largest fields in the downlink control information. In the case of SC-FDMA uplink, the allocated resource blocks need to be contiguous to guarantee single-carrier property. While the contiguous resource allocation can be

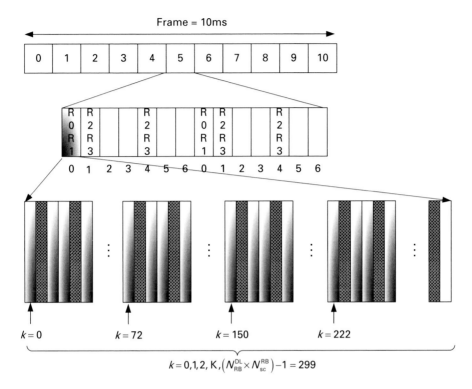

Figure 15.5. PCFICH mapping to resource element groups (REG).

signaled with the minimum number of signaling bits, it also results in limiting the scheduling flexibility. In the case of OFDM, non-contiguous resource blocks can be allocated thus providing maximum scheduling flexibility. However, the signaling overhead also increases for non-contiguous resource block allocations. In order to provide various choices of scheduling performance and signaling overhead, multiple resource allocation types are defined. A contiguous resource allocation scheme is defined for both the uplink and the downlink. As pointed out earlier, a contiguous resource allocation is necessary in the uplink due to single-carrier access. In the downlink, contiguous resource allocation provides a low overhead alternative while limiting scheduling flexibility. In addition to contiguous resource allocation, two types of non-contiguous resource allocations using a bitmap-based signaling are defined for the downlink.

Resource allocation type 0

The type 0 resource allocation allows allocating non-contiguous group of resource blocks referred to as resource block groups (RBGs) using a bitmap where a RBG is a set of consecutive physical resource blocks (PRBs) as shown in Figure 15.6. The bitmap is defined on a group of resource blocks to reduce the signaling overhead. The resource block group size (P) is a function of the system bandwidth as given in Table 15.3. We note that resource block group size P in Table 15.3 is the same as the CQI subband size k in Table 14.17 for $\left(N_{RB}^{DL} \geq 11\right)$. This allows matching the resource allocation to the channel quality feedback. The total number of

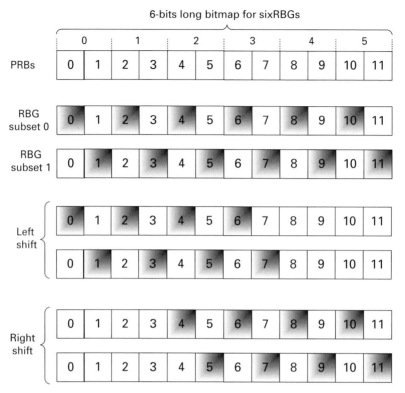

Figure 15.6. An illustration of type 1 resource allocation for $N_{RB}^{DL} = 12, P = 2$ and $\left\lceil N_{RB}^{DL}/P \right\rceil = 6$.

Table 15.3. Type 0 resource allocation RBG size vs. downlink system bandwidth.

System bandwidth	RBG size
N_{RB}^{DL}	(P)
≤ 10	1
$11 - 26$	2
$27 - 63$	3
$64 - 110$	4

RBGs (N_{RBG}) for downlink system bandwidth of N_{RB}^{DL} PRBs is given by:

$$N_{RBG} = \left\lceil \frac{N_{RB}^{DL}}{P} \right\rceil, \tag{15.8}$$

where $\left\lfloor N_{RB}^{DL}/P \right\rfloor$ of the RBGs are of size P and if $\left\lceil N_{RB}^{DL}/P \right\rceil - \left\lfloor N_{RB}^{DL}/P \right\rfloor > 0$ then one of the RBGs is of size $N_{RB}^{DL} - P \cdot \left\lfloor N_{RB}^{DL}/P \right\rfloor$. The bitmap is of size N_{RBG} bits with one bitmap bit per RBG such that each RBG is addressable. We note that for small system bandwidth $\left(N_{RB}^{DL} \leq 10 \right)$, $P = 1$ and hence each resource block can be allocated with full flexibility.

The RBG size is increased with system bandwidth because the signaling overhead would be overwhelming if the resource block level bitmap is used for larger bandwidths.

Resource allocation type 1

The type 1 resource allocation allows allocating resources on an RB level even for larger bandwidths using the same number of resource allocation bits as in type 0 resource allocation. This is achieved by defining RBG subsets where the number of RBG subsets is equal to the resource block group size P in type 0 resource allocation. We note that the number of RBG subsets P is equal to the number of bandwidth parts given in Table 14.6 that are used for subband CQI reports sent in a time cycling fashion on PUCCH. Therefore, type 1 allocations can be used to match the resource allocations on the downlink to the subband CQI feedback provided on PUCCH. Moreover, the number of subsets is picked equal to the resource block group size P to make the total resource allocation bits the same between type 0 and type 1 resource allocation. The resource blocks within each subset can be allocated using a bitmap. A bitmap of size $\lceil N_{RB}^{DL}/P \rceil$ indicates to a scheduled UE the PRBs from the set of PRBs from one of P RBG subsets. The portion of the bitmap used to address PRBs in a selected RBG subset has size N_{RB}^{TYPE1} and is defined as:

$$N_{RB}^{TYPE1} = \lceil N_{RB}^{DL}/P \rceil - \left(\lceil \log_2(P) \rceil + 1 \right), \tag{15.9}$$

where $\lceil N_{RB}^{DL}/P \rceil$ is the overall bitmap size and $\lceil \log_2(P) \rceil$ is the minimum number of bits needed to select one of the P RBG subsets and one additional bit is used to indicate whether the addressable PRBs of a selected RBG subset is left shifted or right shifted. The shift is needed for addressability of all PRBs since the number of PRBs in an RBG subset is larger than the PRB addressing portion of the bitmap as indicated by $N_{RB}^{TYPE1} < \lceil N_{RB}^{DL}/P \rceil$. An example of type 1 resource allocation is depicted in Figure 15.6 for $N_{RB}^{DL} = 12, P = 2$ and $\lceil N_{RB}^{DL}/P \rceil = 6$. In the case of type 0 resource allocation, a 6-bits long bitmap can be used to indicate one of the six resource block groups. With two RBG subsets, we need 1 bit to indicate selection of one of the two RBG subsets this leaves us with only 5 bits to indicate PRBs within a subset. However, the number of PRBs in each subset is six. This problem is solved by using another bit (leaving only 4 bits for the bitmap) to indicate the bitmap shift of the selected RBG subset so that all the PRBs within a subset can be allocated. Note that for resource allocation to one UE, the bitmap shift can be towards the right while for another UE the shift can be towards the left.

Resource allocation type 2

The type-2 resource allocation is used for uplink assignments that require a set of contiguous resource blocks as well as compact downlink assignments. In the downlink, the resource allocation information indicates to a scheduled UE a set of contiguously allocated localized virtual resource blocks or distributed virtual resource blocks depending on the setting of a 1-bit flag carried on the associated PDCCH. Localized VRB allocations for a UE vary from a single VRB up to a maximum number of VRBs spanning the system bandwidth. Distributed VRB allocations for a UE vary from a single VRB up to N_{VRB}^{DL} VRBs if $N_{RB}^{DL} < 50$ and vary from a single VRB upto 16 if $N_{RB}^{DL} \geq 50$, where N_{VRB}^{DL} is defined in Chapter 8. With a total of N_{RB} resource blocks, the number of combinations for resource allocation when the start position is the first RB is N_{RB}. Similarly, when the start position is the second RB, the number of possible end positions is $(N_{RB} - 1)$ and so on. The total number of possible combinations can then be obtained as:

$$N_{RB} + (N_{RB} - 1) + (N_{RB} - 2) + \cdots + 1 + 2 = \frac{N_{RB}(N_{RB} + 1)}{2}. \tag{15.10}$$

This approach for contiguous RB allocation can save some signaling bits in some cases relative to the case of an explicit indication of the start position and the end position (or the number of RBs allocated).

A type-2 resource allocation field consists of a resource indication value (RIV) corresponding to a starting resource block (RB_{start}) and a length in terms of contiguously allocated resource blocks (L_{CRBs}). The resource indication value is defined by:

$$RIV = \begin{cases} N_{RB}^{DL}(L_{CRBs} - 1) + RB_{start} & (L_{CRBs} - 1) \leq \lfloor N_{RB}^{DL}/2 \rfloor \\ N_{RB}^{DL}(N_{RB}^{DL} - L_{CRBs} + 1) + (N_{RB}^{DL} - 1 - RB_{start}) & \text{otherwise.} \end{cases} \tag{15.11}$$

For uplink scheduling using DCI format 0, the resource indication value is defined by:

$$RIV = \begin{cases} N_{RB}^{UL}(L_{CRBs} - 1) + RB_{start} & (L_{CRBs} - 1) \leq \lfloor N_{RB}^{UL}/2 \rfloor \\ N_{RB}^{UL}(N_{RB}^{UL} - L_{CRBs} + 1) + (N_{RB}^{UL} - 1 - RB_{start}) & \text{otherwise.} \end{cases} \tag{15.12}$$

An example of resource allocation type 2 signaling for $N_{RB} = 6$ is illustrated in Figure 15.7. The total number of combinations to signal in this case are:

$$\frac{N_{RB}(N_{RB} + 1)}{2} = \frac{6(6 + 1)}{2} = 21. \tag{15.13}$$

These combinations numbered from zero to 20 can be signaled using a total of 5 bits. It should be noted that the number of signaling bits needed would be six if the explicit start position and end position are indicated using 3 bits each for the case of six resource blocks. The numbering of the combinations in Figure 15.7 is obtained using (15.11).

In the case where PUCCH transmissions use an odd number of resource block pairs, one PRB within a slot on each band edge is not used by PUCCH as shown in Figure 15.8. Therefore, when a PUSCH resource allocation includes PRBs at a carrier band edge then the PRB of the allocated PUSCH band edge PRB pair occupied by the PUCCH resource slot is not used for the PUSCH.

15.5   Downlink control information

The downlink control information (DCI) carries downlink or uplink scheduling information as well as uplink power control commands. A dedicated control (also referred to as separate

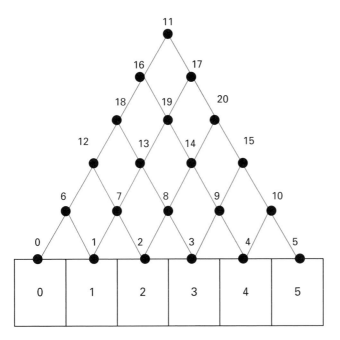

Figure 15.7. An illustration of resource allocation type 2 signaling for $N_{RB} = 6$.

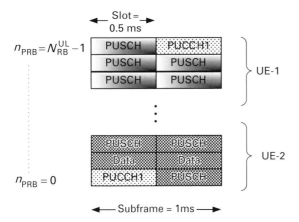

Figure 15.8. An illustration of PUSCH resource allocation where PUCCH transmissions use an odd number of resource block pairs.

coding) approach is used for the physical downlink control channel (PDCCH) in the LTE system. A dedicated control approach was selected over the common control (also referred to as joint coding) approach because it allows one to perform link adaptation and power control according to the channel conditions of the UE addressed in PDCCH. A drawback of this approach, however, is increased number of PDCCH blind decodings that each UE needs to perform. This is because multiple PDCCHs can be present in a subframe and a large number of possible PDCCH transmission formats is supported.

The control information is different between the uplink and the downlink as well as among different downlink transmission modes. For example, a synchronous hybrid ARQ is used in the uplink, which does not require signaling of the hybrid ARQ process number. In contrast, a hybrid ARQ process needs to be indicated in the downlink due to asynchronous hybrid ARQ retransmissions. Additionally, multiple modes such as single antenna port, transmit diversity, spatial multiplexing and closed-loop rank 1 are defined for downlink transmissions. The control information can be different between, for example, closed-loop spatial multiplexing and transmit diversity. This is because a single codeword is carried in transmit diversity mode while two codewords are transmitted in spatial multiplexing. In the case of two codewords, MCS, new data indicator and redundancy version need to be signaled separately for both the codewords. Moreover, precoding information is signaled in the case of closed-loop spatial multiplexing while no precoding needs to be signaled for transmit diversity mode.

A simple DCI design would have been possible by using a single format with the largest amount of control information required by any mode. However, this would be quite inefficient from a control overhead perspective because unnecessary control information would be transmitted in some transmission modes. On the other hand, many different formats complicate the system operation because they require a large number of blind hypotheses testing, which also affects false alarm performance. Therefore, a relatively small set of DCI formats is defined by reusing some of the formats among multiple transmission modes. As an example, the formats used for uplink scheduling grants and compact downlink assignments share the same number of payload bits. A 1-bit flag, however, is required to differentiate between the uplink and downlink assignments.

15.5.1 DCI formats

DCI Format 0

The DCI format 0 carries information for scheduling uplink transmissions on PUSCH. The different fields of format 0 are summarized in Table 15.4. The number of bits required for resource block assignment (type-2 resource allocation) and hopping resource allocation is given as:

$$
N_{RB}^{assign} = \begin{cases} \left(\lceil \log_2(N_{RB}^{UL}(N_{RB}^{UL}+1)/2) \rceil \right) & \text{non-hopping} \\ \left(\lceil \log_2(N_{RB}^{UL}(N_{RB}^{UL}+1)/2) \rceil - N_{UL\_hop} \right) & \text{hopping.} \end{cases} \tag{15.14}
$$

It should be noted that in order to keep the single-carrier property of SC-FDMA, the resources allocations in the uplink are always contiguous.

In (15.14), N_{UL_hop} indicates hopping information and is one or two bits based on the system bandwidth as given in Table 15.5. The number of contiguous RBs that can be assigned to a hopping user is limited to:

$$
\min \left(\lfloor 2^y / N_{RB}^{UL} \rfloor, \lfloor N_{RB}^{PUSCH} / N_{sb} \rfloor \right), \tag{15.15}
$$

Table 15.4. DCI format 0.

Field	Bits
Format0/format1A differentiation flag	1
Hopping flag	1
Resource block assignment and hopping resource allocation	$\left\lceil \log_2 (N_{RB}^{UL}(N_{RB}^{UL}+1)/2) \right\rceil$
Modulation and coding scheme and redundancy version	5
New data indicator	1
TPC command for scheduled PUSCH	2
Cyclic shift for DM RS	3
CQI request	1

Table 15.5. Max PUSCH BW, and number of hopping bits vs. system bandwidth.

System BW N_{RB}^{UL}	Max BW assigned to a hopping user	Number of hopping bits N_{UL_hop}
6–49	$\min\left(\left\lfloor 2^y/N_{RB}^{UL} \right\rfloor, \left\lfloor N_{RB}^{PUSCH}/N_{sb} \right\rfloor\right)$	1
50–110	$\min\left(\left\lfloor 2^y/N_{RB}^{UL} \right\rfloor, \left\lfloor N_{RB}^{PUSCH}/N_{sb} \right\rfloor\right)$	2

where the number of subbands N_{sb} is set by higher layers and N_{RB}^{PUSCH} is given as:

$$
N_{RB}^{PUSCH} = \begin{cases} N_{RB}^{UL} - N_{RB}^{PUCCH} & N_{RB}^{PUCCH} \quad \text{even} \\ N_{RB}^{UL} - N_{RB}^{PUCCH} + 1 & N_{RB}^{PUCCH} \quad \text{odd,} \end{cases} \tag{15.16}
$$

where N_{RB}^{PUCCH} is the number of RBs allocated to PUCCH. When N_{RB}^{PUCCH} is odd, an RB at the edge of the bandwidth in each slot is not used for PUCCH. This is because of slot-level hopping used for PUCCH and the fact that PUCCH is transmitted at the band edge to assure the single-carrier property of SC-FDMA.

For PUSCH hopping type 1, the hopping bits are used to obtain the value of $\tilde{n}_{PRB}(i)$ as indicated in Table 15.6. The lowest index PRB (n_{PRB}^{S1}) of the first slot RA in subframe i is defined as:

$$
n_{PRB}^{S1}(i) = \tilde{n}_{PRB}^{S1}(i) + \left\lceil N_{RB}^{PUCCH}/2 \right\rceil. \tag{15.17}
$$

The lowest index PRB ($n_{PRB}(i)$) of the second slot RA in subframe i is defined as:

$$
n_{PRB}(i) = \tilde{n}_{PRB}(i) + \left\lfloor N_{RB}^{PUCCH}/2 \right\rfloor. \tag{15.18}
$$

In PUSCH hopping type 2 the set of physical resource blocks to be used for transmission in slot n_s is given by the scheduling grant together with a predefined pattern (see Chapter 8, Section 8.6).

Table 15.6. PDCCH DCI format 0 hopping bit definition.

System BW N_{RB}^{UL}	Number of hopping bits	Information in hopping bits	$\tilde{n}_{PRB}(i)$
6 – 49	1	0	$\left(\left\lfloor N_{RB}^{PUSCH}/2\right\rfloor + \tilde{n}_{PRB}^{S1}(i)\right) \bmod N_{RB}^{PUSCH}$,
		1	Type 2 PUSCH hopping
		00	$\left(\left\lfloor N_{RB}^{PUSCH}/4\right\rfloor + \tilde{n}_{PRB}^{S1}(i)\right) \bmod N_{RB}^{PUSCH}$,
50 – 110	2	01	$\left(-\left\lfloor N_{RB}^{PUSCH}/4\right\rfloor + \tilde{n}_{PRB}^{S1}(i)\right) \bmod N_{RB}^{PUSCH}$,
		10	$\left(\left\lfloor N_{RB}^{PUSCH}/2\right\rfloor + \tilde{n}_{PRB}^{S1}(i)\right) \bmod N_{RB}^{PUSCH}$,
		11	Type 2 PUSCH hopping

Table 15.7. DCI format 1.

Field	Bits
Resource allocation type 0 / type 1 flag	1
Resource block assignment	Type 0 $\left\lceil N_{RB}^{DL}/P\right\rceil$ Type 1 $\left(\left\lceil N_{RB}^{DL}/P\right\rceil - \left\lceil\log_2(P)\right\rceil - 1\right)$
Modulation and coding scheme	5
HARQ process number	3
New data indicator	1
Redundancy version	2
TPC command for PUCCH	2

DCI format 1

The DCI format 1 carries information for scheduling transmission of one codeword on PDSCH. The contents of format 1 are given in Table 15.7 which supports both types 0 and 1 resource allocations as described in Section 15.4. For small downlink bandwidths with $N_{DL}^{RB} \leq 10$, there is no resource allocation header and resource allocation type 0 is assumed. For resource allocation type 0, a $\left\lceil N_{RB}^{DL}/P\right\rceil$-bits-long bitmap provides the resource allocation information. For resource allocation type 1, $\left\lceil\log_2(P)\right\rceil$ bits are used to indicate the selected resource blocks subset, 1 bit indicates a shift of the resource allocation span and $\left(\left\lceil N_{RB}^{DL}/P\right\rceil - \left\lceil\log_2(P)\right\rceil - 1\right)$ bits provide the resource allocation information. When the number of information bits in format 1 is equal to that in format 0/1A, one bit of value zero is appended to format 1 to avoid confusing format 1 with format 0/1A.

DCI format 1A

DCI format 1A is used for the compact scheduling of one PDSCH codeword. Since format 0 and format 1A share the same payload size, a 1-bit flag is used to differentiate between the two formats. The other fields in format 1A are summarized in Table 15.8. Similar to format 0, $\left\lceil\log_2(N_{RB}^{DL}(N_{RB}^{DL} + 1)/2)\right\rceil$ bits are used for resource block assignment (type-2

Table 15.8. DCI format 1A.

Field		Bits
Format0/format1A differentiation flag		1
Localized/distributed VRB assignment flag		1
Resource block assignment	localized VRB	$\left\lceil \log_2(N_{RB}^{DL}(N_{RB}^{DL}+1)/2) \right\rceil$
	distributed VRB $N_{RB}^{DL} < 50$	$\left\lceil \log_2(N_{RB}^{DL}(N_{RB}^{DL}+1)/2) \right\rceil$
	distributed VRB $N_{RB}^{DL} \geq 50$	$\left(\left\lceil \log_2(N_{RB}^{DL}(N_{RB}^{DL}+1)/2) \right\rceil - 1 \right)$
Modulation and coding scheme		5
HARQ process number		3
New data indicator		1
Redundancy version		2
TPC command for PUCCH		2

Table 15.9. DCI format 1B.

Field		Bits
Localized/distributed VRB assignment flag		1
Resource block assignment	localized VRB	$\left\lceil \log_2(N_{RB}^{DL}(N_{RB}^{DL}+1)/2) \right\rceil$
	distributed VRB $N_{RB}^{DL} < 50$	$\left\lceil \log_2(N_{RB}^{DL}(N_{RB}^{DL}+1)/2) \right\rceil$
	distributed VRB $N_{RB}^{DL} \geq 50$	$\left(\left\lceil \log_2(N_{RB}^{DL}(N_{RB}^{DL}+1)/2) \right\rceil - 1 \right)$
Modulation and coding scheme		5
HARQ process number		3
New data indicator		1
Redundancy version		2
TPC command for PUCCH		2
Precoding information	2 antenna ports	2
	4 antenna ports	4

resource allocation). However, for distributed VRB allocations for system bandwidths larger than 50 RBs $(N_{RB}^{DL} \geq 50)$, 1 bit among the resource assignment bits is used to indicate $N_{gap} = N_{gap,i}$, $i = 1, 2$, and the remaining $\left(\left\lceil \log_2(N_{RB}^{DL}(N_{RB}^{DL}+1)/2) \right\rceil + 1 \right)$ bits provide the resource allocation information.

DCI format 1B

The DCI format 1B is used for the compact scheduling of one PDSCH codeword with MIMO precoding information, see Table 15.9. The remaining contents of format 1B are the same as format 1A with the difference that a 1-bit flag to differentiate between format 0/1B is not needed. This is because with the precoding information, the size of format 1B becomes greater than that of format 0. The precoding information consists of two and four bits for two and four antenna ports respectively. We note that the precoding information does not include the rank information as the rank is always one for the case of one PDSCH codeword.

Table 15.10. DCI format 1C.

Field	Bits
Gap flag	0 or 1
Resource block assignment	See (15.19)
Transport block size	5

DCI format 1C

DCI format 1C is used for highly compact assignments where transmissions on DL-SCH use QPSK modulation and distributed virtual resource blocks. This format carries scheduling assignment of system information (SI), random access response and paging. For scheduling assignment of system information (SI), the 2-bit redundancy version is implicitly derived by system frame number (SFN) and subframe number. For random access response and paging, no redundancy version signaling is necessary since HARQ combining is not used for these messages. Therefore, there is no need for explicit redundancy version signaling in DCI format 1C. The information size of PDCCH DCI format 1C is same for the scheduling assignment of SI, random access response and paging.

The control information fields for format 1C are summarized in Table 15.10. A 1-bit flag indicates the distributed transmission gap value for $N_{RB}^{DL} \geq 50$, where value 0 indicates $N_{gap} = N_{gap,1}$ and value 1 indicates $N_{gap} = N_{gap,2}$. For $N_{RB}^{DL} < 50$, there is a single gap value and hence the 1-bit gap flag is not defined. The number of bits for resource block assignment is given as:

$$\left[\log_2 \left(\frac{\left[\frac{N_{VRB,gap1}^{DL}}{N_{RB}^{step}} \right] \cdot \left(\left[\frac{N_{VRB,gap1}^{DL}}{N_{RB}^{step}} \right] + 1 \right)}{2} \right) \right], \tag{15.19}$$

where $N_{VRB,gap1}^{DL}$ is given by (8.7). $N_{RB}^{step} = 2,4$ for $N_{RB}^{DL} < 50$ and $N_{RB}^{DL} \geq 50$ respectively. The distributed VRB allocatins in format 1C vary from N_{RB}^{step} VRB(s) up to $([N_{VRB}^{DL}/N_{RB}^{step}].N_{RB}^{step})$ VRBs with an increment step of N_{RB}^{step}.

DCI format 1D

The DCI format 1D is used for compact scheduling of one PDSCH codeword with precoding and power offset information, see Table 15.11. The power offset information is required for multi-user MIMO (see Chapter 7, Section 7.6.2) scheduling in the downlink. This is because when more than one UE is scheduled on the same resource blocks, the total power is shared among the UEs. The two power offset values represented by '1' and '0' are 0 and –3dB respectively relative to the single user transmit power offset signaled by higher layers. The remaining fields of format 1D are the same as those of fomat 1B. The format 1D is used with wideband precoding feedback from the UE and hence subband precoding confirmation is not required.

DCI format 2

The DCI format 2 is used for scheduling PDSCH to UEs configured in closed-loop spatial multiplexing mode. The information transmitted in DCI format 2 is summarized in Table 15.12. The resource allocation information in format 2 is conveyed in the same way as in format 1. In spatial multiplexing mode, two codewords are transmitted. Therefore, the difference relative

Table 15.11. DCI format 1D.

Field		Bits
Localized/distributed VRB assignment flag		1
Resource block assignment	localized VRB	$\left\lceil \log_2(N_{RB}^{DL}(N_{RB}^{DL}+1)/2) \right\rceil$
	distributed VRB $N_{RB}^{DL} < 50$	$\left\lceil \log_2(N_{RB}^{DL}(N_{RB}^{DL}+1)/2) \right\rceil$
	distributed VRB $N_{RB}^{DL} \geq 50$	$\left(\left\lceil \log_2(N_{RB}^{DL}(N_{RB}^{DL}+1)/2) \right\rceil - 1 \right)$
Modulation and coding scheme		5
HARQ process number		3
New data indicator		1
Redundancy version		2
TPC command for PUCCH		2
Precoding information	2 antenna ports	2
	4 antenna ports	4
Downlink power offset		1

Table 15.12. DCI format 2.

Field		Bits
Resource allocation type 0/type 1 flag		1
Resource block assignment	Type 0	$[N_{RB}^{DL}/P]$
	Type 1	$\left(\left\lceil N_{RB}^{DL}/P \right\rceil - \lceil \log_2(P) \rceil - 1 \right)$
HARQ process number		3
Transport block to codeword swap flag		1
TPC command for PUCCH		2
Modulation and coding scheme		5
New data indicator	Transport block 1	1
Redundancy version		2
Modulation and coding scheme		5
New data indicator	Transport block 2	1
Redundancy version		2
Precoding information		3 for $P = 2$
		6 for $P = 4$

to format 1 is that MCS, new data indicator and redundancy version are indicated separately. Additionally, a 1-bit transport block to codeword swap flag allows switching mapping of transport blocks to codewords and hence mapping of transport blocks to MIMO layers. As in DCI format 1, for small downlink bandwidths with $N_{RB}^{DL} \leq 10$, there is no resource allocation header and resource allocation type 0 is assumed.

The transport block (TB) to codeword (CW) mapping is given in Table 15.13. When both transport blocks are enabled, the transport block to codeword swap flag determines the transport block to codeword swap flag is reserved and the transport block to codeword mapping is predetermined as given in Table 15.13. The disabling of a transport block is indicated by a predetermined combination of MCS and redundancy version fields for that transport block. A transport block is disabled if $I_{MCS} = 0$ (see Table 12.1) and if $rv_{idx} = 1$. Otherwise, the transport block is enabled.

Table 15.13. Transport block (TB) to codeword (CW) mapping.

	Transport block to codeword swap flag	CW1	CW2
Two transport blocks enabled	0	TB1	TB2
	1	TB2	TB1
One transport block enabled	Reserved	TB1	NA
		TB2	NA

The contents of the precoding information field for two and four antenna ports are summarized in Tables 15.14 and 15.15 respectively. The interpretation of the precoding information field depends on the number of enabled codewords according to Tables 15.14 and 15.15. A single enabled codeword is transmitted on a single layer with transmission rank 1 (TRI = 1). However, an exception is allowed when a codeword that is transmitted over two layers needs a hybrid ARQ retransmission. In this case a single codeword can be transmitted over two layers with (TRI = 2). It should be noted that this situation can only happen when the rank on initial transmission of two codewords is greater than 2 (TRI > 2). Therefore, the transmission of a single codeword with rank-2 only applies to the case of four antenna ports as indicated in Table 15.15. In short, the combination of a single enabled codeword and (TRI = 2) is only supported for retransmission of the corresponding hybrid ARQ process.

We note from Table 15.12 that multiple PMI is not indicated on the downlink. When multiple subbands PMI mode is configured, eNB uses a confirmation mechanism where some combinations of the precoding field are used to confirm that the eNB is using the PMI as reported by the UE in the most recent feedback message, In these cases, the precoding for the corresponding RB(s) in subframe n is according to the latest PMI(s) reported by the UE on PUSCH, on or before subframe $(n - 4)$. This approach allows one to use subband PMI without additional overhead on the downlink. The same three or six bits precoding field also allows to indicate a single PMI when the precoding is the same for all the resource blocks allocated to a UE.

DCI format 2A

The DCI format 2A is used for scheduling PDSCH to UEs configured in open-loop spatial multiplexing mode without PMI feedback. The information transmitted in DCI format 2A is summarized in Table 15.16. As in DCI format 1 and DCI format 2, for small downlink bandwidths with $N_{RB}^{DL} \leq 10$, there is no resource allocation header and resource allocation type 0 is assumed. The only difference relative to format 2 is that precoding information is zero and two bits for two and four antenna ports respectively. Therefore, in the case of open-loop spatial multiplexing transmission mode with two antenna ports ($P = 2$), the precoding information field is not present. A fixed 2×2 identity matrix is instead used for transmission of two codewords in open-loop spatial multiplexing mode for the case of two antenna ports. The information on the number of transmission layers is derived from the information obtained if one codeword or two codewords are enabled and a single layer transmission is assumed if codeword 1 is enabled while codeword 2 is disabled. A single codeword is transmitted using two antenna ports SFBC transmit diversity scheme.

The transport block (TB) to codeword (CW) mapping is the same as in format 2 given in Table 15.13. When both transport blocks are enabled, the transport block to codeword swap flag determines the transport block to codeword mapping. In the case where one of the

Table 15.14. Precoding information field for two antenna ports.

Number of codewords				
1			**2**	
Bit field mapped to index	Message		Bit field mapped to index	Message
0	TRI = 1: transmit diversity		0	TRI = 2: Precoding corresponding to precoder matrix $\frac{1}{2}\begin{bmatrix} 1 & 1 \\ 1 & -1 \end{bmatrix}$
1	TRI = 1: Precoding corresponding to precoder vector $\begin{bmatrix} 1 & 1 \end{bmatrix}^T / \sqrt{2}$		1	TRI = 2: Precoding corresponding to precoder matrix $\frac{1}{2}\begin{bmatrix} 1 & 1 \\ j & -j \end{bmatrix}$
2	TRI = 1: Precoding corresponding to precoder vector $\begin{bmatrix} 1 & -1 \end{bmatrix}^T / \sqrt{2}$		2	TRI = 2: Precoding according to the latest PMI report on PUSCH, using the precoder(s) indicated by the reported PMI(s)
3	TRI = 1: Precoding corresponding to precoder vector $\begin{bmatrix} 1 & j \end{bmatrix}^T / \sqrt{2}$		3	Reserved
4	TRI = 1: Precoding corresponding to precoder vector $\begin{bmatrix} 1 & -j \end{bmatrix}^T / \sqrt{2}$		4	Reserved
5	TRI = 1: Precoding according to the latest PMI report on PUSCH, using the precoder(s) indicated by the reported PMI(s), if R1 = 2 was reported, using first column of all precoders implied by the reported PMI(s)		5	Reserved
6	TRI = 1: Precoding according to the latest PMI report on PUSCH, using the precoder(s) indicated by the reported PMI(s), if R1 = 2 was reported, using second column of all precoders implied by the reported PMI(s)		6	Reserved
7	Reserved		7	Reserved

Table 15.15. Precoding information field for four antenna ports.

Number of codewords			
1		**2**	
Bit field mapped to index	Message	Bit field mapped to index	Message
0	TRI = 1: transmit diversity	0	TRI = 2: TPMI = 0
1	TRI = 1: TPMI = 0	1	TRI = 2: TPMI = 1
2	TRI = 1: TPMI = 1	⋮	⋮
⋮	⋮	⋮	TRI = 2: TPMI = 15
16	TRI = 1: TPMI = 15	16	TRI = 2: Precoding according to the latest PMI report on PUSCH using the precoder(s) indicated by the reported PMI(s)
17	TRI = 1: Precoding according to the latest PMI report on PUSCH using the precoder(s) indicated by the reported PMI(s)	17	TRI = 1: TPMI = 0
18	TRI = 2: TPMI = 0	18	TRI = 3: TPMI = 1
19	TRI = 2: TPMI = 1	⋮	⋮
⋮	⋮	32	TRI = 3: TPMI = 15
33	TRI = 2: TPMI = 15	33	TRI = 3: Precoding according to the latest PMI report on PUSCH using the precoder(s) indicated by the reported PMI(s)
34	TRI = 2: Precoding according to the latest PMI report on PUSCH using the precoder(s) indicated by the reported PMI(s)	34	TRI = 4: TPMI = 0
35 – 63	Reserved	35	TRI = 4: TPMI = 1
		⋮	⋮
		49	TRI = 4: TPMI = 15
		50	TRI = 4: Precoding according to the latest PMI report on PUSCH using the precoder(s) indicated by the reported PMI(s)

Table 15.15. (Continued).

			Number of codewords		
	1		2		
Bit field mapped to index		Message	Bit field mapped to index	Message	
			51 – 63	TRI = 3: Precoding according to the latest PMI report on PUSCH using the precoder(s) indicated by the reported PMI(s)	
0		TRI = 1: transmit diversity	0	TRI = 2: TPMI = 0	

Table 15.16. DCI format 2A.

Field		Bits
Resource allocation type 0 / type 1 flag		1
Resource block assignment	Type 0	$[N_{RB}^{DL}/P]$
	Type 1	$\left(\left[N_{RB}^{DL}/P\right] - \left[\log_2(P)\right] - 1\right)$
HARQ process number		3
Transport block to codeword swap flag		1
TPC command for PUCCH		2
Modulation and coding scheme		5
New data indicator	Transport block 1	1
Redundancy version		2
Modulation and coding scheme		5
New data indicator	Transport block 2	1
Redundancy version		2
Precoding information		0 for $P = 2$
		2 for $P = 4$

transport blocks is disabled, the transport block to codeword swap flag is reserved and the transport block to codeword mapping is predetermined as given in Table 15.13. Similar to format 2, the combination of a single enabled codeword and (RI = 2) is only supported for retransmission of the corresponding hybrid ARQ process.

The contents of the precoding information field for four antenna ports are summarized in Table 15.17. For rank-1 transmissions, four antenna ports SFBC-FSTD transmit diversity scheme is applied. For rank greater than one, a precoder cycling approach with large delay CDD as described in Chapter 7 (Section 7.6) is employed. It should be noted that precoder cycling is not allowed for the two-antenna ports case as a fixed 2×2 identity matrix is used for transmission of two codewords in open-loop spatial multiplexing mode for the case of two antenna ports.

Table 15.17. Precoding information field for four antenna ports for open-loop spatial multiplexing.

	Number of codewords			
	1		2	
Bit field mapped to index	Message	Bit field mapped to index	Message	
0	RI = 1: transmit diversity	0	RI = 2: precoder cycling with large delay CDD	
1	RI = 2: precoder cycling with large delay CDD	1	RI = 3: precoder cycling with large delay CDD	
2	Reserved	2	RI = 4: precoder cycling with large delay CDD	
3	Reserved	3	Reserved	

DCI format 3 and 3A

The DCI format 3 and 3A are used for the transmission of TPC commands for PUCCH and PUSCH with two bits and a single bit power adjustments respectively. The number of TCP commands transmitted, N, in format 3 and format 3A are given as:

$$N = \begin{cases} \left\lfloor \frac{L_{\text{format }0}}{2} \right\rfloor, & \text{format3} \\ L_{\text{format }0}, & \text{format3A,} \end{cases} \qquad (15.20)$$

where $L_{\text{format }0}$ is equal to the payload size of format 0 before CRC attachment.

15.5.2 PDCCH monitoring

As was pointed out earlier, the dedicated control approach used for PDCCH results in a large number of blind decodings because multiple PDCCHs are transmitted in a subframe possibly using multiple transmission formats. In order to keep the number of blind decoding that a UE needs to make in a subframe manageable, the concept of search spaces and aggregation levels is defined.

The control region consists of a set of control channel elements (CCEs), numbered from 0 to $N_{\text{CCE},k} - 1$, where $N_{\text{CCE},k}$ is the total number of CCEs in the control region of subframe k. The UE monitors a set of PDCCH candidates for control information. The UE is not required to decode control information on a PDCCH if the channel-code rate is larger than 3/4, where channel-code rate is defined as the number of downlink control information bits (including RNTI) divided by the number of physical channel bits on the PDCCH. For example, a PDCCH mapped to a single CCE can carry a maximum of 72 coded bits as given in Table 15.20, below. Therefore, a UE can assume that DCI formats containing more than 54 bits are not transmitted on a single CCE. A control channel element (CCE) that consists of 9 resource element groups can carry a total of 36 modulation symbols and hence a total of 72 coded bits using QPSK modulation.

The set of PDCCH candidates to monitor is defined in terms of search spaces, where a search space $S_k^{(L)}$ at aggregation level $L \in \{1, 2, 4, 8\}$ is defined by a set of PDCCH candidates. The CCEs corresponding to PDCCH candidate m of the search space $S_k^{(L)}$ are given by

$$L \cdot \{(Y_k + m) \bmod [N_{\text{CCE},k}/L]\} + i, \tag{15.21}$$

where Y_k is defined in (15.22), $i = 0, \cdots, L - 1$, and $m = 0, \cdots, M^{(L)}$ is the number of PDCCH candidates to monitor in the given search space.

A UE monitors one common search space at each of the aggregation levels 4 and 8 and one UE-specific search space at each of the aggregation levels 1, 2, 4, 8. The common and UE-specific search spaces may overlap. The aggregation levels defining the search spaces and the corresponding DCI formats that the UE monitors are listed in Table 15.18. A UE monitors either DCI format 3 or format 3A as determined by the configuration. The DCI formats that the UE monitors in the UE specific search spaces is a subset of those listed in Table 15.18 and depends on the configured transmission mode as given in Table 15.19.

An example of PDCCH monitoring for UE-specific search space with aggregation level $L = 2$ in Table 15.8 is shown in Figure 15.9. The total number of CCEs that are monitored is 12 and the number of PDCCH candidates $M^{(2)} = 6$. We would also like to point out that a PDCCH consisting of L consecutive CCEs only starts on a CCE fulfilling $i \bmod L = 0$, where i is the CCE number. Therefore, with $L = 2$, the PDCCH starts at even-numbered CCEs and therefore the UE only needs to monitor six PDCCHs in Figure 15.8. In the case where this condition is not enforced, the number of possible decoding would be larger than six as a PDCCH can start at any CCE number.

Table 15.18. PDCCH candidates monitored by a UE.

Search space $S_k^{(L)}$				
Type	Aggregation level L	Size [in CCEs]	Number of PDCCH candidates $M^{(L)}$	DCI formats
UE-specific	1	6	6	
	2	12	6	
	4	8	2	0, 1, 1A, 1B, 2, 2A
	8	16	2	
Common	4	16	4	
	8	16	2	0, 1A, 1C, 3/3A

Table 15.19. Reference DCI format(s) supported by each transmission mode.

Transmission mode	Scheme	Reference DCI format
1	Single-antenna port ($p = 0$)	1, 1A
2	Transmit diversity	1, 1A
3	Open-loop spatial multiplexing	2A
4	Closed-loop spatial multiplexing	2
5	Multi-user MIMO	1D
6	Closed-loop rank = 1 precoding	1B
7	Single-antenna port; port 5	1, 1A

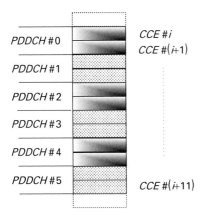

Figure 15.9. PDCCH monitoring for UE-specific search space with aggregation level $L = 2$.

For the common search spaces, $Z_k^{(L)}$ is set to 0 for the two aggregation levels $L = 4$ and 8. For the UE-specific search space $S_k^{(L)}$ at aggregation level L, the variable $Z_k^{(L)}$ is defined by:

$$Z_k^{(L)} = L \cdot \left(Y_k \bmod \lfloor N_{\text{CCE},k}/L \rfloor \right)$$
$$Y_k = (A \cdot Y_{k-1}) \bmod D \tag{15.22}$$

where $Y_{-1} = n_{\text{RNTI}} \neq 0$, $A = 39\,827$ and $D = 65\,537$.

A UE receives PDSCH broadcast control transmissions including paging, RACH response and BCCH associated with DCI formats 1A or 1C signaled by a PDCCH in the common search spaces. In addition, the UE is semi-statically configured to receive PDSCH data transmissions signaled via PDCCH UE-specific search spaces, based on one of the transmission modes given in Table 15.9. A UE not configured to receive PDSCH data transmissions based on one of the transmission modes may receive PDSCH data transmissions with DCI format 1A signaled by a PDCCH in its UE specific search spaces or the common search spaces.

15.5.3 PDCCH formats

A control channel element (CCE) that consists of nine resource element groups is the minimum unit of transmission for PDCCH. A physical control channel is transmitted on an aggregation of one, two, four or eight consecutive control channel elements (CCEs) as given in Table 15.20. A PDCCH consisting of L consecutive CCEs may only start on a CCE fulfilling $i \bmod L = 0$, where i is the CCE number as shown in Figure 15.10. This condition is applied to help reduce the number of blind decoding assumptions. We should point out that there is no explicit relationship between PDCCH formats in Table 15.20 and the DCI formats described in Section 15.5.1. For example, it should not be assumed that DCI format 0 is transmitted using PDCCH format 0 and so on.

15.5.4 PDCCH transmit chain

The PDCCH transmission chain is depicted in Figure 15.11. With dedicated control, multiple PDCCHs may be transmitted in a subframe, one for each UE scheduled for uplink or downlink

Table 15.20. CFI supported PDCCH formats.

PDCCH format	Number of CCEs	Number of resource-element groups	Number of PDCCH bits
0	1	9	72
1	2	18	144
2	4	36	288
3	8	72	576

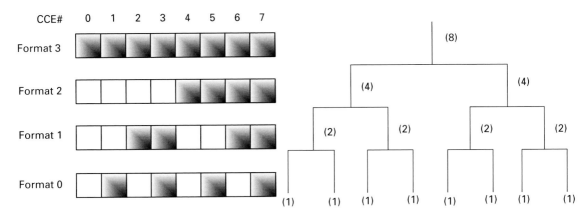

Figure 15.10. PDCCH formats.

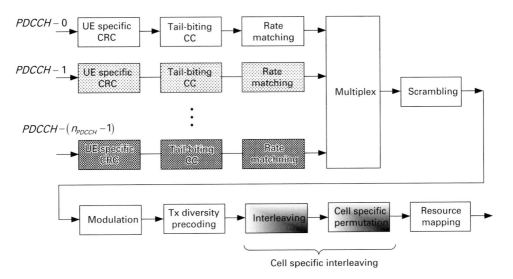

Figure 15.11. PDCCH transmission chain processing.

transmission. The number of PDCCHs transmitted in a subframe is n_{PDCCH} and is numbered from 0 to $(n_{PDCCH} - 1)$. A CRC is attached to PDCCH information and the MAC ID, also referred as RNTI (radio network temporary identifier), is implicitly encoded in the CRC. A tail-biting convolutional coding and rate matching is performed separately on each PDCCH . The coded bits from multiple PDCCHs are then multiplexed. The remaining processing steps including scrambling, modulation, transmit diversity precoding, interleaving, cell-specific permutation and mapping are then performed on the multiplexed stream. The interleaving along with cell-specific permutation effectively results in cell-specific interleaving of the multiplexed PDCCH stream. The cell-specific interleaving helps to randomize interference to the neighboring cells as different PDCCHs can use different power levels as each PDCCH is power controlled to the UE it is addressed to.

A 16-bits-long CRC is used both to carry the RNTI as well as to provide for error detection. The entire PDCCH payload is used to calculate the CRC parity bits. Let us denote the PDCCH payload bits by $a_0, a_1, a_2, a_3, \ldots, a_{A-1}$, and the parity bits by $p_0, p_1, p_2, p_3, \ldots, p_{L-1}$, where A is the PDCCH payload size and $L = 16$ is the CRC size. The CRC parity bits are scrambled with the RNTI $x_{ue,0}, x_{ue,1}, \ldots, x_{ue,15}$ to form the sequence of bits $c_0, c_1, c_2, c_3, \ldots, c_{B-1}$:

$$
c_k = \begin{cases} b_k & k = 0, 1, 2, \ldots, (A-1) \\ (b_k + x_{ue,k-A}) \bmod 2 & k = A, A+1, A+2, \ldots, A+15. \end{cases}
\tag{15.23}
$$

The information on uplink antenna selection (when configured by higher layers and used with format 0 only) is also carried implicitly in the CRC. In this case, the CRC parity bits are scrambled with both the antenna selection mask $x_{AS,0}, x_{AS,1}, \ldots, x_{AS,15}$ from Table 15.21 and the RNTI as below:

$$
c_k = \begin{cases} b_k & k = 0, 1, 2, \ldots, (A-1) \\ (b_k + x_{ue,k-A} + x_{AS,k-A}) \bmod 2 & k = A, A+1, A+2, \ldots, A+15. \end{cases}
\tag{15.24}
$$

For each PDCCH, the block of bits after tail-biting convolutional coding, sub-block interleaving and circular buffer rate matching (see Chapter 11) is represented as $b^{(i)}(0), \ldots, b^{(i)}(M_{bit}^{(i)} - 1)$, where $M_{bit}^{(i)}$ is the number of bits to be transmitted on physical downlink control channel number i. The bits from all the PDCCHs transmitted within a subframe are multiplexed as:

$$
\begin{aligned} &b^{(0)}(0), \ldots, b^{(0)}(M_{bit}^{(0)} - 1), b^{(1)}(0), \ldots, b^{(1)}(M_{bit}^{(1)} - 1), \ldots, \\ &b^{(n_{PDCCH}-1)}(0), \ldots, b^{(n_{PDCCH}-1)}(M_{bit}^{(n_{PDCCH}-1)} - 1), \end{aligned}
\tag{15.25}
$$

where n_{PDCCH} is the number of PDCCHs transmitted in the subframe. The sequence of bits in (15.25) is scrambled with a cell-specific sequence $c(i)$ as below:

$$
\tilde{b}(i) = (b(i) + c(i)) \bmod 2.
\tag{15.26}
$$

Table 15.21. UE transmit antenna selection mask.

UE transmit antenna selection	Antenna selection mask $\langle x_{AS,0}, x_{AS,1}, \ldots, x_{AS,15} \rangle$
UE port 0	$\langle 0, \ 0, \ 0, \ 0, \ 0, \ 0, \ 0, \ 0, \ 0, \ 0, \ 0, \ 0, \ 0, \ 0, \ 0, \ 0 \rangle$
UE port 1	$\langle 0, \ 0, \ 0, \ 0, \ 0, \ 0, \ 0, \ 0, \ 0, \ 0, \ 0, \ 0, \ 0, \ 0, \ 0, \ 1 \rangle$

The scrambling sequence generator is initialized at the start of each subframe with a cell-specific value as:

$$c_{\text{init}} = \lfloor n_s/2 \rfloor 2^9 + N_{\text{ID}}^{\text{cell}}. \tag{15.27}$$

A CCE consists of 36 resource elements (nine REGs). The number of bits that can be carried on a CCE using a QPSK modulation is therefore 72. A CCE number n then corresponds to bits:

$$b(72n), b(72n + 1), \ldots, b(72n + 71). \tag{15.28}$$

The block of scrambled bits $\tilde{b}(0), \ldots, \tilde{b}(M_{\text{tot}} - 1)$ is modulated using QPSK modulation, resulting in another block of complex-valued modulation symbols $d(0), \ldots, d(M_{\text{symb}} - 1)$. This block of modulation symbols is mapped to layers and precoded according to the single antenna, two antenna SFBC or four antenna SFBC-FSTD scheme (see Chapter 6) resulting in a block of vectors

$$y(i) = \left[y^{(0)}(i) \quad \cdots \quad y^{(P-1)}(i) \right]^T \quad i = 0, \ldots, (M_{\text{symb}} - 1), \tag{15.29}$$

where $y^{(p)}(i)$ represents the signal for antenna port $p = 0, \ldots, P - 1$ with $P \in \{1, 2, 4\}$.

The mapping to resource elements is defined by operations on quadruplets of complex-valued symbols $z^{(p)}(i) = \left(y^{(p)}(4i), y^{(p)}(4i + 1), y^{(p)}(4i + 2), y^{(p)}(4i + 3) \right)$. This is because the symbols are mapped to resource element groups each consisting of four resource elements. The block of quadruplets

$$z^{(p)}(0), \ldots, z^{(p)}(M_{\text{quad}} - 1), \quad M_{\text{quad}} = \frac{M_{\text{symb}}}{4} \tag{15.30}$$

is interleaved according to the subblock interleaver described in Chapter 11. The complex-valued symbol quadruplets at the output of the interleaver are denoted as $w^{(p)}(0), \ldots, w^{(p)}(M_{\text{quad}} - 1)$. The interleaving is performed on symbol quadruplets with each entry in the interleaving matrix consisting of a symbol quadruplet.

The block of quadruplets $w^{(p)}(0), \ldots, w^{(p)}(M_{\text{quad}} - 1)$ is shifted cyclically as a function of physical layer cell identity as below:

$$\bar{w}^{(p)}(i) = w^{(p)} \left((i + N_{\text{ID}}^{\text{cell}}) \bmod M_{\text{quad}} \right) \quad i = 0, 1, \ldots, (M_{\text{quad}} - 1). \tag{15.31}$$

We note that a common interleaving based on subblock interleaver followed by cell-specific cyclic shift results in a cell-specific interleaving. The cell-specific interleaving is beneficial for interference randomization to neighboring cells as different PDCCHs may use different transmit power levels. The block of quadruplets $\bar{w}^{(p)}(i)$ is mapped to the OFDM symbols indicated by CFICH. The mapping is done in a time-first fashion with each symbol quadruplet mapped to a resource-element group (REG) that is not used by either PCFICH or PHICH as shown in Figure 15.12.

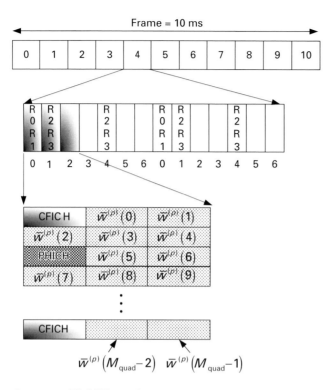

Figure 15.12. PDCCH mapping to resources.

Table 15.22. HI codewords.

HI	HI	HI codeword $\langle b_0, b_1, b_2 \rangle$
ACK	0	$\langle 0, 0, 0 \rangle$
NACK	1	$\langle 1, 1, 1 \rangle$

15.6   Hybrid ARQ indicator

A hybrid ARQ indicator (HI) with corresponding physical channel referred to as physical hybrid ARQ indicator channel (PHICH) provides ACK/NACK feedback for uplink data transmissions on PUSCH. A maximum of a single codeword or transport block can be transmitted on PUSCH and therefore a 1-bit binary ACK/NACK feedback where "1" indicates a positive acknowledgment (ACK) and "0" indicates a negative acknowledgment (NACK) respectively is sufficient. As no sophisticated coding can be applied to a single bit transmission, the bit is simply repeated three times as shown in Table 15.22.

Another challenge with a single-bit transmission is capturing transmit diversity. When a large message size is transmitted, the total number of coded bits is fairly large. These bits then can be distributed over a larger bandwidth to capture frequency diversity via coding. However, a single-bit transmission in a distributed fashion over a large bandwidth would require a very large number of repetitions thus making inefficient use of the resources. Another aspect to consider is interference averaging as PHICH is power controlled to make it reliable

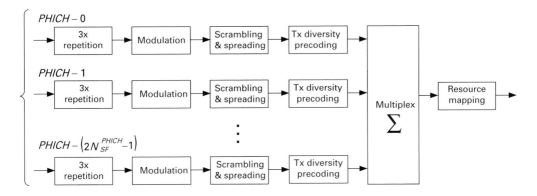

Figure 15.13. HI transmission chain processing.

for the desired user. Therefore, if PHICH is transmitted on a very narrow bandwidth, it will create interference peaks in the neighboring cells. A solution to solve these problems is to spread PHICH using orthogonal sequences. The spreading, however, also results in bandwidth expansion and therefore the HIs for multiple UEs within a PHICH group are code-multiplexed as shown in Figure 15.13. In order to further improve resource utilization, the PHICHs are also multiplexed on in-phase (I) and quadrature (Q) components, which means that a BPSK modulation is used.

The multiple PHICHs mapped to the same set of resource elements using code-multiplexing form a PHICH group. A PHICH resource is identified by the index pair $\left(n_{\text{PHICH}}^{\text{group}}, n_{\text{PHICH}}^{\text{seq}}\right)$, where $n_{\text{PHICH}}^{\text{group}}$ is the PHICH group number and $n_{\text{PHICH}}^{\text{seq}}$ is the orthogonal sequence index within the group. The number of PHICH groups in a cell can be configured in a semi-static fashion to match the load in the cell. The number of PHICH groups $N_{\text{PHICH}}^{\text{group}}$ is constant in all subframes and is given by:

$$N_{\text{PHICH}}^{\text{group}} = \begin{cases} \left\lceil N_g \left(N_{\text{RB}}^{\text{DL}}/8\right)\right\rceil & \text{normal CP} \\ 2 \cdot \left\lceil N_g \left(N_{\text{RB}}^{\text{DL}}/8\right)\right\rceil & \text{extended CP} \end{cases} \qquad N_g \in \{1/6, 1/2, 1, 2\}, \qquad (15.32)$$

where N_g is provided by higher layers. The index $n_{\text{PHICH}}^{\text{group}}$ ranges from 0 to $N_{\text{PHICH}}^{\text{group}} - 1$. The number of groups is two times larger for the extended cyclic prefix than for the normal cyclic prefix because the spreading factor of the orthogonal sequence is two times smaller in the case of the extended cyclic prefix. This means that a PHICH groups can only carry half the PHICHs for the extended cyclic prefix case when compared to the normal cyclic prefix. The number of PHICH groups increases with $N_{\text{RB}}^{\text{DL}}$ because for larger bandwidths, a larger number of UEs can transmit simultaneously and hence require transmission of a larger number of hybrid ARQ ACK/NACK responses. In theory each RB can contain data for a different UE and hence the number of simultaneous ACK/NACKs required is equal to the number of RBs. However, this situation does not occur frequently as some UEs require multiple RBs for their transmission. The term $\left(N_{\text{RB}}^{\text{DL}}/8\right)$ accounts for the fact that in the case of normal cyclic prefix, up to eight PHICHs can be carried in a single PHICH group.

Let us consider an example assuming $N_{RB}^{DL} = 50$. The number of PHICH groups N_{PHICH}^{group} for the normal cyclic prefix in this case is:

$$N_{PHICH}^{group} = \lceil N_g (25/8) \rceil = 2, 4, 7, 13 \quad N_g \in \{1/6, 1/2, 1, 2\}. \quad (15.33)$$

The total number of PHICHs in this case is 16, 32, 56 and 104 for $N_g = 1/6, 1/2, 1, 2$ respectively. It should be noted that the maximum number of PHICHs is approximately two times larger than the number of RBs. This is because in the extreme case one could imagine that two UEs are scheduled per RB in multi-user MIMO transmission mode on PUSCH. Then each of these UEs requires a PHICH for ACK/NACK feedback.

The block of bits $b(0), b(1), b(2)$ transmitted on one PHICH in one subframe is modulated using BPSK modulation generating a block of modulation symbols $z(0), z(1), z(2)$:

$$z(i) = \frac{1}{\sqrt{2}} (1 - b(i)). \quad (15.34)$$

15.6.1 Spreading for hybrid ARQ indicator

The block of modulation symbols $z(0), z(1), z(2)$ is bit-wise multiplied with an orthogonal sequence, resulting in another sequence of modulation symbols $d(0), \ldots, d(M_{symb} - 1)$ according to:

$$d(i) = w \left(i \bmod N_{SF}^{PHICH} \right) \cdot (1 - 2c(i)) \cdot z \left(\lfloor i/N_{SF}^{PHICH} \rfloor \right) \quad i = 0, \ldots, (M_{symb} - 1) \quad (15.35)$$

$$d(i) = w \left(i \bmod N_{SF}^{PHICH} \right) \cdot (1 - 2c(i)) \cdot z \left(\lfloor i/N_{SF}^{PHICH} \rfloor \right),$$

where $M_{symb} = 3N_{SF}^{PHICH}$, $N_{SF}^{PHICH} = 4, 2$ for the normal and extended cyclic prefix respectively and $c(i)$ is a cell-specific scrambling sequence initialized as:

$$c_{init} = \left(\lfloor n_s/2 \rfloor + 1 \right) \cdot \left(2N_{ID}^{cell} + 1 \right) \cdot 2^9 + N_{ID}^{cell}. \quad (15.36)$$

We note that similar to the case of CFICH, the scrambling sequence is initialized at the start of each subframe. The term $(1 - 2c(i))$ converts 0 and 1 in $c(i)$ to $+1$ and -1 respectively. The orthogonal sequences $[w(0) \cdots w(N_{SF}^{PHICH} - 1)]$ are given in Table 15.23 where the sequence index n_{PHICH}^{seq} corresponds to the PHICH number within the PHICH group. We note that the orthogonal sequences are merely Walsh codes on I and Q branches. For example, the orthogonal sequence $[+j \ -j \ +j \ -j]$ is in effect the Q branch equivalent of the Walsh code $[+1 \ +1 \ -1 \ -1]$ on the I branch.

Let us consider an example where the orthogonal sequence with index $n_{PHICH}^{seq} = 6$, $[+j \ -j \ +j \ -j]$, is used to spread ACK/NACK modulation symbols for a UE. Ignoring the scrambling sequence for simplicity, the spread sequences of modulation symbols $d(0), \ldots, d(11)$ for an ACK and a NACK are:

$$d_{ACK} = \frac{1}{\sqrt{2}} [+j \ -j \ +j \ -j \ +j \ -j \ +j \ -j \ +j \ -j \ +j \ -j]$$

$$d_{NACK} = \frac{1}{\sqrt{2}} [-j \ +j \ -j \ +j \ -j \ +j \ -j \ +j \ -j \ +j \ -j \ +j]. \quad (15.37)$$

Table 15.23. Orthogonal sequences $\left[w(0) \cdots w(N_{SF}^{PHICH} - 1) \right]$ for PHICH.

Sequence index	Orthogonal sequence	
n_{PHICH}^{seq}	Normal cyclic prefix $N_{SF}^{PHICH} = 4$	Extended cyclic prefix $N_{SF}^{PHICH} = 2$
0	$[+1 \quad +1 \quad +1 \quad +1]$	$[+1 \quad +1]$
1	$[+1 \quad -1 \quad +1 \quad -1]$	$[+1 \quad -1]$
2	$[+1 \quad +1 \quad -1 \quad -1]$	$[+j \quad +j]$
3	$[+1 \quad -1 \quad -1 \quad +1]$	$[+j \quad -j]$
4	$[+j \quad +j \quad +j \quad +j]$	-
5	$[+j \quad -j \quad +j \quad -j]$	-
6	$[+j \quad +j \quad -j \quad -j]$	-
7	$[+j \quad -j \quad -j \quad +j]$	-

The number of symbols $d(0), \ldots, d(M_{symb} - 1)$ for the normal cyclic prefix is 12, which is obtained by spreading each of the three repeated ACK/NACK modulation symbols by an orthogonal sequence with spreading factor (SF) of four $\left(N_{SF}^{PHICH} = 4\right)$. Therefore, the 12 symbols can be mapped to three resource element groups.

$$d^{(0)}(i) = d(i), \quad i = 0, 1, \ldots, 11. \tag{15.38}$$

However, for the extended cyclic prefix, the spreading factor (SF) is only two $\left(N_{SF}^{PHICH} = 2\right)$. A lower spreading factor is used for the extended cyclic prefix because the channel delay spread is generally larger in deployments using the extended cyclic prefix. A larger delay spread translates into narrower coherence bandwidths, which in turn affect the orthogonality performance of sequences with a larger spreading factor. We should note that Walsh codes are orthogonal as long as the channel gain is constant over the sequence length. With narrower coherence bandwidth, the frequency-domain channel gains can change over four adjacent resource elements (subcarriers). In order to limit the loss of orthogonality in deployments with larger delay spread, a smaller spreading factor is chosen for the case of the extended cyclic prefix. The spreading of three repeated PHICH modulation symbols generates a total of six symbols. In order to evenly map these six symbols to three resource element groups, the block of modulation symbols $d(0), \ldots, d(5)$ is extended to length 12 as below:

$$\begin{bmatrix} d^{(0)}(4i) \\ d^{(0)}(4i+1) \\ d^{(0)}(4i+2) \\ d^{(0)}(4i+3) \end{bmatrix} = \begin{cases} \begin{bmatrix} d(2i) & d(2i+1) & 0 & 0 \end{bmatrix}^T & n_{PHICH}^{group} \bmod 2 = 0 \\ \begin{bmatrix} 0 & 0 & d(2i+0) & d(2i+1) \end{bmatrix}^T & n_{PHICH}^{group} \bmod 2 = 1 \end{cases} \tag{15.39}$$

for $i = 0, 1, 2.$

15.6.2   Transmit diversity for hybrid ARQ indicator

The transmit diversity scheme used on PHICH for the four antenna ports case is different from the balanced SFBC-FSTD scheme used on other downlink channels such as CFICH, PDCCH and PDSCH. The transmit matrix for the balanced SFBC-FSTD scheme is given as

$$
\begin{bmatrix}
y^{(0)}(4i) & y^{(0)}(4i+1) & y^{(0)}(4i+2) & y^{(0)}(4i+3) \\
y^{(1)}(4i) & y^{(1)}(4i+1) & y^{(1)}(4i+2) & y^{(1)}(4i+3) \\
y^{(2)}(4i) & y^{(2)}(4i+1) & y^{(2)}(4i+2) & y^{(2)}(4i+3) \\
y^{(3)}(4i) & y^{(3)}(4i+1) & y^{(3)}(4i+2) & y^{(3)}(4i+3)
\end{bmatrix}
$$

$$
=
\begin{bmatrix}
x^{(0)}(i) & x^{(1)}(i) & 0 & 0 \\
0 & 0 & x^{(2)}(i) & x^{(3)}(i) \\
-\left(x^{(1)}(i)\right)^{*} & \left(x^{(0)}(i)\right)^{*} & 0 & 0 \\
0 & 0 & -\left(x^{(3)}(i)\right)^{*} & \left(x^{(2)}(i)\right)^{*}
\end{bmatrix},
\tag{15.40}
$$

where $y^{(p)}(i)$ represents the signal for antenna port p, $p = 0, \ldots, P-1$ on the ith resource element. The equivalent channel matrix for the balanced SFBC-FSTD scheme H_4-Balanced-SFBC-FSTD can be written as:

$$
H_{4\text{-SFBC-FSTD}} = \frac{1}{\sqrt{4}}
\begin{bmatrix}
h_0 & -h_2^{*} & 0 & 0 \\
h_2 & h_0^{*} & 0 & 0 \\
0 & 0 & h_1 & -h_3^{*} \\
0 & 0 & h_3 & h_1^{*}
\end{bmatrix},
\tag{15.41}
$$

where h_i represents complex channel gain on the ith antenna port. We assume that the channel gain is constant over the four resource elements within a resource element group. Assuming a matched filter receiver, the resulting channel gains matrix can be written as:

$$
=
\begin{bmatrix}
\dfrac{\left(h_0^2 + h_2^2\right)}{2} & 0 & 0 & 0 \\
0 & \dfrac{\left(h_0^2 + h_2^2\right)}{2} & 0 & 0 \\
0 & 0 & \dfrac{\left(h_1^2 + h_3^2\right)}{2} & 0 \\
0 & 0 & 0 & \dfrac{\left(h_1^2 + h_3^2\right)}{2}
\end{bmatrix}.
\tag{15.42}
$$

We note that when this transmit diversity scheme is used on the spread PHICH modulation symbols with spreading using $\left(N_{\text{SF}}^{\text{PHICH}} = 4\right)$, the first half of the symbols experience channel gain of $\frac{(h_0^2 + h_2^2)}{2}$ while the remaining half of the symbols experience channel gain $\frac{(h_1^2 + h_3^2)}{2}$. Since these two channel gains are different the Walsh sequence experiences loss of orthogonality. It should be noted that there is no issue with the spreading factor of two $\left(N_{\text{SF}}^{\text{PHICH}} = 2\right)$ as both the symbols from the spread sequence will experience similar channel gain. The problem with $\left(N_{\text{SF}}^{\text{PHICH}} = 4\right)$ is solved by modifying the four antenna ports transmit diversity precoding as described below.

The block of 12 symbols $d^{(0)}(0), \ldots, d^{(0)}(11)$ is mapped to layers and precoded, resulting in a block of vectors $y(i) = \left[y^{(0)}(i) \quad \ldots \quad y^{(P-1)}(i)\right]^T, i = 0, 1, \ldots, 11$, where $y^{(p)}(i)$ represents the signal for antenna port p, $p = 0, \ldots, P - 1$ with $P \in \{1, 2, 4\}$. The layer mapping and precoding operation depends on the cyclic prefix length and the number of antenna ports used for transmission of the PHICH. The PHICH is also transmitted on the same set of antenna ports as the PBCH. For transmission on a single antenna port, that is $P = 1$, the layer mapping and precoding can be simply represented as:

$$y^{(0)}(i) = x^{(0)}(i) = d^{(0)}(i), \quad i = 0, 1, \ldots, 11. \tag{15.43}$$

For transmission on two antenna ports, that is $P = 2$, even-numbered symbols are mapped to layer 0 while odd-numbered symbols are mapped to layer 1 as below:

$$\begin{aligned} x^{(0)}(i) &= d^{(0)}(2i) \\ x^{(1)}(i) &= d^{(0)}(2i + 1) \end{aligned} \quad i = 0, 1, \ldots, 5. \tag{15.44}$$

The output $\left[y^0(i) \quad y^{(1)}(i)\right]^T$ of the precoding operation for two antenna ports transmit diversity is written as:

$$\begin{bmatrix} y^{(0)}(2i) \\ y^{(1)}(2i) \\ y^{(0)}(2i + 1) \\ y^{(1)}(2i + 1) \end{bmatrix} = \begin{bmatrix} 1 & 0 & j & 0 \\ 0 & -1 & 0 & j \\ 0 & 1 & 0 & j \\ 1 & 0 & -j & 0 \end{bmatrix} \times \begin{bmatrix} x_I^{(0)}(i) \\ x_I^{(1)}(i) \\ x_Q^{(0)}(i) \\ x_Q^{(1)}(i) \end{bmatrix} \quad i = 0, 1, \ldots, 5, \tag{15.45}$$

where $x_I^{(0)}(i)$ and $x_Q^{(0)}(i)$ are respectively real and imaginary parts of the modualtion symbol on layer 0 and $x_I^{(1)}(i)$ and $x_Q^{(1)}(i)$ are respectively real and imaginary parts of the modualtion symbol on layer 1.

$$\begin{bmatrix} y^{(0)}(2i) \\ y^{(1)}(2i) \\ y^{(0)}(2i + 1) \\ y^{(1)}(2i + 1) \end{bmatrix} = \begin{bmatrix} x_I^{(0)}(i) + jx_Q^{(0)}(i) \\ -x_I^{(1)}(i) + jx_Q^{(1)}(i) \\ x_I^{(1)}(i) + jx_Q^{(1)}(i) \\ x_I^{(0)}(i) - jx_Q^{(0)}(i) \end{bmatrix} = \begin{bmatrix} x^{(0)}(i) \\ -\left(x^{(1)}(i)\right)^* \\ x^{(1)}(i) \\ \left(x^{(0)}(i)\right)^* \end{bmatrix} \quad i = 0, 1, \ldots, 5. \tag{15.46}$$

Let us rewrite (15.46) as below:

$$\begin{bmatrix} y^{(0)}(2i) & y^{(0)}(2i + 1) \\ y^{(1)}(2i) & y^{(1)}(2i + 1) \end{bmatrix} = \begin{bmatrix} x^{(0)}(i) & x^{(1)}(i) \\ -\left(x^{(1)}(i)\right)^* & \left(x^{(0)}(i)\right)^* \end{bmatrix}. \tag{15.47}$$

We note that the precoding scheme in (15.47) is simply a Space Frequency Block Coding (SFBC) approach on antenna ports 0 and 1. In this case all symbols experience channel gain of $\frac{(h_0^2+h_1^2)}{2}$ and therefore there is no issue of loss of orthogonality of the Walsh sequence.

The transmit diversity precoding for the case of four antenna ports for PHICH is done slightly differently than the other channels to avoid the problem of loss of orthogonality of the sequence. In particular when $(i + n_{\text{PHICH}}^{\text{group}}) \mod 2 = 0$ for the normal cyclic prefix, or $(i + \lfloor n_{\text{PHICH}}^{\text{group}}/2 \rfloor) \mod 2 = 0$ for the extended cyclic prefix, where $n_{\text{PHICH}}^{\text{group}}$ is the PHICH group number and $i = 0, 1, 2$, the precoding is given as:

$$
\begin{bmatrix}
y^{(0)}(4i) \\
y^{(1)}(4i) \\
y^{(2)}(4i) \\
y^{(3)}(4i) \\
y^{(0)}(4i+1) \\
y^{(1)}(4i+1) \\
y^{(2)}(4i+1) \\
y^{(3)}(4i+1) \\
y^{(0)}(4i+2) \\
y^{(1)}(4i+2) \\
y^{(2)}(4i+2) \\
y^{(3)}(4i+2) \\
y^{(0)}(4i+3) \\
y^{(1)}(4i+3) \\
y^{(2)}(4i+3) \\
y^{(3)}(4i+3)
\end{bmatrix}
=
\begin{bmatrix}
1 & 0 & 0 & 0 & j & 0 & 0 & 0 \\
0 & 0 & 0 & 0 & 0 & 0 & 0 & 0 \\
0 & -1 & 0 & 0 & 0 & j & 0 & 0 \\
0 & 0 & 0 & 0 & 0 & 0 & 0 & 0 \\
0 & 1 & 0 & 0 & 0 & j & 0 & 0 \\
0 & 0 & 0 & 0 & 0 & 0 & 0 & 0 \\
1 & 0 & 0 & 0 & -j & 0 & 0 & 0 \\
0 & 0 & 0 & 0 & 0 & 0 & 0 & 0 \\
0 & 0 & 1 & 0 & 0 & 0 & j & 0 \\
0 & 0 & 0 & 0 & 0 & 0 & 0 & 0 \\
0 & 0 & 0 & -1 & 0 & 0 & 0 & j \\
0 & 0 & 0 & 0 & 0 & 0 & 0 & 0 \\
0 & 0 & 0 & 1 & 0 & 0 & 0 & j \\
0 & 0 & 0 & 0 & 0 & 0 & 0 & 0 \\
0 & 0 & 1 & 0 & 0 & 0 & -j & 0 \\
0 & 0 & 0 & 0 & 0 & 0 & 0 & 0
\end{bmatrix}
\times
\begin{bmatrix}
x_I^{(0)}(i) \\
x_I^{(1)}(i) \\
x_I^{(2)}(i) \\
x_I^{(3)}(i) \\
x_Q^{(0)}(i) \\
x_Q^{(1)}(i) \\
x_Q^{(2)}(i) \\
x_Q^{(3)}(i)
\end{bmatrix}
\quad i = 0, 1, 2
$$

(15.48)

$$
\begin{bmatrix}
y^{(0)}(4i) \\
y^{(1)}(4i) \\
y^{(2)}(4i) \\
y^{(3)}(4i) \\
y^{(0)}(4i+1) \\
y^{(1)}(4i+1) \\
y^{(2)}(4i+1) \\
y^{(3)}(4i+1) \\
y^{(0)}(4i+2) \\
y^{(1)}(4i+2) \\
y^{(2)}(4i+2) \\
y^{(3)}(4i+2) \\
y^{(0)}(4i+3) \\
y^{(1)}(4i+3) \\
y^{(2)}(4i+3) \\
y^{(3)}(4i+3)
\end{bmatrix}
=
\begin{bmatrix}
x_I^{(0)}(i) + jx_Q^{(0)}(i) \\
0 \\
-x_I^{(1)}(i) + jx_Q^{(1)}(i) \\
0 \\
x_I^{(1)}(i) + jx_Q^{(1)}(i) \\
0 \\
x_I^{(0)}(i) - jx_Q^{(0)}(i) \\
0 \\
x_I^{(2)}(i) + jx_Q^{(2)}(i) \\
0 \\
-x_I^{(3)}(i) + jx_Q^{(3)}(i) \\
0 \\
x_I^{(3)}(i) + jx_Q^{(3)}(i) \\
0 \\
x_I^{(2)}(i) - jx_Q^{(2)}(i) \\
0
\end{bmatrix}
=
\begin{bmatrix}
x^{(0)}(i) \\
0 \\
-\left(x^{(1)}(i)\right)^* \\
0 \\
x^{(1)}(i) \\
0 \\
\left(x^{(0)}(i)\right)^* \\
0 \\
x^{(2)}(i) \\
0 \\
-\left(x^{(3)}(i)\right)^* \\
0 \\
x^{(3)}(i) \\
0 \\
\left(x^{(2)}(i)\right)^* \\
0
\end{bmatrix}
\quad i = 0, 1, 2. \quad (15.49)
$$

Let us rewrite (15.49) as:

$$
\begin{bmatrix}
y^{(0)}(4i) & y^{(0)}(4i+1) & y^{(0)}(4i+2) & y^{(0)}(4i+3) \\
y^{(1)}(4i) & y^{(1)}(4i+1) & y^{(1)}(4i+2) & y^{(1)}(4i+3) \\
y^{(2)}(4i) & y^{(2)}(4i+1) & y^{(2)}(4i+2) & y^{(2)}(4i+3) \\
y^{(3)}(4i) & y^{(3)}(4i+1) & y^{(3)}(4i+2) & y^{(3)}(4i+3)
\end{bmatrix}
$$

$$
=
\begin{bmatrix}
x^{(0)}(i) & x^{(1)}(i) & x^{(2)}(i) & x^{(3)}(i) \\
0 & 0 & 0 & 0 \\
-\left(x^{(1)}(i)\right)^{*} & \left(x^{(0)}(i)\right)^{*} & -\left(x^{(3)}(i)\right)^{*} & \left(x^{(2)}(i)\right)^{*} \\
0 & 0 & 0 & 0
\end{bmatrix}.
$$

(15.50)

We note that the scheme in (15.50) is simply an SFBC scheme on antenna ports 0 and 2 with all symbols belonging to an orthogonal sequence experiencing a channel gain of $\frac{(h_0^2+h_2^2)}{2}$. For $(i + n_{\text{PHICH}}^{\text{group}})$ mod $2 = 1$ for the normal cyclic prefix, or $(i + \lfloor n_{\text{PHICH}}^{\text{group}}/2 \rfloor)$ mod $2 = 1$ for the extended cyclic prefix, where $n_{\text{PHICH}}^{\text{group}}$ is the PHICH group number and $i = 0, 1, 2$, the four antenna ports transmit diversity precoding is applied as:

$$
\begin{bmatrix}
y^{(0)}(4i) \\
y^{(1)}(4i) \\
y^{(2)}(4i) \\
y^{(3)}(4i) \\
y^{(0)}(4i+1) \\
y^{(1)}(4i+1) \\
y^{(2)}(4i+1) \\
y^{(3)}(4i+1) \\
y^{(0)}(4i+2) \\
y^{(1)}(4i+2) \\
y^{(2)}(4i+2) \\
y^{(3)}(4i+2) \\
y^{(0)}(4i+3) \\
y^{(1)}(4i+3) \\
y^{(2)}(4i+3) \\
y^{(3)}(4i+3)
\end{bmatrix}
=
\begin{bmatrix}
0 & 0 & 0 & 0 & 0 & 0 & 0 & 0 \\
1 & 0 & 0 & 0 & j & 0 & 0 & 0 \\
0 & 0 & 0 & 0 & 0 & 0 & 0 & 0 \\
0 & -1 & 0 & 0 & 0 & j & 0 & 0 \\
0 & 0 & 0 & 0 & 0 & 0 & 0 & 0 \\
0 & 1 & 0 & 0 & 0 & j & 0 & 0 \\
0 & 0 & 0 & 0 & 0 & 0 & 0 & 0 \\
1 & 0 & 0 & 0 & -j & 0 & 0 & 0 \\
0 & 0 & 0 & 0 & 0 & 0 & 0 & 0 \\
0 & 0 & 1 & 0 & 0 & 0 & j & 0 \\
0 & 0 & 0 & 0 & 0 & 0 & 0 & 0 \\
0 & 0 & 0 & -1 & 0 & 0 & 0 & j \\
0 & 0 & 0 & 0 & 0 & 0 & 0 & 0 \\
0 & 0 & 0 & 1 & 0 & 0 & 0 & j \\
0 & 0 & 0 & 0 & 0 & 0 & 0 & 0 \\
0 & 0 & 1 & 0 & 0 & 0 & -j & 0
\end{bmatrix}
\times
\begin{bmatrix}
x_I^{(0)}(i) \\
x_I^{(1)}(i) \\
x_I^{(2)}(i) \\
x_I^{(3)}(i) \\
x_Q^{(0)}(i) \\
x_Q^{(1)}(i) \\
x_Q^{(2)}(i) \\
x_Q^{(3)}(i)
\end{bmatrix}
\quad i = 0, 1, 2
$$

(15.51)

$$
\begin{bmatrix}
y^{(0)}(4i) \\
y^{(1)}(4i) \\
y^{(2)}(4i) \\
y^{(3)}(4i) \\
y^{(0)}(4i+1) \\
y^{(1)}(4i+1) \\
y^{(2)}(4i+1) \\
y^{(3)}(4i+1) \\
y^{(0)}(4i+2) \\
y^{(1)}(4i+2) \\
y^{(2)}(4i+2) \\
y^{(3)}(4i+2) \\
y^{(0)}(4i+3) \\
y^{(1)}(4i+3) \\
y^{(2)}(4i+3) \\
y^{(3)}(4i+3)
\end{bmatrix}
=
\begin{bmatrix}
0 \\
x_I^{(0)}(i)+jx_Q^{(0)}(i) \\
0 \\
-x_I^{(1)}(i)+jx_Q^{(1)}(i) \\
0 \\
x_I^{(1)}(i)+jx_Q^{(1)}(i) \\
0 \\
x_I^{(0)}(i)-jx_Q^{(0)}(i) \\
0 \\
x_I^{(2)}(i)+jx_Q^{(2)}(i) \\
0 \\
-x_I^{(3)}(i)+jx_Q^{(3)}(i) \\
0 \\
x_I^{(3)}(i)+jx_Q^{(3)}(i) \\
0 \\
x_I^{(2)}(i)-jx_Q^{(2)}(i)
\end{bmatrix}
=
\begin{bmatrix}
0 \\
x^{(0)}(i) \\
0 \\
-\left(x^{(1)}(i)\right)^* \\
0 \\
x^{(1)}(i) \\
0 \\
\left(x^{(0)}(i)\right)^* \\
0 \\
x^{(2)}(i) \\
0 \\
-\left(x^{(3)}(i)\right)^* \\
0 \\
x^{(3)}(i) \\
0 \\
\left(x^{(2)}(i)\right)^*
\end{bmatrix}
\quad i=0,1,2
$$

$$(15.52)$$

$$
\begin{bmatrix}
y^{(0)}(4i) & y^{(0)}(4i+1) & y^{(0)}(4i+2) & y^{(0)}(4i+3) \\
y^{(1)}(4i) & y^{(1)}(4i+1) & y^{(1)}(4i+2) & y^{(1)}(4i+3) \\
y^{(2)}(4i) & y^{(2)}(4i+1) & y^{(2)}(4i+2) & y^{(2)}(4i+3) \\
y^{(3)}(4i) & y^{(3)}(4i+1) & y^{(3)}(4i+2) & y^{(3)}(4i+3)
\end{bmatrix}
$$
$$=$$
$$
\begin{bmatrix}
0 & 0 & 0 & 0 \\
x^{(0)}(i) & x^{(1)}(i) & x^{(2)}(i) & x^{(3)}(i) \\
0 & 0 & 0 & 0 \\
-\left(x^{(1)}(i)\right)^* & \left(x^{(0)}(i)\right)^* & -\left(x^{(3)}(i)\right)^* & \left(x^{(2)}(i)\right)^*
\end{bmatrix}.
$$

$$(15.53)$$

We note that the scheme in (15.53) is simply an SFBC scheme on antenna ports 1 and 3 with all symbols belonging to an orthogonal sequence experiencing a channel gain of $\frac{(h_1^2+h_3^2)}{2}$. We note that the modified four antenna transmit diversity scheme avoids the problem of loss of orthogonality of the Walsh sequence by performing equivalent two-antenna SFBC within a sequence length. However, the pair of antenna ports is changed between repetitions of PHICH as given by (15.50) and (15.53) hence capturing diversity from all the four antenna ports.

15.6.3   Resource mapping for PHICH

The symbol sequences obtained after transmit diversity precoding for each of the PHICH within a PHICH group are added to obtain another symbol sequence $\bar{y}^{(p)}(0),\ldots,\bar{y}^{(p)}(11)$

given as:

$$\bar{y}^{(p)}(n) = \sum_i y_i^{(p)}(n) \quad n = 0, 1, \ldots, 11, \tag{15.54}$$

where the sum is over all PHICHs in the PHICH group and $y_i^{(p)}(n)$ represents the symbol sequence from the ith PHICH in the PHICH group.

Since a resource element group contains four resource elements, the mapping of the symbol sequence $\bar{y}^{(p)}(0), \ldots, \bar{y}^{(p)}(11)$ to resource elements is defined in terms of symbol quadruplets:

$$z^{(p)}(i) = \left\langle \bar{y}^{(p)}(4i), \bar{y}^{(p)}(4i+1), \bar{y}^{(p)}(4i+2), \bar{y}^{(p)}(4i+3) \right\rangle \quad i = 0, 1, 2, \tag{15.55}$$

where $z^{(p)}(i)$ denotes symbol quadruplet i for antenna port p.

The symbol quadruplet i is then mapped to the resource-element group represented by (k_i', l_i'), where the indices k_i' and l_i' depend upon the cyclic prefix length as well as whether normal or extended PHICH duration is used. The normal and extended PHICH durations are one and three OFDM symbols respectively. Moreover, for MBSFN subframes, a PHICH duration of two OFDM symbols is used. The PHICH duration configured puts a lower limit on the size of the control region signaled by the PCFICH. Therefore, when an extended PHICH duration of three OFDM symbols is configured, the control format indicator (CFI) information is redundant as downlink control always spans three OFDM symbols.

Let n_i denote the number of resource element groups not assigned to PCFICH in OFDM symbol i. We number these REGs from 0 to $(n_i - 1)$ starting from the resource-element group with the lowest frequency-domain index. For simplicity, let us focus on resource mapping for the case of normal PHICH duration where the PHICH is carried in the first OFDM symbol with $l_i' = 0$. The frequency-domain index k_i' is set to the resource-element group number $\bar{n}_i$ given by:

$$\bar{n}_i = \begin{cases} \left(\left\lfloor N_{\text{ID}}^{\text{cell}} \cdot n_{l_i'} \big/ n_0 \right\rfloor + m' \right) \bmod n_{l_i'} & i = 0 \\ \left(\left\lfloor N_{\text{ID}}^{\text{cell}} \cdot n_{l_i'} \big/ n_0 \right\rfloor + m' + \left\lfloor n_{l_i'} \big/ 3 \right\rfloor \right) \bmod n_{l_i'} & i = 1 \\ \left(\left\lfloor N_{\text{ID}}^{\text{cell}} \cdot n_{l_i'} \big/ n_0 \right\rfloor + m' + \left\lfloor 2n_{l_i'} \big/ 3 \right\rfloor \right) \bmod n_{l_i'} & i = 2, \end{cases} \tag{15.56}$$

where m' indicates the PHICH group number.

Let us consider an example with $N_{\text{ID}}^{\text{cell}} = 0$, $m' = 0$, $l_i' = 0$ and $n_0 = 50$ then $\bar{n}_0, \bar{n}_1, \bar{n}_2 = 0, 16, 33$. The second PHICH group will use $\bar{n}_0, \bar{n}_1, \bar{n}_2 = 1, 17, 34$ and so on. In the case of extended PHICH over three OFDM symbols, each symbol quadruplet is localized within a REG in an OFDM symbol with the first, second and third symbol quadruplet transmitted on the first, second and third OFDM symbol. The extended PHICH duration helps to improve PHICH coverage as each PHICH can span over three OFDM symbols allowing to transmit three times more energy for the same transmit power. We note that the transmission of different symbol quadruplets is uniformly distributed in frequency with each symbol quadruplet localized within a REG. It should be noted that each symbol quadruplet contains an ACK/NACK modulation symbol spread by an orthogonal sequence and therefore should experience constant channel gain. The localized transmission of the symbol quadruplet within a REG allows it to experience an approximately constant channel gain as the channel on adjacent resource elements in a REG are correlated due to very small bandwidth of the REG. However, different symbol quadruplets can be distributed in frequency.

15.6.4 PHICH assignment procedure

In Section 12.6, we noted that the eNB processing time for PUSCH transmissions in the uplink is equal to three subframes (3 ms). Therefore, for a scheduled PUSCH transmission in subframe n, a UE receives ACK/NACK feedback from the eNB on PHICH in subframe (n + 4). The PHICH resource is determined from the lowest index physical resource block (PRB) of the uplink resource allocation and from the 3-bit uplink demodulation reference symbol (DMRS) cyclic shift associated with the PDCCH with DCI format 0 granting the PUSCH transmission. The PHICH resource is identified by the index pair n_{PHICH}^{group}, n_{PHICH}^{seq}, where n_{PHICH}^{group} is the PHICH group number and n_{PHICH}^{seq}, is the orthogonal sequence index within the group as defined by:

$$n_{PHICH}^{group} = \left(I_{PRB\_RA}^{lowest\_index} + n_{DMRS}\right) \bmod N_{PHICH}^{group}$$

$$n_{PHICH}^{seq} = \left(\left[\frac{I_{PRB\_RA}^{lowest\_index}}{N_{PHICH}^{group}}\right] + n_{DMRS}\right) \bmod \left(2N_{SF}^{PHICH}\right), \qquad (15.57)$$

where n_{DMRS} is related to the cyclic shift of the DMRS (see Table 15.4) used in the uplink transmission as given by Table 15.25. $I_{PRB\_RA}^{lowest\_index}$ is the lowest index PRB of the uplink resource allocation. The modulo $(2N_{SF}^{PHICH})$ operation in (15.57) accounts for the fact that the number of sequences is twice the spreading factor due to I/Q multiplexing as indicated in Table 15.23. The PHICH group number n_{PHICH}^{group} and the sequence index n_{PHICH}^{seq} for N_{RB}^{UL} = 15 and n_{DMRS} = 0 are given in Table 15.24. We note that the maximum number of PHICHs required is equal to the number of uplink PRBs. However, the LTE standard allows scheduling more than one UE on the same uplink resource blocks in a multi-user MIMO configuration.

Table 15.24. PHICH group number n_{PHICH}^{group} and sequence index n_{PHICH}^{seq} for N_{RB}^{UL} = 15 and n_{DMRS} = 0.

Lowest index PRB $I_{PRB\_RA}^{lowest\_index}$	PHICH group number n_{PHICH}^{group}	PHICH sequence index n_{PHICH}^{seq}
0	0	0
1	1	0
2	0	1
3	1	1
4	0	2
5	1	2
6	0	3
7	1	3
8	0	4
9	1	4
10	0	5
11	1	5
12	0	6
13	1	6
14	0	7

Table 15.25. Relationship between n_{DMRS} and cyclic shift field in DCI format 0.

Cyclic shift field in DCI format 0	n_{DMRS}
000	0
001	1
010	2
011	3
100	4
101	5
110	6
111	7

Therefore, the cyclic shift of the DMRS is used to differentiate PHICHs for UEs transmitting using the same lowest index PRB $I_{PRB\_RA}^{lowest\_index}$.

15.7 Summary

The downlink control information consists of control format indication, uplink and downlink scheduling assignments and the ACK/NACK feedback for uplink hybrid ARQ transmissions. The control information is time-multiplexed with data transmissions with the control occupying up to three or four OFDM symbols. The control format indication is used to dynamically indicate the number of OFDM symbols used for control information. This dynamic indication of the control region allows using the OFDM symbols not used for control in a given subframe for downlink data transmission. This way the control resources are not wasted in cases where control information does not need to use the maximum number of three or four OFDM symbols.

The benefits of time-multiplexed control include the possibility of a micro-sleep mode and lower latency. The micro-sleep mode is enabled when a UE decodes the control in the first few OFDM symbols and determines that there are no scheduling assignments for it and turns off its transmitter and receiver until the start of the next subframe. The lower latency comes from the fact that the control information can be decoded before the data decoding starts. Similar to the uplink, a drawback of the time-multiplexed approach is that transmit power cannot be shared between control and data that may limit control channel coverage in certain scenarios.

Three different types of resource allocations are permitted for the downlink. In addition to compact assignments of contiguous resource blocks as in the uplink, a group bitmap approach and a resource block bitmap approach are supported. These two allocations are aligned with the subband channel quality feedback schemes in the uplink. The goal is to best exploit the frequency-selective scheduling gains by matching the resource allocation strategy to the channel quality feedback scheme.

Multiple downlink control formats are supported in terms of both number of payload bits as well as channel coding and resources used. The goal with different payload sizes is to match the control information fields to the downlink transmission modes such as single antenna transmission, transmit diversity open-loop and closed-loop spatial multiplexing. On the other hand, the number of resources used for control is a function of both the control information payload size as well as the channel coding rate determined by the link adaptation according to the channel conditions of the addressed UE. It should be noted that modulation is not adapted as QPSK is always used for PDCCH transmissions.

A large number of blind decodings is required to decode different payload sizes as well as different transmission formats. In order to keep the number of blind hypotheses to a manageable level, a concept of search spaces is defined where a UE only monitors a subset of the total PDCCHs.

The hybrid ARQ ACK/NACK information uses the concepts of groups where ACK/NACKs within a group are code-multiplexed. The code-multiplexing approach allows for the carrying of multiple ACK/NACKS on the same set of frequency resources. The goal of code-multiplexing is to exploit frequency diversity as well as randomize inter-cell interference. The four-antenna-ports transmit diversity scheme used for ACK/NACKs is different from that used for other downlink channels. The modified scheme avoids loss of code orthogonality.

References

[1] 3GPP TSG RAN v8.3.0, Evolved Universal Terrestrial Radio Access: Physical Channels and Modulation (3GPP TS 36.211).
[2] 3GPP TSG RAN v8.3.0, Evolved Universal Terrestrial Radio Access: Multiplexing and Channel Coding (3GPP TS 36.212).
[3] 3GPP TSG RAN v8.3.0, Evolved Universal Terrestrial Radio Access: Physical Layer Procedures (3GPP TS 36.213).

16 Inter-cell interference control

An important requirement for the LTE system is improved cell-edge performance and through-put. This is to provide some level of service consistency in terms of geographical coverage as well as in terms of available data throughput within the coverage area. In a cellular system, however, the SINR disparity between cell-center and cell-edge users can be of the order of 20 dB. The disparity can be even higher in a coverage-limited cellular system. This leads to vastly lower data throughputs for the cell-edge users relative to cell-center users creating a large QoS discrepancy.

The cell-edge performance may be either noise-limited or interference-limited. In a noise-limited situation that typically occurs in large cells in rural areas, the performance can generally be improved by providing a power gain. The power gain can be achieved by using high-gain directional transmit antennas, increased transmit power, transmit beam-forming and receive beam-forming or receive diversity, etc. The total transmit power is generally dictated by regula-tory requirements and hence limits the coverage gains possible due to increased transmit power.

The situation is different in small cells interference-limited cases, where, in addition to noise, inter-cell interference also contributes to degraded cell-edge SINR. In this case, providing a transmit power gain may not help because as the signal power goes up, the interference power also increases. This is assuming that with a transmit power gain all cells in the system will operate at a higher transmit power. However, if we can somehow eliminate or reduce inter-cell interference, the cell-edge SINR can be improved.

16.1 Inter-cell interference

As a UE moves away from the cell-center, SINR degrades due to two factors. Firstly, the received signal strength goes down as the path-loss increases with distance from the serving eNB. Secondly, the inter-cell interference (ICI) goes up because when a UE moves away from one eNB, it is generally getting closer to another eNB as shown in Figure 16.1. We assume that the UE is connected to eNB1 and moving away from eNB1 towards eNB2. Furthermore, we assume a universal frequency reuse, which means that both eNB1 and eNB2 transmit on the same frequency resources. Therefore, the signal transmitted from eNB2 appears as interference to the UE. The SINR experienced by the UE at a distance r from eNB2 can be written as:

$$\rho = \frac{P_1 r^{-\alpha}}{N_0 W + P_2 (2R - r)^{-\alpha}},$$ (16.1)

where α is the path-loss exponent and P_k the transmit power for the kth eNB. Also, R is the cell-radius with $2R$ the distance between eNB1 and eNB2. In general, all the eNBs in a system

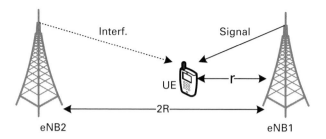

Figure 16.1. Inter-cell interference.

use the same transmit power and therefore we will assume $P_1 = P_2$. In a severely interference-limited scenario, the background noise N_0W can be ignored and the above expression can be simplified as:

$$\rho = \left(\frac{2R}{r} - 1\right)^{\alpha}. \tag{16.2}$$

We note that SINR degrades with increasing r. Also, for a given $r < R$, the SINR is higher for a larger path-loss exponent α. This is because the interference travels a longer distance for $r < R$ and is attenuated more for larger α. We also note that the maximum SINR at the cell-edge with $r = R$ is limited to 0 dB.

Let us assume the path-loss model from Chapter 19 for 2 GHz frequency:

$$PL_s = 128.1 + 37.6 \times \log_{10}(r) \quad \text{dB s}, \tag{16.3}$$

where R is the distance between the UE and Node-B in kilometers. In addition, we assume in-building penetration loss of 20 dB. The same path-loss model is assumed for the interferer eNB2.

$$PL_i = 128.1 + 37.6 \times \log_{10}(2R - r) \quad \text{dB s}. \tag{16.4}$$

The SINR experienced by the UE can be written as:

$$\rho_{\text{ICI}} = \frac{P\left(10^{\frac{PL_s}{10}}\right)}{N_0W + P\left(10^{\frac{PL_i}{10}}\right)}. \tag{16.5}$$

When the ICI is not present, the SINR experienced by the UE can be written as:

$$\rho_{No\text{-}\text{ICI}} = \frac{P\left(10^{\frac{PL_s}{10}}\right)}{N_0W}. \tag{16.6}$$

In Figure 16.2, we plot SINR with and without assuming ICI as a function of distance from the cell-center r for a UE receiving transmission over a 10 MHz bandwidth. The total background noise in a 10 MHz bandwidth is $N_0W = -104$ dBm. Also, we assume eNB transmit power of $P = 43$ dBm. We note that the SINR gain by ICI elimination is larger for lower SINR UEs. The lower SINR happens when r approaches R, which is the case for cell-edge UEs. The relative gains in throughput by ICI eliminations are expected to be even larger for low SINR UEs as the capacity scales almost linearly at lower SINR. For high SINR

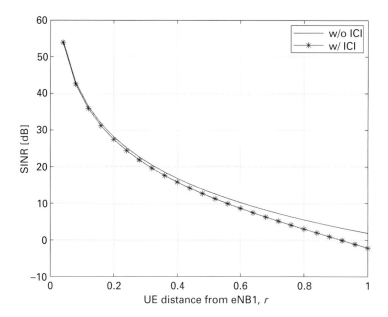

Figure 16.2. SINR as a function of distance from the cell-center.

users, small gains in SINR by ICI elimination do not translate into any meaningful gains in throughput as shown in Figure 16.3. From this discussion, we can conclude that inter-cell interference is more important for cell-edge UEs than for the cell-center UEs.

16.2   Inter-cell interference mitigation

The inter-cell interference (ICI) mitigation techniques can be classified into three categories namely ICI randomization, ICI cancellation and ICI co-ordination or avoidance. Furthermore, the use of beam-forming antennas can be seen as another method for ICI mitigation.

In ICI randomization, the interfering signals are randomized enabling interference suppression at the receiver due to processing gain. The randomization of interfering signals can be achieved by applying pseudo-random scrambling (see Chapter 9, Section 9.1) after channel coding.

The ICI cancellation aims at interference suppression at the receiver beyond what can be achieved by just exploiting the processing gain. The spatial suppression can be achieved by means of multiple antennas at the receiver. Such a scheme, referred to as Interference Rejection Combining (IRC) [1] that can improve system performance in interference-limited scenarios. A more advanced form of interference cancellation can be achieved by first detecting/decoding the interfering signal and canceling it from the overall received signal before proceeding for own signal decoding. This scheme, however, requires knowledge of the transmission format and resource allocation information of the interfering signal, which means that the UE needs to decode the control channels from the neighboring cells. The LTE system does not support such a signaling and hence the post-decoding type of interference cancellation cannot be performed. However, the IRC scheme that requires the knowledge of interference statistics

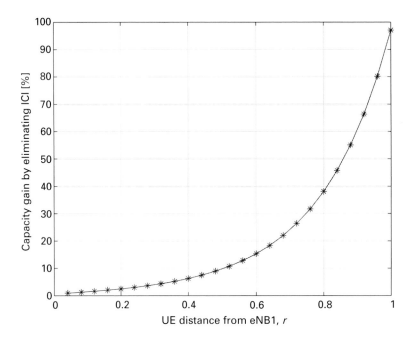

Figure 16.3. Capacity gain by ICI elimination.

can be used for interference suppression when an additional degree of freedom due to multiple receiver antennas is available at the receiver.

The goal of the ICI co-ordination or avoidance scheme is to apply certain restrictions on the resources used in different cells in a co-ordinated way. These restrictions can be in the form of restrictions to what time–frequency resources are available to the resource manager or restrictions on the transmit power that can be applied to certain time–frequency resources. Such restrictions in a cell provide the possibility for improvement in SNR, and cell-edge data-rates on the corresponding time–frequency resources in a neighbor cell. The ICI co-ordination requires certain inter-eNB communication in order to configure the scheduler restrictions. The co-ordination can be static, which means reconfiguration of the restrictions is done on a time scale corresponding to days. In this case, the inter-eNB communication is very limited. The semi-static co-ordination requires configuration of the restrictions on a time scale corresponding to seconds or longer.

In this chapter, we focus on the inter-cell interference co-ordination or avoidance schemes.

16.3   Cell-edge performance

Let us consider a case of hexagonal cell-layout, two tiers of interferers and universal frequency reuse i.e. reuse of one as shown in Figure 16.4. In this case, a UE at the cell edge experiences interference from 11 cells with two interferers at distance R, three interferers at distance $2R$ and six interferers at a distance of $2.7R$, where R is the cell radius. The worst-case signal-to-interference-plus-noise ratio (SINR), ignoring background noise, is then

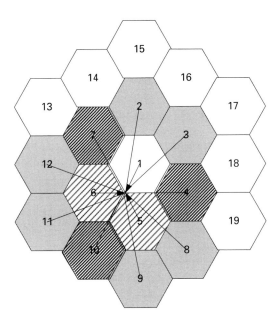

Figure 16.4. An example of interference experienced by a cell-edge UE.

written as [2]:

$$\rho_{\text{reuse-1}} = \frac{R^{-\alpha}}{2 \times (R)^{-\alpha} + 3 \times (2R)^{-\alpha} + 6 \times \left[\left(\sqrt{3}+1\right)R\right]^{-\alpha}}$$

$$= \frac{1}{2 + 3 \times (2)^{-\alpha} + 6 \times \left(\sqrt{3}+1\right)^{-\alpha}}, \qquad (16.7)$$

where α is the path-loss exponent. Also, if we ignore the six interferers at a distance of $2.7R$, the worst-case SINR is given as:

$$\rho_{\text{reuse-1}} = \frac{R^{-\alpha}}{2 \times (R)^{-\alpha} + 3 \times (2R)^{-\alpha}} = \frac{1}{2 + 3 \times (2)^{-\alpha}}. \qquad (16.8)$$

We note that the SINR increases faster with increasing path-loss exponent when a larger number of interferers is assumed. This is explained by the fact that a larger path-loss exponent reduces the effect of interference and the larger the number of interferers considered, the larger the reduction in interference seen. We also note that since two interferers are located at the same distance as the desired cell, that is at distance R, the maximum SINR is upper bounded by -3 dB as given below:

$$\rho_{\text{reuse-1}} = \frac{R^{-\alpha}}{2 \times (R)^{-\alpha}} = \frac{1}{2} = -3 \text{ dB.} \qquad (16.9)$$

Let us now assume a reuse of three for cell numbers 1, 5 and 6 only. This means that all the remaining cells implement a reuse of 1. In this case, the interference from two dominant interferers at distance R (cells 5 and 6) from the UE is eliminated. This results in a worst-case SINR as given below.

$$\rho_{\text{reuse-3}} = \frac{R^{-\alpha}}{\frac{3 \times (2R)^{-\alpha} + 6 \times \left[\left(\sqrt{3}+1\right)R\right]^{-\alpha}}{3}} = \frac{3}{3 \times (2)^{-\alpha} + 6 \times \left(\sqrt{3}+1\right)^{-\alpha}}.$$

(16.10)

If we ignore the six interferers at a distance of $2.7R$, the worst-case SINR with a total of 3 interferers at distance $2R$ is given as:

$$\rho_{\text{reuse-3}} = \frac{R^{-\alpha}}{\frac{3 \times (2R)^{-\alpha}}{3}} = (2)^{\alpha}.$$

(16.11)

Now let us assume a reuse of seven implemented in cell numbers 1–7. In addition to the interference from two dominant interferers at distance R (cells 5 and 6) for the case of reuse of three, the interference from 2 interferers at distance $2R$ (cells 4 and 7) and another 2 interferers at distance $2.7R$ (cells 2 and 3) is eliminated. This results in a worst-case SINR as given below.

$$\rho_{\text{reuse-7}} = \frac{R^{-\alpha}}{\frac{(2R)^{-\alpha}+4\times\left[\left(\sqrt{3}+1\right)R\right]^{-\alpha}}{7}} = \frac{7}{(2)^{-\alpha} + 4 \times \left(\sqrt{3}+1\right)^{-\alpha}}.$$

(16.12)

The cell-edge SINR for various reuse factors is given in Figure 16.5. We note that at a path-loss exponent of 3.6, we observe an approximately 13 dB improvement in cell-edge SINR

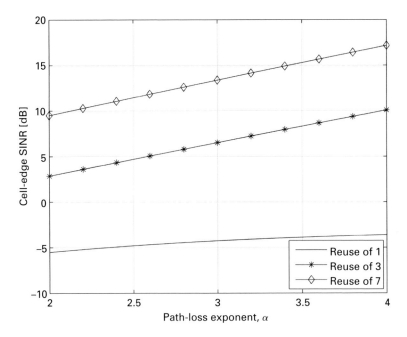

Figure 16.5. Cell-edge SINR for various reuse factors.

by using reuse of three relative to reuse of one. A reuse of seven increases cell-edge SINR by another 8 dB. Note that with a reuse of three and seven, the power spectral density on the transmitted bandwidth increases by a factor of 3 and 7 respectively. This is because, with a higher reuse, the frequency bandwidth used in each cell in the reuse scheme decreases. We have accounted for this increase in power spectral density in the above calculations.

We note that improvements in cell-edge SINR come at the expense of reduced bandwidth available for transmissions in a given cell when higher reuse factors are employed. Therefore, a more meaningful metric to look at is the improvement in spectral efficiency by accounting for the bandwidth loss effect resulting from a reuse of greater than one. The capacity limit for cell-edge users can be approximated as:

$$C_{\text{reuse-1}} = \log_2 (1 + \rho_{\text{reuse-1}}) \ [\text{b/s/Hz}]$$

$$C_{\text{reuse-3}} = \left(\frac{1}{3}\right) \times \log_2 (1 + \rho_{\text{reuse-3}}) \ [\text{b/s/Hz}]$$

$$C_{\text{reuse-7}} = \left(\frac{1}{7}\right) \times \log_2 (1 + \rho_{\text{reuse-7}}) \ [\text{b/s/Hz}]. \tag{16.13}$$

The cell-edge capacity limits for various reuse factors are plotted in Figure 16.6. It can be noted that, at a path-loss exponent of 3.6, a reuse of three provides approximately two times improvement (1 versus 0.5 b/s/Hz) in cell-edge throughput relative to the case of universal frequency reuse. It should be noted that the analysis in this contribution is based on a path-loss model only while ignoring the effects of fading, etc. Therefore, the potential improvement in performance is merely an indication of the gains achievable by inter-cell interference co-ordination for the cell-edge users. We further note that reuse of seven while providing some capacity gains relative to universal frequency reuse performs worse than reuse of three.

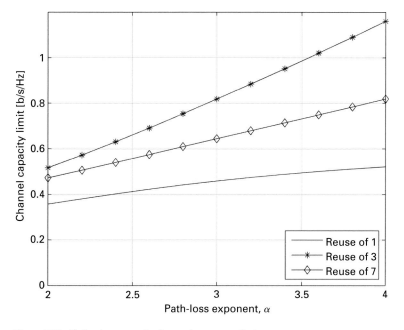

Figure 16.6. Cell-edge capacity for various reuse factors.

This means that the bandwidth penalty resulting from reuse of seven does not justify the improvements in cell-edge SINR.

16.4   Cell-center performance

Let us now assume a UE at the cell-center as shown in Figure 16.7 experiencing interference from six first-tier cells. Assuming that the UE is located near the cell-center, the six first-tier interferers are at a distance of approximately $\sqrt{3}R$. The UE also receives interference from 12 second-tier cells with six interferers at a distance of approximately 3R and remaining six second-tier interferers at a distance of $2\sqrt{3}R$. The SINR for this cell-center UE can be written as:

$$\rho_{reuse-1} = \frac{(\beta R)^{-\alpha}}{6 \times \left(\sqrt{3}R\right)^{-\alpha} + 6 \times (3R)^{-\alpha} + 6 \times \left(2\sqrt{3}R\right)^{-\alpha}}$$

$$= \frac{(\beta)^{-\alpha}}{6 \times \left(\sqrt{3}\right)^{-\alpha} + 6 \times (3)^{-\alpha} + 6 \times \left(2\sqrt{3}\right)^{-\alpha}}, \tag{16.14}$$

where $\beta R \, (\ll R)$ represents the distance of the UE from the own-cell. Assuming frequency reuse of three, the interference from two interferers at a distance of $\sqrt{3}R$ can be eliminated.

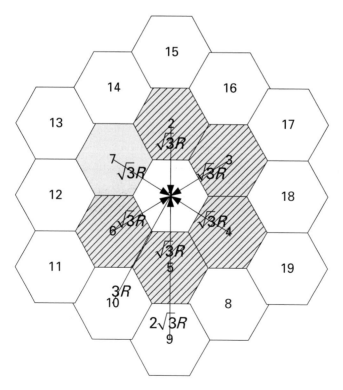

Figure 16.7. An example of interference experienced by a cell-center UE.

The SINR in this case can be written as:

$$\rho_{\text{reuse-3}} = \frac{3 \times (\beta R)^{-\alpha}}{4 \times \left(\sqrt{3}R\right)^{-\alpha} + 6 \times (3R)^{-\alpha} + 6 \times \left(2\sqrt{3}R\right)^{-\alpha}}$$

$$= \frac{3 \times (\beta)^{-\alpha}}{4 \times \left(\sqrt{3}\right)^{-\alpha} + 6 \times (3)^{-\alpha} + 6 \times \left(2\sqrt{3}\right)^{-\alpha}}. \tag{16.15}$$

If we assume a frequency reuse of seven, the interference from all the first-tier interferers can be eliminated resulting in SINR given by the relationship below:

$$\rho_{\text{reuse-7}} = \frac{7 \times (\beta R)^{-\alpha}}{6 \times (3R)^{-\alpha} + 6 \times \left(2\sqrt{3}R\right)^{-\alpha}} = \frac{7 \times (\beta)^{-\alpha}}{6 \times (3)^{-\alpha} + 6 \times \left(2\sqrt{3}\right)^{-\alpha}}. \tag{16.16}$$

These three SINRs as a function of the path-loss exponent are plotted in Figure 16.8 for the case of $\beta = 0.2$. This means that the cell-center user is assumed to be at a distance of $0.2R$ from the cell center. We note that incremental gains in SINR with increasing reuse factor in this case are smaller compared to the case of the cell-edge user. This is explained by the fact that a cell-center UE receives smaller inter-cell interference and eliminating inter-cell interference increases SINR marginally.

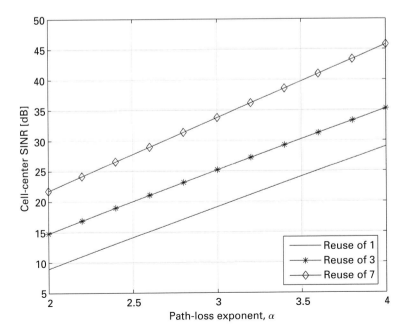

Figure 16.8. Cell-center SINR for various reuse factors.

The capacity limit for cell-center users for reuse of one, three and seven can be approximated as:

$$C_{\text{reuse-1}} = \log_2 (1 + \rho_{\text{reuse-1}}) \; [\text{b/s/Hz}]$$

$$C_{\text{reuse-3}} = \left(\frac{1}{3}\right) \times \log_2 (1 + \rho_{\text{reuse-3}}) \; [\text{b/s/Hz}] \qquad (16.17)$$

$$C_{\text{reuse-7}} = \left(\frac{1}{7}\right) \times \log_2 (1 + \rho_{\text{reuse-7}}) \; [\text{b/s/Hz}].$$

The cell-center capacity results are plotted in Figure 16.9.

It can be noted that reuse of one provides the highest throughput for the cell-center users. This means that the penalty in bandwidth resulting from higher reuse factors does not justify marginal gains in SINR for the cell-center user. Also, the cell-center users experience a larger SINR even for the case of reuse of one and any further gains in SINR result in logarithmic increase in capacity at higher SINRs. In contrast, for the cell-edge users, the SINR for reuse of 1 is very low and SINR gains result in an almost linear increase in throughput.

We discussed the frequency reuse schemes in the context of downlink. The gains in cell-edge performance from frequency reuse schemes can be larger for the uplink due to lower SINR in the uplink. We note from Figure 16.4 that with a reuse of 1, for a user served by cell-1 and located at the cell-edge, there can be two equally strong interferers namely cell-5 and cell-6. We also noted that this limits the maximum SINR at the cell-edge to -3 dB. However in case of uplink using multi-user MIMO, it is possible that many users located at the cell-edge are transmitting to their corresponding cell (eNB1) in the uplink as shown in Figure 16.10. A user transmitting in eNB2 from the cell-edge will see these multiple interferers received at eNB2 with approximately the same power as its own received power at eNB2. When the

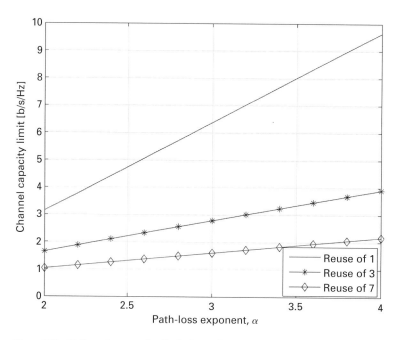

Figure 16.9. Cell-center capacity limit for various reuse factors.

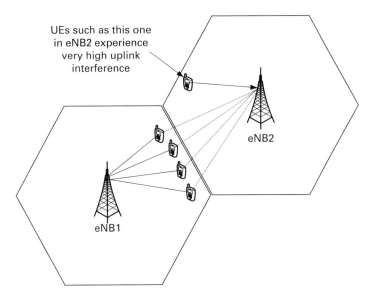

Figure 16.10. An example of high interference experienced by a cell-edge UE transmitting in the uplink.

number of these interfering users is greater than 2, the SINR seen on the uplink can be lower than the downlink SINR in interference-limited scenarios. However, when a frequency reuse scheme, for example with a reuse of 3, is applied, the interference from these multiple users transmitting on the uplink in eNB2 is eliminated. This results in a larger improvement in SINR and correspondingly larger improvements in uplink capacity or throughput. Moreover, since the starting SINR with reuse of 1 is low, the capacity scales approximately linearly with SINR and therefore results in larger gains in uplink capacity for cell-edge users.

16.5   Fractional frequency reuse

A fractional frequency reuse scheme is based on the concept of reuse partitioning [3]. In reuse partitioning, the users with the highest signal quality use a lower reuse factor while users with low SINR use a higher reuse factor. This is in line with the discussions we had earlier in this chapter which showed that reuse of 1 provides the best throughput for users in the cell-center experiencing higher SINR while reuse of 3 provides the highest throughput for the cell-edge users experiencing low SINR. The fractional frequency reuse scheme, for example, uses a universal (reuse of one) frequency reuse for cell-center users while a reuse of three is used for the cell-edge users as shown in Figure 16.11. The total frequency resource is divided into four segments namely (f_1, f_2, f_3, f_4). The frequency resource (f_1) is used in all the cells to serve users experiencing good SINR. A frequency reuse of three is implemented on the remaining three resource segments (f_2, f_3, f_4). In Figure 16.11, the top-right, middle-right and bottom-right cells use frequency resource (f_2), (f_3) and (f_4) respectively. Even though we have discussed the fractional frequency reuse in the context of omni-cells, similar principles apply to the case of a sectorized system. Figure 16.12, for example, shows planning of frequency resources (f_2, f_3, f_4) for cell-edge users between different sectors of a cell. The

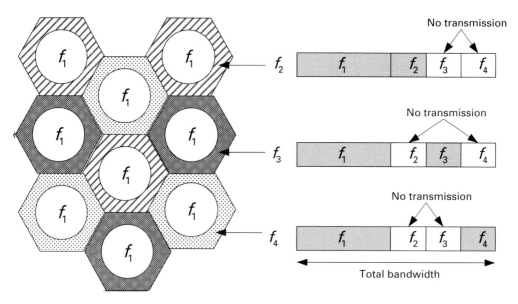

Figure 16.11. An example of fractional frequency reuse.

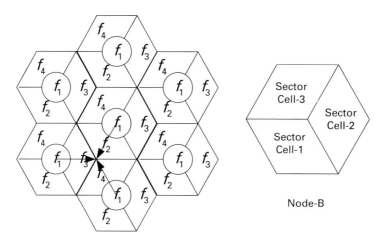

Figure 16.12. An example of fractional frequency reuse in a sectorized system.

frequency resources (f_2), (f_3) and (f_4) are used for sector-1, sector-2 and sector-3 of a Node-B respectively. As for the omni-cell case, cell-center users use frequency resource (f_1) in all the sectors. A user at the cell-edge in sector-1 of the center Node-B, for example, is served on frequency resource (f_2). The neighboring sectors to sector-1 use frequency resources (f_3) and (f_4). Therefore, this user will not experience interference from the two neighboring sectors.

16.5.1   Soft frequency reuse

We noticed that in fractional frequency reuse schemes, the frequency resources used for cell-edge users in the neighboring cells are left empty in a given cell. In soft frequency reuse

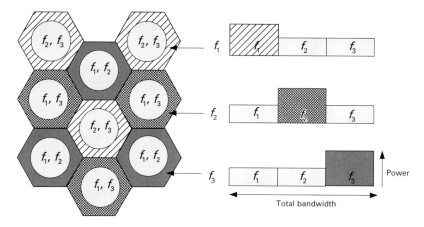

Figure 16.13. An example of soft-frequency reuse.

schemes on the other hand all the frequency resources can be used in all the cells. For a soft-frequency reuse of three, the total frequency resource is divided into three segments (f_1, f_2, f_3) as shown in Figure 16.13. The top-right cell uses frequency resource (f_1) for the cell-edge users while using frequency resources (f_2, f_3) for cell-center users. The middle-right cell uses frequency resource (f_2) for the cell-edge users while using frequency resources (f_1, f_3) for cell-center users. The bottom-right cell uses frequency resource (f_3) for the cell-edge users while using frequency resources (f_1, f_2) for cell-center users. Since all the frequency resource is used in all the cells like universal frequency reuse, the cell-edge users experience high interference. In order to reduce the interference to cell-edge users in the neighboring cells, a lower transmit power is used on the frequency resource used for cell-center users. Note that the frequency resource used for cell-center users in a cell is used for cell-edge users in neighboring cells. This type of power allocation arrangement in soft-frequency reuse improves cell-edge SINR while degrading SINR for users towards the cell-center. However, the expectation is that since cell-edge users experience lower SINR, the throughput would increase almost linearly with SINR. On the other hand, degradation in SINR for a high SINR user would only result in logarithmic reduction of throughput.

The resource allocated for cell-edge and cell-center users in both fractional frequency reuse and soft-frequency reuse schemes can be varied in a semi-static fashion based on traffic load and other network conditions.

16.6   Fractional loading

We noted that the cell-edge performance could be improved by frequency reuse or reuse partitioning. However, in a practical situation, a frequency reuse approach may not deliver its promise due to irregular cell shapes and propagation conditions that makes frequency planning challenging. In this section, we describe an approach referred to as fractional loading that can serve as a simpler alternative to frequency reuse.

In the case of a fractional loading approach, each cell randomly selects a subset of time-frequency resources for transmission as shown in Figure 16.14. In the example of Figure 16.14,

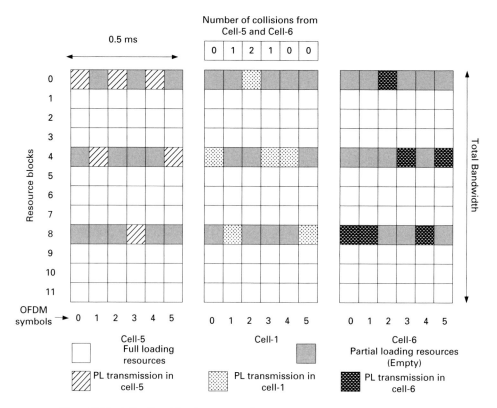

Figure 16.14. A fractional loading approach.

1/4 of the frequency resource (3 out of 12 resource blocks) is reserved for supporting cell-edge or low SINR users with each cell operating at a fractional loading of 1/3 on the reserved frequency resource. This scenario is representative of cell-edge users for unicast traffic, single-cell non-SFN broadcast multicast traffic as well as broadcast control channels. Also, for illustration purposes we assumed that the frequency resource reserved for fractional loading consists of a number of resource blocks. We note that the frequency resource can also be reserved on a subcarrier-level to provide greater granularity in resource allocation as well as increased frequency diversity. Furthermore, in order to randomize the inter-cell interference on the reserved resource due to competing transmissions in the neighboring cells, the selection of time-frequency resources can be independent from cell to cell. This can be achieved by using, for example, a pre-known pseudo-random cell-specific sequence that determines the resources used for transmission in a cell employing partial loading.

With independent transmissions between neighboring cells, the fractional loading approach does not need any co-ordination between cells and therefore, in some cases, the transmissions from neighboring cells can overlap and collide. In the example of Figure 16.14, cell-1 transmissions do not overlap with the transmissions from the neighboring cells 5 and 6 (see Figure 16.4) in OFDM symbols 0, 4 and 5. In OFDM symbols 1 and 3, cell-1 transmission overlaps with one neighboring cell while in OFDM symbol 2 transmissions overlaps with both cell-5 and cell-6. The cell-1 will see highest SINR in OFDM symbols 0, 4 and 5, a relatively lower SINR in OFDM symbol 1 and 3 and an even lower SINR in OFDM symbol 2. In general, a fractional loading approach can provide an overall gain if statistically the gains in capacity are higher benefiting from reduced interference than the loss in bandwidth due to operation at fractional loading.

The average capacity in a fractional loading approach can be approximated as:

$$C_{average} = p_0 c_0 + p_1 c_1 + p_2 c_2 + p_3 c_3 \ [b/s/Hz]$$
$$c_i = \log_2(1 + \rho_i) \ [b/s/Hz], \qquad (16.18)$$

where c_i is the capacity in a time-frequency resource with i number of transmissions among the neighboring cells. In the example where a given time-frequency resource is shared among 3 neighboring cells, there can be 0, 1, 2 or 3 transmissions in a time-frequency resource. Also, ρ_i indicates the SINR experienced with i number of transmissions among the neighboring cells and is approximated as:

$$\rho_0 = 0.0 = -\infty \ \text{dB}$$

$$\rho_1 = \frac{R^{-\alpha}}{\dfrac{3 \times (2R)^{-\alpha} + 6 \times \left[\left(\sqrt{3}+1\right)R\right]^{-\alpha}}{3}} = \frac{3}{3 \times (2)^{-\alpha} + 6 \times \left(\sqrt{3}+1\right)^{-\alpha}}$$

$$\rho_2 = \frac{R^{-\alpha}}{(R)^{-\alpha} + \dfrac{3 \times (2R)^{-\alpha} + 6 \times \left[\left(\sqrt{3}+1\right)R\right]^{-\alpha}}{3}} = \frac{1}{1 + \dfrac{3 \times (2)^{-\alpha} + 6 \times \left(\sqrt{3}+1\right)^{-\alpha}}{3}}$$

$$\rho_3 = \frac{R^{-\alpha}}{2 \times (R)^{-\alpha} + \dfrac{3 \times (2R)^{-\alpha} + 6 \times \left[\left(\sqrt{3}+1\right)R\right]^{-\alpha}}{3}} = \frac{1}{2 + \dfrac{3 \times (2)^{-\alpha} + 6 \times \left(\sqrt{3}+1\right)^{-\alpha}}{3}}.$$

The numerical results for average capacity in a fractional loading approach are provided in Figure 16.15. It can be noted that the capacity is maximized at around 2/3 loading (effective reuse of $3/2 = 1.5$). At lighter loading, the capacity is lower because some time-frequency chunks go wasted where none of the neighboring cells transmit. At very high loading, the capacity is lower again due to increased interference. The fractional loading of 1 corresponds to a universal frequency reuse case where all the cells transmit in all the time-frequency resources. The capacity for reuse of 1 for path-loss exponents of 2, 3, 3.6 and 4 respectively are 0.48, 0.54, 0.56 and 0.57 b/s/Hz. The maximum capacity numbers with fractional loading for path-loss exponents of 2, 3, 3.6 and 4 respectively are 0.55, 0.72, 0.84 and 0.91 b/s/Hz. The fractional loading of 0.6–0.7 appears to maximize the achievable capacity for various path-loss exponents. It can also be noted that for larger path-loss exponents such as $\alpha = 4$, the fractional loading approach can provide around 60% (0.91 vs. 0.57 b/s/Hz) improvement in cell-edge performance relative to reuse of 1. In addition, under an ideal frequency reuse of 3, cell-edge performance can be further improved by approximately 26% (see Figure 16.6, 1.15 vs. 0.91 b/s/Hz) relative to fractional loading. However, we reiterate that a frequency reuse approach might be challenging to implement in practice due to the required frequency planning.

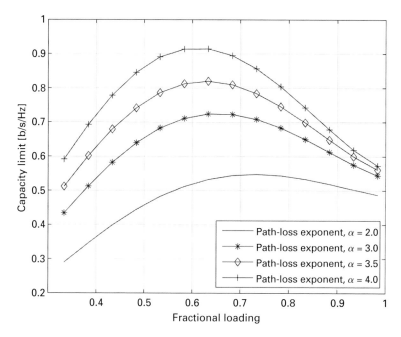

Figure 16.15. Capacity as a function of fractional loading.

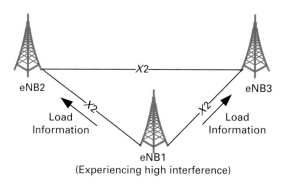

Figure 16.16. Load information exchange.

16.7   ICI co-ordination in the LTE system

In the downlink, ICI co-ordination is enabled by eNB power restriction as described in Chapter 13, Section 13. 2. 3. The power restriction information consists of a bitmap of size equal to the number of physical resource blocks (PRBs) that is exchanged among eNBs on the X2 interface. Each position in the bitmap represents a PRB for which value '1' indicates transmit power exceeding the threshold and value '0' indicates transmit power not exceeding the threshold. The threshold value is also exchanged along with the bitmap.

The LTE system defines a load indication procedure for uplink inter-cell interference co-ordination. This mechanism is used to send an interference overload indication when an eNB experiences too high interference on some resource blocks. The eNB experiencing excessive interference initiates the procedure by sending an uplink interference overload indication (OI) message to intra-frequency neighboring eNBs. The information exchanged consists of high interference, medium interference or low interference indication for each PRB. Additionally, high interference indication (HII) information consisting of a bitmap of size equal to the number of PRBs is exchanged. Each position in the bitmap represents a PRB for which value "1" indicates high interference sensitivity and value "0" indicates low interference sensitivity.

16.8   Summary

The LTE system targets improved cell-edge performance in order to provide service consistency in terms of geographical coverage and throughputs. In interference-limited scenarios, the cell-edge performance can be improved via inter-cell interference co-ordination. We noted that, for a cell-edge user, reuse of seven while providing some data rate gains relative to universal frequency reuse performs worse than reuse of three. This means that the bandwidth penalty resulting from reuse of seven does not justify the improvements in cell-edge SINR. We also observed that the most benefit gained from inter-cell interference control is for the cell-edge users. Therefore, schemes that allow universal frequency reuse for cell-center users and reuse greater than one for cell-edge users such as fractional frequency reuse present a greater potential for cell-edge performance improvement while minimizing the impact on overall system capacity and throughput. However, the inter-cell interference co-ordination schemes based on fractional frequency reuse require careful frequency planning which may be challenging in practice due to irregular cell shapes and varying propagation conditions.

A simpe fractional loading approach can also be considered for improving cell-edge performance without requiring careful frequency planning. In this case, each cell randomly selects a subset of time-frequency resources for transmission. The gains come from the fact that higher SINR can be achieved on resources not instantaneously used in neighboring cells due to random selection of resources for transmission.

The LTE system defines eNB power restriction signaling to support inter-cell interference co-ordination on the downlink. Moreover, two types of inter-eNB communication mechanisms that include overload indication and high interference indication are defined for uplink inter-cell interference control. The load indication procedure is used for sending an interference indication when an eNB experiences too high interference on some resource blocks in the uplink. The receiving eNB may then control the power transmitted on the resource blocks experiencing higher interference in the eNB sending the overload indication signal.

References

[1] Winters, J. H., "Optimum combining in digital mobile radio with cochannel interference," *IEEE Journal on Selected Areas in Communications*, vol. SAC-2, no. 4, July 1984.

[2] Stüber, G. L., *Principles of Mobile Communications*, Dordrecht: Kluwer, 2003.

[3] Halpern, S. W., "Reuse partitioning in cellular systems," *Proceedings of 33rd IEEE Vehicular Technology Conference*, vol. 33, pp. 322–327, 25–27 May 1983.

17 Single frequency network broadcast

Voice communication and download data services such as web browsing are based on point-to-point (PTP) communication. On the other hand, multicast and broadcast services are based on point-to-multipoint (PTM) communication, where data packets are simultaneously transmitted from a single source to multiple destinations. Examples of broadcast services are radio and television services that are broadcast over the air or over cable networks and the content is available to all the users. Multicast refers to services that are delivered to users who have joined a particular multicast group. The service delivery using point-to-multipoint (PTM) communication is generally more efficient when a large number of users is interested in receiving the same content such as a mobile TV channel. This results in efficient transmission not only over the wireless link but also in the core and access networks. This is because a single multicast broadcast packet travels in the core and access networks and is copied and forwarded to multiple Node-Bs in the multicast broadcast area.

The broadcast services can be delivered to mobile devices either via an independent broadcast network such as DVB-H (digital video broadcast-handheld) [1], DMB (digital multimedia broadcast), MediaFLO [2] or over a service provider's cellular network. The DMB is a South Korean standard derived from the digital audio broadcast (DAB) standard [3]. In the case of an independent broadcast network, dual mode UEs capable of receiving service from both the broadcast network and the cellular network are required. In the case where broadcast service is delivered using the service provider's cellular network, a single network is used for delivery of both broadcast and unicast traffic. This allows single mode terminals to receive both the broadcast and unicast traffic from the same network. Another drawback of a separate broadcast network is spectrum availability and the cost of building a new broadcast network infrastructure.

The multicast and broadcast capabilities have already been introduced in the existing 3G wireless systems. In the cdma2000 family of standards, both high rate packet data (HRPD) [4] and cdma2000 revision D [5] support broadcast and multicast applications. The multimedia broadcast multicast service (MBMS) is specified in UMTS release 6 standard [6]. The broadcast services need to be delivered in a cost effective way in order for these services to be popular among consumers. This demands, among other factors, extremely high spectral efficiency for these services due to scarcity of the radio spectrum. Therefore, techniques that provide increased spectral efficiency for broadcast services such as single frequency network (SFN) operation have recently been introduced in the LTE and in other cellular standards.

In this chapter, we first describe SFN and its benefits for broadcast transmission. We then discuss two orthogonal multiplexing approaches for unicast and broadcast namely time-multiplexing and frequency-multiplexing along with their pros and cons. Finally, we expose a non-orthogonal multiplexing approach that is based on unicast and broadcast superposition and the interference cancellation concept.

17.1 Multicast broadcast system

A broadcast or multicast service is delivered in a broadcast multicast service area which refers to the coverage area in which a specific broadcast service is available. A broadcast area is defined on a per-broadcast-service basis which means that different broadcast multicast services may have different broadcast multicast coverage areas. A broadcast service area may represent the coverage area of the entire network or part of the network. A simplified diagram of a cellular broadcast multicast system with broadcast multicast area consisting of six cells is shown in Figure 17.1. A content provider can be a cellular service provider or a third party and acts as the source of multicast broadcast content. The evolved broadcast multicast service center (eBM-SC) acts as an entry point for content-delivery services. The eBM-SC forwards the broadcast multicast packets to the eMBMS gateway from where the packets are distributed to eNode-Bs in the broadcast multicast area. A key point in broadcast multicast transmission over the air is that the same information is transmitted simultaneously from Node-Bs in the broadcast multicast area. This allows UEs to receive broadcast multicast signals from multiple cells.

17.2 Single frequency network

In the case of point-to-point (PTP) unicast communication, different information contents are transmitted to different UEs in different cells in the system. A UE at the cell-edge as shown in the example of Figure 17.2 experiences interference from 11 cells. This is assuming a hexagonal cell-layout, 2-tiers of interferers and universal frequency reuse i.e. reuse of one. In this case, two interferers are at distance R, three interferers at distance $2R$ and six interferers at a distance of $2.7R$, where R is the cell radius. The worst-case signal-to-interference-plus-noise

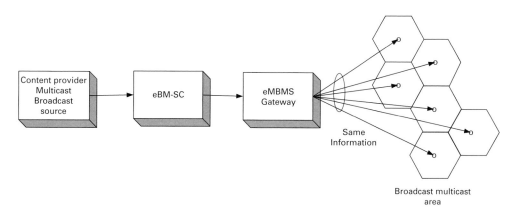

Figure 17.1. A multicast broadcast system.

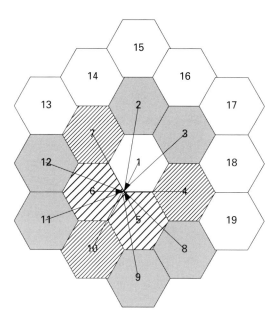

Figure 17.2. An example of interference experienced by a cell-edge UE.

ratio (SINR), assuming a total of 11 interferers, is then written as:

$$\rho_{reuse\text{-}1} = \frac{P\left(R^{-\alpha}\right)}{P\left(2 \times (R)^{-\alpha} + 3 \times (2R)^{-\alpha} + 6 \times \left[\left(\sqrt{3}+1\right)R\right]^{-\alpha}\right) + N_0 W}$$

$$= \frac{1}{2 + 3 \times (2)^{-\alpha} + 6 \times \left(\sqrt{3}+1\right)^{-\alpha} + \frac{N_0 W}{P(R^{-\alpha})}}, \tag{17.1}$$

where α is the path-loss exponent. We assumed that all the base stations transmit at the same constant power P watts. Also, if we ignore the background noise ($N_0 W$), six interferers at a distance of $2.7R$ and three interferers at distance $2R$, the maximum SINR is upper bounded by -3 dB as given below:

$$\rho_{reuse\text{-}1} = \frac{R^{-\alpha}}{2 \times (R)^{-\alpha}} = \frac{1}{2} = -3 \text{ dB}. \tag{17.2}$$

This is because two interferers are located at the same distance as the desired cell, that is, distance R from the UE. A low SINR of -3 dB means that very high data rates cannot be supported at the cell-edge for transmission in unicast mode where different base stations transmit different content.

Let us now assume a broadcast case where all cells in the system transmit the same content. We first consider the case of broadcast transmission using code division multiple access (CDMA). For simplicity of analysis, we assume that the UE receives equal power P from all the base stations transmitting the broadcast content. This may be a reasonable assumption for the broadcast scenario where the cell edge users limit the performance and these users receive signals from multiple base stations. Therefore, in our assumption the signal from a given base station is either received at the same equal power or no signal is received at all. Note that in a practical system, a given UE would only be receiving signals from a subset of base stations in the system. For base stations farther away from the mobile, a very weak or no signal may be received. Furthermore, we assume a single path flat-fading channel i.e. no inter-symbol-interference (ISI).

The signal-to-interference-plus-noise ratio (SINR) for the signal received from base station j is written as:

$$\rho_j = \frac{P}{(J-1)P + N_0}. \tag{17.3}$$

It is assumed that the different copies of the signal received from different base stations are asynchronous, so that copies from transmitters other than the jth base station become interference. Note that even when the cells are synchronized, the signals received at the UE from multiple cells are not guaranteed to be received in a synchronous manner due to different propagation delays from different cells. We further assume that a Rake receiver is used to track and combine signals from a total of J base station transmitters. Using maximum-ratio-combining (MRC) for signals received from different base stations, the average signal-to-noise ratio can be expressed as the sum:

$$\rho = \sum_{j=1}^{J} \rho_j = \sum_{l=1}^{J} \left(\frac{P}{(J-1)P + N_0} \right) = \frac{JP}{(J-1)P + N_0}. \tag{17.4}$$

It can be noted that the maximum achievable SINR is limited to 1 (0.0 dB) when the broadcast signal is received from a large number of base stations ($J \gg 1$). The capacity for broadcast traffic in a CDMA system using Rake receiver can then be written as:

$$C_{\text{Broadcast-CDMA}} = \log_2(1 + \rho) = \log_2 \left(1 + \frac{JP}{(J-1)P + N_0} \right) \quad [\text{b/s/Hz}]. \tag{17.5}$$

We observe that the achievable capacity for broadcast in a CDMA system is limited to 1 b/s/Hz when the broadcast signal is received from a large number of base stations ($J \gg 1$). It should be noted that the above formulation assumes that CDMA Rake receiver has a large number of fingers available to potentially receive signals from a large number of base stations. In practice, lower capacity would be achieved for CDMA due to limitation on the maximum number of Rake fingers. However, we note that advanced receivers employing equalization can be used to improve the performance of a broadcast CDMA system at the expense of additional complexity.

Let us now turn our attention to a broadcast system using OFDM. We assume that cells in the system transmitting broadcast content are synchronized such that the signals arrive at the UE with delays within the OFDM cyclic prefix length. The cyclic prefix length in the OFDM broadcast systems are generally chosen such that propagation delays are within the cyclic prefix length. This generally requires a larger cyclic prefix for a system with larger cells deployment.

With an OFDM system transmitting the same information from multiple cells, the delayed copies of the same signal received at the UE simply result in a phase shift in the frequency. Let D_j denote the delay in samples for the OFDM symbol received from the jth cell. The angle

φ_j for the jth cell can then be written as:

$$\varphi_j = \frac{2\pi}{N} D_j. \tag{17.6}$$

The OFDM subcarrier k received from the jth cell experiences a phase shift of $e^{j\varphi_j k}$.

Let $H_j(k)$ represent the channel gain from cell j on the kth subcarrier, then the composite channel gain $H_c(k)$ experienced by a modulation symbol transmitted on the kth subcarrier is given as:

$$H_c(k) = H_0(k) + H_1(k) \cdot e^{j\phi_1 k} + \cdots + H_{(J-1)}(k) \cdot e^{j\phi_{(J-1)}k}. \tag{17.7}$$

Note that we assumed that the received power from each of the cells is equal. We remark that when the same signal is transmitted from multiple cells using OFDM and received with certain relative delays, the overall composite channel becomes frequency-selective. This is because the delayed signals from different cells just appear as multi-paths carrying the same information signal. The channel gain power $|H_p(k)|^2$ is plotted in Figure 17.3 for the case of three cells. In Figure 17.3, we assumed delay values of 0, 4 and 8 samples from the first, second and third cells respectively. Also, we assumed that the channel is flat-fading from each cell. We note that three cells results in peaks of up to $10 \log_{10}(3) = 4.7$ dB relative to a flat-fading case when the signals transmitted from the three cells combine coherently. At some other subcarriers, the signal power is lower than in the baseline flat-fading case. We also note that there is no loss of power and there is no interference in the system apart from the background noise. The only effect is the introduction of frequency selectivity. Such a broadcast system using OFDM transmission from synchronized base stations is referred to as a single frequency network (SFN).

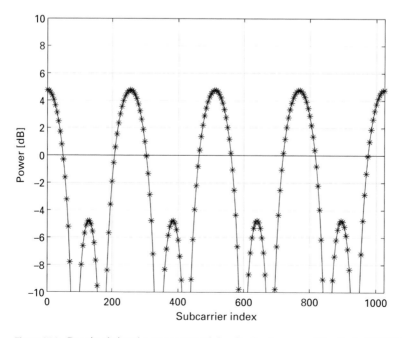

Figure 17.3. Received signal power spectral density for the case of 3 cells in OFDM broadcast.

The channel capacity limit for an SFN system can be written as:

$$C = E[\log_2(1 + |H_c(k)|^2 \, \text{SNR})], \tag{17.8}$$

where $H_c(k)$ is the composite channel gain experienced by a modulation symbol transmitted on the kth subcarrier. The frequency selectivity introduced by SFN transmission provides a frequency diversity benefit. On the other hand, the drawback of frequency selectivity is the impact on channel capacity. Let us consider this effect at both low and high SNR. At low SNR, the capacity scales approximately linearly with SNR as given by:

$$C = E[\log_2(1 + |H_c(k)|^2 \, \text{SNR})] \approx E[|H_c(k)|^2 \, \text{SNR}] \log_2 e = SNR \times \log_2 e. \tag{17.9}$$

This means that at low SNR, capacity is not a concave function of SNR and hence there is no capacity penalty due to frequency selectivity.

At very high SNR, $|H_c(k)|^2 \, \text{SNR} \gg 1$ and the above equation can be simplified as:

$$C \approx E[\log_2(\text{SNR} \times |H_c(k)|^2)] = \log_2 \text{SNR} + E[\log_2(|H_c(k)|^2)], \tag{17.10}$$

where $E[\log_2(|H_c(k)|^2)]$ represents the penalty due to frequency selectivity introduced by SFN which is characterized as -2.5 dB for a Rayleigh fading channel as discussed in Section 3.3. Since the broadcast performance is limited by the outage for the weakest user in the system, the diversity benefit outweighs the capacity loss as is discussed in Chapter 6 on transmit diversity. Note that we assumed a flat-fading channel from each of the base stations. In a more realistic situation, when the channel from each base station is frequency-selective, there is already some frequency diversity present without SFN. In this case, the SFN operation makes the overall channel from multiple base stations even more selective.

This increased frequency selectivity in an SFN environment leads to two consequences. Firstly, channel estimation in a highly frequency selective channel requires a high density of reference signals in the frequency-domain. In the LTE system, the frequency density for Multicast Broadcast SFN (MBSFN) reference signals are three times higher than the unicast reference signals (see Chapter 9, Section 9.6.2). Secondly, high-frequency selectivity results in the puncturing of some of the transmitted modulation symbols, which requires the use of stronger coding. This requires switching to higher-order modulations at a lower coding rate (see Chapter 12, Section 12.4.2). The lower coding rate allows decoding of the multicast broadcast SFN (MBSFN) transmissions even if some of the modulations symbols are punctured. Note that in SFN there is no loss of energy, rather some of the modulations symbols experience very high SINR while some symbols experience a very low SINR. When higher-order modulation is transmitted at a lower coding rate, the transmissions may be decoded because some symbols are received at very high SINR. A lower coding rate assures that there is still sufficient redundancy even when some modulation symbols are erased.

17.3   Multiplexing of MBSFN and unicast

Let us start by looking at an interference-limited case for the unicast traffic. In an interference-limited scenario, we can ignore the background noise ($N_0 W$) in Equation (17.1). The worst-case signal-to-interference-plus-noise ratio (SINR) for the user in Figure 17.2 can then be written as:

$$\rho_{\text{reuse-1}} = \frac{R^{-\alpha}}{2 \times (R)^{-\alpha} + 3 \times (2R)^{-\alpha} + 6 \times \left[\left(\sqrt{3}+1\right)R\right]^{-\alpha}}$$

$$= \frac{1}{2 + 3 \times (2)^{-\alpha} + 6 \times \left(\sqrt{3}+1\right)^{-\alpha}}. \tag{17.11}$$

We assumed that all the base stations transmit at the same constant power. We note that the unicast SINR in an interference-limited case is independent of the base station transmit power.

We noted that for MBSFN, the signals received from multiple synchronized base stations do not interfere with each other as long as the relative delays of the received signals are within the OFDM symbol cyclic prefix length. Therefore, there is no interference when the same broadcast content is transmitted system-wide apart from the background noise.

The signal-to-interference-plus-noise ratio (SINR) for SFN assuming reception from a total of 12 cells as shown in Figure 17.2 can be written as:

$$\rho_{\text{SFN-12}} = \frac{P(3 \times (R)^{-\alpha} + 3 \times (2R)^{-\alpha} + 6 \times [(\sqrt{3}+1)R]^{-\alpha})}{N_0 W}. \tag{17.12}$$

It can be noted that increasing transmit power results in linear increase (within practical receiver limits) of SFN broadcast SINR.

Let us summarize our investigation on the effect of base station transmit power on SINR for a unicast and SFN broadcast traffic. We noted that in an interference-limited scenario, the unicast SINR does not benefit from the increased base station power. This is because when the power is increased in the own cell, the transmit power is also increased in the interfering neighboring cells. However, in the case of SFN broadcast, the increased base station transmit power results in a linear increase in SINR. In the case of an interference-limited unicast scenario if increased transmit power does not help increase SINR, lowering transmit power should not effect SINR either. Therefore, we may say that excess base station power is available in an interference-limited unicast system. This is because base station transmit power is generally dimensioned for the coverage-limited situation and full transit power is generally not needed in interference-limited cases.

Can the excess unicast power somehow be used for SFN broadcast? The answer lies in the SFN broadcast and unicast multiplexing approach. In the case of time-domain sharing of resources between SFN broadcast and unicast, the maximum power spectral density (PSD) between SFN broadcast and unicast is generally the same and is limited by the maximum base station (eNode-B) transmit power. The power cannot be shared across time and therefore there are no benefits in lowering unicast transmit power.

On the other hand, if SFN broadcast and unicast are frequency-multiplexed, the total eNode-B transmit power can be shared between SFN broadcast and unicast. The power spectral density (PSD) for frequency resources used for SFN broadcast can be increased while lowering the PSD on unicast frequency resources.

This way the excess unicast power can be allocated to SFN broadcast traffic resulting in enhanced broadcast SINR and hence improved performance in an SFN-based transmission.

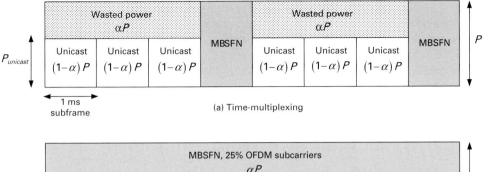

Figure 17.4. An example of power sharing between MBSFN and unicast.

Figure 17.4 displays an example of power sharing between MBSFN and unicast. In unicast slots for the time-multiplexed case, a power of $(1 - \alpha)P$ watts is used for unicast traffic and αP watts is wasted because transmission at increased power does not help to improve unicast SINR. However, during broadcast slots, the transmission happens at the full power P watts because transmitting at a higher power always helps the MBSFN traffic. The total energy transmitted for MBSFN for the time-multiplexed case is then $(P/4)$ joules assuming 1/4 duty cycle for the MBSFN traffic. In the the frequency-multiplexed case, however, the power can be shared between the unicast and broadcast traffic. Therefore, when broadcast traffic is transmitted at the same time as the unicast traffic using orthogonal OFDM subcarriers, the unused power can be allocated to broadcast traffic. This unused power that was not helping the unicast traffic can now help to improve the broadcast performance. In the case of frequency-multiplexing, 1/4 of the subcarrier resource is allocated to broadcast in order to account for the same bandwidth fraction as with 1/4 duty cycle with time-multiplexing. Therefore, the performance of the unicast traffic is unaffected. However, the energy for broadcast traffic is now αP joules. Therefore, the broadcast energy is always equal or better with frequency-multiplexing compared to the time-multiplexing case as long as α is greater than the broadcast duty cycle, that is the power spectral density (PSD) on the broadcast resources is higher than the PSD on unicast resources.

We provide a simple analysis showing potential gains of power sharing between SFN broadcast and unicast by frequency-multiplexing. We noted that excess power is available at the base station carrying only unicast traffic and operating in an interference-limited situation. This excess power can be allocated to SFN broadcast traffic resulting in enhanced broadcast SINR. When SFN broadcast and unicast are frequency multiplexed, the capacity for SFN broadcast can be written as:

$$C_{\text{MBSFN}} = \beta \cdot W \cdot \log_2 \left(1 + \frac{\alpha \cdot P}{\beta \cdot W \cdot N_0} \right) \quad \text{b/s/Hz}, \qquad (17.13)$$

where β is the fraction of the total bandwidth allocated to MBSFN. We assume that the remaining bandwidth $(1 - \beta)$ is allocated to unicast. α represents the total fraction of power allocated to MBSFN. We assume that the remaining power $(1 - \alpha)$ is allocated to unicast. P and W represent the total received power and total system bandwidth respectively. Note that, for simplicity, we ignore the penalty due to frequency selectivity in (17.13). For the case of uniform power spectral density (PSD) between MBSFN and unicast $(\alpha = \beta)$, we can write (17.13) as:

$$C_{\text{MBSFN}} = \beta \cdot W \cdot \log_2 \left(1 + \frac{P}{W \cdot N_0} \right) \quad \text{b/s/Hz.} \qquad (17.14)$$

The above equation also determines the capacity limit when broadcast and unicast traffic is time-multiplexed. In this case, β will represent the fraction of time allocated to the broadcast traffic. The capacity limit for unicast traffic with frequency-multiplexing of broadcast and unicast traffic can be written as:

$$C_{\text{unicast}} = (1 - \beta) \cdot W \cdot \log_2 \left(1 + \frac{(1 - \alpha) \cdot P}{(1 - \beta) \cdot W \cdot N_0 + (1 - \alpha) \cdot f \cdot P} \right) \quad \text{b/s/Hz,}$$
$$(17.15)$$

where f represents the ratio between other-cell and own-cell signal. f is in general larger for users at the cell-edge. For the case of uniform power spectral density (PSD) between MBSFN and unicast, i.e. $\alpha = \beta$, the above equation simplifies to:

$$C_{\text{unicast}} = (1 - \beta) \cdot W \cdot \log_2 \left(1 + \frac{P}{W \cdot N_0 + f \cdot P} \right) \quad \text{b/s/Hz.} \qquad (17.16)$$

This equation also determines the unicast capacity limit when broadcast and unicast traffic is time-multiplexed. In this case, $(1 - \beta)$ will represent the fraction of time allocated to the unicast traffic. In an interference-limited situation (the case for most cellular deployments in urban areas) with $(WN_0 \ll fP)$, the capacity limit can be written as:

$$C_{\text{unicast}} = (1 - \beta) \cdot W \cdot \log_2 \left(1 + \frac{P}{f \cdot P} \right). \qquad (17.17)$$

That is, increasing power P does not help in improving the unicast SINR. This is in line with our earlier conclusion that excess power is available at the eNode-B carrying only unicast traffic and operating in an interference-limited situation. The PSD ratio between SFN broadcast and unicast traffic is given as:

$$\frac{\text{PSD}_{\text{MBSFN}}}{\text{PSD}_{\text{unicast}}} = 10 \log_{10} \left[\frac{\alpha/\beta}{\frac{(1-\alpha)}{(1-\beta)}} \right] = 10 \log_{10} \left[\frac{(1 - \beta) \cdot \alpha}{(1 - \alpha) \cdot \beta} \right] \text{dB.} \qquad (17.18)$$

Figure 17.5 displays the numerical results for the case of 20%, 40% and 60% bandwidth allocated to MBSFN using frequency-multiplexing. We consider a case where the total system bandwidth available for both unicast and MBSFN is 10 MHz. In the case of time multiplexing the numbers 20%, 40% and 60% represent the fraction of time allocated to broadcast traffic.

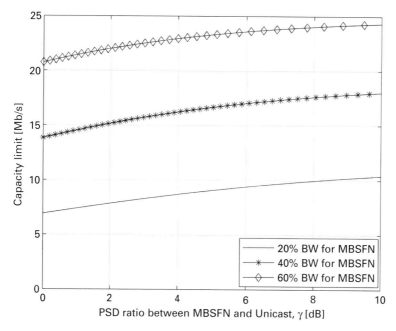

Figure 17.5. Capacity limit for broadcast using frequency multiplexing, $\frac{P}{N_0 W} = 10\,\text{dB}$.

Furthermore, we assumed the MBSFN SNR of ($P/N_0 W = 10\,\text{dB}$). The case of 0.0 dB PSD ratio between broadcast and unicast represents the case of time multiplexing and also the frequency-multiplexing case if the PSD between broadcast and unicast is the same. We note that the frequency-multiplexing approach can provide about 50% gain over the time-multiplexing approach (capacity limit of 10.5 relative to 7.0 Mb/s) when PSD on broadcast resources is 10.0 dB higher than the unicast for the case of 20% broadcast BW allocation. The gains are relatively smaller when a larger bandwidth is allocated to broadcast. This is due to the fact that with larger MBSFN bandwidth, a larger amount of power needs to be borrowed from the unicast to boost the broadcast PSD relative to unicast. However, there is a smaller amount of total unicast power due to smaller bandwidth allocation to unicast. The gains of frequency multiplexing will be larger when broadcast uses a small amount of bandwidth.

Figures 17.6 and 17.7 show MBSFN spectral efficiency for the case of ($P/N_0 W = 0.0\,\text{dB}$) and ($P/N_0 W = 20\,\text{dB}$) respectively. We note that frequency multiplexing of MBSFN and unicast provides larger gains relative to the time-multiplexing approach when the MBSFN SINR is low as depicted by Figure 17.6. We note that for ($P/N_0 W = 0.0\,\text{dB}$) and 20% MBSFN bandwidth allocation, the frequency-multiplexing approach provides two times larger spectral efficiency when PSD on broadcast resources is 8.0 dB higher than the unicast PSD. When the MBSFN SINR is large the gains from frequency multiplexing are small as is shown in Figure 17.7. We note that the frequency-multiplexing approach allows higher PSD for MBSFN that increases SINR. However, when the SINR is already large, broadcast performance is not power-limited and therefore increasing PSD on broadcast resources results in a logarithmic increase in capacity. We remark that for ($P/N_0 W = 20\,\text{dB}$) and 20% MBSFN bandwidth allocation, the frequency-multiplexing approach provides only 22% gain when PSD on broadcast resources is 8.0 dB higher than the unicast PSD.

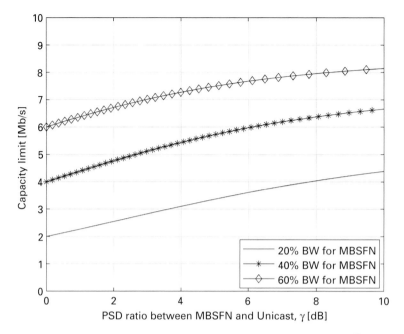

Figure 17.6. Capacity limit for broadcast using frequency multiplexing, $\frac{P}{N_0 W} = 0.0\,\text{dB}$.

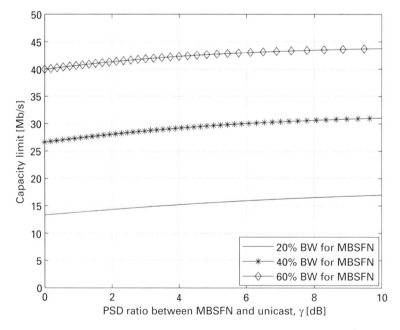

Figure 17.7. Capacity limit for broadcast using frequency multiplexing, $\frac{P}{N_0 W} = 20\,\text{dB}$.

We noted that under SFN operation, broadcast SINR improves when more power is concentrated on broadcast resources in the case of frequency multiplexing. It should be noted that for the case of frequency multiplexing, the broadcast SINR would be higher depending upon the amount of power available for broadcast that is not otherwise used by the unicast traffic. The performance of unicast is unaffected by frequency multiplexing because unicast SINR is unaffected when power is borrowed from unicast in an interference-limited situation. In the performance analysis presented here, we ignored the effect of the cyclic prefix overhead. In some cases with small cell deployments, it is possible to use the normal cyclic prefix for the broadcast traffic as well. Therefore, there would not be any impact on unicast performance. In cases where unicast is forced to use the extended cyclic prefix, the gains from broadcast frequency multiplexing should overcome the loss due to the longer cyclic prefix.

So far we have only considered the power-sharing aspect in MBSFN and unicast multiplexing. Another multiplexing consideration is the impact on UE power consumption. In the case of time multiplexing of MBSFN and unicast, the duty cycle of MBSFN transmissions is minimized. This is because all the resources (bandwidth) within a subframe are used for MBSFN that allows carrying a large amount of data within a short period of time. With a low duty cycle, the UE receiver only needs to be turned on during MBSFN transmissions and hence power savings can be achieved by turning off the receiver during unicast subframes. In the case of frequency multiplexing, the MBSFN transmissions use a fraction of the bandwidth and hence the transmission duration in time needs to be extended in order to carry the same amount of data. In the example of Figure 17.4, time multiplexing requires a duty cycle of 25% while in the frequency multiplexing approach the duty cycle in time is 100% with continuous transmission in time. This means that the UE receiver needs to be turned on for longer periods resulting in increased power consumption in the UE receiver. Therefore, from a UE power saving perspective, time multiplexing of MBSFN and unicast traffic is preferable.

We noted that with the same amount of resources allocated for the MBSFN traffic, the frequency-multiplexing approach could stretch a transmission longer in time therefore benefiting from additional time diversity. We also remark that MBSFN can still benefit from full frequency diversity by distributed allocation of frequency resources to the MBSFN traffic when the total frequency resource in a subframe is shared between unicast and broadcast traffic. Therefore, from a diversity performance point of view, the frequency-multiplexing approach can be better than the time-multiplexing approach.

It can also be argued that since frequency multiplexing increases throughput of MBSFN, it requires fewer resources than the time-multiplexing approach to support the same data rate. This may help to partly offset the transmission expansion in time required for the frequency-multiplexing approach. Moreover, there may be a requirement of a maximum data rate that can be received by certain UE classes. Since the MBSFN transmission data rate is determined by the highest data rate than can be supported by the lowest class UE, this may also limit the maximum data rate that can be used for MBSFN transmission. This means that using full bandwidth in a given subframe for MBSFN may result in underutilization of the system capacity in that subframe. Moreover, a UE may not be interested in receiving all the MBSFN services at a given time. In the case of mobile TV application, for example, a UE may be interested in receiving a single TV channel at a given time. It is therefore possible to time-multiplex different channels to reduce the duty cycle from a given UE perspective. However, these different TV channels can be frequency-multiplexed with the unicast traffic.

Figure 17.8 displays a hybrid time/frequency-multiplexing approach. The MBSFN is frequency multiplexed along with the unicast traffic in the same subframe of 1.0 ms. The

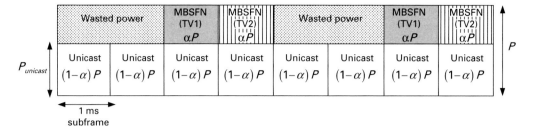

Figure 17.8. An example of hybrid time/frequency-multiplexing between MBSFN and unicast.

hybrid subframes containing MBSFN and unicast traffic are further time-multiplexed with the unicast traffic. In order to allow a low duty cycle of MBSFN from a UE perspective, the TV channel#1 and TV channel#2 are time-multiplexed with respect to each other. This allows UEs interested in a single MBSFN service to turn off their receivers in subframes carrying other MBSFN services. In the example of Figure 17.8, the duty cycle for a UE interested in a single mobile TV channel is only 25% (every fourth subframe). The amount of frequency and time resources needed for MBSFN can be configured by the network based on the MBSFN traffic requirements. A pure time multiplexing of MBSFN and unicast can be realized by allocating the whole frequency resource in the hybrid subframes to the broadcast traffic. This hybrid approach allows exploiting some of the benefits of frequency multiplexing such as power sharing between MBSFN and unicast and also enabling lower data rates transmissions without wasting resources within a subframe. Note that unlike the time-multiplexing approach, the resources not used by MBSFN in hybrid subframes can be used by unicast.

17.4   MBSFN and unicast superposition

So far we have considered multiplexing schemes that require orthogonal resources for MBSFN and unicast. In this section, we discuss a new approach whereby the MBSFN and unicast traffic are carried over the same time and frequency resources in a non-orthogonal fashion [7]. This can be achieved by a simple linear superposition of MBSFN and unicast transmissions as illustrated in Figure 17.9. We note that different information content is transmitted in different cells for the case of unicast while the same information content is simultaneously transmitted from multiple cells for the case of MBSFN. Since MBSFN transmission is targeted to be received by all the UEs in the MBSFN area, MBSFN signals can be successfully decoded and cancelled from the overall composite received signal leaving no MBSFN interference to the unicast traffic. Once the MBSFN signal is detected and cancelled, demodulation and decoding of unicast traffic can start. This way the MBSFN transmissions do not use any orthogonal resources thus greatly improving the overall system capacity and efficiency. This, of course, is achieved at the expense of additional interference cancellation complexity at the UE receiver.

Figure 17.10 displays an MBSFN/unicast superposition transmitter. The control information such as unicast and MBSFN reference signals and scheduling grants are transmitted over orthogonal resources. This is to enable robust system operation not requiring MBSFN interference cancellation for reception of critical control information. In general, the scheduling grants do not use hybrid ARQ and therefore need to be transmitted with very high reliability on the first transmission attempt. Allocation of orthogonal resources to signaling and

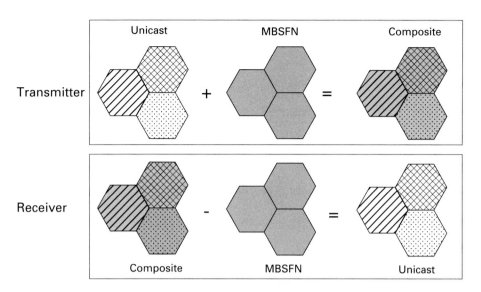

Figure 17.9. An illustration of MBSFN/unicast superposition.

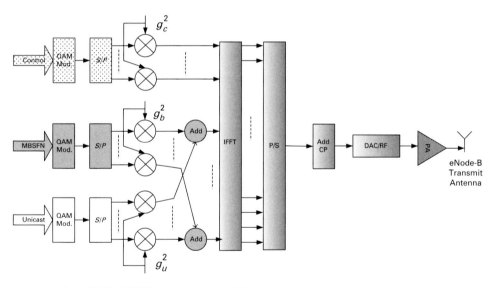

Figure 17.10. MBSFN/unicast superposition transmitter.

control assures high reliability for the critical system control information. The power allocated to control channels can be adjusted dynamically by controlling the control channel gain factor, g_c^2. The MBSFN and unicast signals are linearly superimposed (added) in the frequency-domain before IFFT on the same subcarrier resources. The ratio between MBSFN and unicast power g_b^2/g_u^2 can be selected for the desired MBSFN data rate taking into account total available power for MBSFN and unicast traffic. The superimposed composite signal is then mapped to OFDM subcarriers at the input of IFFT. It should be noted that both MBSFN and unicast traffic use the same set of subcarriers thereby not requiring any orthogonal resources.

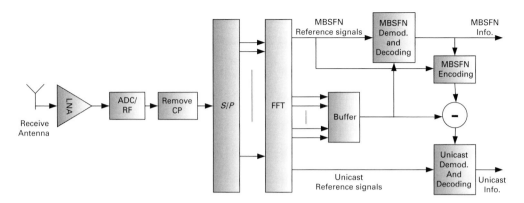

Figure 17.11. Receiver operation for MBSFN/unicast superposition and interference cancellation.

The total power is shared between the MBSFN and the unicast traffic by appropriately selecting the gain factors g_b^2 and g_u^2 respectively.

The receiver operation for MBSFN superposition and interference cancellation scheme is shown in Figure 17.11. The received signal is filtered, amplified and converted from analog to digital. After discarding the cyclic prefix (CP), an FFT operation is performed on the received signal. Since the control information including MBSFN and unicast reference signals are transmitted over orthogonal subcarriers, these signals can be recovered at the output of FFT without requiring any interference cancellation. The frequency domain samples of the composite signal are buffered for further processing. The MBSFN data is first demodulated and decoded using MBSFN channel estimates obtained from MBSFN reference signals. The MBSFN reference signals are transmitted using the same time-frequency positions from all the base stations in an MBSFN zone. This provides for an overall channel estimate for the signal received from multiple base stations transmitting the same content in the MBSFN zone. The MSFN signal is reconstructed using the MBSFN channel estimates and successfully decoded MBSFN information block. The reconstructed MBSFN signal is then cancelled from the overall received signal stored in the buffer. The reconstruction of the MBSFN signal using the overall SFN channel estimate assures that all the MBSFN interference including MBSFN interference from neighboring cells to the unicast traffic is cancelled. This results in a clean unicast signal that is free from any MBSFN interference. This clean unicast signal is then further processed for unicast traffic demodulation and decoding. The channel estimates obtained from unicast reference signals are used for unicast demodulation.

Some of the challenges with interference cancellation in the context of unicast traffic only such as channel estimation error and error propagation are discussed in [9]. However, we will point out here that canceling an MBSFN signal affords certain attractive features from the complexity and robustness point of view. In general, the MBSFN information is available to all the users in the system with very high reliability i.e. the target block error rate on the MBSFN traffic is 0.1–1%. Therefore, the MBSFN signal can be reconstructed and cancelled with a fair degree of accuracy given that good channel estimates are available. It should be noted that MBSFN generally does not use hybrid ARQ and, therefore, 99% (with 1% FER requirement) of the MBSFN transmissions are successful on the first attempt. A successful transmission is a prerequisite for effective reconstruction of the signal and its cancellation from the overall received signal. Since a single MBSFN stream is cancelled, the complexity

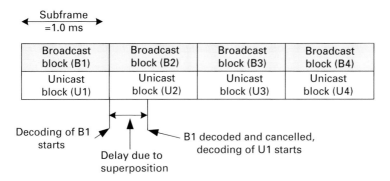

Figure 17.12. MBSFN and unicast information blocks decoding timing.

of the proposed approach is also moderate. Moreover, unlike conventional successive interference cancellation there are no issues with error propagation because of a single stream cancellation.

An MBSFN information block is independently coded along with a CRC and is transmitted within a subframe as shown in Figure 17.12. The subframe duration or transmission time interval in the LTE system is 1 ms. By using the same subframe for both MBSFN and unicast, a small additional delay is introduced for the unicast traffic as shown in Figure 17.12. The unicast decoding starts immediately after decoding of the MBSFN information block. It can be noted that some loss of time-diversity for MBSFN may occur relative to the case where the MBSFN information block is transmitted over multiple subframes. The broadcast traffic is generally not time-sensitive and its transmission can be spread out over multiple subframes. However, in the MBSFN superposition and interference cancellation approach, MBSFN traffic benefits from full frequency diversity because the MBSFN signal is transmitted over almost the whole system bandwidth making it less dependent upon time diversity. We transmit the broadcast information block in the same time interval as the unicast information block to limit the delays introduced for the unicast traffic. It should, however, be noted that when unicast traffic employs the successive-interference cancellation receiver for MIMO spatial multiplexing, similar delays as in the superposition approach are introduced.

In the MBSFN superposition and interference cancellation approach, the unicast decoding only happens after decoding the MBSFN information. This may represent a significant computing burden on the UE processing and battery life if it continuously needs to perform interference cancellation on the MBSFN traffic. However, we know that in packet-based communications a user receives traffic for short periods of time with large inactivity periods in between. During inactivity periods, the UE only needs to listen to control information such as paging messages and scheduling grants transmitted from the network. Therefore, a system design that minimizes the computation effort for the UE requires that the control information be transmitted on orthogonal OFDM subcarriers as shown in Figure 17.10. In this way, a UE who is listening to just the control information does not need to perform interference cancellation of the MBSFN traffic. When a traffic transmission happens for a UE, it is indicated via the scheduling grant. Therefore, a UE needs to perform interference cancellation of the MBSFN signal only when it is receiving unicast traffic transmission. In general, the control and signaling information uses around 10–20% of the system bandwidth leaving the remaining 80–90% of the bandwidth for traffic. Therefore, even by using orthogonal subcarriers for control and signaling, the MBSFN traffic can be superimposed on a large fraction of the traffic bandwidth. For simplicity, in the following discussions, we will assume that all the bandwidth is available for unicast and broadcast traffic.

Similar to the MBSFN/unicast frequency-multiplexing approach, the MBSFN superposition and interference cancellation approach allows to trade off power allocation between unicast and MBSFN traffic for certain performance and capacity targets. In the frequency-multiplexing approach, the total resource needs to be shared between MBSFN and unicast traffic. However, in the MBSFN/unicast superposition approach full frequency resources are available to both MBSFN and unicast. When MBSFN and unicast traffic is superimposed, the capacity for MBSFN can be written as:

$$C_{\text{mbsfn}} = W \cdot \log_2 \left(1 + \frac{\alpha \cdot P}{W \cdot N_0 + (1 - \alpha) \cdot P} \right). \tag{17.19}$$

It should be noted that with the superposition approach, full bandwidth is available to both broadcast and unicast. However, the power allocated to unicast $(1 - \alpha) P$ now appears as interference to broadcast traffic. The broadcast to unicast power ratio can be written as:

$$\frac{g_b^2}{g_u^2} = 10 \cdot \log_{10} \left(\frac{\alpha}{1-\alpha} \right) \text{dB}. \tag{17.20}$$

The capacity limit for unicast traffic with broadcast and unicast superposition is given as:

$$C_{\text{unicast}} = W \cdot \log_2 \left(1 + \frac{(1 - \alpha) \cdot P}{I_{\text{sfn}} + W \cdot N_0 + (1 - \alpha) \cdot f \cdot P} \right), \tag{17.21}$$

where I_{sfn} is the residual interference due to imperfect cancellation of the MBSFN signal. The impact of residual interference on the system capacity is evaluated in [7]. With perfect interference cancellation, $I_{\text{sfn}} = 0$ and capacity for unicast can be written as:

$$C_{\text{unicast}} = W \cdot \log_2 \left(1 + \frac{(1 - \alpha) \cdot P}{W \cdot N_0 + (1 - \alpha) \cdot f \cdot P} \right). \tag{17.22}$$

In an interference-limited situation with $(1 - \alpha) \cdot f \cdot P \gg W \cdot N_0$, the above equation simplifies to:

$$C_{\text{unicast}} = W \cdot \log_2 \left(1 + \frac{P}{f \cdot P} \right). \tag{17.23}$$

It can be noted that the unicast broadcast superposition approach provides $\frac{1}{(1-\beta)}$ times higher unicast capacity relative to the frequency- or time-multiplexing approaches.

The MBSFN capacity results assuming 10 MHz system bandwidth are shown in Figures 17.13–17.15 for 20%, 40% and 80% bandwidth allocation to broadcast in the case of frequency multiplexing. In the case of time-multiplexing these numbers represent the fraction of time allocated to broadcast traffic. With superposition, full bandwidth is available for unicast traffic. In all the cases, we assumed $P/N_0 W = 10$ dB. From Figure 17.13, for example, we can see that the superposition approach provides gains relative to frequency multiplexing when a larger fraction of power can be allocated to broadcast. It can be noted that superposition starts outperforming the frequency-multiplexing approach when more than 50% power ($\alpha = 0.5$) is allocated to broadcast for the case of 20% bandwidth allocation. This corresponds to broadcast to unicast power ratio $\alpha/(1 - \alpha)$ of around 0.0 dB. At very high $\alpha/(1 - \alpha)$ ratios (α approaching 1), superposition provides about three times higher broadcast capacity while at the same time providing a higher unicast capacity. In particular for 20% broadcast bandwidth allocation case ($\beta = 0.2$) in Figure 17.13, the unicast capacity with superposition is 25% (1/0.8) higher than the case of the frequency- or time-multiplexing approach. The unicast capacity in the case of superposition is higher relative to the frequency-multiplexing approach because full bandwidth is now available to unicast.

The time-multiplexing scheme limits the maximum power spectral density to the same value between unicast and broadcast traffic. Therefore, the point where the power fraction and the bandwidth fraction allocated to broadcast are the same can be seen as the performance with a time-multiplexing approach. In Figure 17.13, for example, the 20% power fraction point ($\alpha/(1 - \alpha) = -6$ dB on the horizontal axis) corresponds to performance of the time-multiplexing approach. We also note that frequency multiplexing provides gain over the time-multiplexing approach as the power spectral density on broadcast resources can be increased relative to the unicast traffic. It can also be noted that superposition gains relative to the time-multiplexing approach at higher $\alpha/(1 - \alpha)$ ratios are even greater compared to the frequency-multiplexing approach.

We note from Figures 17.14 and 17.15 that the point at which superposition starts performing better than the frequency-multiplexing scheme moves towards higher $\alpha/(1 - \alpha)$ ratios as the bandwidth allocated to broadcast for the case of frequency multiplexing increases. This is expected because as more bandwidth is allocated to broadcast for the frequency multiplexing

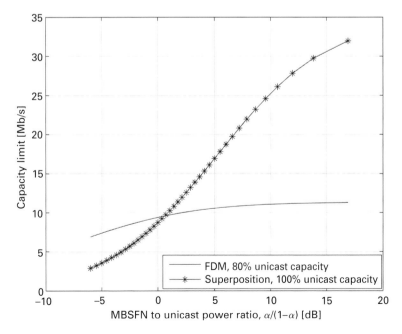

Figure 17.13. Capacity limit for MBSFN using superposition of broadcast and unicast for $\beta = 0.2$, $\frac{P}{N_0 W} = 10$ dB.

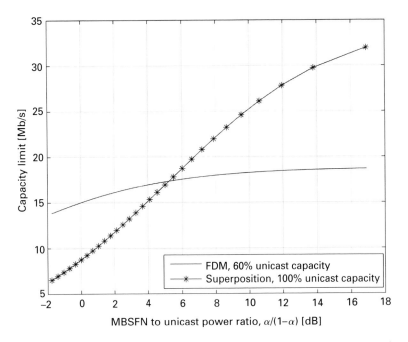

Figure 17.14. Capacity limit for MBSFN using superposition of broadcast and unicast for $\beta = 0.4$, $\frac{P}{N_0 W} = 10$ dB.

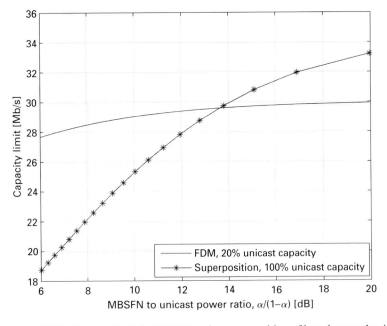

Figure 17.15. Capacity limit for MBSFN using superposition of broadcast and unicast for $\beta = 0.8$, $\frac{P}{N_0 W} = 10$ dB.

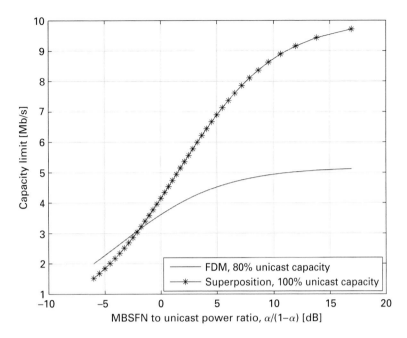

Figure 17.16. Capacity limit for MBSFN using superposition of broadcast and unicast for $\beta = 0.2$, $\frac{P}{N_0 W} = 0\,\mathrm{dB}$.

approach, broadcast capacity increases. However, an important point to note here is that the available bandwidth for unicast reduces as more bandwidth is allocated to broadcast which lowers unicast capacity. We note from Figure 17.15, for example, that superposition outperforms the frequency-multiplexing approach for $\alpha/(1-\alpha)$ ratio of greater than 14 dB while providing five times (100% relative to 20%) larger unicast capacity.

Figures 17.16 and 17.17 show broadcast capacity limits for the case of 20% bandwidth allocation to broadcast and $P/N_0 W$ of 0.0 and 20 dB respectively. For $P/N_0 W$ of 0.0 dB in Figure 17.16, we can see that superposition starts outperforming frequency-multiplexing for broadcast to unicast power ratios $\alpha/(1-\alpha)$ of less than 0.0 dB. At very high $\alpha/(1-\alpha)$ ratios, superposition provides about two times higher broadcast capacity relative to the frequency-multiplexing approach while at the same time providing a higher unicast capacity. The superposition gains relative to the time-multiplexing approach for high $\alpha/(1-\alpha)$ ratios are approximately five times higher capacity. For the case of $P/N_0 W$ of 20.0 dB in Figure 17.17, we can see that superposition starts outperforming frequency multiplexing for $\alpha/(1-\alpha)$ ratios of more than approximately 3.0 dB. At very high broadcast to unicast power ratios $\alpha/(1-\alpha)$, superposition provides more than three times higher broadcast capacity while at the same time providing a higher unicast capacity. The superposition gains relative to the time-multiplexing approach for high $\alpha/(1-\alpha)$ ratios are approximately four times higher capacity.

17.5   Summary

The simultaneous transmission of broadcast information using the same time-frequency resources and the same modulation and coding format from multiple synchronized

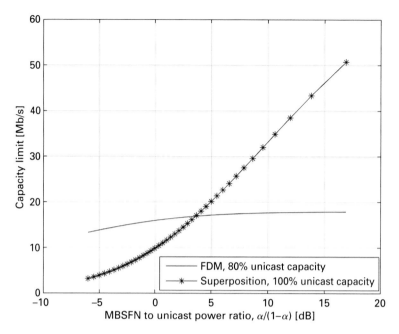

Figure 17.17. Capacity limit for MBSFN using superposition of broadcast and unicast for $\beta = 0.2$, $\frac{P}{N_0 W} = 20$ dB.

Node-Bs results in an interference-free signal at the UE; such an operation or system is commonly referred to as a single frequency network (SFN). When multicast broadcast SFN (MBSFN) and unicast traffic are carried on the same carrier, two orthogonal multiplexing schemes can be considered. The time-multiplexing approach leads to lower MBSFN duty cycles resulting in UE battery power savings as the UE needs to turn on its receiver infrequently. Another benefit of the time-multiplexing approach is that a longer cyclic prefix length can be used for broadcast while using a shorter cyclic prefix for the unicast traffic. This is because transmission of MBSFN and unicast is separated in time. The MBSFN channel delay spread is generally larger than the unicast channel due to reception from multiple cells generally requiring a longer cyclic prefix. The drawback of the time-multiplexing approach, however, is that total transmit power cannot be shared between MBSFN and unicast.

The frequency-multiplexing approach allows power sharing between MBSFN and unicast. With the frequency-multiplexing approach, in interference-limited scenarios, the power allocated to the unicast traffic can be lowered without affecting the unicast performance. The additional available power can then be allocated to MBSFN traffic improving MBSFN spectral efficiency. The frequency-multiplexing approach, however, requires the use of the same cyclic prefix length between unicast and MBSFN. It should be noted that in small cell deployments, it is possible to use the same shorter CP for both MBSFN and unicast transmissions. Therefore, there will not be any impact on unicast performance by frequency multiplexing. In cases where unicast is forced to use a long cyclic prefix, the gains from MBSFN frequency multiplexing should overcome the loss due to use of a longer cyclic prefix. We also noted that TDM multiplexing of MBSFN services is possible in both cases when unicast and MBSFN is time multiplexed or frequency multiplexed. This allows a low duty cycle for UEs interested in a single MBSFN service at a given time, thus prolonging the UE battery life.

We also discussed a non-orthogonal MBSFN and unicast multiplexing approach based on the concept of superposition and interference cancellation. Like the frequency-multiplexing scheme, the superposition approach also allows to share total Node-B power between MBSFN and unicast traffic. The superposition is achieved by a simple linear superposition of MBSFN and unicast transmissions. The superposed MBSFN signal is decoded and cancelled before unicast demodulation and decoding. Using this approach, the MBSFN interference to unicast traffic from own cell and neighboring cells is eliminated. We observed that the superposition scheme could provide a large capacity advantage relative to orthogonal multiplexing schemes because full bandwidth is available to both broadcast and unicast. This is particularly true for the interference-limited cases where excess unicast power is available for use by the superposed MBSFN traffic.

The LTE system employs a simple time-multiplexing approach for MBSFN and unicast multiplexing. The frequency multiplexing and superposition of broadcast and unicast were not included in the standard specifications due to concern on UE complexity. However, both the frequency multiplexing and superposition schemes are currently under investigation for the LTE-advanced system.

References

[1] EN 302 304 v1.1.1 ETSI, Digital Video Broadcasting (DVB); Transmission System for Handheld Terminals (DVB-H).

[2] Chari, M., Ling, F., Mantravadi, A., Krishnamoorthi, R., Vijayan, R., Walker, G. K. and Chandhok, R., "FLO physical layer: an overview," *IEEE Transactions on Broadcasting*, vol. B-53, no. 1, Mar. 2007.

[3] ETS 300 401, Radio Broadcasting Systems; Digital Audio Broadcasting (DAB) to Mobile, Portable and Fixed Receivers.

[4] 3GPP2 C.S-0054-A, cdma2000 High Rate Broadcast–Multicast Packet Data Air-Interface Specification.

[5] 3GPP2 C.S0002-D, Physical Layer Standard for cdma2000 Spread Spectrum Systems – Revision D.

[6] 3GPP TS 22.146, Multimedia Broadcast/Multicast Service (MBMS); Stage 1.

[7] Khan, F., "Broadcast overlay on unicast via superposition coding and interference cancellation," *64th IEEE Vehicular Technology Conference, VTC-2006 Fall*, Sept. 2006.

[8] Kim. D., Khan, F., Rensburg, C., Pi. Z. and Yoon, S., "Superposition of broadcast and unicast in wireless cellular systems," *IEEE Communication Magazine*, pp. 110–117, July 2008.

[9] Andrews, J. G., "Interference cancellation for cellular systems: a contemporary overview," *IEEE Wireless Communications Magazine*, pp. 19–29, Apr. 2005.

18 Spatial channel model

Specification of a propagation channel model is of foremost importance in the design of a wireless communication system. A propagation model is used to predict how the channel affects the transmitted signal so that transmitters and receivers that best compensate for the channel's corrupting behaviors can be developed. A propagation model is also used as a basis for performance evaluation and comparison of competing wireless technologies. An example of such propagation models is ITU-R channel models [1] that were developed for IMT-2000 system evaluation. A wireless propagation channel model needs to be refined as new system parameters (e.g. larger bandwidths and new frequency bands) or radio technologies exploiting new characteristics of the channel such as multi-antenna schemes are introduced. A well-defined channel model allows for the assessing of the system performance under new parameters as well as gains due to introduction of new radio technologies. The performance of multi-antennas technologies, for example, depends upon the spatial correlations between antennas. As ITU-R channel models do not characterize the spatial correlations, using these propagation models may lead to overestimating the gains of multi-antenna techniques. In order to provide a reasonable propagation platform for multi-antenna techniques evaluation, the spatial channel model (SCM) was developed [2]. The SCM defines a ray-based model derived from stochastic modeling of scatters and therefore allows to model spatial correlations required for evaluation of multi-antenna techniques.

18.1 Multi-path fading

A cellular radio system consists of a collection of fixed Node-Bs that define the radio coverage areas or cells. Typically, a non-line-of-sight (NLOS) radio propagation path exists between a Node-B and a UE due to natural and man-made objects that are situated between the Node-B and the UE. As a consequence, the radio waves propagate via reflections, diffractions and scattering. The arriving waves at the UE (at the Node-B in the uplink direction) experience constructive and destructive additions because of different phases of the individual waves. This is due to the fact that at high carrier frequencies typically used in the cellular wireless communication, small changes in the differential propagation delays introduce large changes in the phases of the individual waves. If the MS is moving or there are changes in the scattering environment, then the spatial variations in the amplitude and phase of the composite received signal will manifest themselves as the time variations known as fast fading.

The instantaneous received power at the UE due to a large number of reflected and scattered waves is a random variable, which is dependent upon the location of the UE. Let us consider a case of an unmodulated carrier transmit signal given as:

$$s(t) = \cos(2\pi f_c t + \psi). \tag{18.1}$$

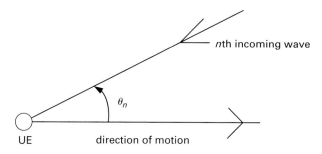

Figure 18.1. Incoming wave at a UE receiver.

Let us assume that the nth wave arrives at an angle θ_n relative to the direction of UE motion as shown in Figure 18.1. The UE movement introduces a frequency shift referred to as Doppler shift $f_{D,n}$ in the incident plane wave given as:

$$f_{D,n} = \frac{v}{\lambda} \cos \theta_n, \qquad (18.2)$$

where v is the velocity of the UE and λ the wavelength given as:

$$\lambda = \frac{C}{f_c}, \qquad (18.3)$$

where C is the speed of light, which is 299 792 458 m/sec in free space. The Doppler shift $f_{D,n}$ can then be written as:

$$f_{D,n} = \frac{f_c v}{C} \cos \theta_n. \qquad (18.4)$$

The received signal $r(t)$ is given as:

$$r(t) = \sum_{n=1}^{N} c_n \cos \left(2\pi \left(f_c + f_{D,n}\right) t + \psi + \phi_n\right), \qquad (18.5)$$

where c_n and ϕ_n are respectively the amplitude and phase of the nth incoming wave. By using the Doppler shift $f_{D,n}$ definition from (18.4), the received signal is written as:

$$r(t) = \sum_{n=1}^{N} c_n \cos \left(2\pi f_c t + \frac{2\pi v f_c t}{C} \cos \theta_n + \psi + \phi_n\right). \qquad (18.6)$$

We can rewrite the received signal $r(t)$ in in-phase and quadrature representation as below:

$$r(t) = I(t) \cos (2\pi f_c t) - Q(t) \sin (2\pi f_c t), \qquad (18.7)$$

where the in-phase $I(t)$ and quadrature $Q(t)$ components are given as:

$$I(t) = \sum_{n=1}^{N} c_n \cos \left(\frac{2\pi v f_c t}{C} \cos \theta_n + \psi + \phi_n\right)$$

$$Q(t) = \sum_{n=1}^{N} c_n \sin \left(\frac{2\pi v f_c t}{C} \cos \theta_n + \psi + \phi_n\right). \qquad (18.8)$$

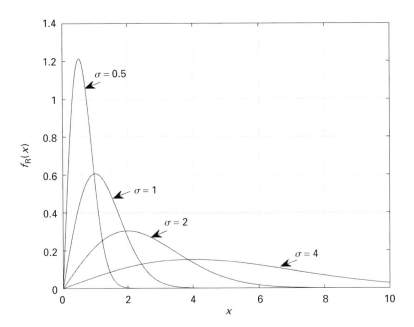

Figure 18.2. Probability density function of Rayleigh distribution.

According to the central limit theorem, for a large number of received waves ($N \to \infty$), the in-phase $I(t)$ and quadrature $Q(t)$ terms are zero-mean *i.i.d.* Gaussian random variables. The received signal amplitude $\sqrt{I^2(t) + Q^2(t)}$ is then Rayleigh distributed with probability distribution function:

$$f_R(x) = \frac{x}{\sigma^2} e^{\left(-x^2/2\sigma^2\right)} \quad x \geq 0, \tag{18.9}$$

where σ^2 is the variance of the in-phase $I(t)$ and quadrature $Q(t)$ components. The probability density function of Rayleigh distribution for various values of the variances σ^2 is shown in Figure 18.2.

The received signal $r(t)$ can then be written as:

$$r(t) = \sqrt{I^2(t) + Q^2(t)} \cos\left(2\pi f_c t + \theta(t)\right), \tag{18.10}$$

where $\theta(t)$ follows a uniform distribution $\theta(t) \in (0, 2\pi)$.

18.2   SCM channel scenarios

In order to provide a system evaluation framework under a wide range of cellular deployments, the SCM channel model describes three channel scenarios namely suburban macro-cell, urban macro-cell and urban micro-cell. In the case of macro-cells, a Node-B serves a large coverage area and the probability of experiencing a line-of-sight (LOS) is generally low. Therefore, for simplicity, an LOS component is not modeled in the macro-cells case. In micro-cell deployments with each Node-B covering a small area, the probability of a LOS channel cannot be neglected. Therefore, the urban micro-cell scenario is differentiated by a non-line-of-sight (NLOS) and a LOS model. The parameters for these channel scenarios are given in Table 18.1. In the case of macro-cell, the Node-B antenna height is assumed above local clutter

Table 18.1. SCM channel scenarios.

	Channel scenario		
Parameter	Suburban macro	Urban macro	Urban micro
Number of paths (N)	6	6	6
Number of subpaths (M) per-path	20	20	20
Mean AS at	$E[\sigma_{AS}] = 5°$	$E[\sigma_{AS}] = 8°, 15°$	NLOS:
Node-B			$E[\sigma_{AS}] = 19°$
AS at Node-B	$\mu_{AS} = 0.69$	For $E[\sigma_{AS}] = 8°$	N/A
modeled as a	$\varepsilon_{AS} = 0.13$	$\mu_{AS} = 0.81$	
lognormal		$\varepsilon_{AS} = 0.34$	
random		For $E[\sigma_{AS}] = 15°$	
variable $\sigma_{AS} = 10 \wedge (\varepsilon_{AS}x + \mu_{AS})$,		$\mu_{AS} = 1.18$	
$x \sim \eta(0, 1)$		$\varepsilon_{AS} = 0.21$	
$r_{AS} = \sigma_{AoD}/\sigma_{AS}$	1.2	1.3	N/A
Per-path AS at BS (Fixed)	2°	2°	5° (for both LOS and NLOS)
BS per-path AoD	$\eta(0, \sigma^2_{AoD})$ where	$\eta(0, \sigma^2_{AoD})$ where	$U(-40°, 40°)$
distribution standard distribution	$\sigma_{AoD} = r_{AS}\sigma_{AS}$	$\sigma_{AoD} = r_{AS}\sigma_{AS}$	
Mean AS at UE	$E[\sigma_{AS}] = 68°$	$E[\sigma_{AS}] = 68°$	$E[\sigma_{AS}] = 68°$
Per-path AS at UE (fixed)	35°	35°	35°
UE per-path AoA distribution	$\eta(0, \sigma^2_{AoA}(\text{Pr}))$	$\eta(0, \sigma^2_{AoA}(\text{Pr}))$	$\eta(0, \sigma^2_{AoA}(\text{Pr}))$
Delay spread	$\mu_{DS} = -6.80$	$\mu_{DS} = -6.18$	N/A
modeled as a lognormal	$\varepsilon_{DS} = 0.228$	$\varepsilon_{DS} = 0.18$	
random variable			
$\sigma_{DS} = 10 \wedge (\varepsilon_{DS}x + \mu_{DS}), x \sim \eta(0, 1)$			
Mean total RMS delay spread	$E[\sigma_{DS}] = 0.17\mu s$	$E[\sigma_{DS}] = 0.65\,\mu s$	$E[\sigma_{DS}] = 0.251\mu s$
$r_{DS} = \sigma_{delays}/\sigma_{DS}$	1.4	1.7	N/A
Distribution for path delays			$U(0, 1.2\mu s)$
Lognormal shadowing standard	8 dB	8 dB	NLOS: 10 dB
deviation, σ_{SF}			LOS: 4 dB
Path-loss model(dB),	$31.5 + 35\log_{10}(d)$	$34.5 + 35\log_{10}(d)$	NLOS: 34.53+
d is in meters			$38\log_{10}(d)$LOS:
			$30.18 + 26\log_{10}(d)$

and, therefore, the angle spread and delay spread are relatively small. The delay spread and angle spread for urban macro-cell are assumed slightly larger than the suburban macro-cell. In the case of the urban micro-cell scenario, Node-B antennas are assumed to be at rooftop height, which results in an even larger angle spread.

18.3 Path-loss models

The received signal power P_R is inversely proportional to the distance between the transmitter and receiver (d) as below:

$$P_R \propto \left(\frac{1}{d}\right)^\alpha,$$

(18.11)

where α is referred to as the path-loss exponent, which is 2 in free space. In cellular systems modeling, the path-loss exponent α is generally assumed between 2 and 4 depending upon the wireless channel environment with smaller values assumed for LOS situations. The macro- and micro-cell path-loss models are described below. Both the models are based on well-known COST231 [3] propagation models.

18.3.1   Macro-cell path-loss

The macro-cell path-loss is based on the modified COST231 Hata urban propagation model [3] given as below:

$$\text{PL [dB]} = \left[44.9 - 6.55 \log_{10}(h_{\text{NB}})\right] \log_{10}\left(\frac{d}{1000}\right) + 45.5 +$$

$$\left[35.46 - 1.1 h_{\text{UE}}\right] \log_{10}(f_c) - 13.82 \log_{10}(h_{\text{NB}}) + 0.7 h_{\text{UE}} + C, \quad (18.12)$$

where

    h_{NB} is the Node-B antenna height in meters
    h_{UE} is the UE antenna height in meters
    f_c is the carrier frequency in MHz
    d is the distance between the Node-B and UE in meters, and
    C is a constant factor ($C = 0$ dB for suburban macro and $C = 3$ dB for urban macro).

Assuming $h_{\text{NB}} = 32\,m, h_{\text{UE}} = 1.5\,\text{m}$ and $f_c = 1.9\,\text{GHz}$, the path-loss in dBs for suburban macro PL_{SM} and urban macro PL_{UM} scenarios are given as:

$$PL_{\text{SM}} = 31.5 + 35 \log_{10}(d) \quad C = 0\,\text{dB}$$

$$PL_{\text{UM}} = 34.5 + 35 \log_{10}(d) \quad C = 3\,\text{dB}. \quad (18.13)$$

Note that the minimum distance between the Node-B and the UE d is required to be at least 35 m.

18.3.2   Micro-cell path-loss

The micro-cell NLOS path-loss is based on the COST 231 Walfish–Ikegami NLOS model. The model parameters assumed are $h_{\text{NB}} = 12.5\,m, h_{\text{UE}} = 1.5\,m$, building height 12 m, building to building distance of 50 m and street width of 25 m. Further we assume orientation of $30°$ for all paths and selection of metropolitan center. With these assumptions, the micro-cell NLOS path-loss $PL_{\text{MI-NLOS}}$ in dBs can be written as:

$$PL_{\text{MI-NLOS}} = -55.9 + 38 \times \log_{10}(d) + \left(24.5 + \frac{1.5 \times f_c}{925}\right) \times \log_{10}(f_c). \quad (18.14)$$

Further assuming a carrier frequency of $f_c = 1.9\,\text{GHz}$ the path-loss equation can be simplified as below:

$$PL_{\text{MI-NLOS}} = 34.53 + 38 \times \log_{10}(d). \quad (18.15)$$

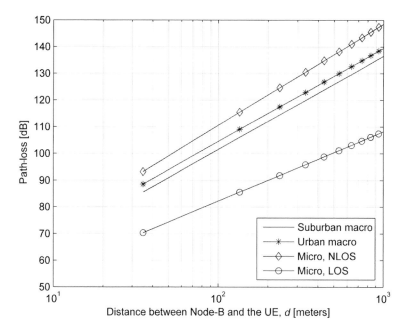

Figure 18.3. Path-loss as a function of distance between the Node-B and the UE, d.

The micro-cell LOS pathloss is based on the COST 231 Walfish–Ikegami street canyon model with the same parameters as in the NLOS case. The micro-cell LOS path-loss PL_{MI-LOS} in dBs can be written as:

$$PL_{\text{MI-NLOS}} = -35.4 + 26 \times \log_{10}(d) + 20 \times \log_{10}(f_c). \tag{18.16}$$

Assuming a carrier frequency of $f_c = 1.9\,\text{GHz}$, the path-loss equation can be simplified as below:

$$PL_{\text{MI-NLOS}} = 30.18 + 26 \times \log_{10}(d). \tag{18.17}$$

Note that the minimum distance between the Node-B and the UE d is required to be at least 20 m for both micro-cell NLOS and LOS cases.

The path-loss as a function of distance between the Node-B and the UE d for suburban macro-, urban macro-, micro-NLOS and micro-LOS scenarios is plotted in Figure 18.3. The pathloss for 1 km separation between Node-B and the UE for suburban macro, urban macro, micro-NLOS and micro-LOS scenarios are 136.5, 139.5, 148.5 and 108 dB respectively. We note that the path-loss difference between micro-NLOS and micro-LOS scenarios at a distance of 1 km is approximately 40 dB.

18.4 SCM user parameters

The SCM parameters for eNode-B and UE are shown in Figure 18.4. For simplicity, a single path between the eNode-B and the UE is shown. In general, the total number of paths can be

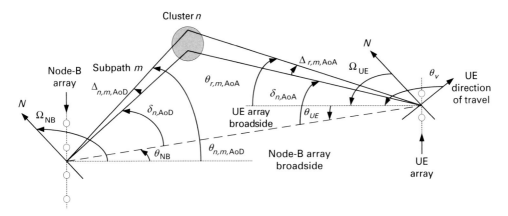

Figure 18.4. SCM parameters for Node-B and UE.

N with each path consisting of M subpaths. The Node-B and UE array orientations defined as the difference between the array broadside and the absolute North (N) reference direction are denoted as Ω_{NB} and Ω_{UE} respectively. The absolute angle of departure (AoD) for the mth subpath within the nth path in relation to Node-B broadside $\theta_{n,m,\mathrm{AoD}}$ is given as:

$$\theta_{n,m,\mathrm{AoD}} = \left(\theta_{\mathrm{NB}} + \delta_{n,\mathrm{AoD}} + \Delta_{n,m,\mathrm{AoD}}\right), \qquad (18.18)$$

where

θ_{NB} LOS AoD direction between Node-B and UE relative to the broadside of the Node-B array

$\delta_{n,\mathrm{AoD}}$ AoD for the nth ($n = 0, 1, 2, \ldots, N-1$) path relative to the LOS AoD θ_{NB}

$\Delta_{n,m,\mathrm{AoD}}$ Offset for the mth ($m = 0, 1, 2, \ldots, M-1$) subpath of the nth path relative to $\delta_{n,\mathrm{AoD}}$.

Similarly, the absolute angle of arrival (AoA) for the mth subpath within the nth path in relation to UE broadside $\theta_{n,m,\mathrm{AoA}}$ is given as:

$$\theta_{n,m,\mathrm{AoA}} = \left(\theta_{\mathrm{UE}} + \delta_{n,\mathrm{AoA}} + \Delta_{n,m,\mathrm{AoA}}\right), \qquad (18.19)$$

where

θ_{UE}:   Angle between Node-B and UE LOS and the broadside of the UE array

$\delta_{n,\mathrm{AoA}}$:   AoA for the nth ($n = 0, 1, 2, \ldots, N-1$) path relative to the LOS AoA θ_{UE}

$\Delta_{n,m,\mathrm{AoD}}$:   Offset for the mth ($m = 0, 1, 2, \ldots, M-1$) subpath of the nth path relative to $\delta_{n,\mathrm{AoA}}$.

The angles measured in a clockwise direction are assumed to be negative. When the UE is assumed as mobile, the angle of the velocity vector v relative to the UE broadside is given as:

$$\theta_v = \arg(v). \qquad (18.20)$$

In contrast to multi-path fast fading, slow-fading or shadow fading (SF) occurs when a large obstruction such as a hill or a large building obscures the main signal path between the transmitter and the receiver. The amplitude changes caused by shadowing is often modeled using a log normally distributed random variable with zero-mean and a standard deviation of 4–10 dB. More recently, it was observed that delay spread (DS) and azimuth or angle spread (AS) are also log normally distributed, which also suggests that DS and AS are correlated with each other as well as with shadow fading [4–8].

These correlations are accounted for in the SCM model by defining an intra-Node-B correlation matrix as below:

$$
A = \begin{bmatrix} \rho_{11} & \rho_{12} & \rho_{13} \\ \rho_{21} & \rho_{22} & \rho_{23} \\ \rho_{31} & \rho_{32} & \rho_{32} \end{bmatrix},
\tag{18.21}
$$

where

$\rho_{12} = \rho_{21}$ represents correlation between DS and AS
$\rho_{13} = \rho_{31}$ represents correlation between DS and SF and
$\rho_{23} = \rho_{32}$ represents correlation between AS and SF.

The ρ_{11}, ρ_{22} and ρ_{33} are DS–DS, AS–AS and SF–SF correlations and are all assumed to be 1, that is $\rho_{11} = \rho_{22} = \rho_{33} = 1$.

The intra-Node-B correlation matrix can then be written as:

$$
A = \begin{bmatrix} 1 & \rho_{12} & \rho_{13} \\ \rho_{12} & 1 & \rho_{23} \\ \rho_{13} & \rho_{23} & 1 \end{bmatrix}.
\tag{18.22}
$$

In order to provide a self-consistent model and that the full correlation matrix is positive-definite, the intra-Node-B correlations are assumed to take the following values:

$$
\begin{aligned}
\rho_{12} = \rho_{21} &= 0.5 \\
\rho_{13} = \rho_{31} &= -0.6 \\
\rho_{23} = \rho_{32} &= -0.6.
\end{aligned}
\tag{18.23}
$$

We note that DS and AS are assumed to be positively correlated, which means that a larger azimuth spread leads to a larger delay spread. On the other hand, DS and AS are assumed to be negatively correlated with the shadow fading. This means that for a larger shadow fading, DS and AS are reduced while for a smaller shadow fading, DS and AS are increased.

The only correlation between Node-Bs is for SF given by the matrix below:

$$
B = \begin{bmatrix} \zeta_{11} & \zeta_{12} & \zeta_{13} \\ \zeta_{21} & \zeta_{22} & \zeta_{23} \\ \zeta_{31} & \zeta_{32} & \zeta_{33} \end{bmatrix},
\tag{18.24}
$$

where

$\zeta_{12} = \zeta_{21}$ represents inter-Node-B correlation between DS and AS
$\zeta_{13} = \zeta_{31}$ represents inter-Node-B correlation between DS and SF and
$\zeta_{23} = \zeta_{32}$ represents inter-Node-B correlation between AS and SF.

The only correlation modeled between Node-Bs is SF–SF correlation and therefore all the elements in matrix B are assumed zero except for $\zeta_{33} = \zeta$:

$$B = \begin{bmatrix} 0 & 0 & 0 \\ 0 & 0 & 0 \\ 0 & 0 & \zeta \end{bmatrix}. \tag{18.25}$$

As we discussed, the DS, AS and SF are modeled as log normal random variables. Therefore for a given UE in the jth Node-B ($j = 1, 2, \ldots, J$), we need to generate values for $\sigma_{DS,j}, \sigma_{AS,j}$ and $\sigma_{SF,j}$. In order to model the correlations, these values are generated by correlated Gaussian random variables α_j, β_j and γ_j for $\sigma_{DS,j}, \sigma_{AS,j}$ and $\sigma_{SF,j}$ respectively. The correlated Gaussian random variables α_j, β_j and γ_j are in turn generated from a set of independent Gaussian random variables for the jth Node-B $[w_{1j}, w_{2j}, w_{3j}]$ and another set of Gaussian random variables $[\xi_1, \xi_2, \xi_3]$ that are common to all the Node-Bs.

The correlated Gaussian random variables α_j, β_j and γ_j are then given as:

$$\begin{bmatrix} \alpha_j \\ \beta_j \\ \gamma_j \end{bmatrix} = \begin{bmatrix} 1 & \rho_{12} & \rho_{13} \\ \rho_{12} & 1 & \rho_{23} \\ \rho_{13} & \rho_{23} & 1 - \zeta \end{bmatrix}^{1/2} \begin{bmatrix} w_{j1} \\ w_{j2} \\ w_{j3} \end{bmatrix} + \begin{bmatrix} 0 & 0 & 0 \\ 0 & 0 & 0 \\ 0 & 0 & \sqrt{\zeta} \end{bmatrix} \begin{bmatrix} \xi_1 \\ \xi_2 \\ \xi_3 \end{bmatrix}, \tag{18.26}$$

where

$$\begin{bmatrix} 1 & \rho_{12} & \rho_{13} \\ \rho_{12} & 1 & \rho_{23} \\ \rho_{13} & \rho_{23} & 1 - \zeta \end{bmatrix}^{1/2} = (A - B)^{1/2}. \tag{18.27}$$

Also $(A - B)^{1/2}$ represents the square root of matrix $(A - B)$. Note that $(A - B)$ is positive-definite and hence the square-root operation is well defined.

It should be noted that different UEs should use independent $[w_{1j}, w_{2j}, w_{3j}]$ triplets as well as independent ξ_3 realizations. In the following discussion, we focus on a link between a Node-B and a single UE.

The value of $\sigma_{DS,j}$ is given as:

$$\sigma_{DS,n} = 10 \wedge \left(\varepsilon_{DS} \alpha_j + \mu_{DS} \right), \tag{18.28}$$

where α_j is the correlated Gaussian random variable derived in Equation (18.26). Also $\varepsilon_{DS} = \sqrt{E\left[\log_{10}^2 (\sigma_{DS})\right] - \mu_{DS}^2}$ and $\mu_{DS} = E[\log_{10} (\sigma_{DS})]$ are respectively the logarithmic standard deviation and logarithmic mean of the distribution of the delay spread (DS). ε_{DS} and μ_{DS} are given in Table 18.1. The variable σ_{DS} is then used to calculate the random delays for each of the N multi-path components as below:

$$\tau_n' = -r_{DS} \, \sigma_{DS,j} \log_e (z_n) \quad n = 1, 2, \ldots, N, \tag{18.29}$$

where z_n ($n = 1, 2, \ldots, N$) are i.i.d. random variables derived from a uniform distribution $U(0, 1)$. The scaling factor r_{DS} is based on measurements and takes into account the statistical relationship between path delays and powers. The path delay variables τ_n' are ordered in ascending order so that $\tau_{(1)}' < \tau_{(2)}' < \cdots < \tau_{(N)}'$ and the smallest delay value is subtracted from each variable that is the delay for the nth path is $\tau_n = \left(\tau'_{(n)} - \tau'_{(1)}\right)$. The ordered path delays $\tau_1 = 0 < \tau_2 < \cdots < \tau_N$ can be quantized to the desired quantization level.

The random average powers for each of the N multi-paths are written as:

$$P'_n = e^{\frac{(1-r_{DS})}{r_{DS}} \cdot \frac{(\tau_n)}{\sigma_{DS}}} \cdot 10^{-\zeta_n/10} \quad n = 0, 1, \ldots, N, \tag{18.30}$$

where ζ_n $(n = 0, 1, \ldots, N)$ are *i.i.d.* Gaussian random variables with standard deviation of 3 dB. This randomization is used to account for the shadowing effect on per path powers. Since $(1 - r_{DS}) < 0$ the average powers per path decay with the path delay, that is, the larger the path delay the smaller the path power.

In the micro-cell scenario, the path delays $(\tau_n, n = 1, 2, \ldots, N)$ are *i.i.d.* random variables drawn from a uniform distribution U $(0, 1.2\,\mu s)$. The power for each path is given as:

$$P'_n = 10^{-\left(\tau_n + \frac{z_n}{10}\right)} \quad n = 0, 1, \ldots, N, \tag{18.31}$$

where $(z_n, n = 1, 2, \ldots, N)$ are *i.i.d.* zero-mean Gaussian random variables with a standard deviation of 3 dB. We note that path powers are exponentially decaying with delay with the addition of a lognormal randomness due to z_n.

The average powers for the paths are normalized so that the total power for the N multi-paths is unity.

$$P_n = \frac{P'_n}{\sum\limits_{i=1}^{N} P'_i} \quad n = 1, 2, \ldots, N. \tag{18.32}$$

The powers are normalized the same way for the macro- and micro-cell NLOS cases. However, when the micro-cell LOS model is considered, the path powers are normalized such that the ratio of the power in the direct path P_{LOS} to the powers in the scattered path is K:

$$P_n = \frac{P'_n}{(K+1)\sum\limits_{i=1}^{N} P'_i} \quad P_{LOS} = \frac{K}{(K+1)}. \tag{18.33}$$

The value of $\sigma_{AS,j}$ for angle spread in the jth Node-B is given as:

$$\sigma_{AS,j} = 10^{(\varepsilon_{AS}\beta_j + \mu_{AS})}, \tag{18.34}$$

where β_j is the correlated Gaussian random variable derived in Equation (18.26). Also $\varepsilon_{AS} = \sqrt{E\left[\log_{10}^2 (\sigma_{AS})\right] - \mu_{AS}^2}$ and $\mu_{AS} = E\left[\log_{10} (\sigma_{AS})\right]$ are respectively the logarithmic standard deviation and logarithmic mean of the distribution of the angle spread (AS).

The distribution for shadow fading (SF) is given as:

$$\sigma_{SF,j} = 10^{\left(\frac{\sigma_{SH}\gamma_j}{10}\right)}, \tag{18.35}$$

where γ_j is the correlated Gaussian random variable derived in Equation (18.26) and σ_{SH} is the shadow fading standard deviation. For the macro-cell scenario, the same shadow fading is applied to all the N multi-path components. However, for the micro-cell scenario, independent shadow fading is assumed for each of the N multi-path components

The AoDs for each of the N multi-paths are *i.i.d.* zero-mean Gaussian random variables as below:

$$\delta_n' \sim \eta\left[0, (r_{AS}\sigma_{AS})^2\right] \quad n = 1, 2, \ldots, N, \qquad (18.36)$$

where r_{AS} equals 1.2 and 1.3 for suburban macro and urban macro environments (see Table 18.1). The scaling factor r_{AS} determines the distribution of powers in angle. A higher value of r_{AS} means more power concentrated in a small AoD or a small number of paths that are closely spaced in angle as is the case for urban macro ($r_{AS} = 1.3$) relative to suburban macro ($r_{AS} = 1.2$) environment.

The AoDs generated for the N multi-paths are ordered in increasing absolute value such that $|\delta_{(1)}'| < |\delta_{(2)}'| < \cdots < |\delta_{(N)}'|$. The AoDs $\delta_{n,AoD}$ are assigned to the ordered variables such that:

$$\delta_{n,\,AoD} = \delta_{(n)}' \quad n = 1, 2, \ldots, N. \qquad (18.37)$$

The path delays $(\tau_n, n = 1, 2, \ldots, N)$ are associated with the AoDs as below:

$$\tau_1 \leftrightarrow \delta_{1,\,AoD}$$

$$\tau_2 \leftrightarrow \delta_{2,\,AoD}$$

$$\vdots$$

$$\tau_N \leftrightarrow \delta_{N,\,AoD}. \qquad (18.38)$$

As both path delays $(\tau_n, n = 1, 2, \ldots, N)$ and AoDs $\left(\delta_{n,\,AoD}, n = 1, 2, \ldots, N\right)$ are ordered in increasing order, a larger path delay is associated with a larger AoD as shown in Figure 18.5.

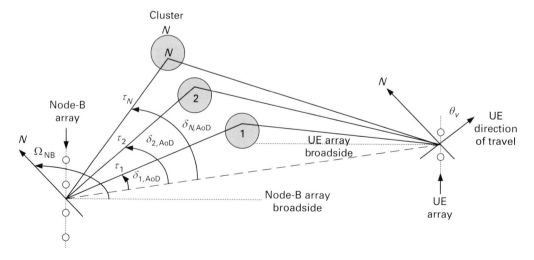

Figure 18.5. Association of path delays and path AoDs.

For the micro-cell scenario, the AoDs for each of the N multi-path components are *i.i.d.* random variables from a uniform distribution:

$$\delta_{n,\text{AoD}} \sim U\left(-40°, 40°\right) \quad n = 1, 2, \ldots, N. \tag{18.39}$$

The AoD $\delta_{n,\text{AoD}}$ of the nth path is associated with the power of the nth path P_n. It should be noted that unlike the macro-cell scenario, the AoDs do not need to be sorted before associating them with the path powers. Also, when the micro-cell LOS model is used, the AoD for the direct path is equal to the line-of-sight path direction. The multi-path delays $(\tau_n, n = 1, 2, \ldots, N)$ are randomly associated with the $(\delta_{n,\text{AoD}}, n = 1, 2, \ldots, N)$ in the case of the micro-cell scenario.

Each multi-path further consists of $(M = 20)$ subpaths. The power for the mth subpath within the nth path $P_{n,m}$ is given as:

$$P_{n,m} = \frac{P_n}{20} \quad n = 1, 2, \ldots, N \quad m = 1, 2, \ldots, M = 20, \tag{18.40}$$

where P_n is derived in Equation (18.32). The powers for all the subpaths within a path are the same. Note that the subpath powers among subpaths belonging to different paths can be different as the path powers are different among different N paths.

The phase for the mth subpath within the nth path $\Phi_{n,m}$ is drawn from a uniform distribution as below:

$$\Phi_{n,m} \sim \text{Uniform} (0, 360) \text{ degrees.} \tag{18.41}$$

The relative offsets for the subpaths $\Delta_{n,m,\text{AoD}}$ are given in Table 18.2. These offsets are selected to obtain the per path angle spread of $2°$ and $5°$ for macro- and micro-cell environments respectively. It should be noted that these offsets result in subpath angle spread within a path. This is in contrast to the paths angle spread σ_{AS}, which refers to the angle spread of the N multi-path components. Also, in the case of the micro-cell LOS model, the direct path component has no per path angle spread.

Table 18.2. Subpath AoD and AoA offsets.

Subpath #(m)	Subpaths offset at Node-B (degrees) $\Delta_{n,m,\text{AoD}}$ (degrees)		Subpaths offset for a 35 deg AS at the UE $\Delta_{n,m,\text{AoA}}$ (degrees)
	2 deg AS (macro-cell)	5 deg AS (micro-cell)	
1, 2	±0.0894	±0.2236	±1.5649
3, 4	±0.2826	±0.7064	±4.9447
5, 6	±0.4984	±1.2461	±8.7224
7, 8	±0.7431	±1.8578	±13.0045
9, 10	±1.0257	±2.5642	±17.9492
11, 12	±1.3594	±3.3986	±23.7899
13, 14	±1.7688	±4.4220	±30.9538
15, 16	±2.2961	±5.7403	±40.1824
17, 18	±3.0389	±7.5974	±53.1816
19, 20	±4.3101	±10.7753	±75.4274

The AoAs at the UE for each of the N multi-paths are *i.i.d.* zero-mean Gaussian random variables as below:

$$\delta'_{n,\text{AoA}} \sim \eta\left[0, \sigma^2_{n,\text{AoA}}\right] \quad n = 1, 2, \ldots N, \tag{18.42}$$

where $\sigma_{n,\text{AoA}}$ for the macro-cell scenario is given as:

$$\sigma_{n,\text{AoA}} = 104.12 \times \left(1 - e^{-0.2175|\log_{10}(P_n)|}\right). \tag{18.43}$$

On the other hand, $\sigma_{n,\text{AoA}}$ for the micro-cell scenario is given as:

$$\sigma_{n,\text{AoA}} = 104.12 \times \left(1 - e^{-0.265|\log_{10}(P_n)|}\right). \tag{18.44}$$

This AoA parameter $\sigma_{n,\text{AoA}}$ is plotted as a function of relative path power in Figure 18.6. We note that for smaller relative path powers, $\sigma_{n,\text{AoA}}$ takes larger values with a maximum of 104.12 degrees. On the other hand, for larger relative path powers $\sigma_{n,\text{AoA}}$ takes smaller values. This means that larger path powers are concentrated in narrower AoA. As for the AoD, when the micro-cell LOS model is used, the AoA for the direct path is equal to the line-of-sight path direction.

The AoA offsets $\Delta_{n,m,\text{AoA}}$ ($n = 1, 2, \ldots, N \quad m = 1, 2, \ldots, M = 20$) for the M subpaths within each of the N paths take fixed values from Table 18.2. The AoA offsets $\Delta_{n,m,\text{AoA}}$ are also shown in Figure 18.7. The offsets are selected such that the resulting per path angle spread is 35 degrees.

We have generated Node-B and UE paths and subpath information. Now we link the Node-B paths and subpaths to the UE paths and subpaths. The nth Node-B path defined by its path delay τ_n, power P_n and AoD $\delta_{n,\text{AoD}}$ is simply associated with the nth UE path defined by its

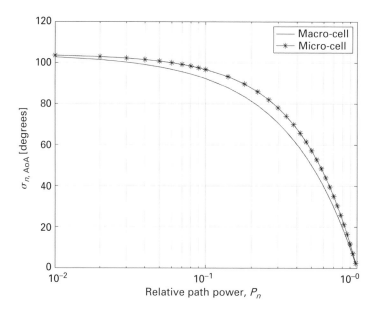

Figure 18.6. $\sigma_{n,AoA}$ as a function of relative path power.

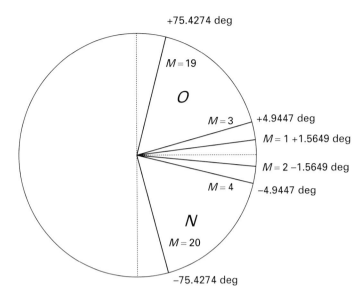

Figure 18.7. Subpath offsets for a 35 deg AS at the UE, $\Delta_{n,m,\text{AoA}}$ (degrees).

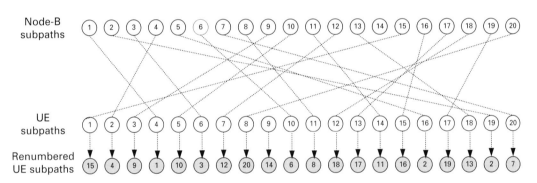

Figure 18.8. Random association of Node-B subpaths to UE subpaths within the n multi-path.

AoA $\delta_{n,\text{AoA}}$. Within each path pair, the M Node-B subpaths defined by their offset $\Delta_{n,m,\text{AoD}}$ are randomly linked to M UE subpaths defined by their offsets $\Delta_{n,m,\text{AoA}}$. In order to simplify the notation, we renumber the M UE subpath offsets with their newly associated Node-B subpaths as shown in Figure 18.8. In this example, the first UE subpath is associated with the 15th Node-B subpath. From Table 18.2, $\Delta_{n,m,\text{AoD}}$ for the first subpath for a 2 degrees angle spread at the Node-B is $\Delta_{n,1,\text{AoD}} = 0.0894°$ while $\Delta_{n,m,\text{AoA}}$ for the 15th subpath for a 35° angle spread at the UE is $\Delta_{n,15,\text{AoA}} = 40.1824°$. Since the first UE subpath is associated with the 15th Node-B subpath, the 15th UE subpath is renumbered as the first UE subpath as $\Delta_{n,1,\text{AoA}} = 40.1824°$. Each subpath pair is combined and phases $\Phi_{n,m}$ defined by Equation (18.41) are applied. The absolute angle of departure (AoD) for the mth subpath within the nth path in relation to Node-B broadside $\theta_{n,m,\text{AoD}}$ is given as:

$$\theta_{n,m,\text{AoD}} = \left(\theta_{\text{BS}} + \delta_{n,\text{AoD}} + \Delta_{n,m,\text{AoD}}\right). \tag{18.45}$$

Similarly, the absolute angle of arrival (AoA) for the mth subpath within the nth path in relation to UE broadside $\theta_{n,m,\text{AoA}}$ is given as:

$$\theta_{n,m,\text{AoA}} = \left(\theta_{\text{UE}} + \delta_{n,\text{AoA}} + \Delta_{n,m,\text{AoA}}\right). \tag{18.46}$$

In the case of 3-sector Node-Bs, the sector antenna pattern is given as:

$$A\left(\theta\right) = -\min\left[12\left(\frac{\theta}{\theta_{3\,\text{dB}}}\right)^2, A_m\right] \quad -180 \le \theta \le 180, \tag{18.47}$$

where $\theta_{3\,\text{dB}}$ is the 3 dB beam-width and A_m is the maximum attenuation. The antenna pattern for UE is considered omni-directional and simply given as $A\left(\theta\right) = 1 = 0\,\text{dB}$. The antenna pattern for 3-sector cell sites and omni-antenna for UE is shown in Figure 18.9. In the case of a 3-sector antenna, we assumed $\theta_{3\,\text{dB}} = 70°$ and $A_m = 20\,dB$. For a 6-sector Node-B scenario $\theta_{3\,\text{dB}} = 35°$ and $A_m = 23$ dB. The antenna gain in linear scale can be written as:

$$G\left(\theta\right) = 10^{\left(\frac{A(\theta)}{10}\right)}. \tag{18.48}$$

The antenna gains are dependent upon the subpath AoDs and AoAs. These gains are given by $G_{\text{Node-B}}\left(\theta_{n,m,\text{AoD}}\right)$ and $G_{\text{UE}}\left(\theta_{n,m,\text{AoA}}\right)$ for Node-B and UE respectively.

$$G_{\text{Node-B}}\left(\theta_{n,m,\text{AoD}}\right) = 10^{\left(\frac{A\left(\theta_{n,m,\text{AoD}}\right)}{10}\right)}$$

$$G_{\text{UE}}\left(\theta_{n,m,\text{AoA}}\right) = 10^{\left(\frac{A\left(\theta_{n,m,\text{AoA}}\right)}{10}\right)}. \tag{18.49}$$

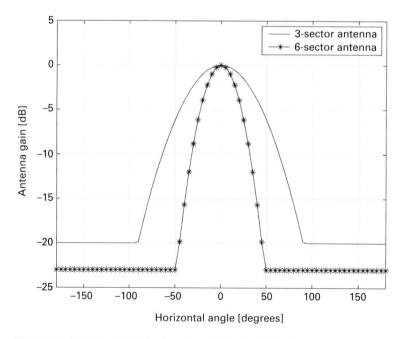

Figure 18.9. Antenna pattern for 3-sector and 6-sector Node-Bs.

18.5 SCM channel coefficients

The user parameters obtained above can now be used to generate the channel coefficients. Let us assume that Node-B has an S-element linear array while the UE has a U-element linear array as shown in Figure 18.10. Then the $U \times S$ channel matrix for the nth $(n = 1, 2, \ldots, N)$ multi-path component is given as:

$$H_n(t) = \begin{bmatrix} h_{1,1,n}(t) & h_{1,2,n}(t) & \cdots & h_{1,S,n}(t) \\ h_{2,1,n}(t) & h_{2,2,n}(t) & \cdots & h_{2,S,n}(t) \\ \vdots & \vdots & \cdots & \vdots \\ h_{U,1,n}(t) & h_{U,2,n}(t) & \cdots & h_{U,S,n}(t) \end{bmatrix} \quad (n = 1, 2, \ldots, N). \quad (18.50)$$

The (u, s)th component of $H_n(t)$ is given as:

$$h_{u,s,n}(t) = \sqrt{\frac{P_n \sigma_{SF}}{M}} \sum_{m=1}^{M} \left(\begin{array}{l} \sqrt{G_{BS}(\theta_{n,m,AoD})} \exp\left(j\left[kd_s \sin(\theta_{n,m,AoD}) + \Phi_{n,m}\right]\right) \times \\ \sqrt{G_{MS}(\theta_{n,m,AoA})} \exp\left(jkd_u \sin(\theta_{n,m,AoA})\right) \times \\ \exp\left(jk\|\mathbf{v}\| \cos(\theta_{n,m,AoA} - \theta_v) t\right) \end{array} \right),$$

$$(18.51)$$

where:
P_n is the power of the nth path from Equation (18.32).
σ_{SF} is the lognormal shadow fading from Equation (18.35), applied as a bulk parameter to the n $(n = 1, 2, \ldots, N)$ paths for a given drop.
M is the number of subpaths per path.
$\theta_{n,m,AoD}$ is the AoD for the mth subpath of the nth path from Equation (18.45).
$\theta_{n,m,AoA}$ is the AoA for the mth subpath of the nth path from Equation (18.46).
$G_{NB}(\theta_{n,m,AoD})$ is the Node-B antenna gain of each array element from Equation (18.49).

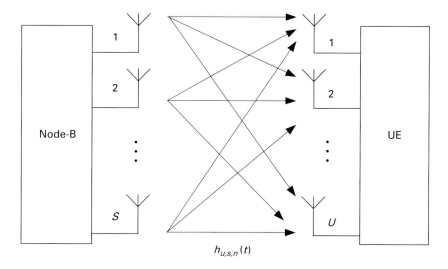

Figure 18.10. A MIMO system employing S transmit and U receive antennas.

$G_{\mathrm{UE}}(\theta_{n,m,\mathrm{AoA}})$ is the UE antenna gain of each array element from Equation (18.49).
$j = \sqrt{-1}$
$k = \frac{2\pi}{\lambda}\lambda$ is the carrier wavelength in meters.
d_s is the distance in meters from Node-B antenna element s to the reference antenna with $(s = 1, d_1 = 0)$.
d_u is the distance in meters from UE antenna element u to the reference antenna with $(u = 1, d_1 = 0)$.
$\Phi_{n,m}$ is the phase of the mth subpath of the nth from Equation (18.41).
$\|\mathbf{v}\|$ is the magnitude of the UE velocity vector.
θ_v is the angle of the UE velocity vector from Equation (18.20).

The SCM channel model is defined in the time-domain. For application to OFDM and also for frequency-domain equalization in SC-FDMA, we need frequency-domain channel coefficients. The equivalent frequency-domain channel coefficients at the kth subcarrier for an $S \times U$ MIMO channel are given as:

$$H(k) = \begin{bmatrix} H_{1,1}(k) & H_{1,2}(k) & \cdots & H_{1,S}(k) \\ H_{2,1}(k) & H_{2,2}(k) & \cdots & H_{2,S}(k) \\ \vdots & \vdots & \cdots & \vdots \\ H_{U,1}(k) & H_{U,2}(k) & \cdots & H_{U,S}(k) \end{bmatrix} \quad (k = 1, 2, \ldots, N_{\mathrm{FFT}}), \qquad (18.52)$$

where N_{FFT} is the FFT size, which also denotes the total number of subcarriers in the OFDM system. The (u, s)th component of $H(k)$ is a function of the multi-path components, $h_{u,s,n}(t), \quad n = 1, 2, \ldots, N$:

$$H_{u,s}(k) = \mathrm{FFT}\left[h_{u,s,1}(t), h_{u,s,2}(t), \ldots, h_{u,s,N}(t)\right], \quad (k = 1, 2, \ldots, N_{\mathrm{FFT}}). \qquad (18.53)$$

For example in the LTE system, the sample period equals $T_s = 1/f_s$, where $f_s = 30.72$ Msamples/sec. This sampling rate applies to the largest FFT size in the LTE system, which is ($N_{\mathrm{FFT}} = 2048$). The multi-path components, $h_{u,s,n}(t), n = 1, 2, \ldots, N$ are mapped to the time-domain samples according to their delays. For example, the six multi-paths for the (u, s)th MIMO channel component shown in Figure 18.11 would be mapped to six out of 2048 samples. In case a multi-path delay does not exactly match the samples delay, the path is mapped to the nearest delay. A size 2048 FFT operation is then performed to obtain 2048 frequency-domain channel coefficients for each of the 2048 subcarriers. The same process is repeated to obtain other components in the frequency-domain channel matrix in (18.52).

The SCM model also provides modeling approaches for special cases of polarized arrays, far scatterer clusters and urban canyons. These special cases are not discussed here and the interested reader is referred to [2] for details.

18.6   SCM extension

The SCM model was defined for a maximum bandwidth of 5 MHz and a carrier frequency of 1.9 GHz for a WCDMA system. The limitation mainly comes from the fact that the maximum number of multi-paths in the SCM model is limited to six. A backward compatible extension of the SCM model (SCME) to larger bandwidths is proposed in [9]. The model is extended by incorporating an intra-path delay spread. When the intra-path delay is zero, the SCME

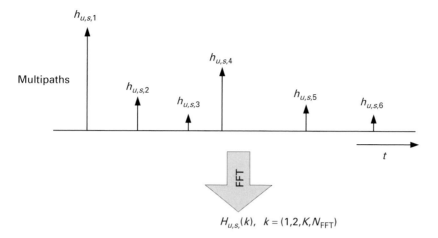

Figure 18.11. The relationship between time-domain and frequency-domain channel coefficients.

Table 18.3. Intra-cluster parameters for 20 MHz.

Mid-path $i =$	Number of sinusoids and power	Delay	Subpaths	AS_i / AS_n
1	10 (of 20)	0	1,2,3,4,5,6,7,8,19,20	0.9865
2	6 (of 20)	12.5 ns	9,10,11,12,17,18	1.0056
3	4 (of 20)	25 ns	13,14,15,16	1.0247

model collapses to the SCM model. The approach of intra-path delay spread was originally proposed for indoor propagation modeling in [10]. A similar model was specified for outdoor propagation modeling in [5].

We now discuss an example of the SCME implementation that was proposed for LTE performance evaluation in 20 MHz bandwidth. In this model, the nth ($n = 1, 2, \ldots, N$) multipath is further decomposed into three mid-paths ($i = 1, 2, 3$) providing a maximum of 18 mid-paths ($N = 6$) as shown in Figure 18.12. Each mid-path then consists of a number of subpaths corresponding to that multi-path. The number of subpaths for the first, second and third mid-paths within the nth multi-path is 10, 6 and 4 respectively as shown in Table 18.3.

Since the power for all the subpaths within a multi-path is the same, the power for the ith mid-path ($i = 1, 2, 3$) within the nth path $P_{n,i}$ is given as:

$$P_{n,1} = \frac{P_n}{2}, \quad P_{n,2} = \frac{3P_n}{10} \quad P_{n,3} = \frac{P_n}{5}. \tag{18.54}$$

This means that the first, second and the third mid-path contains 50%, 30% and 20% of the multi-path power respectively. The relative delay offset for the first, second and third mid-paths within the nth ($n = 1, 2, \ldots, N$) multi-path are 0, 12.5 and 25 ns respectively. This assumes a delay resolution of 1/4 of the sampling rate for 20 MHz bandwidth $\left(\frac{1}{4 \times 20\,\text{MHz}} = 12.5\,\text{ns} \right)$.

In the SCM model, each subpath ($m = 1, 2, \ldots, M = 20$) has a fixed angle relative to the path mean angle assigned to it. By perturbing the set of subpaths assigned to a mid-path, the angular spread (AS) of that mid-path can be varied. The mid-path angle spreads AS_i were

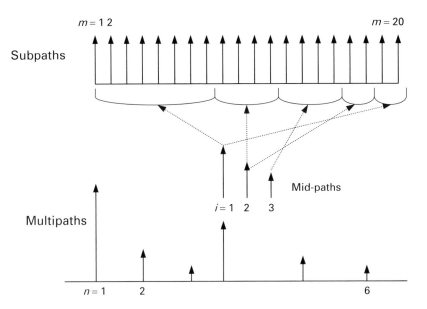

Figure 18.12. Multi-paths, mid-paths and subpaths.

optimized such that the deviation from the path angle spread AS_n that is the angle spread of all mid-paths combined is minimized.

18.7   Summary

The ITU-R channel models that have been widely used in cellular systems performance evaluation are not suitable for multi-antenna technologies. This is because these models primarily developed for single-antenna transmission scenarios assume independent channels between each pair of transmit and receive antennas. Since multi-antenna system performance heavily depends upon the channel correlations, it became obvious that new propagation models that accurately capture the correlations are needed. This led to formation of a joint ad hoc group in 3GPP and 3GPP2, which was chartered to develop the multi-antenna spatial channel model (SCM) [2]. The SCM group defined a ray-based model derived from stochastic modeling of scatters and therefore allows to model spatial correlations required for evaluation of multi-antenna techniques. Since its publication in 2003, SCM has been widely used for MIMO systems performance evaluation.

The SCM model was defined for a maximum bandwidth of 5 MHz and a carrier frequency of 1.9 GHz for a WCDMA system. The limitation mainly comes from the fact that the maximum number of multi-paths in the SCM model is limited to 6. For larger system bandwidths, the SCM model can be extended by incorporating an intra-path delay spread with up to a maximum of 18 multi-paths.

References
[1] Recommendation ITU-R M.1225, Guidelines for Evaluation of Radio Transmission Technologies for IMT-2000.
[2] 3GPP TR25.996 v7.0.0, Spatial Channel Model for MIMO Simulations.

[3] COST Action 231, "Digital radio mobile towards future generation systems, final report," Techincal Report, European Communities, EUR 18957, 1999.

[4] Greenstein, L. J. *et al.,* "A new path-gain/delay-spread propagation model for digital cellular channels," *IEEE Transactions on Vehicular Technology*, vol. 46, no. 2, pp. 477–485, May 1997.

[5] Correia, L. M., *Wireless Flexible Personalized Communications, COST 259: European Cooperation in Mobile Radio Research*, New York: Wiley, 2001.

[6] Cheon, C., Liang, G. and Bertoni, H. L., "Simulating radio channel statistics for different building environments," *IEEE Journal on Selected Areas in Communications*, vol. 19, no. 11, pp. 2191–2200, Nov. 2001.

[7] Algans, A., Pedersen, K. I., Mogensen, P.E., "Experimental analysis of the joint statistical properties of azimuth spread, delay spread, and shadow fading," *IEEE Journal on Selected Areas in Communications*, vol. 20, no. 3, pp. 523–531, Apr. 2002.

[8] Galcev, G., *et al.*, "A wideband spatial channel model for system-wide simulations," *IEEE Transactions on Vehicular Technology*, vol. 56, no. 2, pp. 389–403, Mar. 2007.

[9] Baum, D. S., Hansen, J., Salo, J., Del Galdo, G., Milojevic, M. and Kyosti, P., "An interim channel model for beyond-3G systems," *Proceedings of IEEE Vehicular Technology Conference 2005*, Stockholm, Sweden, May 2005.

[10] Saleh, A. and Valenzuela, R.A., "A statistical model for indoor multi-path propagation," *IEEE Journal on Selected Areas in Communications*, vol. SAC-5, no. 2, pp. 128–137, Feb. 1987.

19 LTE performance verification

The LTE system requirements mandate significant improvement in performance relative to the Release 6 HSPA system. In particular, the spectrum efficiency improvement targets for the downlink are three to four times that of the Release 6 HSPA system. The spectral efficiency improvement targets for the uplink are relatively modest with two to three times improvement over Release 6 HSPA. One of the reasons for lower improvement targets for the uplink is that the same antenna configuration is assumed for the LTE system and Release 6 HSPA system. On the other hand for downlink, LTE assumes two transmit antennas while Release 6 HSPA baseline system assumes only one transmit antenna at the Node-B. Similar targets are set for the peak data rates and also cell-edge performance improvements. The spectral efficiency target for the MBSFN, which is a downlink only service, is set at an absolute number of 1 bps/Hz.

An evaluation methodology specifying the traffic models and simulation parameters was developed for assessing the performance of the LTE and Release 6 HSPA systems. The goal of the evaluation methodology is to provide a fair comparison as all the parties participating in the simulations campaign can evaluate performance under the same set of assumptions. In this chapter, we describe LTE simulations methodology and provide relative performance of the LTE system and Release 6 HSPA system.

19.1 Traffic models

In this section, we discuss various traffic models considered in the performance verification. The traffic mix scenarios are given in Table 19.1. The traffic mix considers different categories of traffic including real-time, best effort, interactive, streaming and interactive real-time traffic.

19.1.1 Voice-over-IP (VoIP) traffic model

A simple two-state voice activity model as shown in Figure 19.1 is considered. The probability of transitioning from state 0 (silence or inactive state) to state 1 (talking or active state) is α while the probability of staying in state 0 is $(1 - \alpha)$. On the other hand, the probability of transitioning from state 1 to state 0 is denoted β while the probability of staying in state 1 is $(1 - \beta)$. The updates are made at the speech encoder frame rate $R = 1/T$, where T is the encoder frame duration (typically 20 ms).

The probabilities of being in state 0 and state 1 denoted as P_0 and P_1 respectively are given as:

$$P_0 = \frac{\beta}{\alpha + \beta}$$

$$P_1 = \frac{\alpha}{\alpha + \beta}.$$

(19.1)

Table 19.1. Traffic models mix.

Application	Traffic category	Percentage of users
VoIP	Real-time	30%
FTP	Best effort	10%
Web browsing / HTTP	Interactive	20%
Video streaming	Streaming	20%
Gaming	Interactive real-time	20%

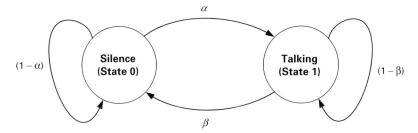

Figure 19.1. Two-state voice activity model.

The voice activity factor (VAF) is the probability of being in the talking state, that is, state 1:

$$\text{VAF} = P_1 = \frac{\alpha}{\alpha + \beta}. \tag{19.2}$$

The mean silence duration and mean talking duration in terms of number of voice frames can be written as:

$$E[\tau_s] = \frac{1}{\alpha}$$
$$E[\tau_t] = \frac{1}{\beta}. \tag{19.3}$$

The probabilities that a silence duration or a talking duration is n voice frames long is given by:

$$P_{\tau_s = n} = \alpha(1 - \alpha)^{n-1} \quad n = 1, 2, \ldots$$
$$P_{\tau_t = n} = \beta(1 - \beta)^{n-1} \quad n = 1, 2, \ldots . \tag{19.4}$$

The distribution of the time period τ_{AE} (in voice frames) between successive transitions into the talking state is the convolution of the distributions of τ_s and τ_t. The probability that this duration is n voice frames $P_{\tau_{\text{AE}}} = n$ is given as:

$$P_{\tau_{\text{AE}}=n} = \frac{\alpha}{\alpha - \beta}\beta(1 - \beta)^{n-1} + \frac{\beta}{\beta - \alpha}\alpha(1 - \alpha)^{n-1} n = 1, 2, \ldots . \tag{19.5}$$

Since the state transitions from state 0 to state 1 and vice versa are independent, the mean time between successive transitions into the talking state is simply the sum of the mean time in each state.

$$E[\tau_{AE}] = E[\tau_s] + E[\tau_t] = \frac{1}{\alpha} + \frac{1}{\beta}. \tag{19.6}$$

Table 19.2. VoIP traffic model parameters.

Parameter	Value
Voice codec	RTP AMR 12.2, Source rate 12.2 Kb/s
Encoder frame length	20 ms
Voice activity factor (VAF)	50% $\alpha = \beta = 0.01$
SID payload	SID packet every 160 ms during silence 15 bytes (5 bytes + header)
Protocol overhead with header compression	10 bit + padding (RTP pre-header) 4 byte (RTP/UDP/IP) 2 byte (RLC/security)16 bits (CRC)
Total voice payload on air interface	40 bytes

The mean rate of arrivals into the active state is simply $1/E\,[\tau_{AE}]$.

The rate of arrivals into the active state can serve as a guide on the number of resource requests needed for persistent allocation of resources for VoIP traffic. In general, it is expected that a single resource request and/or scheduling grant will be required for persistent allocation of resources when a VoIP user moves from the inactive to the talking state.

The VoIP traffic model parameters are given in Table 19.2. At a voice source rate of 12.2 Kb/s, a voice frame generated every 20 ms consists of 244 bits. The total protocol overhead per voice frame includes 10-bits of RTP pre-header and 2-bits padding resulting in a total of 236 bits (32 bytes). Furthermore, a compressed RTP/UDP/IP header consisting of 4 bytes is attached to the packet making the total size of 36 bytes. With 2 bytes of Layer 2 overhead consisting of RLC and security header and 16 bits (2 bytes) CRC, the total VoIP payload size transmitted over the air interface becomes 40 bytes.

A Silence Insertion Descriptor (SID) packet consisting of a total of 15 bytes is transmitted every 160 ms (or equivalent of 8 voice frames) during silence periods.

Let us assume a simple case with 20 ms voice frame duration and a desired VAF of 40%. We further assume a desired mean talking duration of 2 seconds (100 voice frames). Using Equation (19.3), we calculate $\beta = 1/100 = 0.01$. Furthermore with 40% VAF assumed, α is given as:

$$\alpha = \frac{v}{(1-v)}\beta = \frac{0.4}{(1-0.4)} \times 0.01 = 0.006\,67. \tag{19.7}$$

The distributions of silence and active state occupancy duration are given in Figure 19.2. The mean talking duration is 100 voice frames or 2 seconds while the mean silence duration is 150 voice frames or 3 seconds. The distribution of active state re-entry duration is given in Figure 19.3. The resulting mean time between successive transitions to the active state from Equation (19.6) is 5 seconds (mean silence duration + mean active duration). Then the mean rate of arrivals into the active state is $1/5 = 0.2$ talk-spurts per second.

We noted that for AMR 12.2 Kb/s voice source, total protocol overhead is approximately 20% (8 bytes out of a total of 40 bytes). This overhead can be reduced if multiple voice frames can be aggregated into a single packet for transmission. In this way, a single set of overheads is introduced for multiple voice frames. For example, if four voice frames containing 80 ms worth of speech can be aggregated or bundled into a single packet, the protocol overhead is reduced to 5%. However, packet aggregation leads to certain drawbacks such as increased

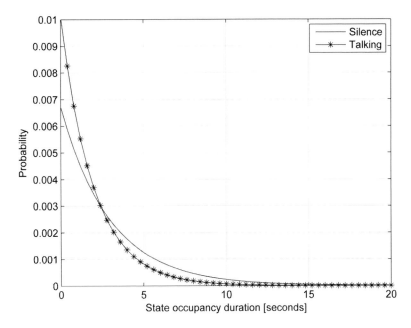

Figure 19.2. Distribution of silence and active state occupancy duration.

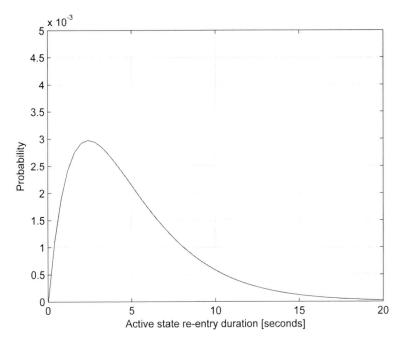

Figure 19.3. Distribution of active state re-entry duration.

delay and also increased sensitivity to packet loss. Note that with packet aggregation, when a single packet is lost it results in loss of multiple consecutive voice frames degrading the quality of voice service.

Table 19.3. FTP traffic model parameters.

Parameter	Statistical characterization
File size S	Truncated lognormal distribution mean = 2 Mbytes, standard deviation = 0.722 Mbytes, maximum size = 5 Mbytes (before truncation) PDF: $f_x = \dfrac{1}{\sqrt{2\pi}\sigma x} e^{\dfrac{-(\ln x - \mu)^2}{2\sigma^2}}$ $x > 0$ $\sigma = 0.35, \mu = 14.45$
Reading time D	Exponential distribution with mean = 180 seconds PDF: $f_x = \lambda e^{-\lambda x}$ $x \geq 0$ $\lambda = 0.006$

The LTE evaluations assumed an end-to-end delay of below 200 ms for mobile-to-mobile communications. Under this assumption, the delay budget available for radio interface is calculated as 50 ms. The system capacity for VoIP is defined as the number of users supported in the cell when more than 95% of the users are satisfied. A VoIP user is satisfied if 98% of its packets experience a delay of less than 50 ms.

19.1.2 Best effort FTP traffic model

A file transfer protocol (FTP) is considered as the best effort traffic, see Table 19.3. An FTP session is a sequence of file transfers separated by reading times. The two main FTP session parameters are the size S of a file to be transferred and the reading time D, i.e. the time interval between the end of the download of the previous file and the user request for the next file. The FTP traffic model is described assuming transmission on the downlink. However, the model can easily be extended for applicability to uplink.

19.1.3 Web browsing HTTP traffic model

A packet trace of a typical HTTP (hypertext transfer protocol) web browsing session is shown in Figure 19.4. The session is divided into active and inactive periods representing web-page downloads and the intermediate reading times. The web-page downloads are generally referred to as packet calls. These active and inactive periods are a result of human interaction where the packet call represents a web user's request for information and the reading time identifies the time required to digest the web-page. It has been suggested that web traffic exhibits self-similar behavior, which means that the traffic statistics on different timescales are similar [1–2]. Therefore, a packet call, like a packet session, is divided into active/inactive periods. Unlike a packet session, the active/inactive periods within a packet call are attributed to machine interaction rather than human interaction. A web-browser will begin serving a user's request by fetching the initial HTML page using an HTTP GET request. The retrieval of the initial page and each of the embedded objects (e.g. pictures, advertisements, etc.) is represented by the active period within the packet call while the parsing time and protocol overhead are represented by the inactive periods within a packet call. The parsing time refers to the time the browser spends in parsing for the embedded objects in the packet call or the web-page.

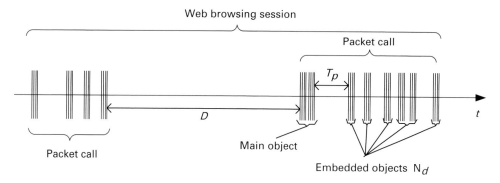

Figure 19.4. Packet trace of a web-browsing session.

Table 19.4. HTTP traffic model parameters.

Parameter	Statistical characterization
Main object size S_M	Truncated lognormal distribution, mean = 10710 bytes, standard deviation = 25032 bytes, minimum = 100 bytes, maximum = 2 Mbytes (before truncation) $$\text{PDF:} f_x = \frac{1}{\sqrt{2\pi}\sigma x} e^{\dfrac{-(\ln x - \mu)^2}{2\sigma^2}} \quad x > 0 \quad \sigma = 1.37, \mu = 8.37$$
Embedded object size S_E	Truncated lognormal distribution, mean = 7758 bytes, standard deviation = 126168 bytes, minimum = 50 bytes, maximum = 2 Mbytes (before truncation) $$\text{PDF:} \quad f_x = \frac{1}{\sqrt{2\pi}\sigma x} e^{\dfrac{-(\ln x - \mu)^2}{2\sigma^2}} \quad x > 0 \quad \sigma = 2.36, \mu = 6.17$$
Number of embedded objects per page N_D	Truncated Pareto distribution, mean = 5.64, maximum = 53 (before truncation) $$\text{PDF:} \quad f_x = \frac{\alpha_k^\alpha}{\alpha + 1}, k \leq x < m \quad f_x = \left(\frac{k}{m}\right)^\alpha, x = m$$ $\alpha = 1.1, \quad k = 2, \quad m = 55$ Note: subtract k from the generated random value to obtain N_D
Reading time D	Exponential distribution with a mean = 30 seconds $\text{PDF:} f_x = \lambda e^{-\lambda x} \quad x \geq 0 \quad \lambda = 0.033$
Parsing time T_p	Exponential distribution with mean = 0.13 seconds $\text{PDF:} f_x = \lambda e^{-\lambda x} \quad x \geq 0 \quad \lambda = 7.69$

The main parameters to characterize for the web-browsing traffic are the main object size S_M, the size of an embedded object in a web-page S_E, the number of embedded objects N_D, reading time D and parsing time T_p. These parameters are given in Table 19.4.

HTTP/1.1 [3] persistent mode transfer is assumed for downloading the objects serially over a single TCP connection. Based on observed packet size distributions [4], 76% of the packet calls use a maximum transmission unit (MTU) of 1500 bytes while the remaining 24% of the packet calls use an MTU of 576 bytes. These packet sizes also include a

40 byte TCP/IP packet header thereby resulting in useful data payloads of 1460 and 536 bytes respectively.

19.1.4 Video streaming traffic model

We assume that each frame of video data arrives at a regular interval T determined by the number of frames per second. Each video frame is decomposed into a fixed number of slices, each transmitted as a single packet. The size of these packets/slices is modeled as a truncated Pareto distribution. The video encoder introduces encoding delay intervals between the packets of a frame. These intervals are also modeled by a truncated Pareto distribution. The video streaming traffic model parameters are given in Table 19.5. In this model, the video source rate is assumed at 64 Kb/s.

19.1.5 Interactive gaming traffic model

The interactive gaming traffic model parameters for the uplink are given in Table 19.6. An initial packet arrival time is uniformly distributed between 0 and 40 ms. This initial time was considered to model the random timing relationship between client traffic packet arrival and uplink frame boundary in cdma2000 systems [5]. In the LTE systems with subframe duration of only 1.0 ms, this initial time to account for the resource request and scheduling grant is

Table 19.5. Video streaming traffic model parameters.

Parameter	Statistical characterization
Inter-arrival time between the beginning of each frame	Deterministic at 100 ms (10 frames per second)
Number of packets (slices) in a frame	Deterministic, 8 packets per frame
Packet (slice) size	Truncated Pareto distribution, mean = 10 Bytes, maximum = 250 bytes (before truncation)
	PDF: $f_x = \dfrac{\alpha_k^\alpha}{\alpha+1}, k \leq x < m \quad f_x = \left(\dfrac{k}{m}\right)^\alpha, x = m$
	$\alpha = 1.2, k = 20$ bytes, $m = ??$
Inter-arrival time between packets (slices) in a frame	Truncated Pareto distribution, mean $= m = 6$ ms, maximum $= 12.5$ ms (before truncation)
	PDF: $f_x = \dfrac{\alpha_k^\alpha}{\alpha+1}, k \leq x < m \quad f_x = \left(\dfrac{k}{m}\right)^\alpha, x = m$
	$\alpha = 1.2, k = 2.5$ ms, $m = ??$

Table 19.6. Interactive gaming traffic model parameters for the uplink.

Parameter	Statistical characterization
Initial packet arrival	Uniform distribution
	$f_x = \dfrac{1}{b-a} \quad a \leq x \leq b \quad a = 0 \quad b = 40$ ms
Packet arrival	Deterministic, 40 ms
Packet size	Largest extreme value distribution (also known as Fisher–Tippett distribution)
	$f_x = \dfrac{1}{b} e^{-\frac{x-a}{b}} e^{-e^{-\frac{x-a}{b}}} \quad a = 45$ bytes $\quad b = 5.7$

expected to be very small. The packet inter-arrival time is deterministic with a packet appearing every 40 ms. The packet size is assumed to follow the largest extreme value distribution, which is also known as the Fisher–Tippett distribution or the log–Weibull distribution. The values for this distribution can be generated by the following procedure:

$$x = a - b \ln (- \ln y), \tag{19.8}$$

where y is drawn from a uniform distribution in the range [0, 1]. Since the packet size needs to be an integer number of bytes, the largest integer less than or equal to x is used as the actual packet size. A compressed UDP header consisting of 2 bytes is added to each packet.

A maximum delay of 160 ms is applied to all uplink packets, i.e. a packet is dropped by the UE if any part of the packet has not started physical layer transmission, 160 ms after entering the UE buffer. The packet delay of a dropped packet is counted as 180 ms. A mobile network gaming user is in outage if the average packet delay is greater than 60 ms. The average delay is the average of the delays of all packets, including the delay of packets delivered and the delay of packets dropped.

The interactive gaming traffic model parameters for the downlink are given in Table 19.7. An initial packet arrival time is uniformly distributed between 0 and 40 ms. The packet inter-arrival times as well as the packet size on the downlink are modeled using the largest extreme value distribution.

19.2 System simulations scenarios and parameters

19.2.1 Deployment scenarios

A macro-cell reference system deployment is considered for the performance verification. In the macro-cell reference case, four different cases representing a mix of carrier frequency, inter-site distance, system bandwidth, penetration loss and UE speed as shown in Table 19.8 are simulated [6]. Since the LTE system requires optimization at low to medium speeds, higher UE speeds were not mandated in the performance verification. In addition to the macro-cell reference scenario, optional micro-cell simulation cases are specified for MIMO simulations.

Table 19.7. Interactive gaming traffic model parameters for the downlink.

Parameter	Statistical characterization
Initial packet arrival	Uniform distribution $f_x = \dfrac{1}{b-a} \quad a \le x \le b \quad a = 0 \quad b = 40\,\text{ms}$
Packet arrival	Largest Extreme Value Distribution (also known as Fisher–Tippett distribution) PDF: $f_x = \dfrac{1}{b} e^{-\frac{x-a}{b}} e^{-e^{-\frac{x-a}{b}}} \quad a = 55\,\text{ms}, \quad b = 6$
Packet size	Largest extreme value distribution (also known as Fisher–Tippett distribution) PDF: $f_x = \dfrac{1}{b} e^{-\frac{x-a}{b}} e^{-e^{-\frac{x-a}{b}}} \quad a = 120\,\text{bytes}, \quad b = 36$

Table 19.8. HSPA and LTE simulation case minimum set.

	Simulation Cases	Carrier frequency (GHz)	Inter-site distance (meters)	System bandwidth (MHz)	Penetration loss (dB)	Speed (km/h)
Macro-cell	1	2.0	500	10	20	3
	2	2.0	500	10	10	30
	3	2.0	1732	10	20	3
	4	0.9	1000	1.25	10	3
Micro-cell (only for MIMO)	Outdoor-to-outdoor	2.0	130	10	NA	3/30
	Outdoor-to-indoor	2.0	130	10	NA	3

The penetration loss for the micro-cell scenarios is included in the distance-dependent path-loss model. All simulation cases are simulated separately, which means that all the users in a given simulation run use parameters from the same case.

19.2.2 Cell layout

The cellular layout considered for macro-cell system simulations consists of a hexagonal grid assuming 19 cell sites and three sectors per site with a total of 57 sectors as shown in Figure 19.5. In the case of micro-cell simulations, a single sector per site consisting of the whole hexagon is considered. For the 3-sector sites the antenna bore sight points toward the flat side of the cell. The cell radius is calculated as $ISD/\sqrt{3}$, where ISD is the inter-site distance with hexagonal cell layout. The users are dropped uniformly in the entire cell. The minimum distance between the UE and the Node-B is assumed as 35 m. Therefore, if a user is dropped within 35 m from the Node-B, the drop is cancelled and a new drop is attempted.

19.2.3 Simulation parameters

The system simulation baseline parameters for the macro-cell and micro-cell deployment model are given in Tables 19.9 and 19.10 respectively.

The long-term shadow or log normal fading in the logarithmic scale around the mean path-loss PL dB is characterized by a Gaussian distribution with zero mean and a standard deviation of 8 and 10 dB for macro-cell and micro-cell scenarios respectively. In macro-cells, the shadow fading is caused by large terrain features such as buildings and hills. In micro-cells, smaller objects such as vehicles etc. cause shadow fading. Therefore, shadow fading is a function of the UE location in the cell and there should also be some correlation between shadow fading experienced by users in the vicinity of each other. Therefore, a distance separation (Δx) dependent correlation is introduced. The normalized autocorrelation function $R(\Delta x)$ can be described by an exponential function [7].

$$R(\Delta x) = e^{-\frac{|\Delta x|}{d_{\text{corr}}} \ln 2}, \tag{19.9}$$

where d_{corr} represents decorrelation length, which is dependent on the environment considered. The above expression indicates that the shadow fading values decorrelate exponentially

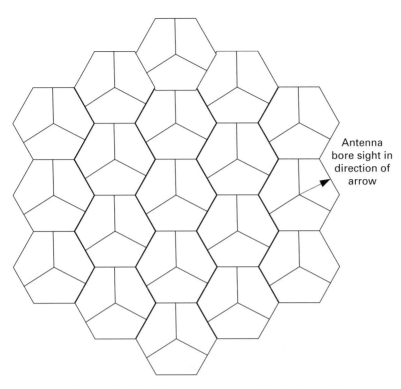

Antenna
bore sight in
direction of
arrow

Figure 19.5. Hexagonal grid cellular layout assuming 19 cell sites and three sectors per site (total of 57 sectors).

Table 19.9. Macro-cell system simulation baseline parameters.

Parameter		Assumption
Distance-dependent path-loss	Carrier freqeuncy = 2 GHz	$PL = 128.1 + 37.6 \times \log_{10}(R)$ dBs where R is distance between the UE and Node-B in kilometers
	Carrier freqeuncy = 900 MHz	$PL = 120.9 + 37.6 \times \log_{10}(R)$ dBs
Minimum distance between UE and Node-B		≥ 35 meters
Shadow fading standard deviation		8 dB
Decorrelation distance of Shadowing, d_{corr}		50 m
Shadow fading correlation	Between Node-Bs	0.5
	Between sectors	1.0
Antenna pattern for 3-sector cell sites		$A(\theta) = -\min\left[12\left(\dfrac{\theta}{\theta_{3\,dB}}\right)^2, A_m\right]$ $\times -180 \leq \theta \leq 180$ $\theta_{3\,dB} = 70$ degrees, $A_m = 20$ dB
Total BS Tx power	1.25 and 5 MHz BW	43 dBm
	10 MHz BW	46 dBm
UE power class		21 dBm (125 mW) and 24 dBm (250 mW)

Table 19.10. Micro-cell system simulation baseline parameters.

Parameter	Assumption	
	Outdoor to indoor	Outdoor to outdoor
Distance-dependent path-loss	$L\,[dB] = 7 + 56\log_{10}(d\,[m])$	$L\,[dB] = \begin{cases} 39 + 20\log_{10}(d\,[m]) & 10m < d \le 45m \\ -39 + 67\log_{10}(d\,[m]) & d > 45m \end{cases}$
Minimum distance between UE and Node-B	$>= 10m$ (and minimum coupling loss of -53 dB) The distance dependent path-loss + shadow fading is lower limited to free-space distance-dependent path-loss	
Shadowing standard deviation	10 dB	10 dB
Correlation distance of shadowing	10 m	25 m
Shadowing correlation	Between cells	0.0
	Between sectors	NA
Antenna pattern (horizontal)	$A\,(\theta) = 1$	
Channel model	According to Table A.2.1.2-1	
UE speeds of interest	3 km/h	3 km/h, 30 km/h
Total BS TX power (Ptotal)	38 dBm assuming 10 MHz BW	
UE power class	21 dBm (125 mW) and 24 dBm (250 mW)	

with distance. The decorrelation length considered in the simulations is 50, 25 and 10 m respectively for macro-cell, outdoor-to-outdoor micro-cell and outdoor-to-indoor micro-cell scenarios. Note that a larger decorrelation length is considered for macro-cells because shadow-fading in this case is generally caused by large structures such as buildings. A smaller decorrelation length is considered for the outdoor-to-indoor micro-cell scenario. We also note that the concept of decorrelation length may not be realistic to consider for this environment. However, the model is used for this case as well with a smaller decorrelation length for consistency.

The shadow fading correlation is modeled as follows. Assume that the shadow fading component of the path loss at the first position P_1 is L_1. Now assume that we want to compute the lognormal component L_2 at the next position P_2. Let us further assume that P_1 and P_2 are separated by Δx meters. Then L_2 follows a normal distribution with mean $R\,(\Delta x) \times L_1$ and variance $\left(1 - R\,(\Delta x)^2\right) \times \sigma^2$, where σ is the standard deviation, 8 and 10 dB for micro-cell and macro-cell scenarios respectively. This interpretation is based on the assumption that the successive path-loss components L_1 and L_2 are jointly normally distributed, each with zero mean and with correlation $R\,(\Delta x)$. The distribution quoted for L_2 is then the conditional distribution of L_2 given the value of L_1.

19.2.4 Antenna pattern

The antenna pattern (horizontal) for 3-sector cell sites using fixed pattern is given by:

$$A\,(\theta) = -\min\left[12\left(\frac{\theta}{\theta_{3\,dB}}\right)^2, A_m\right] \qquad -180 \le \theta \le 180, \qquad (19.10)$$

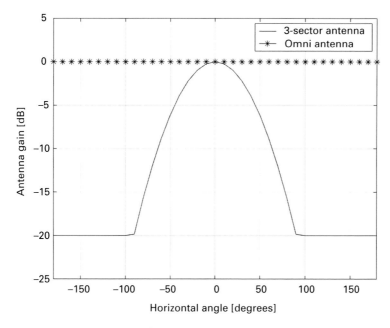

Figure 19.6. Antenna pattern for 3-sector cell sites used in the macro-cell scenario and omni-antenna for the micro-cell scenario.

where $\theta_{3\,dB}$ is the 3 dB beam-width and A_m is the maximum attenuation. The actual antenna pattern for multi-antenna transmission depends upon the MIMO precoding and beam-forming scheme used. The antenna pattern for micro-cell scenario considering omni-cells is simply given as $A(\theta) = 1 = 0\,dB$. The antenna pattern for 3-sector cell sites used in the macro-cell scenario and omni-antenna for the micro-cell scenario is shown in Figure 19.6. In the case of the 3-sector antenna, we assumed $\theta_{3\,dB} = 70°$ and $A_m = 20\,dB$. The antenna pattern for the micro-cell scenario considering omni-cells is simply given as $A(\theta) = 1 = 0\,dB$.

19.2.5 Reference UE and Node-B parameters

The reference HSPA Release 6 and LTE parameters for UE and Node-B are given in Tables 19.11 and 19.12 respectively. We note that the major difference between HSPA Release 6 and LTE is that the latter allows using 2×2 MIMO on the downlink. Also, the UE and Node-B receivers with two receive antennas in the case of HSPA Release 6 are assumed to employ maximum ratio combining (MRC) receive diversity. On the other hand, the LTE receiver is allowed to assume interference rejection combining (IRC) [8], which can result in improved performance in interference-limited scenarios.

19.3   Link to system performance mapping

A key issue for accurate system-level evaluations is to be able to map an instantaneous channel state, such as the instantaneous SNR for each subcarrier in the case of OFDM, to a corresponding block-error rate (BLER). Instead of directly mapping the channel state to a BLER,

Table 19.11. Reference UE parameters.

Parameter	HSPA Release 6	LTE
Receiver	2 antennas (Rx diversity only)	2 antennas
Transmitter	1 antenna	1 antenna
Antenna gain	0 dBi	0 dBi
Noise figure	9 dB	9 dB
MIMO	No MIMO	Support for 2×2 downlink MIMO

Table 19.12. Reference Node-B parameters.

Parameter	HSPA Release 6	LTE
Receiver	2 antennas (Rx diversity only)	2 antennas
Transmitter	1 antenna	2 antenna
Antenna gain	14 dBi for both micro- and macro-cell cases	14 dBi for both micro- and macro-cell cases, 6 dBi for micro-cell case with omni-antennas
Noise figure	5 dB	5 dB
MIMO	No MIMO	Support for 2×2 downlink MIMO

link-to-system mapping methods first map the instantaneous channel state, e.g. the set of subcarrier SNRs $\{\gamma_k, k = 0, 1, 2, \ldots, (N-1)\}$, into an instantaneous effective SNR γ_{eff} (a scalar value). The effective SNR is then used to find an estimate of the BLER from an AWGN link-level performance.

19.3.1 Channel capacity based mapping

A simple link-to-system mapping can be achieved by using the channel capacity formula. An average channel capacity is first computed based on the subcarrier SNRs $\{\gamma_k, k = 0, 1, 2, \ldots, (N-1)\}$ as below:

$$C_{\text{avg}} = \frac{1}{N} \left(\sum_{k=0}^{(N-1)} \log_2 \left(1 + \frac{\gamma_k}{Q} \right) \right), \tag{19.11}$$

where Q represents the gap to capacity of the actual AWGN performance of the code with $Q = 1$ denoting no gap. The average capacity C_{avg} is then converted back to an instantaneous effective SNR γ_{eff}:

$$\gamma_{\text{eff}} = Q(2^{C_{\text{avg}}} - 1). \tag{19.12}$$

The gap factor Q is obtained from actual link simulations for each set of modulation and coding rates. A more accurate approach will be to use the modulation-constrained capacity to arrive at the effective SNR γ_{eff}:

$$C_{\text{avg}} = \frac{1}{N} \left(\sum_{k=0}^{(N-1)} C_m \left(1 + \frac{\gamma_k}{Q} \right) \right), \tag{19.13}$$

where $C_m(\cdot)$ is the modulation-constrained capacity function which depends upon the modulation used such as QPSK, 16-QAM or 64-QAM.

19.3.2 Exponential Effective SNR Mapping (EESM)

A second link-to-system performance mapping method referred to as exponential effective SIR mapping (EESM) [9–11] is derived based on the Union–Chernoff bound of error probabilities. For a BPSK modulation, instantaneous effective SNR γ_{eff} for an N-state channel can be written as:

$$\gamma_{\text{eff}} = -\ln\left(\sum_{k=0}^{(N-1)} p_k e^{-\gamma_k}\right), \tag{19.14}$$

where γ_k is the SNR experienced at the kth channel state and occurs with probability p_k. In the case of OFDM with N subcarriers and SNR γ_k on the kth subcarrier:

$$\gamma_{\text{eff}} = -\ln\left(\frac{1}{N}\sum_{k=0}^{(N-1)} e^{-\gamma_k}\right). \tag{19.15}$$

For QPSK modulation with 2-bits per modulation symbol, the EESM expression becomes:

$$\gamma_{\text{eff}} = -2 \cdot \ln\left(\frac{1}{N}\sum_{k=0}^{(N-1)} e^{-\gamma_k/2}\right). \tag{19.16}$$

For higher-order modulation, such as 16-QAM and 64-QAM, it is not as straightforward to determine the exact expression for the EESM. The reason is that higher-order modulation can itself be seen as a multi-state channel from a binary-symbol transmission point of view. Therefore a generalized EESM that can be applied to higher order modulations needs to include a parameter β that can be adjusted to match a specific modulation scheme or more generally a specific combination of modulation scheme and a coding rate. The generalized EESM expression is then stated as:

$$\gamma_{\text{eff}} = -\beta \ln\left(\frac{1}{N}\sum_{k=1}^{N} e^{-\gamma_k/\beta}\right). \tag{19.17}$$

Note that when a user is scheduled transmission on a subset of the total N subcarriers, only the subcarriers used for the user's transmission need to be used for effective SNR γ_{eff} calculation.

Then suitable values for the parameter β for each modulation and coding scheme derived from link-level simulations using a random OFDM subcarrier interleaver are given in Table 19.13 [12].

The LTE evaluation allows using either of the constrained-capacity or EESM methods for downlink link-to-system performance mapping. In the uplink, the situation is different as all the modulation symbols transmitted using SC-FDMA experience similar SNR. This is because modulation symbols are spread using DFT and transmitted over all the subcarriers allocated

Table 19.13. Estimated β parameter values.

Modulation	Code rate	β value
QPSK	1/3	1.49
	2/5	1.53
	$^1/_2$	1.57
	3/5	1.61
	2/3	1.69
	$^3/_3$	1.69
	4/5	1.65
16-QAM	1/3	3.36
	$^1/_2$	4.56
	2/3	6.42
	$^3/_3$	7.33
	4/5	7.68
64-QAM	1/3	9.21
	2/5	10.81
	$^1/_2$	13.76
	3/5	17.52
	2/3	20.57
	17/24	22.75
	$^3/_3$	25.16
	4/5	28.38

for transmission. The symbol SNR in SC-FDMA depends upon the receiver type and can be derived from subcarrier SNRs $\{\gamma_k, k = 0, 1, 2, \ldots, (N-1)\}$. For example, for the MMSE receiver, SNR in SC-FDMA can be derived as (see Section 3.4.2):

$$\gamma_{\text{SC-FDMA}} = \frac{1}{\frac{1}{N} \sum_{k=0}^{(N-1)} \frac{1}{\gamma_k + 1}} - 1. \tag{19.18}$$

19.4   System performance

The LTE performance evaluation was carried out for the case of full buffers traffic model and VoIP separately. The simulations for the mixed traffic case and also other traffic types were not conducted during the evaluation phase. The performance for the broadcast traffic was evaluated assuming SFN operation with transmission from multiple-synchronized Node-Bs. A number of companies participated in the evaluation campaign and provided the performance results. The simulations were performed assuming either a typical urban (TU) [13] channel model or a spatial channel model (SCM) [14]. When a TU channel is assumed for MIMO simulations, the antennas are assumed to be perfectly decorrelated, which favors MIMO spatial multiplexing. The interference in the downlink is modeled either by assuming that neighboring cells transmit at full power or explicitly performing scheduling in the neighboring cells. In the downlink, assuming that all cells transmit at full power with uniform power spectral density (PSD), assumption of full power in the neighboring cells is generally reasonable. In the uplink, however, since the interfering UEs in the neighboring cells can be at different locations from

Table 19.14. Uplink and downlink spectral efficiency.

	System	Case 1		Case 3	
		[bps/Hz/ cell]	x HSPA	[bps/Hz/ cell]	x HSPA
Uplink	Baseline 1 × 2	0.332	1	0.316	1
	1 × 2	0.735	2.2	0.681	2.2
	1 × 4	1.103	3.3	1.038	3.3
Downlink	Baseline 1 × 2	0.53	1.0	0.52	1.0
	2 × 2 SU-MIMO	1.69	3.2	1.56	3.0
	4 × 2 SU-MIMO	1.87	3.5	1.85	3.6
	4 × 4 SU-MIMO	2.67	5.0	2.41	4.6

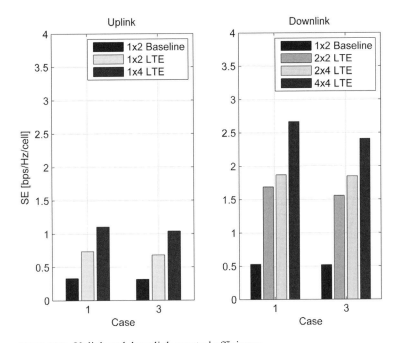

Figure 19.7. Uplink and downlink spectral efficiency.

the UE of interest, explicit modeling of interference is generally required. Explicit modeling of interference is also necessary for the downlink when power control resulting in different PSD on different parts of the transmitted band is employed.

The full buffers traffic spectral efficiency performance for the LTE system relative to a Release 6 HSPA baseline system is summarized in Table 19.14 [15] and Figure 19.7. The cases 1 and 3 refer to the simulation scenarios in Table 19.8. The difference between the two cases is in cell sizes, with case 1 representing inter-site distance of 500 meters and case 3 representing inter-site distance of 1732 m. In a hexagonal cell-layout, cell radii for cases 1 and 3 are 288.7 and 1000 m respectively.

We note that uplink performance for case 3 is slightly worse than for case 1 due to the larger cell radius of case 3 which leads to uplink transmit power and hence coverage limitation for users towards the cell-edge. However, relative gains of LTE over Release 6 HSPA are unchanged between cases 1 and 3 as both systems suffer similarly from uplink coverage limitation in larger cells. The LTE performance gain for the uplink mainly comes from SC-FDMA-based orthogonal access relative to a WCDMA-based non-orthogonal access in the case of Release 6 HSPA (see Chapter 4 for performance comparison of orthogonal and non-orthogonal access). It should be noted that a Rake receiver is assumed for Release 6 HSPA. By using a more advanced receiver such as a successive interference cancellation (SIC) receiver, the uplink performance of WCDMA can also be improved. Also, we note that doubling the number of receive antennas at the Node-B from two to four provides about 50% improvement in spectral efficiency. In general, a 3 dB increase in received SINR is expected by doubling the number of receive antennas. However, as the throughput increase for good users at high SNR is not linear with SNR, the overall gain in throughput is less than two times. The uplink performance can further be improved by employing multi-user MIMO in the uplink by scheduling up to two (four) users simultaneously on the same time-frequency resources with two (four) receive antennas at the Node-B. It should be noted that uplink multi-user MIMO (MU-MIMO) requires only one transmit antenna at the UE. This is in contrast to single-user MIMO (SU-MIMO), which requires at least two transmit antennas at the UE. The uplink SU-MIMO is not supported in the LTE standard due to the complexity of implementing multiple transmit chains in the UE transmitter.

We noted that the SC-FDMA-based uplink in the LTE system provides more than two times improvement in spectral efficiency relative to the HSPA baseline system. These gains can be attributed purely to SC-FDMA as both LTE and HSPA assume 1×2 antenna configuration. In the downlink, however, the baseline antenna configuration for HSPA and LTE is different with LTE assuming 2 transmit antennas relative to one transmit antenna for HSPA. This way LTE can use 2×2 SU-MIMO giving LTE an unfair advantage over HSPA. Therefore, spectral efficiency gains in the downlink are a result of both OFDMA over WCDMA and also 2×2 MIMO over no MIMO. We note that LTE with 2×2 MIMO provides more than three times improvement in spectral efficiency over the 1×2 baseline HSPA system. When larger numbers of antennas are permitted at either the transmitter or receiver or both, the performance improves further. In case 1, for example, 4×2 and 4×4 MIMO provides 10% and 58% spectral effieicny gains relative to a 2×2 MIMO system.

The MBSFN spectral efficiency performance is summarized in Table 19.15 and Figure 19.8. We note that for small-cell scenarios, MBSFN can provide very high spectral efficiency. This is because for MBSFN, with the same content transmitted from all the cells, there is no

Table 19.15. Spectral efficiency and cell range for MBSFN.

Deployment	Spectrum efficiency [bps/Hz]	Inter-site distance @ 1bps/Hz [km]
Case 1	3.13	1.619
Case 2	3.02	2.310
Case 3	0.99	1.619
Case 4	3.18	4.375

Table 19.16. Uplink and downlink VOIP capacity.

	Average VoIP capacity (UEs/cells)	
Deployment scenario	Uplink	Downlink
Case 1	241	317
Case 3	123	289

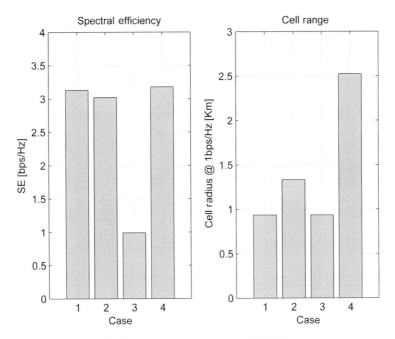

Figure 19.8. Spectral efficiency and cell range for MBSFN.

interference in the system apart from the background noise. For larger cells, as in case 3, the signals received from multiple cells become weaker due to larger path-loss lowering the SNR for the cell-edge users, which degrades achievable spectral efficiency. We also provide inter-site distance (or equivalently cell radius) achievable for an MBSFN spectral efficiency of 1bps/Hz. We note that case 4 provides the largest range. This is because case 4 assumes a lower carrier frequency of 900 MHz relative to 2GHz for the other cases and no penetration loss. Also, the bandwidth assumed for case 4 is 1.25 MHz relative to 10 MHz in the other cases, which provides higher power spectral density assuming the same Node-B total transmit power. The range for case 2 is larger than cases 1 and 3 because penetration loss for case 2 is only 10 dB relative to 20 dB in cases 1 and 3.

The VoIP capacity results are summarized in Table 19.16 and Figure 19.9. As for the data traffic full buffers traffic, the downlink capacity is generally higher than the uplink capacity. An important observation is that uplink VoIP capacity for case 3 is much smaller than the downlink capacity. A reason for this is that the uplink in case 3 is severely power limited due to a larger cell radius of 1000 meters and 20 dB penetration loss assumed. The downlink is also affected by larger cell radius and penetration loss but, since the Node-B transmit power is generally much larger than the UE transmit power, more power can be allocated to power-limited UEs at the cell-edge than the UEs in good channel conditions. This allows for the lowering of the VoIP outage in the downlink.

19.5 Summary

A set of simulation models and parameters was developed to assess the performance of the LTE and Release 6 HSPA system. The performance evaluation of Release 6 HSPA system

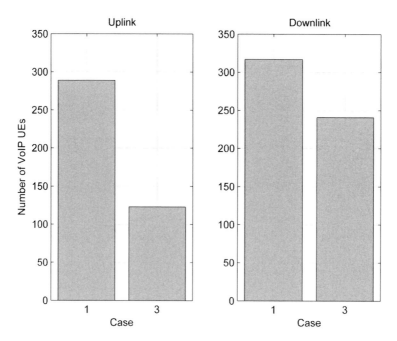

Figure 19.9. Uplink and downlink VoIP capacity.

was necessary as the LTE performance targets were set relative to Release 6 HSPA system performance. The goal of performance evaluation is to verify if the LTE system meets the expected performance targets or not.

The performance evaluation was carried out for full buffers traffic model and VoIP separately. The simulations for mixed traffic cases and for other traffic types such as web browsing, video streaming and mobile gaming were not carried out due to lack of time. For the full buffers traffic case, LTE provides about two times improved spectral efficiency in the uplink and three times improvement in the downlink. The uplink gains can be attributed purely to SC-FDMA as both LTE and HSPA assumes 1×2 antenna configuration. In the downlink, however, the baseline antenna configuration for HSPA and LTE is different with LTE assuming 2 transmit antennas relative to one transmit antenna for HSPA. This way LTE can use 2×2 SU-MIMO giving LTE an unfair advantage over HSPA. Also, for the LTE system, an interference rejection combining (IRC) receiver at the UE is assumed, further helping improve LTE performance in interference-limited scenarios.

With the baseline LTE antenna configurations of 1×2 in the uplink and 2×2 in the downlink, LTE meets the lower-end spectral efficiency performance targets. However, by assuming 1×4 configuration in the uplink with four receive antennas at the Node-B, LTE performance is about three times better than Release 6 HSPA 1×2 system. Similarly, assuming 4×4 MIMO in the downlink, LTE performance is more than four times better than 1×2 Release 6 HSPA 1×2 system. Therefore, LTE can meet the higher-end spectral efficiency targets by assuming more antennas. However, this comparison is not fair as we can argue that the performance of the Release 6 HSPA system would also improve when more antennas are assumed.

An area where LTE performance was significantly higher than the set targets is MBSFN. In small cells, MBSFN achieves more than 3 bps/Hz spectral efficiency. A prime reason for such high spectral efficiency is that for MBSFN, with the same content transmitted from all the cells, there is no interference in the system apart from the background noise. For larger cells, however, the signals received from multiple cells become weaker due to larger path-loss lowering the SNR for the cell-edge users, which degrades achievable spectral efficiency. We noted that even for cell radius of 1000 m and 20 dB building penetration loss, MBSFN achieves 1 bps/Hz spectral efficiency.

References

[1] Leland, W., Taqqu, M., Willinger, W. and Wilson, D., "On the self-similar nature of ethernet traffic (extended version)," *IEEE/ACM Transactions on Networking*, vol. 2, no. 1, pp. 1–15, Feb. 1994.

[2] Crovella, M. E. and Bestavros, A., "Self-similarity in world wide web traffic: evidence and possible causes," *IEEE/ACM Transactions on Networking*, vol. 5, no. 6, pp. 835–846, Dec. 1997.

[3] IETF RFC 2616, Hypertext Transfer Protocol – HTTP/1.1.

[4] Cao, J., Cleveland, W. S., Lin, Dong, Sun., Don X., "On the nonstationarity of internet traffic," *Proceedings ACM SIGMETRICS 2001*, pp. 102–112, 2001.

[5] 3GPP2 C.R1002-0 vl. 0, cdma2000 Evaluation Methodology.

[6] 3GPP RAN WG1 TR 25.814 v7.1.0, Physical Layer Aspects for Evolved UTRA (Release 7).

[7] Gudmundson, M., "Correlation model for shadow fading in mobile radio systems," *Electronics Letters*, vol. 27, no. 23, pp. 2145–2146, Nov.1991.

[8] Winters, J. H., "Optimum combining in digital mobile radio with cochannel interference," *IEEE Journal on Selected Areas in Communications*, vol. SAC-2, no. 4, July 1984.

[9] 3GPP Tdoc R1-031303, System-Level Evaluation of OFDM – Further Considerations.

[10] Brueninghaus, K. *et al.*, "Link performance models for system level simulations of broadband radio access systems," *Proceedings of IEEE Personal Indoor and Mobile Radio Communications (PIMRC) Conference*, 2005.

[11] 3GPP RAN TR25.892 v6.6.0, Feasibility Study for Orthogonal Frequency Division Multiplexing (OFDM) for UTRAN enhancement.

[12] 3GPP Tdoc R1-061506, System Analysis of the Impact of CQI Reporting Period in DL SIMO OFDMA.

[13] ETSI TR 101 112 V3.1.0, Universal Mobile Telecommunications System (UMTS); Selection Procedures for the Choice of Radio Transmission Technologies of the UMTS (UMTS 30.03) v6.1.0.

[14] 3GPP TR 25.996 v6.1.0, "Spatial Channel Model for MIMO simulations."

[15] 3GPP, TR 25.912 v7.2.0, Feasibility Study for Evolved Universal Terrestrial Radio Access (UTRA) and Universal Terrestrial Radio Access Network (UTRAN).

Index